Organosilicon Derivatives of Phosphorus and Sulfur

MONOGRAPHS IN INORGANIC CHEMISTRY

Editor: Eugene G. Rochow
Department of Chemistry, Harvard University

I. I. Vol'nov—Peroxides, Superoxides, and Ozonides of Alkaline Earth Metals—1966

M. Tsutsui, M. N. Levy, A. Nakamura, M. Ichikawa, and K. Mori—Introduction to Metal π-Complex Chemistry—1970

S. N. Borisov, M. G. Voronkov, and E. Ya. Lukevits—Organosilicon Heteropolymers and Heterocompounds—1970

S. N. Borisov, M. G. Voronkov, and E. Ya. Lukevits—Organosilicon Derivatives of Phosphorus and Sulfur—1971

Organosilicon Derivatives of Phosphorus and Sulfur

S. N. Borisov
Director, Laboratory of Elastomer Synthesis
All-Union Synthetic Rubber Research Institute
Leningrad

M. G. Voronkov
Director, Laboratory of Heteroorganic Compounds
Institute of Organic Synthesis
Academy of Sciences of the Latvian SSR, Riga

and

E. Ya. Lukevits
Group Leader, Laboratory of Heteroorganic Compounds
Institute of Organic Synthesis
Academy of Sciences of the Latvian SSR, Riga

Translated from Russian by
C. Nigel Turton and Tatiana I. Turton

PLENUM PRESS · NEW YORK–LONDON · 1971

Sergei Nikolaevich Borisov (1928-1967) was head of the section on heat-resistant elastomers and director of the Laboratory of Elastomer Synthesis, All-Union Synthetic Rubber Research Institute, Leningrad. A graduate student at Leningrad University from 1953 to 1956, in 1958 he defended his Candidate's dissertation on conjugated hydrogenation-halogenation of organosilicon compounds. He was the author of more than 50 articles and reviews on the chemistry and physics of organosilicon, organometallic, and high-molecular compounds and held a number of patents in these fields. Dr. Borisov drowned accidentally on August 6, 1967.

Mikhail Grigor'evich Voronkov was born in 1921. In 1947 he defended his Candidate's dissertation on the action of sulfur on unsaturated compounds. From 1947 to 1954 he was senior scientist of the Organic Chemistry Department, Faculty of Chemistry, Leningrad University. In 1954 he transferred to the Institute of Silicate Chemistry, Academy of Sciences of the USSR, where he worked as senior scientist, and in 1959 he became director of the Inorganic Polymer Laboratory. In 1961 he went to Riga, where he set up the Laboratory of Heteroorganic Compounds, Institute of Organic Synthesis, Academy of Sciences of the Latvian SSR, which he presently directs. In 1961 he was also awarded the degree of Doctor of Chemical Sciences for his work on the heterolytic cleavage of the siloxane bond. With E. Ya. Lukevits, Dr. Voronkov is author of *Organic Insertion Reactions of Group IV Elements* (Consultants Bureau, 1966).

Edmund Yanovich Lukevits was born in 1936. In 1966 he defended his Candidate's dissertation on organosilicon derivatives of furan. He currently directs a group of laboratories devoted to the study of heteroorganic compounds in the Institute of Organic Synthesis, Academy of Sciences of the Latvian SSR, Riga.

The original Russian text, published by Khimiya Press in Leningrad in 1968, has been corrected by the authors for the present edition. The English translation is published under an agreement with Mezhdunarodnaya Kniga, the Soviet book export agency.

KREMNE-ORGANICHESKIE PROIZVODNYE FOSFORA I SERY
КРЕМНЕ-ОРГАНИЧЕСКИЕ ПРОИЗВОДНЫЕ ФОСФОРА И СЕРЫ
С. Н. БОРИСОВ, М. Г. ВОРОНКОВ, Э. Я. ЛУКЕВИЦ

Library of Congress Catalog Card Number 74-159028
SBN 306-30511-9

A Division of Plenum Publishing Corporation
227 West 17th Street, New York, N.Y. 10011

United Kingdom edition published by Plenum Press, London
A Division of Plenum Publishing Company, Ltd.
Davis House (4th Floor), 8 Scrubs Lane, Harlesden, NW10 6SE, England

Printed in the United States of America

Preface to American Edition

We are very pleased that Plenum Press is continuing to publish an English translation of our multivolume monograph on organosilicon heterocompounds. At the same time we regret that our coauthor Sergei Borisov, who died tragically in August 1967, is unable to share our pleasure.

We recently received the American edition of the first volume of our monograph, in which monomeric and polymeric organosilicon heterocompounds containing inorganic elements are examined. We now present to English speaking chemists the second volume, which is devoted to organosilicon derivatives of phosphorus and sulfur.

In this volume we attempted to systematize and to examine critically all available literature data on the synthesis, physical, chemical, and biological properties, analysis, and practical application of monomeric and polymeric phosphorus- and sulfur-containing organosilicon compounds. In the year and a half since the manuscript of the Russian edition was sent to the publishers, the number of publications on these two classes of organosilicon heterocompounds has increased by more than 25%. This vigorous progress compelled us to supplement the American edition with literature published up to June 1, 1969 (in the abstract journal "Khimiya") and to introduce the necessary additions into the text.

We hope that Plenum Press will introduce our readers to the next two volumes of our monograph, which are devoted, respectively, to organosilicon heterocompounds containing nitrogen and compounds containing fluorine, bromine, and iodine, and which are now being prepared for publication.

It is our pleasant duty here to thank all our colleagues who have sent remarks and comments on the Russian and American editions of the first volume and the Russian edition of the second.

Riga, June 1969 M. G. Voronkov and E. Lukevits

Preface

One of the characteristics of the development of chemical science in the middle of the present century is the vigorous progress of the "third chemistry," which is often named now the chemistry of heteroorganic compounds. Then in the last decade, among specialists in this field there has been a marked increase in interest in heteroatomic organic derivatives of silicon, i.e., heteroorganic silicon compounds. However, until recently this new class of chemical substances, which is extremely interesting theoretically and practically, has been without a single specialized monograph which systematizes and generalizes all progress in the heteroorganic chemistry of silicon. The first attempt in this direction was our book "Heteroorganic Compounds of Silicon" [42 (F), 17 (S)*], which appeared at the end of 1966 and was published as an English translation in the USA in 1969. However, as follows from its subtitle "Derivatives of Inorganic Elements," this monograph could not cover the whole broad field of the chemistry of heteroorganic compounds of silicon. The main reason for this was above all the abundance and variety of original investigations of organosilicon derivatives of inorganic elements, which was unexpected even to the authors themselves. As a result of this the planned length of the book compelled us to omit the sections on organosilicon compounds of phosphorus and sulfur, which had already been prepared for publication. The authors conceived the idea of publishing a series of monographs on heteroorganic com-

*(F) – see literature list for the first part; (S) – see literature list for the second part.

pounds of silicon containing not only inorganic elements, but also sulfur, phosphorus, nitrogen, fluorine, bromine, and iodine. Thus, these monographs would exclude only silicohydrocarbons and organosilicon compounds containing oxygen and chlorine, which have already been examined in some detail in monographs and reviews.

This book is the result of the gradual realization of this project.

The generalization of the results of investigations of organophosphorus compounds of silicon (the number of publications in this field had already reached 600 by the middle of 1967) seemed particularly vital to the authors. Compounds of this type form a connection between the chemistry of organosilicon compounds and the chemistry of organophosphorus compounds, i.e., the two branches of heteroorganic chemistry which are growing most rapidly and which are of the greatest practical value to the national economy.

Until now the only literature on organosilicon derivatives of phosphorus has been extremely poor review publications. The well-known monographs on organosilicon chemistry [7 (F), 179, 287, 530 (S)] contain only isolated or disconnected factual data on monomeric organophosphorus derivatives of silicon, fixing the attention of the reader only on individual substances of this type and some of the most typical methods of synthesizing them. It seems to us that the only complete review of this field for its time was the section in the article by Borisov [41 (F)]. In recent years classes of organophosphorus compounds of silicon have been examined in the literature reviews of dissertations [46, 55, 57, 93, 94, 136, 148, 204, 206, 214, 230 (F)], but these are not available to a wide range of readers. In recent years published attempts to summarize progress in this field have been made only for compounds containing Si – P [310, 348, 350, 549a (F); 179 (S)]. The only real review among these was the publication of Fritz [350 (F)], which summarizes information published up to the middle of 1965.

More often in monographs and reviews we find data, which unfortunately are far from exhaustive, on polymers containing the groups Si – O – P [4, 8, 119, 420, 421 (F); 2, 11, 530 (S)], low-molecular cyclic compounds [8 (F); 2, 11 (S)], and triorganosiloxy derivatives of phosphorus [8, 420 (F); 2 (S)], and also on polymers which contain silicon, phosphorus, and aluminum atoms at the same time [8 (F); 2 (S)]. In the publications listed there are also scant mentions of polymeric compounds containing a phosphorus atom in organic radicals attached to a silicon atom [119 (F)], and of polyphosphazenes with organosilicon substituents [8 (F)]. When this book had already been prepared for publication we were kindly provided with the manuscript of the review [223 (F)], which is remarkable for its completeness and quality. However, it was intended to review monomeric materials and covers less than half the references known to us.

The state of generalized information on organosilicon compounds of sulfur is equally unsatisfactory. In monographs and reviews of organosilicon chemistry available up to now, there has been scattered through a mass of other materials pieces of information on individual classes of monomeric and polymeric compounds of this type – organosilanethiols [287, 530 (S)], organothiosilanes [7 (F); 287 (S)], isothiocyanatosilanes [7 (F); 179, 530 (S)], silyl sulfates [420 (F); 179, 530 (S)], thienylsilanes, and other compounds containing the grouping Si–C–S [7 (F); 179, 346 (S)], cyclic silthianes and siloxythianes [420, 421 (F); 11 (S)], and linear polysilthianes [2, 287, 530 (S)]. In only two of the publications listed [7 (F); 346 (S)] were organosilicon compounds containing sulfur examined in a separate section, and the data in them are more than ten years old. In more recent work [179 (S)] there is a brief analysis of the character of the Si–S bond and the characteristics of its formation and cleavage. Finally, in monographs on hydrides of group IV elements [310, 549a (F)] five or six pages each are devoted to compounds containing the grouping H_3Si-S.

The review of Haas [374 (S)], which appeared at the end of 1965, should be regarded as a well-intentioned attempt to fill the gap in the field of organosilicon compounds of sulfur. However, it is only an outline in which the properties of compounds containing the Si–S bond are discussed and a considerable number of these have no organic groups. When this monograph had already been written we learned of the fifth volume of the series "Advances in Organometallic Chemistry," which had only just been published. It contains a relatively incomplete review of sulfur-containing organic derivatives of Group IV elements, where compounds containing the Si–S bond (Si–S–H, Si–S–C, and Si–S–Si) are examined [132a (S)]. Organosilicon isothiocyanates are examined in another review contained in this volume [682a (S)].

While it is clear that none of these sources gives a complete picture of the situation, during work on the first book [42 (F), 17 (S)] we had at our disposal more than 600 original publications containing data on organosilicon compounds of sulfur.

All this led us to the conclusion that there was a pressing need for a separate examination of organosilicon compounds of phosphorus and sulfur, and this monograph is devoted to this. In it we strived for an exhaustive exposition of the contemporary state of the chemistry of these two classes of heteroorganic compounds of silicon (using literature published up to the beginning of 1967 with additions from Chemical Abstracts up to August 1, 1967). The make-up and coverage of the book are the same as in the previous one with the only difference that the literature cited, including patent literature, is given separately here at the end of each of the parts. Unfortunately, the very limited length of this monograph compelled us to give a very brief examination of many of the problems considered.

The first part of the monograph was written by S. N. Borisov, M. G. Voronkov, and É. Ya. Lukevits, and the second part by É. Ya. Lukevits and M. G. Voronkov. M. G. Voronkov was the over-all editor.

We would like to express our thanks to all our Soviet and foreign friends and colleagues who helped in some form or other with the publication of our work. The comments we received on the first book were found to be particularly valuable. We hope that they have made it possible for us to reduce the number of deficiencies in this monograph. We will be very grateful to readers for all critical comments and wishes concerned with this monograph and also previous and subsequent books in the series we plan.

S. N. Borisov,* M. G. Voronkov, and É. Ya. Lukevits

Riga – Leningrad
August, 1967

*Deceased.

Contents

Chapter II. Organosilicon Derivatives of Sulfur

Chapter I

Organosilicon Derivatives of Phosphorus

1. COMPOUNDS CONTAINING THE GROUPING Si – O – P*

1.1. Preparation Methods

1.1.1. Reactions of Halosilanes with Oxygen Acids of Phosphorus and Their Acid and Neutral Esters and Salts

When a mixture of trimethylchlorosilanes and anhydrous orthophosphoric acid (molar ratio 3 : 1) is heated to boiling [62a] or when H_3PO_4 is added gradually to trimethylchlorosilane [430, 525, 526], hydrogen chloride is liberated and tris(trimethylsilyl) phosphate is formed:

$$3(CH_3)_3SiCl + H_3PO_4 \longrightarrow [(CH_3)_3SiO]_3PO + 3HCl \quad (1.1)$$

Trimethylchlorosilane reacts analogously with butylphosphinic acid, forming bis(trimethylsilyl) butylphosphonate [424]:

$$2(CH_3)_3SiCl + (HO)_2P(O)C_4H_9 \longrightarrow [(CH_3)_3SiO]_2P(O)C_4H_9 + 2HCl \quad (1.2)$$

---

* The following nomenclature is used below for oxygen-containing organophosphorus compounds: derivatives of orthophosphoric acid $(HO)_3PO$ are named phosphates, those of phosphorous acid $(HO)_2P(O)H$ – phosphites, those of organophosphinic acids $RP(O)(OH)_2$ and $R_2P(O)(OH)$ – phosphonates, and those of organophosphinous acids $RP(OH)_2$ and R_2POH – phosphinates.

The reaction of trialkylchlorosilanes with phosphorous acid (molar ratio 3 : 1) leads to the formation of bis(trialkylsilyl) phosphites [63], confirming the hypothesis on the dibasicity of phosphorous acid, which reacts with the formation of diorganyl phosphites:

$$2R_3SiCl + (HO)_2P(O)H \longrightarrow (R_3SiO)_2P(O)H + 2HCl \qquad (1.3)$$

Diorganophosphinous acids, although according to P^{31} NMR data having the structure of diorganophosphine oxides $R_2P(O)H$, still react with triorganochlorosilanes in benzene in the presence of triethylamine with the formation of derivatives, namely, triorganosilyl diorganophosphinates (80% yield) [417]:

$$R_3SiCl + HP(O)R'_2 + (C_2H_5)_3N \longrightarrow R_3SiOPR'_2 + (C_2H_5)_3N \cdot HCl \qquad (1.4)$$

The reaction of diorganodichlorosilanes and organotrichlorosilanes with orthophosphoric acid leads to the formation of polymers which contain the grouping P–O–Si–O–P. Thus, when a mixture of dimethyldichlorosilane and orthophosphoric acid (molar ratio 3 : 2) is heated until the evolution of HCl ceases, there is formed a highly viscous, clear polymer, which is soluble in the usual solvents, but is hydrolytically unstable. It contains ~18.9% P, which is 2% greater than that calculated for the "stoichiometric polymer," i.e., $[3(CH_3)_2SiO \cdot P_2O_5]_n$ or $\{[(CH_3)_2Si]_3(PO_4)_2\}_n$ [62a]. The value of n and the structure of the polymer have not been determined, nor has the character of the reaction products with different reagent ratios. Heating a mixture of dimethyldichlorosilane, methyltrichlorosilane, and H_3PO_4 (molar ratio ~1 : 2 : 4) in dioxane yields phosphorus-containing organosilicon resins, which are resistant to the action of water and dilute mineral acids, but dissolve in hot alkalis [504, 505]. Polymers are also formed as a result of the reaction of diorganodichlorosilanes with derivatives of cyclotriphosphazene containing several hydroxyl groups [85, 307-309]:

$$P_3N_3R_3Cl(OH)_2 + R'_2SiCl_2 \xrightarrow[-HCl]{} [-R'_2SiO(P_3N_3R_3Cl)O-]_n \qquad (1.5)$$

When R = CH_3 the polymer softens at 320°C, while when R = C_6H_5 it softens at about 390°C. It is believed that n equals at least 30.

Analysis of the product of the reaction, to constant weight, of $SiCl_4$ and H_3PO_4 at 1200°C indicates that silicon orthophosphate, $Si_3(PO_4)_4$, is formed.

The reaction of $SiCl_4$ in benzene with dibutylphosphinous acid in the presence of triethylamine does not lead to the formation of phosphorus-containing organosilicon compounds:

$$4(C_4H_9)_2P(O)H + SiCl_4 + 4(C_2H_5)_3N \longrightarrow$$
$$\longrightarrow 2(C_4H_9)_2PP(O)(C_4H_9)_2 + SiO_2 + 4(C_2H_5)_3N \cdot HCl \quad (1.6)$$

Partial acid chlorides [338] and acid esters [320, 327] of phosphoric acid react with trialkylchlorosilanes like H_3PO_4:

$$R_3SiCl + HOP(O)Cl_2 \longrightarrow R_3SiOP(O)Cl_2 + HCl \quad (1.7)$$

$$nR_3SiCl + (HO)_nP(O)(OR')_{3-n} \longrightarrow (R_3SiO)_nP(O)(OR')_{3-n} + nHCl \quad (1.8)$$

It is recommended that reaction (1.8) is carried out in the presence of HCl acceptors to suppress side reactions caused by the hydrogen chloride formed. Thus, for example, the yield of methyl bis(trimethylsilyl) phosphate obtained in the presence of pyridine is twice as great as from a synthesis without an acceptor (55% and 28%, respectively). However, in the reaction of trialkylhalosilanes with acid esters of phosphoric acid, in principle the Si–O–P group may also be formed by another scheme:

$$\gt SiX + ROP(O) \longrightarrow \gt SiOP(O) + RX \quad (1.9)$$

i.e., by condensation with elimination of alkyl halide.

The reaction of α,ω-dichloropolydimethylsiloxanes with diethyl phosphate gives a 60-80% yield of linear polymers with terminal diethyl phosphate groups [525]:

$$Cl[(CH_3)_2SiO]_nSi(CH_3)_2Cl + 2(C_2H_5O)_2P(O)OH \xrightarrow{acc}$$
$$\longrightarrow 2\,acc \cdot HCl + (C_2H_5O)_2P(O)O[(CH_3)_2SiO]_nSi(CH_3)_2OP(O)(OC_2H_5)_2 \quad (1.10)$$
$$(n = 0\text{—}4)$$

Combination of the methods illustrated schematically by Eqs. (1.8) and (1.10) makes it possible to construct molecules with alternating silicon- and phosphorus-containing groups [321]:

$$2R_2SiHCl + (HO)_2P(O)OR \xrightarrow[-HCl]{}$$
$$\longrightarrow (R_2SiHO)_2P(O)(OR) \xrightarrow{Cl_2} (R_2SiClO)_2P(O)(OR) \xrightarrow{2(RO)_2P(O)OH}$$
$$\longrightarrow (RO)_2P(O)OR_2SiOP(O)(OR)OSiR_2OP(O)(OR)_2 \quad (1.11)$$

In the reactions of diorganodichlorosilanes with dibasic phosphoric acids and their derivatives, i.e., when all the reagents are bifunctional, there is the possibility of the formation of both linear and cyclic compounds. According to data in [390, 655], the reaction of diorganodichlorosilanes with compounds of the type $(RO)_2P(O)R'$ ($R = H$ or alkyl, $R' = H, CH_3, C_6H_5O, OH, CH_2=CH$) is accompanied by the formation of compounds containing an eight-membered ring:

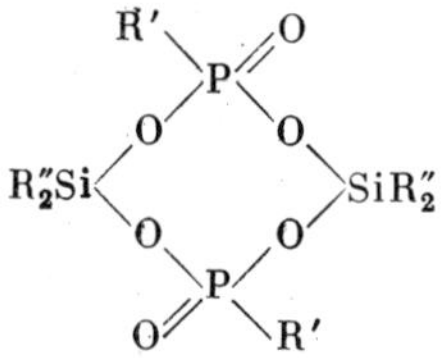

The reaction of $SiCl_4$ with methyl octadecyl phosphate forms a soft resinous brown polymer, while in the case of ethyl isoamyl phosphate a gel is obtained from which it is possible to isolate a solid phase [555].

Triorganohalosilanes react with trialkyl phosphates with the elimination of an alkyl halide and the formation of triorganosilyl alkyl phosphates:

$$nR_3SiX + (R'O)_3PO \longrightarrow nR'X + (R_3SiO)_nP(O)(OR')_{3-n} \quad (1.12)$$

With a molar ratio of reagents of 3:1, reaction (1.12) forms tris(trialkylsilyl) phosphates. Thus, for example, tris(triethylsilyl) phosphate was obtained in 64% yield by simple distillation of a mixture of triethylbromosilane with triethyl phosphate [62a]. Triethylchlorosilane requires the use of a catalyst ($FeCl_3$). However, in this case the yield of tris(triethylsilyl) phosphate (41%) is found to be lower than when X = Br. The use of trimethyl phosphate instead of triethyl phosphate makes it possible to obtain tris(triethylsilyl) phosphate in 64% yield starting from triethylchlorosilane (without the use of a catalyst).

By varying the molar ratio of the reagents it is possible to limit reaction (1.12) to partial replacement of the groups R' by R_3Si [320, 339]. Incidentally, in [430] it is stated that the mixed ester $[(CH_3)_3SiO]_2P(O)OCH_3$ is the main reaction product regardless of the molar ratio of the starting trimethylchlorosilane and trimethyl phosphate (from 1:1 to 1:3). The authors do not report the reaction conditions.

Alkyl organophosphonates react with trialkylhalosilanes like trialkyl phosphates [57, 152, 153, 200, 277, 430, 498, 530]:

$$(3-n)R_3SiCl + R'_nP(O)(OR'')_{3-n} \longrightarrow R'_nP(O)(OSiR_3)_{3-n} + (3-n)R''Cl \quad (n=1 \text{ or } 2) \tag{1.13}$$

The reaction of dialkyldichlorosilanes with trialkyl phosphates forms polymers [276, 277, 469]. However, from the reaction products of dimethyldichlorosilane and triethyl phosphate it was also possible to isolate the monomeric compound $(CH_3)_2 \cdot Si[OP(O)(OC_2H_5)_2]_2$ (yield ~40%) [320]. Silicon- and phosphorus-containing resins were also obtained by boiling dialkyldichlorosilanes and alkyltrichlorosilanes with diethyl ethylphosphonate in xylene [282]. Triphenyl phosphate and ethyl dichlorophosphate do not react with silicon tetrachloride [291, 339, 371]. At the same time, the reaction of $SiCl_4$ with trimethyl phosphate (molar ratio 3:4) is vigorous even at 20°C. In the case of $(C_3H_7O)_3PO$ and $(C_4H_9O)_3PO$ the liberation of alkyl chloride is complete in 24 h. Mixing $SiCl_4$ with triethyl phosphate (molar ratio 3:4) even at room temperature produces the evolution of ethyl chloride. After 2.5 h a gel is formed and after 22 h, a solid product which still contains 17.5% of hydrolyzable chlorine. Heating the reaction mixture to 55°C is accompanied by further evolution of C_2H_5Cl (total yield 96.3%), while stepwise calcination of the residue (180-280; 360-500; 1000°C) gives silicon phosphate [370, 371]. With a reagent ratio of 1:2 the reaction of $SiCl_4$ with trialkyl phosphate may be described by the equation [289, 389, 390, 425, 530]

$$SiCl_4 + 2(RO)_3PO \longrightarrow \frac{1}{n}[SiP_2O_6(OR)_2]_n + 4RCl \tag{1.14}$$

The most suitable medium for this process is provided by higher aliphatic ethers. The reaction product is a macromolecular substance, probably with a cyclo-chain structure, consisting of eight-membered rings connected through spiro atoms of silicon [390]:

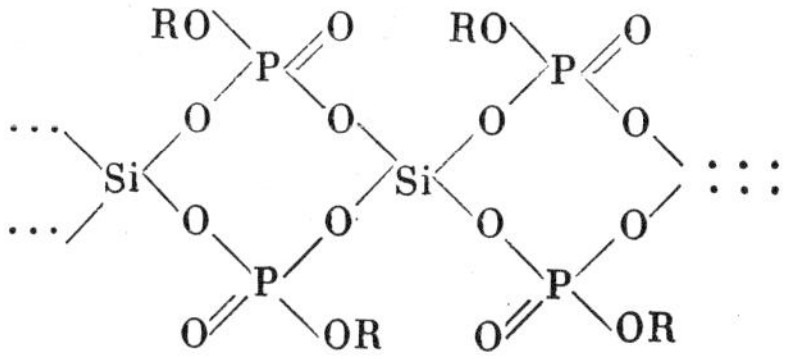

In the opinion of the authors of [390] the configuration of the rings is close to planar, while the arrangement of the substituents at the phosphorus atoms of the rings may correspond to both cis- and trans-configurations.*

The reactions of $SiCl_4$ with diethyl ethylphosphonate [282] and dimethyl cyclohexylphosphinate [530] also form silicon- and phosphorus-containing polymers. In contrast to this, tributyl phosphite exchanges butoxyl groups for chlorine atoms with $SiCl_4$ without the formation of silicon- and phosphorus-containing organic compound [291, 339, 371].

The character of the interaction of triorganohalosilanes with esters of phosphorous acid is of particular interest.

In 1948 B. A. Arbuzov and Pudovik [36] came to the conclusion that the reaction of triorganobromosilanes with triethyl phosphite leads to the formation of an organosilicon compound of phosphorus, containing the Si−P bond:

$$R_3SiX + (R'O)_3P \longrightarrow R_3SiP(O)(OR')_2 + R'Br \qquad (1.15)$$
$$(R = C_2H_5, C_2H_5O;\ X = Cl, Br;\ R' = C_2H_5\ \dagger)$$

The fact that the compounds they obtained were triorganosilyl phosphonates was not doubted originally.‡ In actual fact, the reaction of diethyl sodium phosphite, obtained by the action of sodium on diethyl phosphite, with triethoxychlorosilane yielded a compound identical to the product of the reaction of triethoxychlorosilane with triethyl phosphite. Chlorination of this compound leads to the formation of triethoxychlorosilane and $(C_2H_5O)_2P(O)Cl$.

Nonetheless, two years later Malatesta [453] revised scheme (1.15). In his opinion the ideas of Arbuzov and Pudovik were incorrect, and in the case where R = R' = C_2H_5 the reaction product is not diethyl triethylsilylphosphonate $(C_2H_5)_3SiP(O)(OC_2H_5)_2$, but

*Other authors [530] do not draw such radical conclusions, suggesting only that the spirocyclic compound formed initially is then converted into the polymer $[Si(CH_3PO_4)_2]_n$ (insoluble white powder which is hydrolyzed relatively readily).

†According to data in [335], heating a mixture of trimethyl chlorosilane and $(C_4H_9O)_3P$ (3:1) for 5 h does not result in their reaction.

‡Arbuzov and Pudovik previously described the reaction of trialkyliodostannane with trialkyl phosphites by a scheme analogous to (1.15) [35].

ethyl triethylsilyl ethylphosphonate $C_2H_5P(O)(OC_2H_5)[OSi(C_2H_5)_3]$.*
The formation of the latter may be explained in two ways:

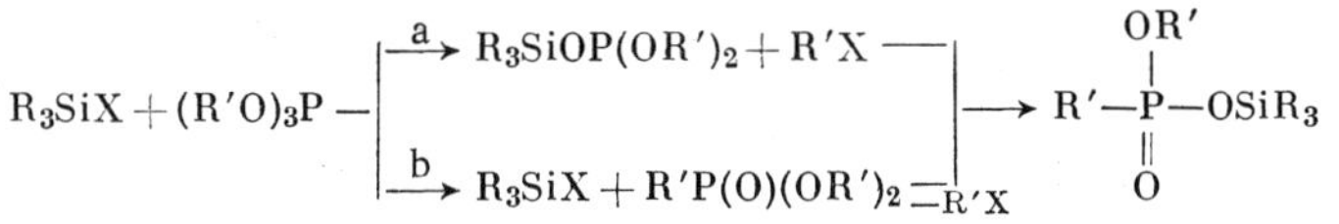

$$R_3SiX + (R'O)_3P - \begin{cases} \xrightarrow{a} R_3SiOP(OR')_2 + R'X - \\ \xrightarrow{b} R_3SiX + R'P(O)(OR')_2 \underset{=R'X}{} \end{cases} \longrightarrow \underset{O}{\overset{OR'}{R'-P-OSiR_3}} \quad (1.16)$$

Both routes involve an Arbuzov rearrangement, but are not
mutually exlusive of each other since in neither case does the rearrangement involve the system Si −O− P. The whole problem is
whether there is initially isomerization of triethyl phosphite into
diethyl ethylphosphonate or whether the primary product from the
reaction of the triorganohalosilane with the trialkyl phosphite re-
arranges.

As is well known, the Arbuzov rearrangement is catalyzed by
alkyl halides and this supports direction (1.16a). On the other
hand, the reaction of triethylbromosilane with diethyl ethylphosphonate proceeds at a higher rate than with triethyl phosphite,
though the reaction products in the two cases are the same, namely, ethyl triethylsilyl ethylphosphonate and bis(triethylsilyl) ethylphosphonate. The formation of the latter may also be explained
in two ways − it is either the product from the reaction of triethylbromosilane with ethyl triethylsilyl ethylphosphonate or the product from the direct reaction of diethyl ethylphosphonate with two
molecules of triethylbromosilane.

Malatesta [454] also revised the point of view of Arbuzov and
Pudovik on the character of the interaction of triethyliodostannane
with triethyl phosphite. The conclusion was analogous: the product
from the reaction of $(C_2H_5)_3SnI$ with triethyl phosphite and diethyl
ethylphosphonate is ethyl triethylstannyl ethylphosphonate, and this
may be explained by a scheme analogous to (1.16) and not (1.15).
The same conclusion was subsequently reached by one of the authors of scheme (1.15), who observed in [34] that triethyl phosphite
isomerizes during the reaction with triethyliodostannane and its product is a derivative of diethyl phosphonate. However, after this
there appeared [286], whose author adheres to scheme (1.15).

---

*The hydrolysis of this compound yielded ethylphosphinic acid and not phosphoric
or phosphorous acids, whose formation might have been expected in the case of
cleavage of an Si−P bond.

The work of Chernyshev and Bugerenko [46, 52, 222] sheds light upon the prolonged argument on the structure of the compounds obtained by the reaction of R_3SiX with $(R'O)_3P$. According to their data, the Arbuzov rearrangement occurs only at moderate temperatures (70-120°C), corresponding to the conditions in [36, 453]. At 160-180°C the reaction of triethylbromosilane with triethyl phosphite proceeds according to the scheme

$$(C_2H_5)_3SiBr + (C_2H_5O)_3P \longrightarrow (C_2H_5)_3SiOP(OC_2H_5)_2 + C_2H_5Br \quad (1.17)$$

$$(C_2H_5)_3SiOP(OC_2H_5)_2 + (C_2H_5)_3SiBr \longrightarrow [(C_2H_5)_3SiO]_2POC_2H_5 + C_2H_5Br \quad (1.18)$$

$$[(C_2H_5)_3SiO]_2POC_2H_5 + (C_2H_5)_3SiBr \longrightarrow [(C_2H_5)_3SiO]_3P + C_2H_5Br \quad (1.19)$$

i.e., with the formation of derivatives of trivalent phosphorus containing the grouping Si–O–P.

However, in this connection there again arises the problem of the identity of the products of the reaction of triethoxychlorosilane with triethyl phosphite and sodium diethyl phosphite [36]. If scheme (1.15) is incorrect, then doubt naturally arises on scheme (1.20), to which a series of investigators adhere [36, 324, 424, 470, 471]*:

$$R_3SiX + NaP(O)(OR')_2 \longrightarrow R_3SiP(O)(OR')_2 + NaCl \quad (1.20)$$

Therefore it is pertinent to mention that the direction of the reaction of triarylhalomethanes Ar_3CX with Ag salts of dialkyl phosphites is determined by the character of the halogen atom in them and the conditions of the process [431]. Thus, when X = Cl in an inert solvent, mixed esters of the type $Ar_3COP(OR)_2$ are obtained. In the absence of solvent or when X = Br, $Ar_3CP(O)(OR)_2$ are usually formed. However, the Ag salt of diisopropyl phosphite reacts with triphenylbromomethane with the formation of $(C_6H_5)_3 \cdot COP(OC_3H_7\text{-iso})_2$.

The investigations of Bugerenko and Chernyshev [52] demonstrated that the reaction of trialkylhalosilanes with sodium diethyl phosphite forms derivatives of trivalent phosphorus containing the grouping Si–O–P and not isomeric derivatives of pentavalent

*In the introduction to the article [424] the data of Arbuzov and Pudovik [36] are distorted somewhat and incorrect conclusions are given on the results of Malatesta's work [453]. The authors of [424] evidently did not have the original articles but only abstracts, since the experimental results described in the abstract of Malatesta's article [C.A., 46, 4472 (1952)] and their interpretation are mainly incorrect.

phosphorus containing the Si – P bond:

$$R_3SiX + NaP(O)(OR')_2 \longrightarrow R_3SiOP(OR')_2 + NaCl \qquad (1.21)$$

Analogously, triorganosilyl diorganophosphinates were obtained from potassium diorganophosphinates and triorganochlorosilanes [417]:

$$R_3SiCl + KOPR'_2 \longrightarrow R_3SiOPR'_2 + KCl \qquad (1.22)$$

The reaction of triorganohalosilanes with salts of acids of pentavalent phosphorus are described by the scheme [29, 30, 98, 100, 101, 103, 319, 320, 525, 526]

$$nR_3SiX + (MO)_nP(Y)R'_{3-n} \longrightarrow (R_3SiO)_nP(Y)R'_{3-n} + nMX \qquad (1.23)$$

$$\left(R = \text{alkyl, alkoxyl, } R''_3SiO;\ X = Cl, Br;\ Y = O \text{ or } S;\right.$$

$$\left.R' = F, OH, OAlk;\ M = Ag, NH_4, Na, K, \tfrac{1}{2}Ba;\ n = 1-3\right)$$

When R' = OH and Y = O there are formed partially substituted trimethylsilyl phosphates [29, 319], which cannot be obtained by other methods. The reaction proceeds at room temperature, but the yield of the trialkylsilyl phosphate depends substantially on the nature of the solvent: 80-90% in ether but only 11-12% in benzene or xylene.

The products from the reaction of dimethyldichlorosilane with sodium methyl octadecyl phosphate or ethyl isoamyl phosphate are white, pasty polymers [554, 555]. In all probability the alkoxyl groups are not touched in this process. In any case the reaction of diorganodichlorosilanes with KH_2PO_4 proceeds in the following way [103, 115, 214]:

$$R_2SiCl_2 + 2KOP(O)(OH)_2 \longrightarrow R_2Si[OP(O)(OH)_2]_2 + 2KCl \qquad (1.24)$$

Ammonium dialkyl thiophosphates may be used with equal success for reaction (1.24) [319], and in the case of diethoxydichlorosilane it is possible to obtain an almost quantitative yield of $(C_2H_5O)_2Si[OP(S)(OC_2H_5)_2]_2$.

The reaction of dialkyldichlorosilane with ammonium difluorophosphate proceeds according to the scheme [102]

$$R_2SiCl_2 + 2NH_4OP(O)F_2 \longrightarrow R_2Si[OP(O)F_2]_2 + 2NH_4Cl \qquad (1.25)$$

Together with the main product of reaction (1.25) there are formed dialkyldifluorosilanes and also compounds of the type $R_2Si(F)OP(O)F_2$.

The reaction of dimethyldichlorosilane with sodium ethyl phosphate (disubstituted) yields a labile brown liquid containing 19.4% P [554].

The reaction of alkyltrichlorosilanes with KH_2PO_4 (molar ratio 1:3) is analogous in general traits to (1.24). However, with the procedure adopted by the authors, namely, prolonged (8–9 h) boiling of the reaction products with anhydrous alcohol, there is partial ethoxylation of them with the formation of $RSi[OP(O)(OH)_2]_2 \cdot [OP(O)(OH)(OC_2H_5)]$ [103, 115].

1.1.2. Reactions of Alkoxysilanes and Acyloxysilanes with Oxygen Acids of Phosphorus and Their Esters and Anhydrides

Trialkylalkoxysilanes react with orthophosphoric acid predominantly according to the scheme [62a]

$$3R_3SiOR' + (HO)_3PO \longrightarrow (R_3SiO)_3PO + 3R'OH \quad (1.26)$$

In the reaction of triethylmethoxysilane with H_3PO_4, almost the theoretical amount of methanol is liberated, while the yield of tris(triethylsilyl) phosphate is 74%. By-products of reaction (1.26) are hexaalkyldisiloxanes, $R_3SiOSiR_3$, which are formed as a result of pyrolysis of $(R_3SiO)_3PO$ or hydrolysis by traces of water present in the H_3PO_4.

Alkylphosphinic acids react with triorganoalkoxysilanes with the formation of bis(triorganosilyl) phosphonates [62, 161, 162, 530]:

$$2R_3SiOR' + R''P(O)(OH)_2 \longrightarrow (R_3SiO)_2P(O)R'' + 2R'OH \quad (1.27)$$

In the case of hydroxymethylphosphinic acid, products of various degrees of substitution are formed, depending on the reaction conditions. Thus, heating a mixture of hydroxymethylphosphinic acid with triorganoalkoxysilanes (molar ratio 1:3) at 150–210°C and atmospheric pressure, with continuous distillation of the alcohol liberated, leads to the formation of bis(triorganosilyl) triorganosiloxymethylphosphonates with 45-55% yield [136, 177, 178, 186]:

$$3R_3SiOR' + HOCH_2P(O)(OH)_2 \longrightarrow (R_3SiO)_2P(O)CH_2OSiR_3 + 3R'OH \quad (1.28)$$

Carrying out the reaction of hydroxymethylphosphinic acid with triorganoalkoxysilanes under milder conditions (distillation of the alcohol in vacuum at 50-80°C) gave quantitative yields of bis(triorganosilyl) hydroxymethylphosphonates [136, 171, 178]:

$$2R_3SiOR' + HOCH_2P(O)(OH)_2 \longrightarrow (R_3SiO)_2P(O)CH_2OH + 2R'OH \quad (1.29)$$

The reaction of trialkylalkoxysilanes with alkylphosphinic acids evidently proceeds through a six-membered activated complex [62]:

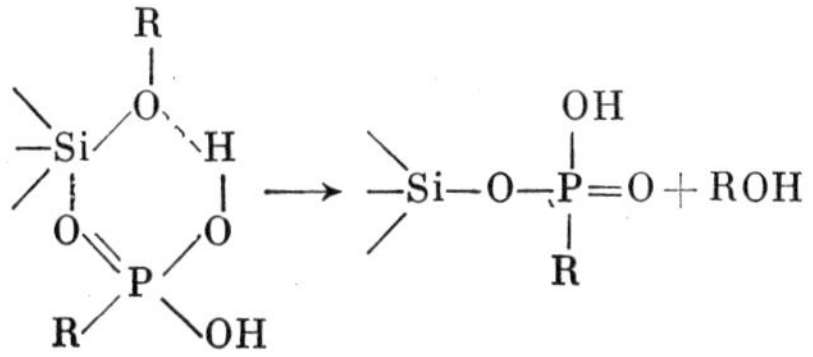

(1.30)

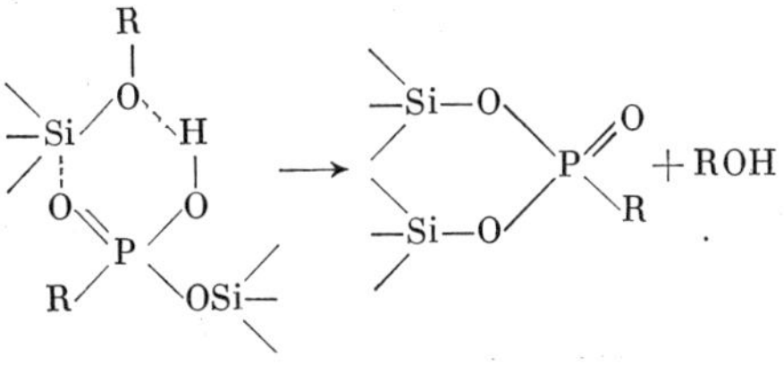

(1.31)

Heating trialkylacetoxysilanes with hydroxymethylphosphinic acid (molar ratio 2:1 or 3:1) at 150-210°C, with continuous distillation of the acetic acid liberated, forms bis(trialkylsilyl) acetoxymethyl phosphonates. The intermediate products of this reaction with a molar ratio of reagents of 3:1 are bis(trialkylsilyl) trialkylsiloxymethylphosphonates, which may be isolated if the process is carried out at 80-120°C with vacuum distillation of the acetic acid [136, 183]:

$$3R_3SiOCOCH_3 + (HO)_2P(O)CH_2OH \xrightarrow[-3CH_3COOH]{} (R_3SiO)_2P(O)CH_2OSiR_3 \quad (1.32)$$

At a higher temperature, there is cleavage of the intermediate trimethylsiloxymethyl derivative by the acetic acid liberated, leading to the bis(trialkylsilyl) acetoxymethylphosphonate:

$$(R_3SiO)_2P(O)CH_2OSiR_3 + CH_3COOH \longrightarrow (R_3SiO)_2P(O)CH_2OCOCH_3 + R_3SiOH \quad (1.33)$$

With a ratio of the starting reagents of 2:1, the initial product of reaction (1:32) is evidently the bis(trialkylsilyl) hydroxymethylphosphonate, which is then acetylated by acetic acid:

$$2R_3SiOCOCH_3 + (HO)_2P(O)CH_2OH \longrightarrow (R_3SiO)_2P(O)CH_2OH \xrightarrow[-H_2O]{+CH_3COOH} (R_3SiO)_2P(O)CH_2OCOCH_3 \quad (1.34)$$

The water liberated in reaction (1.34), and also as a result of condensation of trialkylsilanols formed by reaction (1.33), produces partial hydrolysis of the reaction products, lowering the yield to only 30-50%. The addition of acetic anhydride to the starting reaction mixture raises the yield to 80%, since in this case the reaction proceeds only according to scheme (1.35) without formation of water or silanols in the acetylation [136, 179]:

$$2R_3SiOCOCH_3 + (HO)_2P(O)CH_2OH + (CH_3CO)_2O \longrightarrow$$
$$\longrightarrow (R_3SiO)_2P(O)CH_2OCOCH_3 + 3CH_3COOH \quad (1.35)$$

The reaction of trialkylalkoxysilanes with phosphorous acid does not differ in the character of the products from scheme (1.3) even with a molar ratio of the reagents of 4:1 [63]:

$$2R_3SiOR' + P(OH)_3 \longrightarrow (R_3SiO)_2P(O)H + 2R'OH \quad (1.36)$$

The structure of the bis(trialkylsilyl) phosphites formed is confirmed by the absence from their Raman spectra of lines in the region of 2550-2700 cm^{-1}, which are characteristic of the valence oscillations of P–O–H, and the presence of lines corresponding to the characteristic frequencies of the bond oscillations of P–H (2420-2430 cm^{-1}) and P=O bonds (1230-1260 cm^{-1}).

With rapid removal of the acetic acid formed from the mixture, the reaction of phosphorous acid with a trialkyl(acetoxy)silane proceeds according to a scheme analogous to that for trialkylalkoxysilanes, with the formation of bis(trialkylsilyl) phosphites, whose yield is 50-80% [156, 157]:

$$2R_3SiOCOCH_3 + P(OH)_3 \longrightarrow (R_3SiO)_2P(O)H + 2CH_3COOH \quad (1.37)$$

Boiling a mixture of trialkylacetoxysilanes with triethyl phosphite for 6-10 h at 150-160°C does not lead to an appreciable change in the starting compounds [94].

The reaction of trialkylalkoxysilanes with phosphorus pentoxide may also be used as a method of synthesizing tris(triorganosilyl) phosphates from triorganoalkoxysilanes [68, 91, 93, 107, 109-111, 215, 228]:

$$6R_3SiOR' + P_2O_5 \longrightarrow 2(R_3SiO)_3PO + 3R'_2O \qquad (1.38)$$

The yield of tris(trimethylsilyl) phosphate depends substantially on the nature of the radical R, and is 10, 40-50, and 60-70% on the R_3SiOR' taken when $R' = CH_3$, C_2H_5, and C_4H_9, respectively. Equation (1.38) does not give a complete description of the process, since under the reaction conditions the ratio 6:1 is not optimal, as trimethylalkoxysilane is recovered unchanged to a considerable extent. Therefore, for preparative purposes it is recommended [107] that a molar ratio of $(CH_3)_3SiOR:P_2O_5 = 2:1$ be used.

Together with tris(trimethylsilyl) phosphate, reaction (1.38) yields the polymers $(SiP_2C_3H_9O_{5.5})_n$, where n = 11 and $(Si_2P_2C_6H_{18}O_6)_n$ with n = 9. Their formation is explained [91a, 111] by the interaction of $[(CH_3)_3SiO]_3PO$ which is obtained initially with phosphorus pentoxide, e.g.:

$$2n[(CH_3)_3SiO]_3PO + mnP_2O_5 \begin{cases} \xrightarrow{m=5} 6[(CH_3)_3SiOP_2O_{4.5}]_n \\ \xrightarrow{m=2} 3\{[(CH_3)_3SiO]_2P_2O_4\}_n \end{cases} \qquad (1.39)$$

According to data in [121], the action of orthophosphoric acid on diethyldiethoxysilane regardless of the molar ratio (from 2:1 to 2:3) forms exclusively polydiethylsiloxanes, which do not contain residual ethoxyl groups.* These data correspond to the patent [278] on the preparation of polysiloxanes by heating $R_nSi(OR)_{4-n}$ (n = 1-3) with H_3PO_4. However, the authors of [121] consider that the orthophosphoric acid does not act only as a condensation catalyst, but that the process proceeds through the formation of "polydiethylsiloxane phosphates"

$$(C_2H_5O)_2\underset{\underset{O}{\|}}{P}-O-[(C_2H_5)_2SiO]_x-\underset{\underset{O}{\|}}{P}(OC_2H_5)_2 \quad \text{and} \quad \left[-\overset{\overset{OC_2H_5}{|}}{\underset{\underset{O}{\|}}{P}}-O-Si(C_2H_5)_2O-\right]_x$$

which decompose during distillation of the reaction mixture in high vacuum. Unfortunately, there are no direct demonstrations

*In the case of a mixture of $(C_2H_5)_2Si(OC_2H_5)_2$ and $(C_2H_5)_3SiOC_2H_5$, linear polymers with triethylsilyl groups at the ends of the chains are obtained predominantly.

of the existence of such compounds, however, "polymethylvinyl-
siloxane phosphate" is obtained in ~90% yield by the reaction of
methylvinyldiacetoxysilane with anhydrous phosphoric acid (3.2)
at 180°C [112, 113]. The sticky polymer formed contains 15.7% P
and solidifies on cooling. It is readily soluble in alcohol, somewhat less soluble in chloroform, and is hydrolyzed by water and
alkalis.

Heating a mixture of $C_2H_5O[(CH_3)_2SiO]_nC_2H_5$ with methylphosphinic acid at 100-110°C is accompanied by the liberation of
alcohol and the formation of cyclic phosphonasiloxanes* [12, 14]:

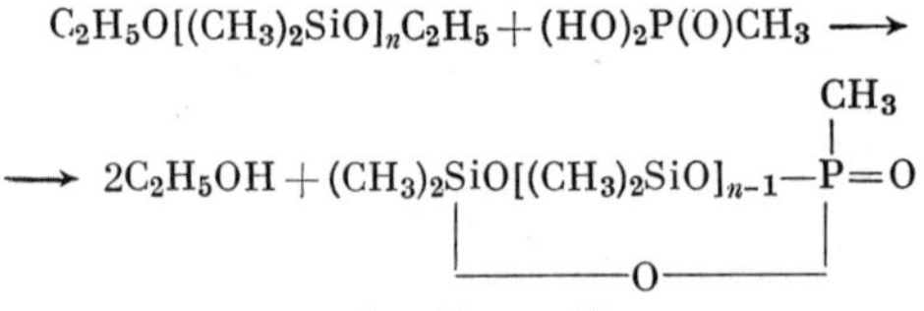

$$C_2H_5O[(CH_3)_2SiO]_nC_2H_5 + (HO)_2P(O)CH_3 \longrightarrow$$

$$\longrightarrow 2C_2H_5OH + \underset{\underset{\displaystyle O}{|\rule{3.5cm}{0pt}|}}{(CH_3)_2SiO[(CH_3)_2SiO]_{n-1}\overset{\displaystyle \overset{CH_3}{|}}{P}{=}O} \qquad (1.40)$$

$$(n = 2 \text{ or } 3)$$

When n = 1, instead of a four-membered ring an eight-membered ring is formed, i.e., $[-(CH_3)_2SiOP(O)(CH_3)O-]_2$. When n = 5
the corresponding twelve-membered ring is obtained in only 15% yield,
while the phosphonasiloxane with n = 3 is formed to a considerably
greater extent (55%).

Data on the reaction of dimethyldiethoxysilane with methylphosphinic acid, which proceeds with the liberation of 85% of the
theoretically possible amount of alcohol [14, 161], are somewhat
divergent. According to [14] this yields the cyclic condensation
product. However, in other work [62, 161, 162] heating the reaction mixture (3 h at 3 mm) at 160°C gave a polymer, which was a
viscous resin with the composition $[-(CH_3)_2SiOP(O)(CH_3)O-]_n$.

The reaction of dimethyldibutoxysilane with phosphorus pentoxide (3:1) is accompanied by the liberation of dibutyl ether and
the formation of a polymer, which corresponds in composition to
$[3(CH_3)_2SiO \cdot P_2O_5]_n$ to a greater extent than the product from the
reaction of dimethyldichlorosilane with H_3PO_4 [107, 108, 111, 118,
228]. The molecular weight of this polymer, 2550, determined by
Rast's method in benzoic acid, indicates that n ≈ 7. However, it

---

*Phosphonasiloxanes are oligo- and polysiloxanes in which one or more Si

atoms are replaced by the group $R-P(O){\Large<}$ (R is a hydrocarbon radical, halogen,

alkoxyl, etc.).

changes with time and two months after preparation it had increased to 2970 (n ≈ 8). The following structure is possible for the poly-mer:

$$\left[\begin{array}{ccccc} & CH_3 & O & CH_3 & \\ & | & \| & | & \\ -O- & Si-O-P-O-Si & -O- & \\ & | & | & | & \\ & CH_3 & O & CH_3 & \\ & & | & & \\ & H_3C-Si-CH_3 & & \\ & & | & & \\ & & O & & \\ & & | & & \\ & -O-P-O- & & \\ & & \| & & \\ & & O & & \end{array}\right]_n$$

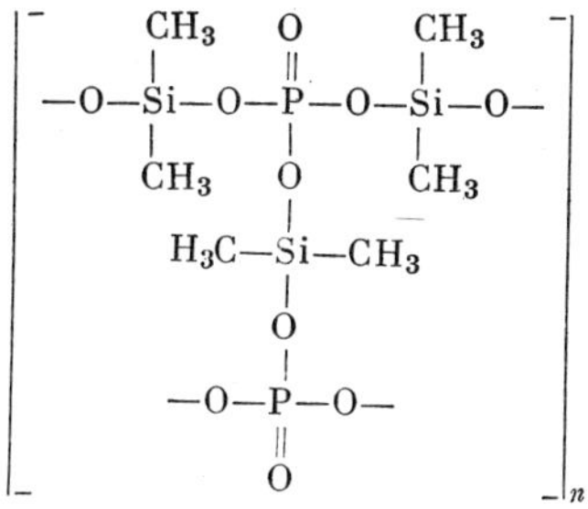

An analogous macromolecular product is obtained by the reaction of P_2O_5 with dimethyldiethoxysilane. Above 200°C it decomposes with the formation of $Si(PO_3)_4$* (cf. [91, 108]). X-ray structural analysis of the latter indicates a crystalline structure [107]. In the case of diethyldiethoxysilane, the polymer $[3(C_2H_5)_2 SiO \cdot P_2O_5]_n$ is formed with $n \approx 4$ [91, 108]. Its solubility in organic solvents is somewhat higher than for the methyl analog.

When alkyltriethoxysilanes are boiled with phosphorus pentoxide (molar ratio 2:1), there precipitates from the reaction mixture a white phosphorus-containing organosilicon polymer with the composition $[Si_2P_2O_6R_2(OR')_4]_n$ [108, 111, 113]. Data on the structure of such polymers are contradictory. The following structural formula is proposed in [108] and its logicality is obvious:

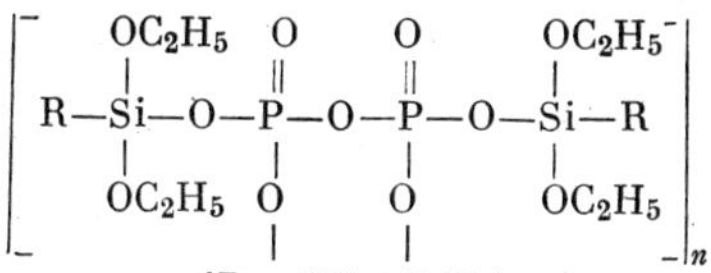

$$\left[\begin{array}{ccccc} OC_2H_5 & O & O & OC_2H_5 \\ | & \| & \| & | \\ R-Si-O-P-O-P-O-Si-R \\ | & | & | & | \\ OC_2H_5 & O & O & OC_2H_5 \\ & | & | & \end{array}\right]_n \qquad (1.41)$$

$$(R = CH_3, C_2H_5)$$

At the same time, a completely different formula is given in [111] without any reservation:

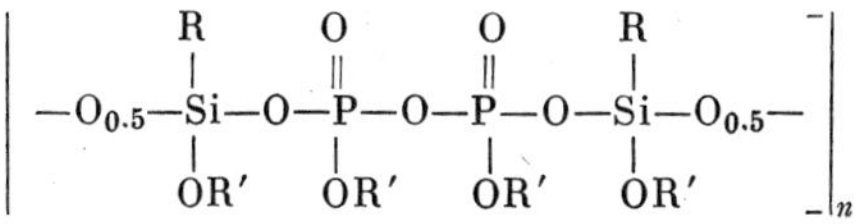

$$\left|\begin{array}{cccc} R & O & O & R \\ | & \| & \| & | \\ -O_{0.5}-Si-O-P-O-P-O-Si-O_{0.5}- \\ | & | & | & | \\ OR' & OR' & OR' & OR' \end{array}\right|_n \qquad (1.42)$$

$$(R = CH_3, C_2H_5, C_6H_5; \quad R' = C_2H_5)$$

_____
*More probably $[Si(PO_3)_4]_n$ [93, 107, 111]. According to [107] the substance corresponds in composition to $Si(PO_3)_4 \cdot 2H_2O$.

In formula (1.42) the ethoxyl groups are at the phosphorus atoms and this naturally makes it necessary to treat the reaction mechanism from different points of view. However, the matter is not limited to this. If structure (1.42) is correct, the polymer is a poly-3,5-diphosphona-1,7-tetrasiloxane and not an organosilyl substituted polyphosphonate as follows from formula (1.41). Without giving any discussion, the authors complete the "evolution" of their views in the communication [113], where a third formula is presented:

$$\left[-O_{0.5}-\overset{\overset{\large R}{|}}{\underset{\underset{\large OR'}{|}}{Si}}-O-\overset{\overset{\large O}{\|}}{\underset{\underset{\large OR'}{|}}{P}}-O_{0.5}- \right]_{2n} \qquad (1.43)$$

$$(R = CH = CH_2,\ R' = C_2H_5)$$

This is the structure of a regular phosphonasiloxane, since the introduction of the subscript 2n does not convert formula (1.43) into (1.42), which corresponds to a "head-to-head" and not a "head-to-tail" structure. The authors give no concrete demonstration of the character of the alternation of the Si–O and P–O groups in the polymer. At the same time, there is an indication [113] that a polymer of analogous structure is obtained by the reaction of phosphorus pentoxide with $[(CH_2{=}CH)(C_2H_5O)_2Si]_2O$.

The reaction of phosphoric acid with tetraethoxysilane leads to silicophosphoric anhydride [62, 62a], while heating tetramethoxysilane with trimethyl phosphate at 200-250°C liberates dimethyl ether and forms polyphosphonasiloxanes [318].

Since the cleavage of disiloxanes by phosphorus compounds will be examined in detail in Section 1.1.5, here we will only report that ether is not liberated in the reaction of sym-divinyltetraethoxydisiloxane with P_2O_5. There is no doubt that in this case the introduction of phosphorus into the structure proceeds through cleavage of the siloxane bond. However, the possibility of simultaneous ethoxylation of the phosphorus in this process is still more doubtful and the description of the structure of the polymer by formula (1.41) could hardly be completely logical.

In the opinion of Kreshkov [105, 228], all reactions of $Si(OR)_4$ with phosphorus pentoxide proceeds through a transesterification stage. With a molar ratio of the reagents of 3:1 in the case of tetraaryloxysilanes, for example, $Si(OC_6H_4CH_3)_4$, the reaction is complete at this. With tetraalkoxysilanes there are further conversions with the formation of solid phosphonasiloxanes with P−O−R bonds. The composition and structure of the polymers are not the same with different tetraalkoxysilanes [93, 105, 118]. Only polymers in which R = C_2H_5 and C_4H_9 have the ratio Si:P = 1. The presence of hydroxyl groups and water of crystallization were explained by the authors of [105] by the strongly hygroscopic nature of the reaction products and hence are not connected with the main polymer formation process.

According to patent data [298, 300, 410], the interaction of the reaction products of polydiorganosiloxanes with HPO_3, H_3PO_4, $H_4P_2O_7$, or P_2O_5, with a viscosity no less than 2000 centistokes, with alkoxysilanes of the type $R_nSi(OR')_{4-n}$ (n = 0 or 1) may be used as a method of "cold vulcanization." Thus, when tetraethoxysilane is added to the product from the polymerization of octamethylcyclotetrasiloxane by phosphorus pentoxide (molar ratio 1000:1), after 2 days there is obtained a rubber having a tensile strength of 17 kg/cm^2 and a relative elongation of 200%.

The reaction of tetraaryloxysilanes with phosphorus pentasulfide proceeds extremely peculiarly. Thus, hydrogen sulfide is liberated when P_4S_{10} is heated with $(C_6H_5O)_4Si$. Hydrolysis of the reaction product forms hydroxyphenylphosphinic acid [39, 40, 120]:

$$Si(OC_6H_5)_4 + P_4S_{10} \longrightarrow 2H_2S + Si(OC_6H_4PS_2)_4 \xrightarrow{H_2O} \longrightarrow 4HOC_6H_4P(O)(OH)_2 + 8H_2S + Si(OH)_4 \quad (1.44)$$

Hydroxycresylphosphinic acids may be obtained analogously.

In the interaction of tetrakis(vinylphenylphosphonyloxy) titanium with excess triethylbutoxysilane or tetrabutoxytitanium, there is the replacement of the hydrogen atom in the phosphonyl group by a triethylsilyl or tributoxytitanyl group [26].

The reaction of tetrakis(methylphosphonyl)titanium with triethylbutoxysilane evidently proceeds by a different scheme, which includes not only replacement of the hydroxyl group, but also rupture of the Ti−O−P bond, since bis(triethylsilyl) methylphosphonate is found in the reaction products in addition to butanol [26]:

$$Ti\left[\begin{matrix} OH \\ | \\ OP(O)R \end{matrix}\right]_4 + 4(C_2H_5)_3SiOC_4H_9 \xrightarrow[-4C_4H_9OH]{} Ti\left[\begin{matrix} OSi(C_2H_5)_3 \\ | \\ OP(O)R \end{matrix}\right]_4 \quad (1.45)$$

1.1.3. Reactions of Alkoxysilanes and Acyloxysilanes with Phosphorus Halides and Acid Halides of Phosphorus Acids

In the patent [459], which was issued in 1945, it is reported that the reaction of trimethylethoxysilane with PCl_5 (1:1; 70°C) is accompanied by the replacement of the ethoxyl group by chlorine. According to data in [320], the reaction between trimethylalkoxylsilanes and PCl_3 has an analogous nature and proceeds in stages:

$$PCl_3 \xrightarrow[-R_3SiCl]{+R_3SiOR'} R'OPCl_2 \xrightarrow[-R_3SiCl]{+R_3SiOR'} (R'O)_2PCl \xrightarrow[-R_3SiCl]{+R_3SiOR'} (R'O)_3P \qquad (1.46)$$

Stopping the reaction at a particular stage depends on the ratio of the components, the experimental conditions, and the character of the radical R'. Thus, when R' = C_4H_9 with a molar ratio of the reagents of 1:1, butoxydichlorophosphine (89%) and trimethylchlorosilane (97%) are obtained predominantly. With a ratio of 2:1, all possible forms of $(RO)_xPCl_{3-x}$ are present in the reaction products.* However, to obtain a satisfactory yield of tributyl phosphite it is necessary to have a ratio $R_3SiOR':PCl_3 \geq$ 1:1 and also to heat the reaction mixture for 3 h at 70°C, even though reaction (1.46) begins even at −10 to +15°C.

The stepwise nature of reaction (1.46) is confirmed by the fact that if an equimolecular mixture of trimethylbutoxysilane and butoxydichlorophosphine is left for 12 h at 15°C, there are formed trimethylchlorosilane (96%) and also dibutoxychlorosilane and tributyl phosphite. Correspondingly, the reaction of trimethylbutoxysilane with $(C_4H_9O)_2PCl$ for 5 h at 65°C gives ~91% of trimethylchlorosilane; tributyl phosphite is also formed in this case.

The replacement of R' = n-C_4H_9 by sec-C_4H_9 in R_3SiOR' changes hardly anything, and trimethylchlorosilane (97%) and sec-butoxydichlorophosphine (79%) are formed. However, $(CH_3)_3CCl$ is formed when R = tert-C_4H_9. In the reaction of PCl_3 with benzhydryloxytrimethylsilane not even traces of an alkoxychlorophosphine are detected and only $(C_6H_5)_2CHCl$ is isolated. With (1-phenylethoxy)-trimethylsilane the formation of $C_6H_5(CH_3)CHOPCl_2$ is observed, although this decomposes even at 18°C.

*The yields are 19.4, 27.3, and 19%, respectively, with x = 1, 2, and 3.

When phosphorus trichloride is replaced by the tribromide, the reaction with butoxy-, sec-butoxy-, and isobutoxytrimethylsilanes proceeds generally in accordance with scheme (1.46), stopping at the second stage. A reaction for 6 h at room temperature gives 90-97% of trimethylbromosilane, 77-81% $R'OPBr_2$, and 66-69% $(R'O)_2PBr$ [335]. In comparison, (1-cyclohexylethoxy)-trimethylsilane forms only $R'OPBr_2$, and in reduced yield (51%). When $R' = (CH_3)_3CCH_2$ (reagent ratio 1:2, 173°C, 16 h), trimethylbromosilane and neopentyl bromide (85%) were isolated [544]. The paper [544] is remarkable in that the view was put forward for the first time that there is the possibility of the intermediate formation of a compound with the group Si−O−P in the reaction of trialkoxysilanes with PX_3:

$$R_3SiOR' + PBr_3 \longrightarrow R'Br + R_3SiOPBr_2 \longrightarrow R_3SiBr + [OPBr] \qquad (1.47)$$

However, this possibility remained a hypothesis until 1957, when it was established [62, 65, 205] that the use of excess R_3SiOR', removal of the RX from the reaction zone, and the use of catalysts* (Lewis acids) direct the reaction along a radically different course:

$$3R_3SiOR' + PBr_3 \longrightarrow (R_3SiO)_3P + 3R'Br \qquad (1.48)$$

When $R = R' = C_2H_5$ only 75% of the theoretically possible amount of ethyl bromide is liberated and together with tris(triethylsilyl) phosphite (32% yield), 14% of bis(triethylsilyl) ethylphosphonate is formed, and this is explained by an Arbuzov side reaction:

$$(R_3SiO)_3P + R'Br \longrightarrow (R_3SiO)_2P(O)R' + R_3SiBr \qquad (1.49)$$

In the case of $(CH_3)_3SiOC_4H_9$ the isomerization product is not detected in the reaction mixture, while butyl bromide is liberated in almost the theoretical amount.

The mechanism of reaction (1.48) was first represented by the following scheme [205]:

$$\begin{gathered} 2PX_3 + ZnCl_2 \longrightarrow 2PX_2^+ + [ZnCl_2X_2]^{2-} \\ R_3SiOR' + PX_2^+ \longrightarrow R_3Si{-}\overset{+}{\underset{\large R}{\overset{|}{O}}}\!{-}PX_2 \longrightarrow R_3SiOPX_2 + R^+ \end{gathered} \qquad (1.50)$$

* 0.5-0.6 mole % of anhydrous $ZnCl_2$, $SnCl_2$, or $FeCl_3$. Reaction (1.48) does not proceed with PCl_3.

$$2R^+ + [ZnCl_2X_2]^{2-} \longrightarrow 2RX + ZnCl_2$$
$$R^+ + PX_3 \longrightarrow RX + PX_2^+ \text{ etc.}$$

A new sequence begins with R_3SiOPX_2.

Later [62, 66] it was concluded that the mechanism of the formation of the group Si–O–P in the catalytic reaction of trialkylalkoxysilanes with PBr_3 is based on the intermediate formation of a six-membered cyclic active complex:

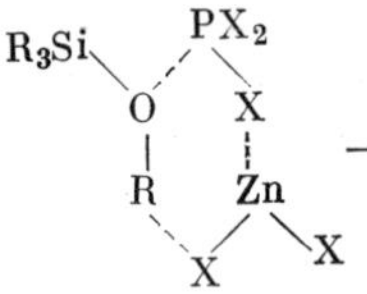

$$\longrightarrow R_3SiOPX_2 + R'X + ZnX_2 \qquad (1.50a)$$

The investigations of Voronkov, Zgonnik, and Skorik [62, 62a, 66, 205] show that specific conditions are needed for the formation of the group Si–O–P in the reaction of phosphorus halides with alkoxysilanes. Therefore it is not surprising that in the reaction of dimethyldiethoxysilane with PCl_5 (1:2) only dimethyldichlorosilane was isolated and in addition, $POCl_3$ and C_2H_5Cl were formed [440]. The formation of analogous compounds was reported in 1872 by Ladenburg [437], who described the reaction of PCl_5 with diethyldiethoxysilane, but in 1956 the preparation of 3-cyanopropyldiethoxychlorosilane from 3-cyanopropyltriethoxysilane and PCl_5 was patented [419a]. According to patent data in [316] polymeric products containing the group N=P–O–Si are formed by heating trichlorocyclotriphosphazene with phenyltriethoxysilane and diphenoxydiethoxysilane at 200-240°C.

The reaction of trialkylacetoxysilanes with PCl_3 (molar ratio 3:1) yields bis(trialkylsilyl) acetylphosphonates, triethylchlorosilane, and acetyl chloride [94, 164, 166, 168, 169]. The initial products of this reaction are tris(trialkylsilyl) phosphites, which are then converted into bis(trialkylsilyl) acetylphosphonates and trialkylchlorosilanes by the acetyl chloride formed [94]:

$$3R_3SiOCOCH_3 + PCl_3 \longrightarrow (R_3SiO)_3P + 3CH_3COCl \qquad (1.51)$$

$$(R_3SiO)_3P + CH_3COCl \longrightarrow (R_3SiO)_2P(O)COCH_3 + R_3SiCl \qquad (1.52)$$

The cleavage of bis(trimethylsilyl) sulfates by PCl_5 in cyclohexane at 80°C proceeds with the formation of trimethylsilyl chlorosulfate, trimethylchlorosilane, and $POCl_3$ [305, 306]:

$$(R_3SiO)_2SO_2 + PCl_5 \longrightarrow R_3SiOSO_2Cl + R_3SiCl + POCl_3 \qquad (1.53)$$

As regards full esters of orthosilicic acid, Mendeleev's description in 1860 of the reaction of tetraethoxysilane with PCl_5 is the first report in the history of organosilicon chemistry of the reaction of $(RO)_4Si$ with phosphorus compounds [135]. According to Mendeleev's data, it was found that even with a molar ratio of $PCl_5 : (C_2H_5O)_4Si = 4:1$ half of the phosphorus pentachloride does not react and $SiCl_4$ is absent from the reaction products. Mendeleev proposed the following reaction scheme:

$$(RO)_4Si + 2PCl_5 \longrightarrow (RO)_2SiCl_2 + 2RCl + 2POCl_3 \qquad (1.54)$$
$$(R = C_2H_5)$$

Later investigations [440] led to analogous results. It should be noted that according to Friedel and Crafts [341] the main product of reaction (1.54) is not diethoxydichlorosilane, but triethoxychlorosilane.

In the opinion of Friedel and Ladenburg [342], at low temperatures the reaction of PCl_3 with ethyltriethoxysilane proceeds by an exchange mechanism, i.e., with the formation of ethyltrichlorosilane and triethyl phosphite. These data are hypothetical to a considerable degree, since no separation and identification of the products were carried out. At temperatures of $\geq 180°C$ there are formed ethyl chloride and a yellow solid which is pyrophoric due to free phosphorus in it. Removal of the latter leads to a compound which is stable in air and gives phosphoric acid on hydrolysis.

Data on the conversion of tetraethoxysilane under the action of PCl_3 are contradictory. According to Stokes [547, 548], their reaction forms only a white solid of indefinite character. But in [33] it is reported that the reaction proceeds without the liberation of ethyl chloride, while the reaction products are ethoxychlorosilanes, Menshutkin's acid chloride $C_2H_5OPCl_2$ (ethoxydichlorophosphine), and a liquid which contains 23% Si and 8.5% P and whose structure was not determined. According to data in [494, 495] tetraethoxysilane does not react with PCl_3 at 20°C, but when a mixture of them is heated to boiling the reaction described by the following scheme occurs:

$$(C_2H_5O)_4Si + PCl_3 \rightleftarrows (C_2H_5O)_nPCl_{3-n} + (C_2H_5O)_{4-n}SiCl_n \quad (1.55)$$

In the case of the reaction of PCl_3 with tetraphenoxysilane, there is complete exchange with the formation of $SiCl_4$ and triphenyl phosphite [91, 92, 104, 106]:

$$3(C_6H_5O)_4Si + 4PCl_3 \longrightarrow 4(C_6H_5O)_3P + 3SiCl_4 \quad (1.56)$$

The reaction of tetraalkoxysilanes with PBr_3 forms ethyl bromide and a polymer [494, 495, 499]. According to [501] the process may be divided into four stages: 1) spontaneous heat evolution by the reaction mixture immediately after mixing of the reagents; 2) evolution of the alkyl bromide; 3) formation of the polymer and its separation from the reaction mixture; 4) a vigorous exothermic reaction. The first stage is due to the formation of a complex and the second is due to its decomposition:

$$(RO)_4Si + PBr_3 \longrightarrow (RO)_3SiOR \cdot PBr_3 \longrightarrow (RO)_3SiOPBr_2 + RBr \quad (1.57)$$

Then the trialkoxysiloxydibromophosphine formed reacts with excess tetraethoxysilane [499]. Naturally this scheme is inadequate to describe the process as a whole, particularly since the liquid polymer obtained by the reaction of $(C_2H_5O)_4Si$ with PBr_3 (3:1) gives a positive reaction for trivalent phosphorus, while polymers obtained with a different ratio of the reagents or with $R = C_3H_7$, C_4H_9 contain pentavalent phosphorus.

The products from the reaction of tetraphenoxysilane with PI_3 (3:2) are triphenyl phosphite and triphenoxyiodosilane [499]:

$$3(C_6H_5O)_4Si + PI_3 \longrightarrow (C_6H_5O)_3P + 3(C_6H_5O)_3SiI \quad (1.58)$$

In the case of tetraalkoxysilanes $(RO)_4Si$ with $R = CH_3$ or C_2H_5, the reaction has a different character and an alkyl iodide, red phosphorus, a hexaalkoxydisiloxane, and a polymer containing the Si–O–P(O) group are obtained [495, 499]. A multistage scheme has been proposed to explain the formation of these products:

$$(RO)_3SiOR + PI_3 - \begin{cases} \longrightarrow [(RO)_3SiO]_3P + RI \\ \longrightarrow (RO)_3SiI + (RO)_3P \end{cases} \quad (1.59)$$

$$[(RO)_3SiO]_3P \longrightarrow 3(RO)_3SiO\cdot + P\cdot; \qquad 4P\cdot \longrightarrow P_4$$

$$2(RO)_3SiO\cdot + (RO)_3P \longrightarrow [(RO)_3Si]_2O + (RO)_3PO$$

$$(RO)_3SiI + (RO)_3PO \longrightarrow [(RO)_3SiO]_3PO$$

$$[(RO)_3SiO]_3PO + PI_3 - \begin{cases} \longrightarrow \{[(RO)_3SiO]_2P(O)OSi(OR)_2O\}_3P + RI \\ \longrightarrow [(RO)_3SiO]_2P(O)OSi(OR)_2I + (RO)_3P \end{cases}$$

The reaction of triorganoorganoxysilanes with phosphorus oxyhalides differs appreciably from a reaction involving PX_3. Thus, for example, the reaction of $(CH_3)_3SiOR$ with $POCl_3$ (2:1) was described in [334, 335] by scheme (1.60), which is shorter than (1.46):

$$POX_3 \xrightarrow[-R_3SiX]{+R_3SiOR'} X_2POR' \xrightarrow[-R_3SiX]{+R_3SiOR'} XP(OR')_2 \qquad (1.60)$$

It should be noted that in the case of trimethylbutoxy- and trimethylisobutoxysilanes the second stage proceeds extremely slowly. While the first stage is complete in 16 h at 20°C, heating an equimolar mixture of trimethylisobutoxysilane and isobutoxydichlorophosphine at 170°C for 5 h is necessary to obtain diisobutoxychlorophosphine [335]. Trimethyl-sec-butoxysilane, which behaves completely analogously to the butoxy isomer in reaction (1.46), gives only sec-butoxydichlorophosphine (90%) when treated with $POCl_3$. Trimethyl-tert-butoxysilane does not undergo this reaction even at 110°C. In this case there is obviously an effect from steric hindrance. With the molar ratio $(CH_3)_3SiOR:POCl_3$ = 1:1 reaction (1.60) is limited to the first stage regardless of the chain length of the alkoxyl radical [526]. The structure of the radical attached to the oxygen has a greater effect with a change to an aliphatic-aromatic derivative [335, 526]:

$$R_3SiOR' + POCl_3 \longrightarrow R_3SiOP(O)Cl_2 + R'Cl$$
$$[R' = (C_6H_5)_2CH, \; C_6H_5(CH_3)CH] \qquad (1.61)$$

The use of "pyrophosphoryl chloride" in reaction (1.61), even in the case of trimethylmethoxysilane, leads to the formation of trimethylsilyl dichlorophosphate [526]:

$$(CH_3)_3SiOCH_3 + Cl_2P(O)OP(O)Cl_2 \longrightarrow (CH_3)_3SiOP(O)Cl_2 + CH_3OP(O)Cl_2 \qquad (1.62)$$

The reaction of trimethylbutoxysilane with $POBr_3$ proceeds more readily than with phosphorus oxychloride. With a molar ratio of 1:1, $C_4H_9OP(O)Br_2$ and trimethylbromosilane (85%) are obtained. A component ratio of 3:1 leads to tributyl phosphate [335].

According to data in [93, 107] the addition of phosphorus oxychloride to dimethyldiethoxysilane produces heat evolution by the mixture, accompanied by the liberation of ethyl chloride. After the reaction mixture had been boiled in an oil bath (1-1.5 h), the diethyl ester and chloroester of phosphoric acid were isolated.

Finally, above 200 °C a precipitate is formed, which is analogous to the product obtained by the reaction of dimethyldiethoxysilane with phosphorus pentoxide.

In 1957–1960 Kreshkov and Karateev gave reports at several conferences and published in a series of publications [91–93, 104–106, 111, 228] practically the same material on the action of $POCl_3$ on tetraalkoxysilanes. However, numerous repetitions of this report did not shed light on the problem. The reaction between tetraethoxysilane and $POCl_3$ is described by the following scheme in [91]:

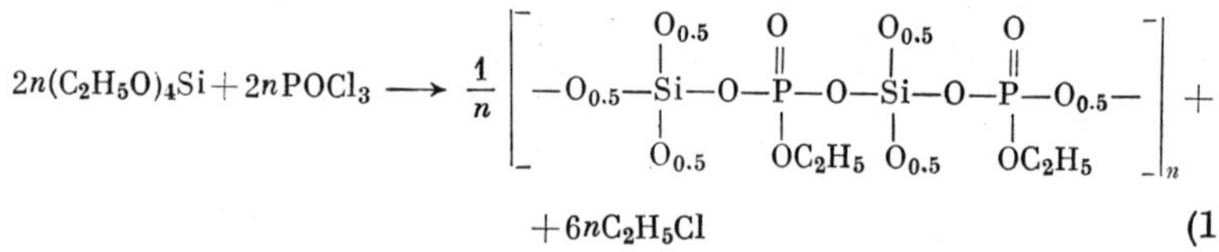

$$2n(C_2H_5O)_4Si + 2nPOCl_3 \longrightarrow \frac{1}{n} \left[-O_{0.5} \begin{matrix} O_{0.5} & O & O_{0.5} & O \\ | & || & | & || \\ -Si-O-P-O-Si-O-P-O_{0.5}- \\ | & | & | & | \\ O_{0.5} & OC_2H_5 & O_{0.5} & OC_2H_5 \end{matrix} \right]_n +$$

$$+ 6nC_2H_5Cl \qquad (1.63)$$

However, the structure of the polymer formed from these two reactants is given in [106] as

$$\left[\begin{matrix} | & & | & & | & & | \\ O & O & O & O \\ | & || & | & || \\ HO-Si-O-P-O-Si-O-P-OH \cdot H_2O \\ | & | & | & | \\ O & OH & O & OC_2H_5 \\ | & & & | \end{matrix} \right]_n$$

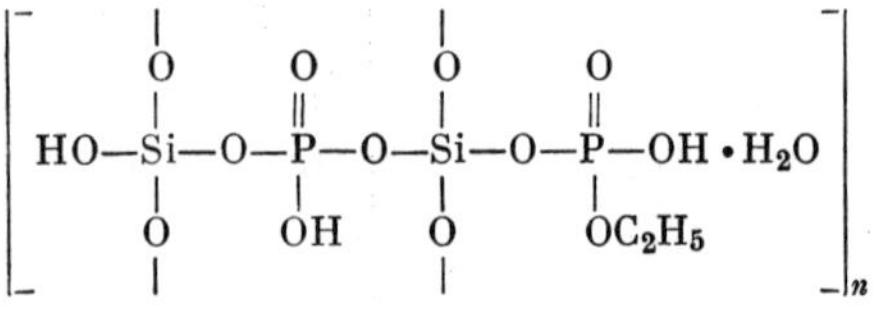

To explain the formation of these polymers, Kreshkov and Karateev started from the assumption that there is essentially transesterification in the reaction of $(RO)_4Si$ and $POCl_3$ and that the formation of the macromolecule is stepwise in nature.* Unfortunately, none of the schemes they presented [106, 111, 228] were substantiated and the structure of the polymers was not established unequivocally and indisputably.

At the same time, according to data in [429, 430], the reaction of tetraethoxysilane with phosphorus oxychloride at ~150°C, with molar ratios of the components from 5:1 to 1:10, leads mainly to a solid, vitreous, amorphous polymer, whose composition is practically independent of the starting ratio of Si:P and is close to $[SiP_2ClO_{5.5}(OC_2H_5)_2]_x$. The highest polymer yield is observed with

---

* This is contradictory to the data of Stokes [547, 548], who showed that $SiCl_4$ is not formed when tetraethoxysilane is treated with $POCl_3$.

a starting ratio $(C_2H_5O)_4Si:POCl_3$ = 1:1. In this case the conversion of phosphorus oxychloride reaches 94%, while 54% of the tetraethoxysilane participates in the formation of the polymer. Silicon is also present in the volatile by-products of the reaction, and the authors of [430] ascribe their formation to transesterification. At the same time the formation of the polymer is associated with condensation processes, which determine its structure:

$$\cdots-\underset{\underset{\displaystyle OR}{|}}{\overset{\overset{\displaystyle OR}{|}}{Si}}-O-\underset{\underset{\displaystyle O}{|}}{\overset{\overset{\displaystyle O}{\|}}{P}}-O-\underset{\underset{\displaystyle Cl}{|}}{\overset{\overset{\displaystyle O}{\|}}{P}}-O-\underset{\underset{\displaystyle Cl}{|}}{\overset{\overset{\displaystyle O}{\|}}{P}}-O-\underset{\underset{\displaystyle O}{|}}{\overset{\overset{\displaystyle O}{\|}}{P}}-O-\underset{\underset{\displaystyle OR}{|}}{\overset{\overset{\displaystyle OR}{|}}{Si}}-\cdots$$

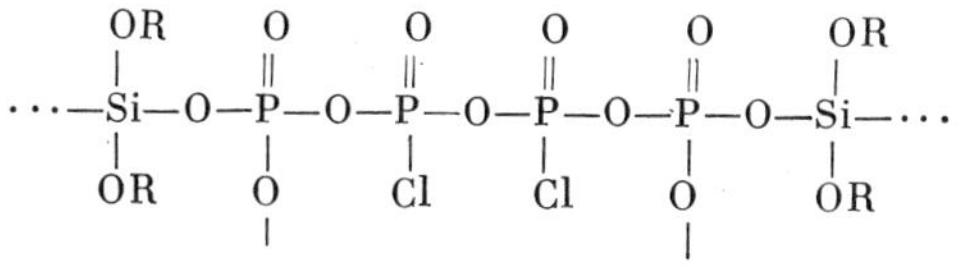

In the opinion of the authors of [430], this structure is demonstrated by the formation of pyrophosphoric acid when the polymer is hydrolyzed by cold water; in contrast to this, orthophosphoric acid is obtained on heating.

Although the data of Kohlschutter and Simoleit [429, 430] differ from the results of Kreshkov and Karateev [106, 111, 228], primarily with respect to the composition of the polymers obtained (this may be ascribed to the difference in the experimental conditions), the main discrepancy lies in the approach to the explanation of the reaction mechanism. While in one case [111, 228], transesterification is a side reaction, in the other [429, 430] it is one of the intermediate stages in the process. The latter is more probable in our opinion.

Heating tetramethoxysilane with $POCl_3$ (1:2) in an autoclave at 180°C leads to quantitative elimination of the methoxyl groups in the form of CH_3Cl and the formation of the acid chloride of silicophosphoric acid $SiP_2O_6Cl_2$ [390]. In this connection it may be remembered that Stokes [547, 548] reported that the reaction of ethoxychlorisilanes with excess phosphorus oxychloride occurs above 175°C and with values of n in $(C_2H_5O)_n SiCl_{4-n}$ from 1 to 3 it gives the same products, namely, $SiCl_4$, C_2H_5Cl, and $SiP_2O_6Cl_2$ (the acid chloride of silicophosphoric acid or, in Stokes terminology, "silicopyrophosphoryl chloride"). The action of PCl_5 on the latter leads to the formation of $POCl_3$ and $SiCl_4$. The silicopyrophosphoryl chloride decomposes at red heat and Stokes described this by the highly debatable scheme

$$6SiP_2O_6Cl_2 \longrightarrow 3SiO_2+3(SiO_2\cdot P_2O_5)+4POCl_3+P_2O_5 \qquad (1.64)$$

When a mixture of phosphorus oxychloride and tetra-o-cresoxysilane in benzene is boiled for 15 h, the only reaction products are found to be tri-o-cresyl phosphate and silicon tetrachloride [91, 92, 106]:

$$3(o\text{-}CH_3C_6H_4O)_4Si + 4POCl_3 \longrightarrow 4(o\text{-}CH_3C_6H_4O)_3PO + 3SiCl_4 \quad (1.65)$$

On the other hand, prolonged heating of tetraphenoxysilane with $POCl_3$ at 240°C gives $SiCl_4$ and phenoxychlorophosphates [549]:

$$(C_6H_5O)_4Si + POCl_3 \longrightarrow SiCl_4 + (C_6H_5O)_nPOCl_{3-n} \quad (1.66)$$
$$(n = 1\text{—}3)$$

It is not possible to obtain phenylated "silicophosphoric acid" by the reaction of $SiCl_4$ with triphenyl phosphate and phenyl dimethyl phosphate [390]. However, it was obtained in accordance with the scheme

$$(CH_3O)_4Si + 2Cl_2P(O)(OC_6H_5) \longrightarrow [(C_6H_5)_2SiP_2O_8] + 4CH_3Cl \quad (1.67)$$

Yet another example of the reaction of acid halides of phosphorus acids is reaction (1.68) [320]:

$$(CH_3O_4)Si + ClP(O)(OC_2H_5)_2 \longrightarrow (CH_3O)_3SiOP(O)(OC_2H_5)_2 + CH_3Cl \quad (1.68)$$

The yield of trimethoxysilyl diethyl phosphate (10%) in this case is still lower than the yield (26%) of the reaction of trimethoxychlorosilane with triethyl phosphate in accordance with scheme (1.12). This low yield is due to decomposition of the $ClP(O) \cdot (OC_2H_5)_2$ with the elimination of ethyl chloride and the formation of polyphosphates.

Triorganoacetoxysilanes react with dichlorides of organophosphinic acids with the formation of acetyl chloride and bis(triorganosilyl) organophosphonates [136, 160, 184):

$$2R_3SiOCOCH_3 + Cl_2P(O)R' \longrightarrow (R_3SiO)_2P(O)R' + 2CH_3COCl \quad (1.69)$$

In the case of organo dichlorophosphites and diorgano chlorophosphites the reaction with triorganoacetoxysilanes proceeds by other routes:

$$2R_3SiOCOCH_3 + Cl_2POR' \longrightarrow R_3SiOP(O)(OR')COCH_3 + R_3SiCl + CH_3COCl \quad (1.70)$$

$$R_3SiOCOCH_3 + ClP(OR')_2 \longrightarrow (R'O)_2P(O)COCH_3 + R_3SiCl \quad (1.71)$$

1.1.4. Reactions of Silanols and Sodium Silanolates with Phosphorus Compounds

The corresponding trialkylsilyl esters are obtained smoothly by azeotropic distillation of water from a mixture of a trialkylsilanol and oxygen acids of phosphorus with an inert solvent (benzene or toluene) [62, 148, 158, 161, 163, 308]:

$$n\mathrm{R_3SiOH} + (\mathrm{HO})_n\mathrm{P(O)R'_{3-n}} \rightleftarrows (\mathrm{R_3SiO})_n\mathrm{P(O)R'_{3-n}} + n\mathrm{H_2O} \qquad (1.72)$$

The mechanism of the reaction, which proceeds in stages, is represented as follows [148, 161]:

$$\mathrm{R_3SiOH} + \mathrm{H_3PO_4} \rightleftarrows [\mathrm{R_3SiOH_2^+} + {}^{-}\mathrm{O{-}P(O)(OH)_2}] \rightleftarrows \mathrm{R_3SiOP(O)(OH)_2} + \mathrm{H_2O}$$

$$\mathrm{R_3SiOH} + \mathrm{R_3SiOP(O)(OH)_2} \rightleftarrows \left[\mathrm{R_3SiOH_2^+} + {}^{-}\mathrm{O{-}\underset{\displaystyle OH}{\underset{|}{P}}(O)OSiR_3}\right] \rightleftarrows$$

$$\rightleftarrows (\mathrm{R_3SiO})_2\mathrm{P(O)OH} + \mathrm{H_2O} \text{ etc.} \qquad (1.73)$$

The first stage, the formation of a siloxonium ion, proceeds through the donation of a proton of the phosphoric acid to the silanol. By subsequent nucleophilic attack a water molecule is liberated and the primary substitution product is formed and this is capable of reacting further with the silanol. This course of the reaction is favored both by the tendency of silicon compounds to undergo S_N2 reactions readily and the shift in the equilibrium of the process with continuous removal of the water from the reaction zone (in principle, reaction 1.72 is reversible).

A clearer representation of the mechanism of reaction (1.72) is given by the following scheme [62]:

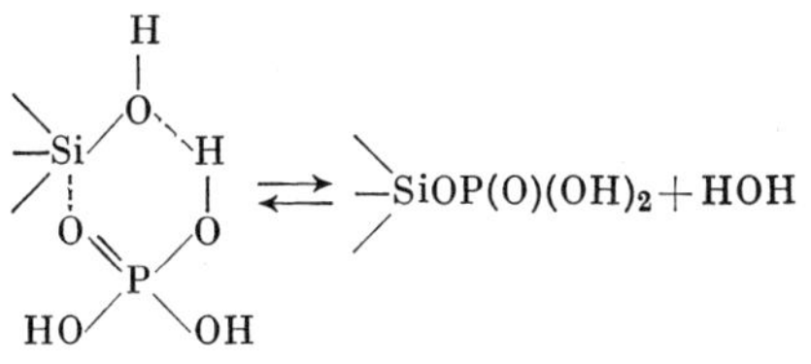
H
O
H
Si
O
O
P
HO
OH
SiOP(O)(OH)₂+HOH

(1.74)

The reaction of silanols (1.72) is accompanied to some extent by homocondensation, but with methylphosphinic acid this process is of less importance than it is with phosphoric acid.

Anhydrocondensation complicates the reaction of tripropylsilanol with esters of phosphoric acids to a considerable degree [529, 530]:

$$(C_3H_7)_3SiOH + (CH_3O)_nP(O)(CH_3)_{3-n} \longrightarrow$$
$$\longrightarrow (C_3H_7)_3SiOP(O)(OCH_3)_{n-1}(CH_3)_{3-n} + CH_3OH \qquad (1.75)$$
$$(n = 1 \text{ or } 3)$$

In the reaction of tripropylsilanol with dimethyl methylphosphonphate at ~150°C, the dehydration of the silanol actually eliminates the phosphorus derivative from the reaction, since the phosphorus derivative dissolves in the water liberated in the condensation and the reaction mixture is converted to a two-phase system. As a result there is only 5% of reaction (1.75) in this case. In the reaction of tripropylsilanol with trimethyl phosphate (27 h at 210°C) hexapropyldisiloxane and dimethyl tripropylsilyl phosphate are formed in a molar ratio of 69:31.

In the reaction of low molecular weight polydimethylsiloxanediols with oxygen acids of pentavalent phosphorus at room temperature there is the formation of heteroxiloxane coordination polymers, which undergo further conversions, namely, polycondensation and oxidation [122-127, 393, 412]. In the case of dialkylphosphinic acids the polycondensation process has been represented by the following scheme [127]:

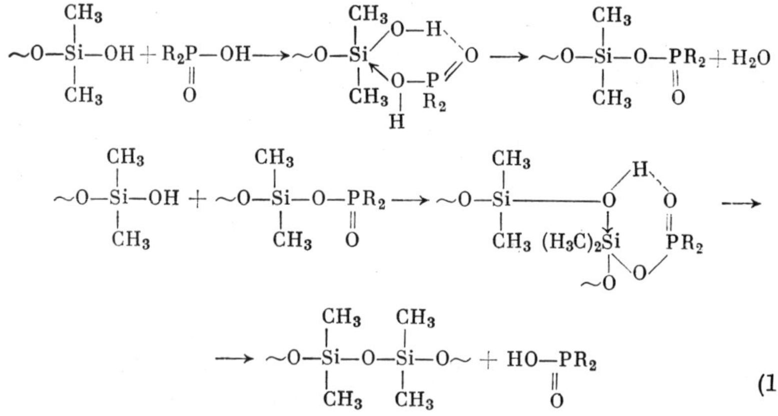

(1.76)

(The symbol ~ denotes the remaining part of the polymer chain)

However, it is more probable that the active cyclic transition complex has a different structure:

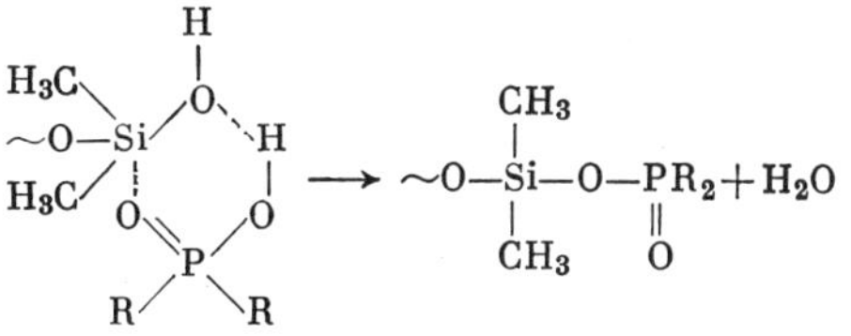
(1.76a)

In contrast to acids of pentavalent phosphorus, esters of phosphoric acid, triorganophosphine oxides, and esters of phosphorous acid do not produce an increase in the viscosity of polydimethylsiloxane-α,ω-diols (P:Si = 1:200) even with prolonged interaction (in an inert gas atmosphere)* [127].

The reaction of triorganosilanols with phosphorus pentoxide may be represented by the simplified scheme [148, 158, 163]

$$6R_3SiOH + P_2O_5 \longrightarrow 2(R_3SiO)_3PO + 3H_2O \qquad (1.77)$$

The possibility of reaction (1.77) indicates the error of the assumption [501-503] that phosphorus atoms are not incorporated into the polymer structure in polycondensations of hydrolyzates of chlorosilanes and that the role of the phosphorus pentoxide is limited to a condensing-dehydrating action on the silanols and siloxanols.† However, a series of experimental data [208, 237, 460, 461] indicates that depending on the amount of phosphorus pentoxide used, the formation of cross-linked polymers competes to some extent with polycondensation of the siloxanols formed by hydrolysis [237]:

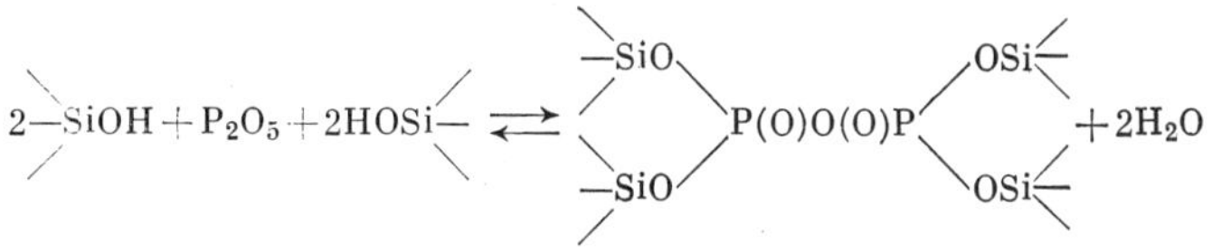
(1.78)

*In [122] and [126], where an increase in the viscosity of the polymer was observed, impure triphenyl phosphate was used [127].

†In the patents [531, 532] it is reported that treatment of hydrolyzates of chlorosilanes containing OH groups with 70-85% phosphoric acid leads to elimination of the OH groups without polymerization or an increase in the viscosity of the product (?!).

Hydrolysis is accompanied by the rupture of Si–O–P bonds with a return to the starting siloxanol or the product of its partial condensation.

The reaction of mixed silyl alkyl esters of phosphoric acid containing highly reactive chlorine atoms at the silicon with trialkylsilanols makes it possible to "grow" the siloxane part of the molecule [321]:

$$(ClR_2SiO)_nP(O)(OR)_{3-n} + nR'_3SiOH \longrightarrow (R'_3SiOSiR_2O)_nP(O)(OR)_{3-n} + nHCl \quad (1.79)$$

$$(n=1 \text{ or } 2)$$

The reaction is carried out in the presence of pyridine in ether. At the same time attempts to obtain tris(triethylsilyl) phosphate by the reaction of triethylsilanol with phosphorus oxychloride in the presence of pyridine [148, 158] and bis(triethylsilyl) chlorophosphinate from triethylsilanol and PCl_3 in the presence of dimethylaniline [94] were unsuccessful since the phosphorus halides were bound in complexes by the amines. In contrast to this, it is reported in [284, 285] that bis(triphenylsilyl) phenylphosphinate was obtained from triphenylsilanol and phenyldichlorophosphine in the presence of pyridine:

$$2(C_6H_5)_3SiOH + Cl_2PC_6H_5 \xrightarrow[-2C_5H_5N\cdot HCl]{+2C_5H_5N} [(C_6H_5)_3SiO]_2PC_6H_5 \quad (1.80)$$

It is possible to obtain trialkylsiloxy derivatives of phosphorus by the reaction of a sodium silanolate instead of the silanol with phosphorus trichloride or $POCl_3$ [231]:

$$3R_3SiONa + Cl_3M \longrightarrow (R_3SiO)_3M + 3NaCl \quad [M = P \text{ or } P(O)] \quad (1.81)$$

There is a serious discrepancy between the constants of tris(triethylsilyl) phosphite obtained by schemes (1.81) [231] and (1.48) [205].

The reaction

$$R_3SiONa + ClP(OR')_2 \longrightarrow R_3SiOP(OR')_2 + NaCl \quad (1.82)$$

was used as an alternative synthesis to prove the structure of the products from the reaction of trialkylhalosilanes with sodium dialkyl phosphites [46, 52].

According to patent data [222, 545, 546] the addition of 0.5-5% of compounds of the type $ROP(O)X_2$ or $RP(O)X_2$ catalyzes the polycondensation of methylsiloxanols. At 20°C the process is complete in 2-20 days and after several hours at 50-150°C. Unfortunately there are no data on the character of the reaction. In the opinion of Rochow [501-503], phosphorus halides and oxyhalides are similar condensing agents to P_2O_5 with respect to siloxanols. In this connection we should mention data [430] which demonstrate the formation of bridges –SiOP(O)(Cl)OSi– when silica gel is treated with $POCl_3$.

There are interesting reactions of silanols and sodium silanolates with phosphonitrilic chlorides $(-PCl_2=N-)_n$ [84-88, 213, 309, 558]. Triethylsilanol does not react with hexachlorocyclotriphosphazene, the trimer of phosphonitrilic chloride $(-PCl_2=N-)_3$, on boiling in benzene [88]. Raising the temperature by replacing the benzene by toluene or xylene leads to reaction, but it is not triethylsiloxy derivatives which are formed, but metapolyphosphimic acid, triethylchlorosilane, and hexaethyldisiloxane. With polyphosphonitrilic chlorides the reaction also proceeds in ether, but the result is the same. In the case of triphenylsilanol hardly any reaction is observed even in boiling o-xylene, and there is only condensation of triphenylsilanol with the elimination of benzene and the formation of polydiphenylsiloxane. A certain amount of a substance containing a small amount of phosphorus, silicon, and chlorine is still observed, but its formation may be connected with the secondary reactions of cleavage of the polysiloxane by the phosphonitrilic chloride.

When the triethylsilanol is replaced by sodium triethylsilanolate there is exchange with hexachlorocyclotriphosphazene, but not condensation. As a result, the disodium salt of phosphonitrilic acid $(-P(ONa)_2=N-)_3$ and the same triethylsilanol and hexaethyldisiloxane are formed. The exchange proceeds at a high rate in the case of sodium triphenylsilanolate, condensation products are not obtained here either.

When hexachlorocyclotriphosphazene is heated with diphenylsilanediol (molar ratio from 1:1 to 1:6) in boiling o-xylene or dichlorobenzene (145-175°C) for 6-9 h, water and a small amount of hydrogen chloride are liberated and polyphosphonitrilic acid $[-P(OH)_2=N-]_n \rightleftharpoons [-P(O)(OH)-NH-]_n$ and polydiphenylsiloxane are formed. Analogous results were obtained by fusing the reagents. When the reaction between hexachlorocyclotriphosphazene and diphenylsilanediol was carried out in the presence of quinoline (1.7–100%) at 90–140°C for 4–9 h, polymetaphosphimic acid and a resinous substance containing 1.5% Si and 11.5% P were obtained [86].

1.1.5. Cleavage of Siloxanes by Phosphorus Compounds

In a series of publications it is reported that the action of phosphorus trichloride, tribromide, or pentachloride, or a mixture of them with $FeCl_3 \cdot 6H_2O$ on hexaorganodisiloxanes, results in cleavage of the Si–O–Si group with almost quantitative formation of triorganohalosilanes [133, 145, 227, 389, 458]. The power of phosphorus (and aluminum) halides to cleave siloxane bonds has even provided a basis for patents [418, 419] aimed at the utilization of waste raw siloxane rubbers and vulcanized mixtures obtained from them.

However, a detailed investigation showed that with a large excess of hexaalkyldisiloxane and in the presence of anhydrous ferric chloride or zinc chloride the reaction with PX_3 may be directed toward the formation of organic compounds of silicon and phosphorus [62, 65, 66, 205]:

$$3R_3SiOSiR_3 + PX_3 \longrightarrow (R_3SiO)_3P + 3R_3SiX \qquad (1.83)$$

When X = Br, distillation of the mixture of reagents gives 25% of tris(trialkylsilyl) phosphite. With phosphorus trichloride, reaction (1.83) is achieved with more difficulty and the yield of $(R_3SiO)_3P$ is only about 7%. In all cases the synthesis must be carried out in an inert gas atmosphere to avoid oxidation of the phosphite to phosphate.

Comparison of the data on the cleavage or organosiloxanes electrophilic halides of various elements (MX_n) leads to the conclusion that the activity of these "Lewis acids" falls in the series B > Al > Ti > Si > P, and this corresponds to the fall in the electron acceptor properties of the central atom of the element M in the halide molecule [62].

Hexaethoxydisiloxane reacts with phosphorus oxychloride with the formation of an organic compound of silicon and phosphorus of undetermined structure [548, 549].

Even when excess hexamethyldisiloxane is used, it reacts with $POCl_3$ at 100°C to give only a monotrimethylsiloxy derivative [525, 526]:

$$R_3SiOSiR_3 + POCl_3 \longrightarrow R_3SiOP(O)Cl_2 + R_3SiCl \quad (1.84)$$

The degree of conversion of $POCl_3$ in this reaction is low and the yield of trimethylsilyl dichlorophosphate is only 11%, though it is quantitative calculated on the amount of phosphorus oxychloride reacting.

Another possibility of preparing the acid chloride of trimethylsilylphosphoric acid is provided by coproportionation:

$$(R_3SiO)_3PO + 2POCl_3 \longrightarrow 3R_3SiOP(O)Cl_2 \quad (1.85)$$

However, scheme (1.85) shows the ideal and not the actual course of the reaction, since the yield of trimethylsilyl dichlorophosphate is only 28% due to the formation of polymeric trimethylsilyl metaphosphate $[(CH_3)_3SiOPO_2]_n$.

The cleavage of hexamethyldisiloxane by the acid chloride of pyrophosphoric acid (pyrophosphoryl chloride) proceeds quantitatively [527]:

$$R_3SiOSiR_3 + Cl_2P(O)OP(O)Cl_2 \longrightarrow 2R_3SiOP(O)Cl_2 \quad (1.86)$$

When hexamethyldisiloxane is replaced in reactions (1.84) and (1.86) by a germanium analog, the character of the reaction does not change. In the reaction of phosphorus oxychloride with trimethyl(trimethylsiloxy)germane, cleavage of the Ge—O bond predominates and not that of the Si—O bond [527]:

$$R_3Si{-}O{-}GeR_3 + POCl_3 \longrightarrow \begin{cases} \xrightarrow{80\%} R_3GeCl + R_3SiOP(O)Cl_2 \\ \xrightarrow{20\%} R_3SiCl + R_3GeOP(O)Cl_2 \end{cases} \quad (1.87)$$

At the same time, the exothermic reaction of trimethyl(trimethylsiloxy)germane with the acid chloride of pyrophosphoric acid gives an equimolecular mixture of trimethylsilyl and trimethylgermyl dichlorophosphates [527]:

$$R_3SiOGeR_3 + [Cl_2P(O)]_2O \longrightarrow R_3SiOP(O)Cl_2 + R_3GeOP(O)Cl_2 \quad (1.88)$$

The acid fluoride of methylphosphinic acid cleaves the siloxane bond in hexaalkyldisiloxanes when a mixture of them is heated to 95°C for 2 h, forming trialkylsilyl methylfluorophosphonates and trialkylfluorosilanes [76, 207]:

$$R_3SiOSiR_3 + CH_3POF_2 \longrightarrow CH_3PF(O)(OSiR_3) + R_3SiF \qquad (1.89)$$

The acid chloride of methylphosphinic acid does not react with hexamethyldisiloxane under these conditions. Their reaction has been observed under more drastic conditions (heating for 10 h in a sealed ampoule at 160°C), but only trimethylchlorosilane was isolated from the reaction mixture [76].

While it is not possible to obtain trialkylsiloxy derivatives by the action of triethylsilanol on hexachlorocyclotriphosphazene, heating $(PNCl_2)_3$ with hexaethyldisiloxane (molar ratio 1:6) at 230°C gives triethylchlorosilane and an oily liquid [87, 88], whose composition corresponds to the replacement of two chlorine atoms in the $(PNCl_2)_3$ molecule by triethylsiloxy groups. It is impossible to distill the substance in vacuum: heating to 260-270°C (1 mm) produces polymerization of it with elimination of hexaethyldisiloxane and the formation of an insoluble white solid condensation product:

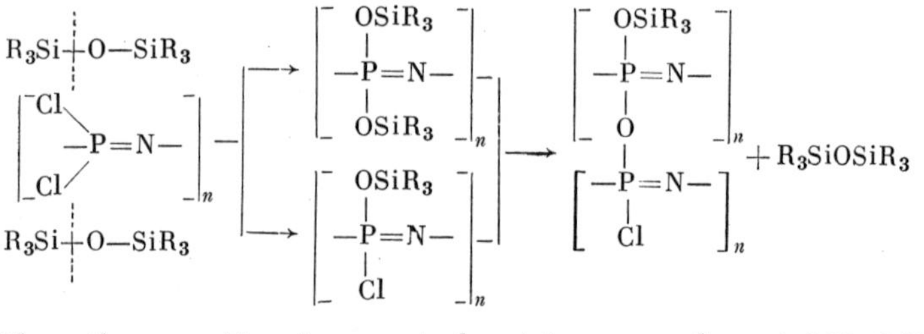

$$+ R_3SiOSiR_3 \qquad (1.90)$$

When the reaction is carried out in ampoules at 260-300°C, powdery polymers are obtained and their silicon content increases, while the phosphorus content falls with a rise in the heating temperature. At lower temperatures there is the possibility of the formation of rubbery polymers which swell in organic solvents and are hydrolyzed by atmospheric moisture. Their composition corresponds to the general formula [88]:

$$\{PN[OSi(C_2H_5)_3]_2\}_x(PON)_y(PNCl_2)_z$$

According to patent data [478], when 0.001-1 wt.% of octachlorocyclotetraphosphazene $(-PCl_2=N-)_4$ is added to a mixture of the hydrolysis products of dimethyldichlorosilane and linear polymethylsiloxanes $(CH_3)_3SiO[(CH_3)_2SiO]_nSi(CH_3)_3$ (viscosity 100 centistokes) after prolonged flushing of the system with air (24 h at 20°C and 24 h at 150°C), a clear oil is obtained with a viscosity of 330,000 centistokes and a high stability. Similar results are observed when the hydrolyzates of other bifunctional diorganosilanes are used [478, 558].

In the patent literature there is a description of the cleavage of cyclosiloxanes by phosphorus trichloride to the corresponding dialkyldichlorosilanes [243] and copolymerization of various cyclosiloxanes by $POCl_3$ [268] and PCl_5 [266, 267] in the presence of small amounts of butanol or water. The reaction of polymethylhydrocyclosiloxanes, obtained by hydrolysis of methyldichlorosilane with PCl_3, is accompanied by reduction of the phosphorus trichloride and the formation of a solid polymeric hydride $(PH)_n$ [477].

While hexaalkyldisiloxanes react readily with sulfuric acid to form bis(trialkylsilyl) sulfates (see Part II, Section 4.1.1), orthophosphoric acid hardly reacts with them under analogous conditions [62].* Tris(trialkylsilyl)phosphates can be synthesized only by azeotropic distillation of the water (with benzene, toluene, or xylene) from a mixture of $R_3SiOSiR_3$ with H_3PO_4 in the presence of catalytic amounts of sulfuric acid or zinc chloride [62, 148, 155, 157, 158, 163]:

$$3R_3SiOSiR_3 + 2H_3PO_4 \longrightarrow 2(R_3SiO)_3PO + 3H_2O \qquad (1.91)$$

The mechanism of this reaction in the presence of sulfuric acid is represented as follows [148, 158]:

$$R_3SiOSiR_3 + H^+ \rightleftarrows R_3Si-\underset{\substack{|\\ H}}{\overset{+}{O}}-SiR_3 \underset{}{\overset{H_3PO_4}{\rightleftarrows}} \underset{\substack{\uparrow\downarrow \\ R_3SiOH + H^+}}{R_3Si\overset{+}{O}H_2} + R_3SiOP(O)(OH)_2 \text{ etc.} \qquad (1.92)$$

A hypothesis was put forward subsequently [62] that this reaction proceeds through the intermediate formation of six-membered cyclic active complexes.

The catalytic action of $ZnCl_2$ is evidently connected with the fact that as an ansolvo acid it formed with orthophosphoric acid a stronger complex ansolvo acid $H^+[(HO)_2P(O)OZnCl_2]^-$, which is

*According to data in [488] the reaction of hexamethyldisiloxane with H_3PO_4 gave an oil which could be used for flameproofing textile fabrics.

actually the catalyst for reaction (1.91) [62, 148, 158]:

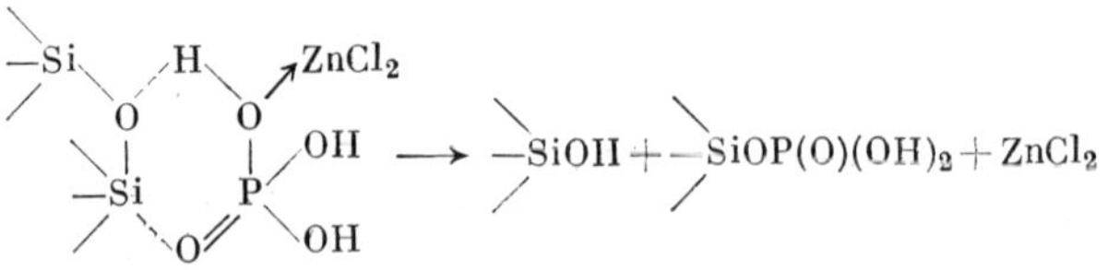

$$\begin{array}{c}\text{—Si} \diagdown \text{H} \diagdown \diagup \text{ZnCl}_2 \\ \diagup \quad \text{O} \quad \text{O} \\ \diagdown \quad | \quad | \diagup \text{OH} \\ \text{—Si} \quad \text{P} \\ \diagup \diagdown \text{O} \diagup \diagdown \text{OH}\end{array} \longrightarrow \text{—SiOH} + \text{—SiOP(O)(OH)}_2 + \text{ZnCl}_2 \qquad (1.93)$$

There are reports of the possibility of the reaction of cyclic polydimethylcyclosiloxanes with orthophosphoric [62, 148, 409, 488] and methylphosphinic [122, 126] acids, with the formation of silicon- and phosphorus-containing organic polymers.[*] However, quaternary phosphonium bases are of much greater value as active agents for the polymerization or copolymerization of cyclosiloxanes for practical purposes [299, 367, 374, 378, 414, 422, 423, 427, 428, 438], and the most convenient of these is tetrabutylphosphonium hydroxide in the form of the siloxanolate [378]:

$$\frac{m}{4}[(CH_3)_2SiO]_4 + [(C_4H_9)_4P]OH \longrightarrow (C_4H_9)_4PO[(CH_3)_2SiO]_mH$$

$$2(C_4H_9)_4PO[(CH_3)_2SiO]_mH \xrightarrow{-H_2O} (C_4H_9)_4PO[(CH_3)_2SiO]_{2m}P(C_4H_9)_4 \qquad (1.94)$$

$$(C_4H_9)_4PO[(CH_3)_2SiO]_mH \rightleftarrows [(C_4H_9)_4P]^+ + {}^-O[(CH_3)_2SiO]_mH$$

The first stage readily proceeds to completion, leaving hardly any unreacted compounds. The elimination of water proceeds much more slowly and the catalyst functions mainly in the form of a mixture of mono- and disiloxanolates. It is assumed that the active part of the catalyst is the siloxanolate anion, but by some mechanism long-chain molecules, with $-OP(C_4H_9)_4$ at the ends, are formed in the polymerization. The use of tetrabutylphosphonium siloxanolates as catalysts is promoted by their high activity, which makes it possible to obtain macromolecular (rubbery) polysiloxanes at moderate temperature with very small amounts of catalyst; thus, for example, from octamethylcyclotetrasiloxane at 110°C in the presence of 0.01 wt.% of catalyst

---

[*]According to data in [409], 85% (syrupy) H_3PO_4 is inactive toward octamethylcyclotetrasiloxane. However, it is reported in the patents [411, 413] that a polymer with a molecular weight of 1310 may be prepared by the reaction of 85% H_3PO_4 with $[CH_3(C_6H_5)SiO]_3$. This should evidently be ascribed to the strain of the six-membered siloxane ring. The ease of interaction of hexamethylcyclotrisiloxane with orthophosphoric acid in comparison with hexaalkyldisiloxanes is also reported in [62, 148].

containing 13.4 mg of $[(C_4H_9)_4P]OH$/ml $[(CH_3)_2SiO]_4$, a polymer is obtained after 3 min with a viscosity of $9.3 \cdot 10^6$ centipoise and after 15 min, $29 \cdot 10^6$ centipoise [378]. It is particularly important that when the polymer is heated at 145-150°C the terminal groups decompose with the formation of products which are inert toward the polymer at high temperatures. As a result rubbers obtained with quaternary phosphonium bases have an improved thermal stability (in comparison with polymers synthesized, for example, with potassium siloxanolates).* Moreover, by utilizing the decomposition of $[R_4P]OH$ at elevated temperatures, it is possible to terminate the polymerization at a given stage and to devise a continuous polymerization process [427].

It is reported that high molecular weight materials, including rubbery ones, may be obtained by polymerization of cyclosiloxanes with $C_6H_5POCl_2$ [510, 511] and $(RO)_2POH$ [236]. In the latter case it is reported that vulcanizates with a tensile strength of 51-54 kg/cm^2 and a relative elongation of 350-500% have been obtained.

Cyclic phosphonasiloxanes of the type

$$\underbrace{(CH_3)_2SiO[(CH_3)_2SiO]_{n-1}P(O)CH_3}_{O},$$

with n = 2 or 3, are very inert toward such active polymerization agents for polymethylcyclosiloxanes as H_2SO_4 or KOH, and are also stable toward $SnCl_4$ and $CH_3PO(OH)_2$ [15]. In mixtures with octamethylcyclotetrasiloxane, polymerization begins only in the presence of small amounts of water, so that we can talk of the polymerization of $[(CH_3)_2SiO]_4$ by the product from hydrolytic cleavage of cyclophosphonasiloxane, $HO[(CH_3)_2SiO]_nP(O)(CH_3)OH$.

Phosphoric acid (and phosphorus pentoxide) cleave siloxane bonds even in silica. Amorphous silicic acid dissolves in orthophosphoric acid on heating [402-404]. However, at 260°C silicon pyrophosphate SiP_2O_7 or, alternatively, $SiO_2 \cdot P_2O_5$, separates from the melt. Above 400°C phosphoric acid reacts with any form of silicic acid [444]. If the starting material be the gel obtained by hydrolysis of SiF_4, then at 260, 360, 700, and 1000°C it is pos-

*Quaternary ammonium bases have similar activity and behavior at different temperatures. However, the complete removal from the polymer of residues of trialkylamines formed in the decomposition, which give the polymer an unpleasant odor, is a difficult task. Polymerization by quaternary phosphonium bases does not suffer from this drawback.

sible to obtain four crystalline modifications of SiP_2O_7. It was shown in [273, 274] that in the system $SiO_2-H_3PO_4$, depending on the ratio of the components and the temperature, it is possible to form two products, namely, the orthophosphate $Si_3(PO_4)_4$ or $3SiO_2 \cdot 2P_2O_5$ and the pyrophosphate SiP_2O_7 or $SiO_2 \cdot P_2O_5$. However there is also the possibility that the primary reaction product is $Si(HPO_4)_2$, which is converted into the pyrophosphate on heating.

According to data in [202], in the reaction of silica with phosphoric acid at 90°C, ternary solid phases of variable composition $nSiO_2 \cdot mP_2O_5 \cdot qH_2O$ are formed. Their decomposition at 325-350°C gives the crystalline silicophosphate $3.5SiO_2 \cdot P_2O_5$.

In 1944, Sauer reported the isolation of tris(trimethylsilyl) phosphate (with no indication of yield) from the products of slow distillation of a large excess of hexamethyldisiloxane over phosphorus pentoxide [509]:

$$3R_3SiOSiR_3 + P_2O_5 \longrightarrow 2(R_3SiO)_3PO \qquad (1.95)$$

Modification of the reaction conditions (stoichiometric amount or excess of P_2O_5, rapid heating, and distillation of the mixture) made reaction (1.95) a convenient preparative method for tris-(triorganosilyl) phosphates (yield ~80%) [61, 62, 320]. With a very large excess of phosphorus pentoxide a polymeric vitreous metaphosphate is formed, and this may be converted into the orthophosphate, i.e., into the tris(triorganosilyl) phosphate, by heating with excess hexaorganodisiloxane [320]:

$$R_3SiOSiR_3 + P_2O_5 \longrightarrow \frac{2}{n}\left[R_3SiOP(O)O-\right]_n \xrightarrow{+(R_3Si)_2O} (R_3SiO)_3PO \qquad (1.96)$$

Attempts to cleave hexachlorodisiloxane by means of phosphorus pentoxide were unsuccessful [62].

The polymerization of octamethylcyclotetrasiloxane by phosphorus pentoxide at 100-150°C has been described [298, 300, 410]. The polymers obtained are solidified in the cold by means of alkylalkoxysilanes due to the terminal phosphorus-containing groups.

1.1.6. Reactions of Phosphorus Compounds with Aminosilanes and Silazans

The reaction of orthophosphoric acid with trialkyl(organoamino)silanes or hexaalkyldisilazans forms tris(trialkylsilyl) phosphates [10, 258]:

$$3R_3SiNHR + H_3PO_4 \longrightarrow (R_3SiO)_3PO + 3R'NH_2 \qquad (1.97)$$

$$3R_3SiNHSiR_3 + 3H_3PO_4 \longrightarrow 2(R_3SiO)_3PO + (NH_4)_3PO_4 \qquad (1.98)$$

Reaction (1.98) is used for the separation of sugar phosphates and mixtures of nucleotides by gas chromatography [398, 400, 401, 563, 653]. For this purpose a solution or suspension of the nucleotides in pyridine is treated with hexamethyldisilazan (normally with the addition of trimethylchlorosilane) and the trimethylsilyl esters obtained (considerably more volatile than the starting phosphates) are chromatographed. Subsequent hydrolysis makes it possible to isolate pure nucleotides. Artificial mixtures containing up to eight different (including isomers) nucleotides were separated in this way [400].

1.1.7. Reaction of Hydrosilanes with Phosphorus Acids

Heating triorganosilanes with phosphorus acids in the presence of colloidal nickel results in their dehydrocondensation [158, 159, 161, 162, 204]:

$$nR_3SiH + (HO)_nP(O)R'_{3-n} \xrightarrow{Ni} (R_3SiO)_nP(O)R_{3-n} + nH_2 \qquad (1.99)$$

$$(n = 2 \text{ or } 3; \quad R = \text{organic}; \quad R' = OH, CH_3, CH_2Cl)$$

The course of the reaction is readily followed through the liberation of hydrogen. In the case of methylphosphinic acid the corresponding organosilicon derivatives are obtained in 75-90% yield. When orthophosphoric acid is used the reaction is complicated by the formation of hexaalkyldisiloxanes, and this reduces the yield of tris(triethylsilyl) phosphate, for example, to 55%.

Scheme (1.99) has no analogy in organic chemistry and results from the hydride character of a hydrogen atom attached directly to a silicon atom. The mechanism of the reaction may be interpreted from the point of view of heterogeneous catalysis [158, 162]. Nucleophilic attack on the surface compound, formed as a result of the capture of the hydride hydrogen by the positively charged surface, by hydroxyl-containing substances with subsequent decomposition of the chemisorbed complex, leads to the formation of triorganosiloxy derivatives:

$$\begin{array}{c} R_3Si - H \cdots Ni \\ \uparrow \\ M{-}O{-}H \end{array} \longrightarrow \begin{array}{c} R_3Si \\ | \\ M{-}O \end{array} + H_2 + Ni \qquad (1.100)$$

Unfortunately up to now there are no kinetic data confirming this mechanism, so that it is hypothetical.*

When the reaction with methylphosphinic acid is carried out with sym-tetraethyldisiloxane, i.e., a compound with two hydride hydrogen atoms in the molecule, there is a vigorous reaction with the formation of a viscous, sticky, soluble polymer of linear structure $[(C_2H_5)_2SiOSi(C_2H_5)_2OP(O)(CH_3)O]_n$ [159, 161, 162].

In the reaction of trialkylsilanes with hydroxymethylphosphinic acid in the presence of colloidal nickel, there is: a) dehydrocondensation with the formation of bis(trialkylsilyl) trialkylsiloxymethylphosphonates (45-60% yield); b) alcoholysis by such intermediate products of dehydrocondensation as $(R_3SiO)_2P(O)CH_2OH$ (the yield of silicon-containing organophosphorus esters is 18-25%); and c) dehydrocondensation of the starting trialkylsilanes with trialkylsilanols liberated as a result of alcoholysis with the formation of hexaalkyldisiloxanes (20-30%). This may be represented by the following scheme [136, 173, 180, 185, 187]:

$$2R_3SiH + (HO)_2P(O)CH_2OH \xrightarrow{Ni} (R_3SiO)_2P(O)CH_2OH + 2H_2$$

$$R_3SiH + (R_3SiO)_2P(O)CH_2OH \xrightarrow{Ni} (R_3SiO)_2P(O)CH_2OSiR_3 + H_2$$

$$(R_3SiO)_2P(O)CH_2OH + (R_3SiO)_2P(O)CH_2OSiR_3 \longrightarrow$$

$$\longrightarrow R_3SiOH + (R_3SiO)_2P(O)CH_2OP(O)(OSiR_3)CH_2OSiR_3 \qquad (1.101)$$

$$R_3SiH + R_3SiOH \xrightarrow{Ni} R_3SiOSiR_3 + H_2$$

The reaction of triorganosilanes with phosphorous acid, with a molar ratio of reagents of 2:1 in the presence of colloidal nickel, leads to the formation of bis(triorganosilyl) phosphites (60-90% yield). According to NMR and IR spectral data, the form with a pentavalent phosphorus atom predominates in their structure [94, 167, 172]:

$$2R_3SiH + H_3PO_3 \xrightarrow{Ni} (R_3SiO)_2PHO + 2H_2 \qquad (1.102)$$

Bis(triorganosilyl) phosphites are capable of reacting further with triorganohydrosilanes with the formation of tris(triorganosilyl) phosphites [94, 172]:

$$(R_3SiO)_2PHO + R_3SiH \xrightarrow{Ni} (R_3SiO)_3P + H_2 \qquad (1.103)$$

* A study of the kinetics of this reaction was begun recently [606].

If the organic radicals attached to the silicon atoms in the molecules of the bis(triorganosilyl) phosphites and triorganosilanes are different, then together with dehydrocondensation according to scheme (1.103) there is also the possibility of transsilylation, e.g. [204]:

$$[(CH_3)_3SiO]_2PHO + (C_2H_5)_3SiH \xrightarrow{Ni} [(CH_3)_3SiO][(C_2H_5)_3SiO]_2P + \\ + [(C_2H_5)_3SiO]_3P + (CH_3)_3SiH + H_2 \quad (1.104)$$

The action of excess of a triorganosilane on phosphorous acid in the presence of colloidal nickel forms tris(triorganosilyl) phosphites in 75-80% yield [94, 149, 150, 172]:

$$3R_3SiH + H_3PO_3 \xrightarrow{Ni} (R_3SiO)_3P + 3H_2 \quad (1.105)$$

The by-products of this reaction are bis(triorganosilyl) phosphites, confirming that the process proceeds through stages (1.102) and (1.103).

At the same time, the reaction

$$R_3SiH + (RO)_2PHO \xrightarrow{Ni} (RO)_2POSiR_3 + H_2 \quad (1.106)$$

forms only a small amount (~3%) of triorganosilyl dialkyl phosphites.

It is assumed [94] that reaction (1.103) is not due to disruption of the tautomeric equilibrium $(R_3SiO)_2PHO \rightleftharpoons (R_3SiO)_2POH$, but due to the catalytic reaction of $(R_3SiO)_2PHO$ with triorganosilanes [94]:

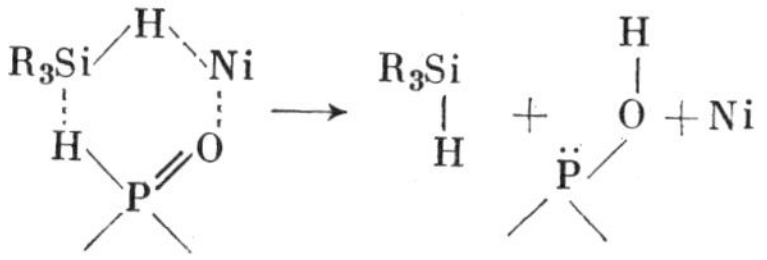

(1.107)

However, it is more probable that reaction (1.103) proceeds as a result of addition of the R_3SiH molecule at the P=O bond:

$$(R_3SiO)_2P(H)=O + R_3SiH \xrightarrow{Ni} (R_3SiO)_3PH_2 \longrightarrow (R_3SiO)_3P + H_2 \quad (1.107a)$$

In this connection it should be noted that hydrosilanes reduce triorganophosphine oxides to triorganophosphines [362-364]. There is no evidence of the possible intermediate formation of organophosphorus compounds of silicon here. However, in this case it may be surmised that the reaction proceeds by a scheme similar to (1.107a):

$$R_3P{=}O + HSiR'_3 \longrightarrow R_3PHOSiR'_3 \longrightarrow R_3P + HOSiR'_3 \qquad (1.107b)$$

The reaction of triorganosilanes with alkyl phosphites in the presence of colloidal nickel at 100-120°C proceeds with the formation of triorganosilyl alkyl phosphites in 25-40% yield [94, 171]:

$$R_3SiH + (HO)_2POR' \xrightarrow{Ni} (R'O)(R_3SiO)PHO + H_2 \qquad (1.108)$$

By further dehydrocondensation with trialkylsilanes the trialkylsilyl alkyl phosphites are converted into bis(trialkylsilyl) alkyl phosphites [94]:

$$R_3SiH + (R'O)(R_3SiO)PHO \xrightarrow{Ni} (R'O)(R_3SiO)_2P + H_2 \qquad (1.109)$$

Reaction (1.109) proceeds with much more difficulty than (1.103). Thus, the reactivity of disubstituted phosphites in their interaction with triorganosilanes in the presence of colloidal nickel falls in the following series:

$$(R_3SiO)_2PHO > (RO)(R_3SiO)PHO > (RO)_2PHO$$

1.1.8. Synthesis of Alumaphosphonasiloxanes

Heating tris(triethylsiloxy)alumane with equimolar amounts of organophosphinic acids or trialkyl phosphates at 150-250°C forms polymers, which are alumaphosphonoxanes or, what is the same, phosphonaalumoxanes, with the main chain framed by triorganosiloxy groups [17-19]:

$$n(R_3SiO)_3Al + n[R_2SiO]_3PO \longrightarrow$$

$$\longrightarrow R_3SiO\text{—}\left[\text{—}\underset{\substack{|\\ OSiR_3}}{Al}\text{—}O\text{—}\underset{\substack{|\\ OSiR_3}}{\overset{\substack{O\\ \|}}{P}}\text{—}O\text{—}\right]_n\text{—}SiR_3 + (2n-1)(R_3Si)_2O \qquad (1.110)$$

$$n(R_3SiO)_3Al + nR'P(O)(OH)_2 \longrightarrow$$

$$\longrightarrow R_3SiO\text{—}\left[\text{—}\underset{\substack{|\\ OSiR_3}}{Al}\text{—}O\text{—}\underset{\substack{|\\ R'}}{\overset{\substack{O\\ \|}}{P}}\text{—}O\text{—}\right]_n\text{—}H + (n-0.5)(R_3Si)_2O + (n-0.5)H_2O \qquad (1.111)$$

Judging by the kinetics of the process, in any case the condensation proceeds in stages. While the reaction of organophosphinic acids with tris(triorganosiloxy)alumanes proceeds with simple heating of the mixture to 170-230°C, to achieve a reaction in accordance with scheme (1.110) it is necessary to simultaneously flush the system with moist air. In this case it is not the tris-(triorganosiloxy) derivatives which actually react, but their partial hydrolysis products.

The character of the polymers formed depends primarily on the nature of the substituents at the silicon atom.

Condensation of tris(triethylsiloxy)alumane with tris(triethylsilyl) phosphate forms a solid vitreous polymer, which is soluble in toluene, benzene, and alcohol. Further heating converts it into an infusible, insoluble form. Under the same conditions the condensation of $[(CH_3)_3SiO]_3Al$ with tris(trimethylsilyl) phosphate or trimethylsilyl phosphate leads immediately to the formation of infusible and insoluble resins. In the reaction of tris(triethylsiloxy)-alumane with triethylsilyl phosphate (2:1) the process may be limited to the formation of a clear, viscous liquid of low molecular weight:

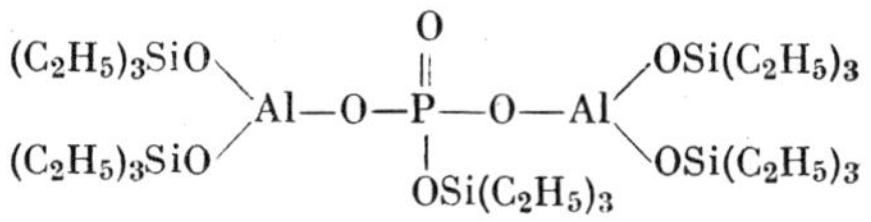

A friable, but soluble, polymer with a molecular weight of 2550 is obtained by the reaction of equimolar amounts of tris(triethylsiloxy)alumane with methylphosphinic acid [20]. Heating for 15-20 min at 200°C is sufficient to convert it into an infusible and insoluble form. The derivative of vinylphosphinic acid (mol. wt. 2700) obtained freshly under analogous conditions is soluble in toluene and styrene, swells only in CCl_4 and benzene, and is inert toward the action of alcohol, acetone, and cellosolve. The polymer is converted to a completely insoluble form even by storage for 25-30 h under normal conditions; then it does not soften up to 450°C [20].

Polymers containing the group Si−O−Al−O−P are also obtained by the reaction of alkoxyalumasiloxanes with diorganophosphinic acids [551];

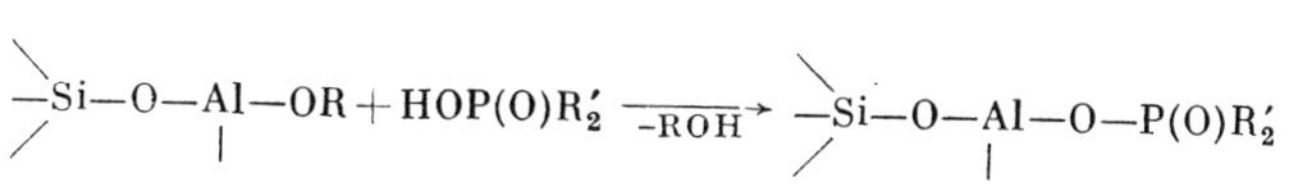

$$\geq Si-O-\underset{|}{Al}-OR + HOP(O)R'_2 \xrightarrow[-ROH]{} \geq Si-O-\underset{|}{Al}-O-P(O)R'_2 \qquad (1.112)$$

the reaction of alkylalumaphosphonoxanes [249] and alkoxyalumaphosphonoxanes [138] with polysiloxanols;

$$\gt Si{-}OH + R{-}\underset{|}{Al}{-}O{-}P(O)\lt \xrightarrow{-RH} \gt Si{-}O{-}\underset{|}{Al}{-}O{-}P(O)\lt, \qquad (1.113)$$

and alkoxyalumaphosphonoxanes with diacyloxysilanes:

$$\gt SiOCOCH_3 + RO{-}\underset{|}{Al}{-}O{-}P\lt \xrightarrow{-CH_3COOR} \gt Si{-}O{-}\underset{|}{Al}{-}O{-}P\lt \qquad (1.114)$$

1.2. Physical Properties

Tris(triorganosilyl) phosphates and bis(triorganosilyl) phosphates (see Table 2) are colorless liquids with a pleasant odor, are readily soluble in most organic solvents, and are quite stable thermally. Among the known substances of this type the only solids are triarylsiloxy derivatives. Mixed silyl alkyl esters of phosphoric acids are appreciably associated. In particular, the fact that replacement of the P(O) group in Si – O – P(O) by P(S) produces hardly any change in the boiling point and density of the compounds is related to the association [319].

Bis(triorganosilyl) acetylphosphates and triorganosilyl alkyl acetylphosphonates are colorless liquids with a weak characteristic odor, which are insoluble in, but readily hydrolyzed by, water; they are readily soluble in organic solvents [94].

Tris(triorganosilyl) phosphites, colorless liquids with a weak camphor-like odor, are readily soluble in organic solvents [94].

A characteristic feature of the polymers of the type $(C_2H_5O)_2P(O)O[(CH_3)_2SiO]_nP(O)(OC_2H_5)_2$ is the decrease in their density with an increase in the molecular weight, which is evidently related to a decrease in the intermolecular interaction (association) with an increase in the number of $(CH_3)_2SiO$ units in the chain.

The products from the copolymerization of cyclodimethylsiloxanes with cyclophosphonasiloxanes [15] are clear and soluble in organic solvents, but in air they become turbid and lose their solubility. If the contact with air is brief (0.5–1 h) this process is reversible, as if the polymer be heated to 130–150°C its original properties are restored. Longer contact with moist air leads to irreversible changes. In the opinion of the authors this may result from hydrolytic cleavage of the group $Si-O-P$ (see Section 1.3) with the subsequent formation of a structure involving hydrogen bonds:

$$
\begin{array}{cc}
\text{O} & (CH_3)_2Si{\sim} \\
\| & | \\
CH_3-P-OH & \cdots \text{O} \\
| & | \\
& (CH_3)_2Si{\sim}
\end{array}
$$

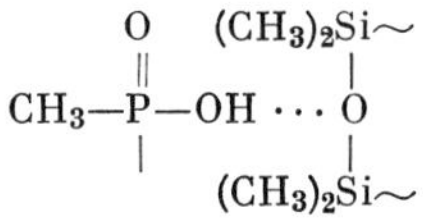

Organic derivatives of silicophosphoric acid with the composition $[SiP_2O_6(OR)_2]_n$ are white hygroscopic powders, which are insoluble in nonpolar solvents, but are decomposed by hydroxyl-containing solvents. They do not have a well-defined melting point, but decompose at elevated temperatures [390]. The degree of crystallization falls in the series $R = H > CH_3 > C_2H_5 > C_4H_9 > C_6H_5$, while $SiP_2O_6Cl_2$ is completely amorphous. The assignment of the characteristic frequencies in the infrared and Raman spectra of organosilicon esters of phosphorous acids is given in Table 1.

The data in Table 1 show that it is still difficult to use the spectral characteristics as decisive evidence in demonstrating the structure of polymers, for example, obtained by the reaction of tetraalkoxysilanes with P_2O_5 and $POCl_3$ [105, 106]. However, they may be used successfully to establish the structure of bis(trialkyl-silyl) phosphites [63, 94].

The physicochemical constants of organosilicon esters of oxygen acids of phosphorus are given in Table 2.

TABLE 1. Spectral Characteristics of Organosilicon Esters of Oxygen Acids of Phosphorus

Frequency, cm^{-1}	Spectrum	Assignment	Literature
590	Raman	$\nu Si-C$	63
610—650	Raman	$\nu Si-O$	63
700—760	Raman	$\nu_s P-O$	63
760	IR	$Si-CH_3$	107
810—840	IR	$\nu_s P-O,\ P-C$	390
870—1070	Raman	$\delta H-P-O,\ \nu C-C$	63
930—980	IR	$P-O$	106, 107, 215, 390
1000—1100	IR	$Si-O,\ Si-O-Si$	14, 80, 94, 101, 102, 106
1090	IR	$P-O-C$	106
1050—1200	IR, Raman	$\nu_{as} P-O$	63, 102, 390
1200—1350	IR, Raman	$\nu P=O,\ \delta C-H,\ Si-C$	14, 63, 106, 107, 289, 338, 389, 390, 417, 425

TABLE 2. Organosilicon Esters of Oxygen Acids of Phosphorus

Empirical formula	Compound	B.p., °C (mm)	n_D^{20}	d_4^{20}	Literature
$C_2H_5ClF_3O_2PSi$	$CH_3(CH_2Cl)SiF[OP(O)F_2]$	38 (92)	1.3610	1.3910	102
$C_2H_5ClF_4O_4P_2Si$	$CH_3(CH_2Cl)Si[OP(O)F_2]_2$	83 (10)	1.3772	1.5232	102
$C_2H_6F_3O_2PSi$	$(CH_3)_2SiF[OP(O)F_2]$	55 (105)	1.3270	1.3692	102
$C_2H_6F_4O_4P_2Si$	$(CH_3)_2Si[OP(O)F_2]_2$	72 (10)	1.3438	1.4702	102
$C_3H_9Cl_2O_2PSi$	$(CH_3)_3SiOP(O)Cl_2$	40 (0.1)	—	—	335
		63—64 (10)	1.4290	1.2105	525
		62—67 (11)	—	—	526
		m.p. —20.5	—	—	335
$C_3H_9F_2O_2PSi$	$(CH_3)_3SiOPOF_2$	36 (36)	1.3428	1.1139	101
		125			
$C_4H_{10}F_3O_2PSi$	$(C_2H_5)_2SiF[OP(O)F_2]$	53 (52)	1.3510	1.2592	102
$C_4H_{10}F_4O_4P_2Si$	$(C_2H_5)_2Si[OP(O)F_2]_2$	72 (6)	1.3622	1.3602	102
$C_4H_{11}F_2O_2PSi$	$(CH_3)_2C_2H_5SiOP(O)F_2$	52 (30)	1.3581	1.1356	101
$C_4H_{12}FO_2PSi$	$CH_3PF(O)OSi(CH_3)_3$	62 (20)	1.3871	1.0525	76
$C_4H_{14}O_6P_2Si_2$	$[(CH_3)_2SiOP(O)(H)O]_2$	129 (0.1)	—	—	469
$C_5H_{15}O_3PSSi$	$(CH_3)_3SiOP(S)(OCH_3)_2$	83—84 (12)	1.4460	1.071	319
$C_5H_{15}O_4PSi$	$(CH_3)_3SiOP(O)(OCH_3)_2$	87—88 (12)	1.4054	1.0705	320
$C_5H_{15}O_4PSi_2$	$(CH_3)_2SiO[(CH_3)_2SiO]P(O)CH_3$	141—143 (1—2)	1.4232	1.1263	14
	|——O——|				
$C_5H_{15}O_6PSi$	$(CH_3O)_3SiOP(O)(CH_3)OCH_3$	61 (7)	—	—	470
$C_5H_{15}O_7PSi$	$(CH_3O)_3SiOP(O)(OCH_3)_2$	—	—	—	320
$C_6H_{15}F_2O_2PSi$	$(C_2H_5)_3SiOP(O)F_2$	72 (20)	1.3890	1.0635	101
$C_6H_{17}O_4PSi$	$(C_2H_5)_3SiOP(O)(OH)_2$	176	—	—	30
		decomp. >200			
$C_6H_{18}FO_3PSi_2$	$[(CH_3)_3SiO]_2POF$	60 (17)	1.3920	1.0160	100
$C_6H_{18}O_6PSi$	$[(CH_3)_2SiOP(O)(CH_3)O]_2$	123 (0.1)	—	—	469
		173—175 (1—2)	1.4415	1.2484	14
$C_6H_{18}O_6P_2S_2Si$	$(CH_3O)_2P(S)OSi(CH_3)_2OP(S)(OCH_3)_2$	90—92 (10^{-3})	1.4690	1.2570	319
$C_6H_{19}O_3PSi_2$	$[(CH_3)_3SiO]_2PHO$	74—75 (3)	1.4145	0.9661	63
		100—102 (5)	1.4140	1.0020	155, 156
		100—102 (6)			

$C_7H_8F_3O_2PSi$	$CH_3(C_6H_5)SiFOP(O)F_2$	57 (7)	1.4392	1.3042	102
$C_7H_8F_4O_4P_2Si$	$CH_3(C_6H_5)Si[OP(O)F_2]_2$	98 (2)	1.4428	1.4212	102
$C_7H_{17}F_2O_2PSi$	$(C_2H_5)_2(C_3H_7)SiOP(O)F_2$	75 (14)	1.3901	1.0569	101
$C_7H_{17}O_4PSi$	$(CH_3)_3SiOP(O)(OC_2H_5)COCH_3$	82—83 (4.5)	1.4212	1.0374	94, 165, 167
$C_7H_{18}FO_2PSi$	$(C_2H_5)_3SiOPCH_3(O)F$	62—64 (2)	1.4175	1.0266	76, 211
$C_7H_{19}O_3PSi$	$(CH_3)_3SiOP(OC_2H_5)_2$	60—62 (11)	1.4116	0.9485	52, 223
$C_7H_{19}O_3PSi$	$C_2H_5P(O)[OSi(CH_3)_3]OC_2H_5$	85—88 (11)	1.4155	0.9947	52
$C_7H_{19}O_3PSSi$	$(CH_3)_3SiOPS(OC_2H_5)_2$	56—57 (1.5)	1.4433	1.0261	52, 223
		96—97 (12)	1.4430	1.027	319
$C_7H_{19}O_3PSSi$	$(CH_3)_3SiOP(O)(OC_2H_5)SC_2H_5$	118.5—119.5 (14)	1.4495	1.0505	52
$C_7H_{19}O_4PSi$	$(CH_3)_3SiOP(O)(OC_2H_5)_2$	97—98 (12)	1.4070	1.018	320
$C_7H_{19}O_6PSi$	$(CH_3O)_3SiOP(O)(C_2H_5)OC_2H_5$	100—103 (12)	1.4053	1.100	320
$C_7H_{19}O_7PSi$	$(CH_3O)_3SiOP(O)(OC_2H_5)_2$	67—70 (10^{-2})	—	—	320
$C_7H_{21}O_3PSi_2$	$[(CH_3)_3SiO]_2P(O)CH_3$	105—107.5 (27)	—	—	498
$C_7H_{21}O_3PSSi_2$	$[(CH_3)_3SiO]_2P(S)OCH_3$	86—88 (12)	1.4483	1.025	319
$C_7H_{21}O_4PSi_2$	$[(CH_3)_3SiO]_2P(O)OCH_3$	98—99 (12)	1.4081	1.001	320
		100—101 (14)	—	—	
$C_7H_{21}O_5PSi_3$	$(CH_3)_2SiO[(CH_3)_2SiO]_2P(O)CH_3$ |——O——|	134—136 (1—2)	1.4198	1.0925	14
$C_8H_7F_3O_2PSi$	$(CH_3)_2(p\text{-}FC_6H_4)SiOP(O)F_2$	93 (4)	1.4467	1.2682	101
$C_8H_{11}F_2O_2PSi$	$(CH_3)_2(C_6H_5)SiOPOF_2$	80 (3)	1.4541	1.1833	101
$C_8H_{19}O_4PSi$	$(CH_3)_2(C_2H_5)SiOP(O)(OC_2H_5)COCH_3$	76.5 (1.5)	1.4270	1.0370	94, 168
	$(CH_3)_3SiOP(O)(OC_3H_7)COCH_3$	86.5—88 (3)	1.4248	1.0245	94, 165, 167
$C_8H_{21}O_3PSi$	$(C_2H_5)_3SiO(C_2H_5O)P(O)H$	82—83 (2)	1.4320	0.9883	94, 171
$C_8H_{21}O_3PSi_2$	$[(CH_3)_3SiO]_2P(O)CH{=}CH_2$	102—104 (14)	—	—	498
$C_8H_{21}O_4PSi$	$H(C_2H_5)SiOP(O)(OC_2H_5)_2$	65—67 (10^{-2})	1.4186	1.018	321
$C_8H_{21}O_4PSi_2$	$[(CH_3)_3SiO]_2P(O)COCH_3$	76.5—77.1 (2.5)	1.4214	1.0048	94, 166
$C_8H_{21}O_7PSi$	$(C_2H_5O)_3SiOP(O)(OCH_3)_2$	72—76 (10^{-2})	1.4025	1.131	320
$C_8H_{22}FO_3PSi_2$	$[(CH_3)_2(C_2H_5)SiO]_2P(O)F$	100 (7)	1.4057	0.9909	100
$C_8H_{23}O_3PSi_2$	$[(CH_3)_2(C_2H_5)SiO]_2P(O)H$	103—104 (3)	1.4230	0.9655	156
		104—106 (4)	1.4235	0.9657	155, 156, 157
$C_8H_{23}O_3PSi_2$	$[(CH_3)_3SiO]_2P(O)C_2H_5$	104—106 (25)	1.4111	0.9718	66, 205
$C_8H_{23}O_3PSSi_2$	$[(CH_3)_3SiO]_2P(S)OC_2H_5$	106—108 (12)	1.4427	1.003	319

TABLE 2 (Continued)

Empirical formula	Compound	B.p., °C (mm)	n_D^{20}	d_4^{20}	Literature
$C_8H_{23}O_4PSi_2$	$[(CH_3)_3SiO]_2P(O)OC_2H_5$	102—103 (12)	1·4062	0·981	320
$C_9H_{19}O_3PSSi_2$	$[(CH_3)_3SiO]_2P(S)OH$	93—95 (3)	1·4200	0·9778	59
$C_9H_{21}O_4PSi$	$CH_3(C_2H_5)_2SiOP(O)(OC_2H_5)COCH_3$	100—101 (4)	1·4340	1·0264	94, 168
	$(CH_3)_2(C_2H_5)SiOP(O)(OC_3H_7)COCH_3$	85—86·5 (1·5)	1·4293	1·0197	94, 168
$C_9H_{22}NO_3PSi_2$	$[(CH_3)_3SiO]_2P(O)CH_2CH_2CN$	108—110 (5)	1·4100	0·9618	59
$C_9H_{23}O_3PSi$	$(C_2H_5)_3SiO(C_3H_7O)P(O)H$	83—84 (1·5)	1·4338	0·9745	94, 171
$C_9H_{23}O_4PSi_2$	$(C_2H_5)_2SiOSi(C_2H_5)_2OP(O)CH_3$ |——O——|	168—170 (2)	1·4490	1·091	12
$C_9H_{24}ClO_3PSi_2$	$[C_2H_5(CH_3)_2SiO]_2P(O)CH_2Cl$	106—107 (3)	1·4400	1·0662	183, 184
$C_9H_{27}O_3PSi_3$	$[(CH_3)_3SiO]_3P$	102 (7)	1·4100	0·9618	66, 205
		124—127 (20)	1·4105	0·9603	66, 205
		127—127·5 (23)	1·4128	0·9588	66
		129—130 (25)	1·4120	0·9594	66, 205
$C_9H_{27}O_4PSi_3$	$[(CH_3)_3SiO]_3PO$	56—57 (1·1)	1·4090	—	526
		m.p. 3.5	—	—	526
		57—59 (1·5)	—	—	526
		67—68 (0·5)	1·4092	—	259, 320
		77 (4)	1·4095	—	19
		85—87 (4)	1·4090	—	510
		91 (5)	—	—	400
		91—93 (5)	1·4082	0·9650	62a
		96—97 (6)	1·4083	0·962	107
		97 (6)	1·4089	0·9591	61
		98—104 (6)	1·4082	—	63
		104—106 (10·5)			
		108—109 (12)			
		110—112 (10)	1·4080	0·9602	158, 163
		161—163 (80)	1·4075	—	158, 163
		228—229 (720)			
		231·8 (749·5)	—	—	158, 163
		235—236 (760)	—	—	158, 163
$C_9H_{27}O_6PSi$	$(CH_3)_2SiO[(CH_3)_2SiO]_3P(O)CH_3$ |——O——|	114—116 (1—2)	1·4189	1·0750	14

$C_{10}H_{17}O_3PSi$	$(CH_3)_2(C_6H_5)SiO(C_2H_5O)P(O)H$	111—112 (2)	1.4900	1.0817	94, 171
$C_{10}H_{23}O_4PSi$	$(C_2H_5)_3SiO(OC_2H_5)P(O)COCH_3$	98—101 (2—2.5)	1.4392	1.0188	94
		102—103 (3)	1.4390	1.0216	165, 167
		102—103 (3)	1.4390	1.0195	94
		103—105 (2)	1.4370	1.0168	170
	$CH_3(C_2H_5)_2SiO(OC_3H_7)P(O)COCH_3$	102—103 (3)	1.4348	1.0141	94, 168
$C_{10}H_{25}O_3PSi$	$(C_2H_5)_3SiOP(OC_2H_5)_2$	95—97 (7)	1.4333	0.9332	52
		105—106 (10)	1.4332	0.9340	222
	$C_2H_5P(O)[OSi(C_2H_5)_3]OC_2H_5$	117—121 (2)	1.4394	0.9678	52
		158—159 (10)	1.4390	0.9659	36
$C_{10}H_{25}O_3PSSi$	$(C_2H_5)_3SiOP(S)(OC_2H_5)_2$	89—90 (2)	1.4540	1.0129	52, 222
$C_{10}H_{25}O_4PSi$	$(C_2H_5)_3SiOP(O)(OC_2H_5)_2$	67—68 (10^{-2})	1.4273	1.004	320
		134—139 (12)			
$C_{10}H_{25}O_4PSi_2$	$[(CH_3)_3SiO]_2P(O)CH_2CH_2OCH=CH_2$	124—126 (5)	—	—	498
	$[C_2H_5(CH_3)_2SiO]_2P(O)COCH_3$	92—93 (2)	1.4319	0.9962	94, 166
$C_{10}H_{25}O_6PSi$	$(C_2H_5O)_3SiOP(O)(C_2H_5)OC_2H_5$	113 (12)	1.4075	0.9282	36
		114 (12)	1.4080	—	36
		114—118 (12)	1.4085	1.034	320
$C_{10}H_{25}O_6PSSi$	$(C_2H_5O)_3SiOP(S)(OC_2H_5)_2$	79—81 (10^{-3})	—	—	319
$C_{10}H_{25}O_7PSi$	$(C_2H_5O)_3SiOP(O)(OC_2H_5)_2$	80—85 (10^{-3})	1.4048	1.090	320
		90—94 (0.3)			
$C_{10}H_{26}O_6P_2Si_2$	$[—P(O)(CH_3)OSi(C_2H_5)_2O—]_2$	226—229 (5)	1.4540	1.174	12
$C_{10}H_{26}O_6P_2S_2Si$	$(C_2H_5O)_2P(S)OSi(CH_3)_2OP(S)(OC_2H_5)_2$	98—102 (10^{-3})	1.4626	1.153	319
$C_{10}H_{26}O_8P_2Si$	$(C_2H_5O)_2P(O)OSi(CH_3)_2OP(O)(OC_2H_5)_2$	101—104 (10^{-3})	1.4180	1.1536	424
$C_{10}H_{27}O_3PSi_2$	$[CH_3(C_2H_5)_2SiO]_2P(O)H$	93—95 (1.5—2)	1.4348	0.9685	58, 94, 167, 172
		136—138 (3)	1.4320	0.9672	58; 156
		136—137 (3)			155, 156, 157
	$[(CH_3)_3SiO]_2P(O)C_4H_9$	138—145 (6)	1.4299	0.9498	63
$C_{10}H_{27}O_4PSi_2$	$[H(C_2H_5)_2SiO]_2P(O)OC_2H_5$	107—108 (5.5)	1.4168^{25}	0.9356^{25}	424
$C_{11}H_{25}O_4PSi$	$(C_2H_5)_3SiOP(O)(OC_3H_7)COCH_3$	81—84 (10^{-2})	1.4282	0.991	321
		112—113 (2)	1.4410	1.0109	94, 165, 167
$C_{11}H_{27}OPSi$	$(C_4H_9)_2POSi(CH_3)_3$	99—101 (12)	—	—	417
$C_{11}H_{27}O_2PSi$	$(C_3H_7)_3SiOP(O)(CH_3)_2$	127—128 (5)	—	—	530
		127 (5)	—	—	529

TABLE 2 (Continued)

Empirical formula	Compound	B.p., °C (mm)	n_D^{20}	d_4^{20}	Literature
$C_{11}H_{27}O_3PSi$	$(CH_3)_3SiOP(O)(C_4H_9)OC_4H_9$	93—94 (4.5)	1.4220[25]	0.9194[25]	424
$C_{11}H_{27}O_4PSi$	$(C_3H_7)_3SiOP(O)(OCH_3)_2$	99—120 (0.5)	—	0.999	530
$C_{11}H_{27}O_5PSi_2$	$[(CH_3)_2(C_2H_5)SiO]_2P(O)CH_2OCOCH_3$	112 (2.5)	1.4332	1.0237	179
$C_{11}H_{27}O_6PSi$	$(CH_3O)_3SiOP(O)(C_4H_9)OC_4H_9$	93—94 (4.5)	—	—	470
$C_{11}H_{28}ClO_3PSi_2$	$[(C_2H_5)_2(CH_3)SiO]_2P(O)CH_2Cl$	112 (3)	1.4485	1.0512	183, 184
$C_{11}H_{29}O_3PSi_2$	$[CH_3(C_2H_5)_2SiO]_2P(O)CH_3$	122 (3)	1.4328	0.9591	161, 162
$C_{11}H_{29}O_4PSi_2$	$[CH_3(C_2H_5)_2SiO]_2P(O)CH_2OH$	—	1.4423	1.0067	177, 178
$C_{11}H_{33}O_7PSi_5$	$(CH_3)_2SiO[(CH_3)_2SiO]_4P(O)CH_3$	119—121 (1—2)	1.4227	1.0721	14
	|——O——|				
$C_{12}H_{19}O_4PSi$	$(CH_3)_2(C_6H_5)SiOP(O)(OC_2H_5)COCH_3$	128—131 (1.5)	1.4852	1.0981	94, 168
$C_{12}H_{23}O_4PSi_2$	$[(CH_3)_3SiO]_2P(O)OC_6H_5$	133—134 (4)	1.4590[19]	—	400
$C_{12}H_{29}O_3PSi_2$	$[(CH_3)_3SiO]_2P(O)C_6H_{11}$	98.5 (0.6)	—	0.9822[23]	530
$C_{12}H_{29}O_4PSi_2$	$[(C_2H_5)_2(CH_3)SiO]_2P(O)COCH_3$	102—103 (1.5—2)	1.4390	0.9869	94, 166
$C_{12}H_{30}FO_3PSi_2$	$[(C_2H_5)_3SiO]_2(O)F$	147 (7)	1.4272	0.9976	100
$C_{12}H_{30}O_8P_2S_2Si$	$(C_2H_5O)_2P(S)OSi(OC_2H_5)_2OP(S)(OC_2H_5)_2$	114—118 (10^{-3})	1.4531	1.155	319
$C_{12}H_{31}O_3PSi_2$	$[(C_2H_5)_3SiO]_2P(O)H$	118—119 (1)	1.4439	0.9625	58, 94, 156, 167, 172
		153—154 (3)	1.4402	0.9692	156, 157
		160—162 (5)	1.4402	0.9668	155, 156
		162—164 (5)	1.4402	0.9690	58, 157
		—	1.4412	—	63
$C_{12}H_{32}O_9P_2Si_2$	$(C_2H_5O)_2P(O)O[Si(CH_3)_2O]_2P(O)(OC_2H_5)_2$	122 (10^{-3})	1.4158	1.1197	424
			1.4223	0.9632	158, 163
$C_{12}H_{33}O_4PSi_3$	$[(CH_3)_2(C_2H_5)SiO]_3PO$	117—121 (3)	1.4023[30]	0.928	442
$C_{13}H_{13}F_2O_2PSi$	$(C_6H_5)_2(CH_3)SiOP(O)F_2$	123 (1)	1.5102	1.1776	101
$C_{13}H_{21}O_4PSi$	$(CH_3)_2(C_6H_5)SiOP(O)(OC_3H_7)COCH_3$	143—145 (2)	1.4830	1.0814	94; 165; 167
$C_{13}H_{25}O_3PSi_2$	$[(CH_3)_3SiO]_2P(O)CH_2C_6H_5$	96.5—98 (8)	—	—	498
$C_{13}H_{31}O_5PSi_2$	$[CH_3(C_2H_5)_2SiO]_2P(O)CH_2OCOCH_3$	138—139 (5.5)	1.4402	1.0125	179
$C_{13}H_{32}ClO_3PSi$	$[(C_2H_5)_3SiO]_2P(O)CH_2Cl$	130—131 (1.5)	1.4568	1.0288	183, 184
		162—165 (4)	1.4585	1.0343	204
$C_{13}H_{33}O_3PSi_2$	$[(C_2H_5)_3SiO]_2P(O)CH_3$	145 (4)	1.4410	0.9611	161, 162

		150 (8)	1.4432	0.9579	188, 204
		151 (7)	1.4432	0.9579	188
		155 (5)	1.4449	0.9610	161, 162
		158 (5)	1.4431	0.9596	160
			1.4438	0.9577	188, 204
			1.4447	0.9557	
$C_{13}H_{33}O_4PSi_2$	$[(C_2H_5)_3SiO]_2P(O)CH_2OH$	—	1.4472	0.9956	177, 178
$C_{13}H_{33}O_5PSi_3$	$(C_2H_5)_2Si[OSi(C_2H_5)_2]_2OP(O)CH_3$	163—165 (1)	1.4450	1.069	12
$C_{13}H_{34}ClO_5PSi_3$	$[(CH_3)_3SiO]_2P(O)CH_2OCH_2CH[OSi(CH_3)_3]CH_2Cl$	153—154 (1)	1.4395	1.0335	153
		153—154 (1)	1.4405	1.0445	154
$C_{13}H_{37}O_6PSSi_4$	$[(CH_3)_3SiO]_3SiOP(S)(OC_2H_5)_2$	82—84 (10^{-3})	1.4205	0.990	14
$C_{13}H_{37}O_7PSi_4$	$[(CH_3)_3SiO]_3SiOP(O)(OC_2H_5)_2$	79—80 (10^{-3})	1.4032	0.982	322
$C_{14}H_{12}FO_2PSi$	$(CH_3)_3SiOP(O)FCH_3$	62 (20)	1.3871	1.0525	207
$C_{14}H_{23}O_4PSi$	$(C_2H_5)_3SiO(OC_6H_5)P(O)COCH_3$	143—145 (1.5)	1.4875	1.0751	94, 168
$C_{14}H_{33}O_4PSi_2$	$[(C_2H_5)_3SiO]_2P(O)COCH_3$	130—131 (1.5)	1.4468	0.9850	166
		130—131 (1.5)	1.4477	0.9904	94, 166, 170
			1.4468	0.9850	94
		135—137 (1.5—2)	1.4478	0.9875	166
		134.5—135 (2)	1.4480	0.9921	94
		142.5—143 (4—4.5)	1.4478	0.9893	166
$C_{14}H_{34}ClO_3PSi_2$	$[(C_2H_5)_3SiO]_2P(O)CH_2CH_2Cl$	145—147 (2)	1.4543	1.0260	183, 184
		156—159 (4)	1.4548	1.0296	204
$C_{14}H_{34}O_6P_2Si_2$	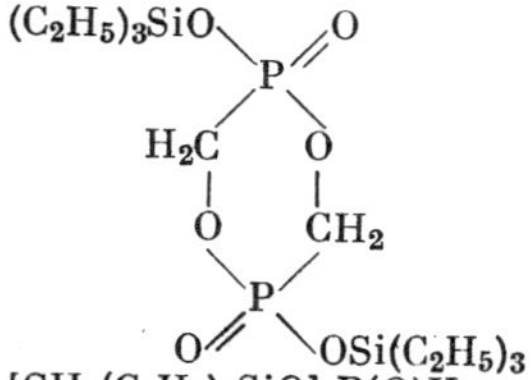	m.p. 85—87	—	—	180
$C_{14}H_{35}O_3PSi_2$	$[CH_3(C_3H_7)_2SiO]_2P(O)H$	185—186 (17)	1.4385	0.9373	63
	$[(C_2H_5)_3SiO]_2P(O)C_2H_5$	106.5—107 (1.5)	1.4448	0.9304	94, 171
$C_{14}H_{35}O_4PSi_2$	$[(C_2H_5)_3SiO]_2P(O)OC_2H_5$	165—170 (8)	1.4460	0.9616	204
		167—169 (8)	1.4438	0.9636	66, 205
$C_{14}H_{35}O_5PSi_2$	$(C_2H_5)_3SiOSi(C_2H_5)_2OP(O)(OC_2H_5)_2$	85—87 (10^{-3})	1.4372	0.972	320
		90—93 (10^{-2})	1.4296	0.992	321

TABLE 2 (Continued)

Empirical formula	Compound	B.p., °C (mm)	n_D^{20}	d_4^{20}	Literature
$C_{14}H_{38}O_{10}P_2Si_3$	$(C_2H_5O)_2P(O)O[Si(CH_3)_2O]_3P(O)(OC_2H_5)_2$	130 (10^{-3})	1.4145	1.1012	424
$C_{15}H_{19}OPSi$	$(C_6H_5)_2POSi(CH_3)_3$	103—106 (0.5)	—	—	417
$C_{15}H_{19}O_4PSi_2$	$CH_3(C_6H_5)SiOSi(CH_3)(C_6H_5)OP(O)CH_3$ (with O bridge between the Si and P(O) groups)	242—244 (2)	—	—	12
$C_{15}H_{28}ClO_3PSi_2$	$[(CH_3)_2(C_2H_5)SiO]_2P(O)CHClC_6H_5$	160—161 (4)	1.4870	1.0792	184
$C_{15}H_{35}O_5PSi_2$	$[(C_2H_5)_3SiO]_2P(O)CH_2OCOCH_3$	144—145 (2.5)	1.4470	1.0094	176
		155 (5.5)	1.4470	1.0103	178, 179
		156—157 (5)	1.4473	1.0120	178
$C_{15}H_{37}O_3PSi_2$	$[(C_2H_5)_3SiO]_2P(O)C_3H_7$	107—109 (1.5—2)	1.4419	0.9404	171
		123.5—124 (3)	1.4441	0.9233	94
	$[CH_3(C_3H_7)_2SiO]_2P(O)CH_3$	160 (4)	1.4372	0.9333	161, 162
$C_{15}H_{37}O_4PSi_2$	$[CH_3(C_3H_7)_2SiO]_2P(O)CH_2OH$	—	1.4465	0.9804	177, 178
$C_{15}H_{39}O_3PSi_3$	$[CH_3(C_2H_5)_2SiO]_3P$	122—124 (1.5)	1.4471	0.9221	94
$C_{15}H_{39}O_4PSi_3$	$[CH_3(C_2H_5)_2SiO]_3PO$	145—147 (1)	1.4302	0.9474	62a
$C_{15}H_{41}O_6PSSi_2$	$[(CH_3)_3SiO]_3SiOP(S)(OC_3H_7)_2$	91—94 (10^{-3})	1.4240	0.982	319
	$[(CH_3)_3SiO]_3SiOP(S)(OC_3H_7\text{-iso})_2$	84—86 (10^{-3})	1.4200	0.973	319
$C_{15}H_{41}O_7PSi_4$	$[(CH_3)_3SiO]_3SiOP(O)(OC_3H_7)_2$	90—92 (10^{-3})	1.4076	0.974	322
	$[(CH_3)_3SiO]_3SiOP(O)(OC_3H_7\text{-iso})_2$	77—79 (10^{-3})	1.4040	0.976	322
$C_{16}H_{20}F_3O_3PSi_2$	$[(CH_3)_2(p\text{-}FC_6H_4)SiO]_2P(O)F$	170 (1)	1.4864	1.1924	100
$C_{16}H_{22}FO_3PSi_2$	$[(CH_3)_2C_6H_5SiO]_2P(O)F$	188 (11)	1.4902	1.0839	100
$C_{16}H_{22}O_6P_2Si_2$	$[CH_3(C_6H_5)Si(O)OP(O)CH_3]_2$	252—255 (2)	—	—	12
$C_{16}H_{22}O_8P_2Si_2$	$[(CH_3)_2SiOP(O)(OC_6H_5)O]_2$	m.p. 135	—	—	469
$C_{16}H_{23}O_3PSi_2$	$[(CH_3)_2(C_6H_5)SiO]_2P(O)H$	169—171 (1)	1.5208	1.0899	58, 94, 167, 172
		206—208 (4)	1.5160	1.0999	156, 157
$C_{16}H_{29}O_4PSi_2$	$CH_3COP(O)[OSi(C_2H_5)_3][OSi(CH_3)_2C_6H_5]$	163—165 (3—3.5)	1.4652	1.0147	94
$C_{16}H_{37}O_7PSi$	$(C_4H_9O)_3SiOP(O)(OC_2H_5)_2$	148—150 (0.1)	—	—	339
$C_{16}H_{41}O_4PSi_3$	$[CH_3(C_2H_5)_2SiO]_2P(O)CH_2OSi(CH_3)(C_2H_5)_2$	133 (2.5)	1.4420	0.9649	185
		133 (3)	1.4436	0.9649	185, 187
		135 (3)	1.4423	0.9647	185, 186
$C_{16}H_{44}O_{11}P_2Si_4$	$(C_2H_5O)_2P(O)O[Si(CH_3)_2O]_4P(O)(OC_2H_5)_2$	~140 (10^{-3})	1.4140	1.0853	424
$C_{17}H_{32}ClO_3PSi_2$	$[CH_3(C_2H_5)_2SiO]_2P(O)CHClC_6H_5$	170 (4)	1.4910	1.0679	184
$C_{17}H_{39}O_5PSi_2$	$[CH_3(C_3H_7)_2SiO]_2P(O)CH_2OCOCH_3$	166 (4.5)	1.4430	0.9752	176, 179

$C_{18}H_{25}O_4PSi_2$	$[C_6H_5(CH_3)_2SiO]_2P(O)COCH_3$	218—220 (4—4.5)	1.5178	1.1073	166
		218—220 (5—5.5)	1.5178	1.1073	94
$C_{18}H_{35}O_4PSi_2$	$[(C_2H_5)_3SiO]_2P(O)OC_6H_5$	164 (1)	1.4709	1.0223	160
$C_{18}H_{39}O_3PSi_3$	$(CH_3)_3SiO[(C_2H_5)_3SiO]_2P$	145—147 (5)	1.4350	0.9232	204
$C_{18}H_{45}O_3PSi_3$	$[(C_2H_5)_3SiO]_3P$	135—136 (1)	1.4542	0.9293	94, 172
		144—145 (1.5—2)	1.4547	0.9293	94
		164—166 (3)	1.4518^{21}	0.9308^{18}	232
		174—176 (6)	1.4540	0.9302	204
		202—205 (8)	1.4473	0.9678	66, 205
		203—206 (10)	1.4480	0.9739	66, 205
$C_{18}H_{45}O_4PSi_3$	$[CH_3(C_2H_5)O]_2P(O)CH_2OSi(CH_3)(C_3H_7)_2$	155—156 (3)	1.4446	0.9609	178, 185
$C_{18}H_{45}O_4PSi_3$	$[(C_2H_5)_3SiO]_3PO$	106—108 (10^{-3})	1.4475	—	320
		134—135 (0.3)	—	—	530
		156—163 (5)	1.4462	—	19
		166.5 (1)	1.4457	0.9670	61
		178—181 (4)	1.4452	0.9663	158, 163
			1.4455	0.9663	58
		180 (4)	1.4400	0.9700	10
		180—181 (3.5)	1.4455	0.9663	158, 161, 163
		182 (4)	1.4429	0.9675	158, 159, 161
		181—183 (3)	1.4460^{21}	$0.9658^{20.5}$	231
		183 (6)	1.4455	0.9668	62a
		193 (8)	1.4480	0.9651	188, 204
		194—196 (8)	1.4467	0.9676	160
		200.5 (11)	—	—	—
		202—205 (10)	1.4445	0.9668	158, 163
		208 (10)	1.4452	0.9668	63
		220—225 (13)	1.4458	—	62a
		240—243 (20)	1.4461	0.9669	62a
$C_{18}H_{45}O_6PSi_3$	$C_2H_5[(C_2H_5)_3SiO]_2SiOP(O)(OC_2H_5)_2$	107—110 (10^{-2})	1.4335	0.987	321
$C_{18}H_{45}O_{12}P_3Si_2$	$C_2H_5OP(O)[OSi(C_2H_5)_2OP(O)(OC_2H_5)_2]_2$	175—180 ($5\cdot10^{-3}$)	1.4330	1.150	321
$C_{18}H_{50}O_{12}P_2Si_5$	$(C_2H_5O)_2P(O)O[Si(CH_3)_2O]_5P(O)(OC_2H_5)_2$	~152 (10^{-3})	1.4127	1.0691	424
$C_{19}H_{35}O_4PSi_2$	$[(C_2H_5)_3SiO]_2P(O)COC_6H_5$	183 (2)	1.4900	1.0332	94, 170
$C_{19}H_{36}ClO_3PSi_2$	$[(C_2H_5)_3SiO]_2P(O)CHClC_6H_5$	175—176 (2)	1.4930	1.0572	183, 184

TABLE 2 (Continued)

Empirical formula	Compound	B.p., °C (mm)	n_D^{20}	d_4^{20}	Literature
$C_{19}H_{37}O_3PSi_2$	$[(C_2H_5)_3SiO]_2P(O)CH_2C_6H_5$	194—195 (4)	1.4797	0.9967	188, 204
		205—207 (7)	1.4785	1.0023	188
		214—216 (7)	1.4788	0.9979	188
$C_{19}H_{47}O_4PSi_3$	$[(C_2H_5)_3SiO]_2P(O)CH_2OSi(C_2H_5)_3$	160—162 (2)	1.4520	0.9640	179, 185
		162 (2.5)	1.4512	0.9653	185, 186, 187
		163 (3)	1.4510	0.9650	135
$C_{20}H_{21}O_3PSSi$	$(C_6H_5)_3SiOP(S)(OCH_3)_2$	m.p. 155	—	—	319
$C_{20}H_{41}O_3PSi_3$	$[(C_2H_5)_3SiO]_2P(O)Si(C_6H_5)(CH_3)_2$	165—170 (2.5)	1.4830	0.9777	94
		168—171 (2—2.5)	1.4831	0.9809	94
$C_{20}H_{45}O_7PSi$	$(C_4H_9O)_3SiOP(O)(OC_4H_9)_2$	168—172 (0.1)	—	—	339
$C_{20}H_{50}O_{10}P_2Si_2$	$(C_2H_5O)_2P(O)OSi[OSi(C_2H_5)_3]_2OP(O)(OC_2H_5)_2$	167—170 ($5 \cdot 10^{-3}$)	1.4325	1.069	321
$C_{21}H_{51}O_4PSi_3$	$[CH_3(C_3H_7)_2SiO]_3PO$	221—224 (10)	1.4380	0.9326	62a
$C_{22}H_{27}O_5PSi_3$	$CH_3(C_6H_5)Si[OSi(CH_3)(C_6H_5)]_2OP(O)CH_3$ (Si and P bridged by —O—)	267—268 (2)	—	—	12
$C_{22}H_{37}O_3PSi_3$	$(C_2H_5)_3SiOP[OSi(CH_3)_2C_6H_5]_2$	188—190 (1.5—2)	1.5100	1.0208	94
$C_{22}H_{52}BrO_5PSi_3$	$[(C_2H_5)_3SiO]_2P(O)CH_2OCH_2CHCH_2Br$ $\vert$ $OSi(C_2H_5)_3$	204—205 (1)	1.4660	1.0877	153, 154
		205—206 (1)	1.4675	1.0977	154
$C_{22}H_{52}ClO_5PSi_3$	$[(C_2H_5)_3SiO]_2P(O)CH_2CHCH_2Cl$ $\vert$ $OSi(C_2H_5)_3$	190—191 (1)	1.4590	1.0115	153
$C_{22}H_{53}O_4PSi_3$	$[CH_3(C_3H_7)_2SiO]_2P(O)CH_2OSi(CH_3)(C_3H_7)_2$	170 (2.5)	1.4469	0.9337	185
			1.4469	0.9464	185, 187
$C_{22}H_{55}O_6PSi_4$	$[(C_2H_5)_3SiO(C_2H_5)_2SiO]_2P(O)OC_2H_5$	150—155 (10^{-2})	1.4385	0.982	321
$C_{22}H_{55}O_7PSi_4$	$[(C_2H_5)_3SiO]_3SiOP(O)(OC_2H_5)_2$	m.p. 44—46			
		128—131 ($5 \cdot 10^{-3}$)	—	—	321
$C_{22}H_{56}O_7P_2Si_4$	$[CH_3(C_2H_5)_2SiO]_2P(O)CH_2OP(O)CH_2OSi(CH_3)(C_2H_5)_2$ $\vert$ $OSi(CH_3)(C_2H_5)_2$	216—218 (3)	1.4510	1.0200	180
$C_{24}H_{33}O_3PSi_3$	$[(CH_3)_2(C_6H_5)SiO]_3P$	201—203 (1.5)	1.5297	1.0526	94; 172
		201—202 (1.5)	1.5298	1.0510	94
$C_{24}H_{53}O_3PSi_2$	$[(C_3H_7)_3SiO]_2P(O)C_6H_{11}$	168 (0.1)	—	0.943^{22}	530

$C_{25}H_{57}O_3P_3Si$	$[(C_4H_9)_2PO]_3SiCH_3$	128—130 (0.01)	—	—	417
$C_{26}H_{26}O_2P_2Si$	$[(C_6H_5)_2PO]_2Si(CH_3)_2$	215—218 (0.5)	—	—	417
$C_{26}H_{26}FO_3PSi_2$	$[CH_3(C_6H_5)_2SiO]_2P(O)F$	243 (4)	1.5612	1.1705	100
$C_{26}H_{33}OPSi$	$[(C_4H_9)_2POSi(C_6H_5)_3$	165—167 (0.01) m.p. 41—42	—	—	417
$C_{26}H_{58}O_8P_2Si$	$[(C_4H_9O)_2PO]_2Si(OC_5H_{11}\text{-tert})_2$	127—129 (0.1)	1.4280	—	233
$C_{26}H_{59}O_{10}PSi_2$	$[(C_4H_9O)_3SiO]_2P(O)OC_2H_5$	160—166 (0.1)	—	—	339
$C_{26}H_{64}O_7P_2Si_4$	$[(C_2H_5)_3SiO]_2P(O)CH_2OP(O)CH_2OSi(C_2H_5)_3$ [on P: $OSi(C_2H_5)_3$]	238—240 (3)	1.4570	1.0166	180
$C_{27}H_{29}O_3PSi_2$	$[CH_3(C_6H_5)_2SiO]_2P(O)CH_3$	270 (3)	1.5781	1.1567	161, 162
$C_{27}H_{29}O_4PSi_2$	$[CH_3(C_6H_5)_2SiO]_2P(O)CH_2OH$	—	1.5755	—	177, 178
$C_{27}H_{63}O_4PSi_3$	$[(C_3H_7)_3SiO]_3PO$	215—225 (5)	—	—	61
$C_{28}H_{63}O_{10}PSi_2$	$[(C_4H_9O)_3SiO]_2P(O)OC_4H_9$	170—185 (0.1)	—	—	339
$C_{29}H_{33}O_3PSi_2$	$[C_2H_5(C_6H_5)_2SiO]_2P(O)CH_3$	300 (6)	1.5701	1.1381	161, 162
$C_{29}H_{35}O_9PSi_4$	$CH_3(C_6H_5)SiO[CH_3(C_6H_5)SiO]_3P(O)CH_3$ [ring: first Si —O— P]	275—276 (1)	—	—	12
$C_{30}H_{25}OPSi$	$(C_6H_5)_2POSi(C_6H_5)_3$	m.p. 85—88	—	—	417
$C_{30}H_{72}O_7P_2Si_4$	$[CH_3(C_3H_7)_2SiO]_2P(O)CH_2OP(O)CH_2OSi(CH_3)(C_3H_7)_2$ [on P: $OSi(CH_3)(C_3H_7)_2$]	244—246 (3)	1.4520	0.9878	183
$C_{36}H_{30}O_8P_2Si_2$	$[(C_6H_5)_2SiOP(O)(OC_6H_5)O]_2$	m.p. 165	—	—	469
$C_{36}H_{81}O_4PSi_3$	$[(C_4H_9)_3SiO]_3PO$	260—270 (5)	—	—	61
$C_{37}H_{33}O_3PSi_2$	$[(C_6H_5)_3SiO]_2P(O)CH_3$	m.p. 196	—	—	162
$C_{37}H_{33}O_3P_3Si$	$[(C_6H_5)_2PO]_3SiCH_3$	—	—	—	417
$C_{38}H_{95}O_{10}PSi_8$	$\{[(C_2H_5)_3SiO]_3SiO\}_2P(O)OC_2H_5$	m.p. 45—50 210 (10^{-3})	—	—	321
$C_{40}H_{41}O_4PSi_3$	$[CH_3(C_6H_5)_2SiO]_2P(O)CH_2OSi(CH_3)(C_6H_5)_2$	306 (1.5)	1.5808^{40}	1.1520^{40}	185, 187
$C_{42}H_{35}O_2PSi$	$[(C_6H_5)_3SiO]_2PC_6H_5$	m.p. 188—191	—	—	285
$C_{54}H_{45}O_4PSi_3$	$[(C_6H_5)_3SiO]_3PO$	m.p. 224	—	—	158
$C_{55}H_{47}O_4PSi_3$	$[(C_6H_5)_3SiO]_2P(O)CH_2OSi(C_6H_5)_3$	m.p. 241—242	—	—	185, 186

TABLE 3. NMR Spectra of Organosilicon Esters of Phosphorus Acid [64] 25% Solutions in CCl_4 to 40 MHz with Internal Standard $[(CH_3)_3Si]_2O$

Compound	Proton chemical shift τ, ppm					Spin-spin coupling constants, Hz		
	P—H	alkoxyl group		triorganosiloxyl group				
		τ_α	τ_β	τ_{CH_3}	τ_{CH_2}	$^1J(P—H)$	$^3J(P—H)$	$^3J(H—H)$
$(C_2H_5O)_2PHO$	3.30	5.93	8.67	—	—	688	8.7	7.1
$C_2H_5O[(C_2H_5)_3SiO]PHO$	3.29	5.97	8.68	9.00	9.28	685	8.9	7.1
$C_3H_7O[(C_2H_5)_3SiO]PHO$	3.29	6.13	8.40	9.00	9.24	683	9.7	6.2
$[(C_2H_5)_3SiO]_2PHO$	3.22	—	—	9.00	9.26	690	—	—
$C_2H_5O[C_6H_5(CH_3)_2SiO]PHO$	3.50	6.27	9.02	9.72	*	695	9.1	6.8
$(C_2H_5O)_3P$	—	6.20	8.78	—	—	—	7.3	7.1
$(C_2H_5O)_2POSi(C_2H_5)_3$	—	5.65	8.58	9.00	9.23	—	7.4	7.4
$C_2H_5OP[OSi(C_2H_5)_3]_2$	—	5.65	8.63	8.95	9.19	—	8.5	6.8
$[(C_2H_5)_3SiO]_3P$	—	—	—	8.98	9.19	—	—	—

* $\tau_{C_6H_5} = 2.80$ ppm.

TABLE 4. P^{31} Magnetic Resonance Chemical Shifts of Organosilyl Phosphates* [417]

Compound	δ, ppm	Compound	δ, ppm	Compound	δ, ppm
$(CH_3)_3SiOP(C_4H_9)_2$	—116	$(C_6H_5)_3SiOP(C_6H_5)_2$	-98.0 ± 1.0	$(CH_3)_2Si[OP(C_6H_5)_2]_2$	-98.1 ± 0.5
$CH_3Si[OP(C_4H_9)]_3$	—119	$(CH_3)_3SiOP(C_6H_5)_2$	-94.1 ± 1.0	$CH_3Si[OP(C_6H_5)_2]_3$	-97.7 ± 1.5
$(C_6H_5)_3SiOP(C_4H_9)_2$	-122.8 ± 0.5				

*Values of δ in ppm relative to the signal of an 85% aqueous solution of H_3PO_4. In the P^{31} magnetic resonance spectrum of tris(trimethylsilyl) phosphate, a signal is observed at +26.5 ppm [514] and +27 ppm [443], taking $\delta = 0$ ppm for H_3PO_4.

There has been little study of the proton magnetic resonance spectra of organosilicon derivatives of phosphorus acids. Among the derivatives of orthophosphoric acid only one compound has been investigated, namely, tris(trimethylsilyl) phosphate, and a signal was observed in its NMR spectrum at −15.3 Hz on the low field side of the internal standard $(CH_3)_4Si$ (at 60 MHz) with a constant J (H^1-C^{13}) of 119.5 Hz [514]. The NMR spectra of organosilicon esters of phosphorous acid have been investigated on a large number of examples (Table 3) [64].

When one alkoxyl group is replaced by a triethylsiloxyl group in fully esterified phosphorus acid a shift is observed in the signal of the α-protons of the alkoxyl group to lower fields; this may be explained by a fall in electron density at the phosphorus atom due to weakening of the $p_\pi-d_\pi$ interaction between the oxygen and phosphorus atoms as a result of a possible analogous interaction between the oxygen and silicon atoms. However, further replacement of an alkoxyl group by a triethysiloxyl group does not produce an additional shift in the proton resonance signal of the methylene group [64].

The P^{31} magnetic resonance chemical shifts of organosilyl phosphonates are given in Table 4 [417].

1.3. Chemical Properties

On heating, trialkoxysilyl alkyl phosphates partly decompose with the liberation of tetraalkoxysilanes and the formation of highly viscous or vitreous polymers. This property is shown most clearly by dimethyl trimethoxysilyl phosphate, which cannot be isolated by distillation even in high vacuum [320]. On heating, triethoxysilyl diethyl thiophosphate also loses tetraethoxysilane. Assuming that the condensation proceeded in accordance with the scheme

$$(C_2H_5O)_2P(S)OSi(OC_2H_5)_3 + C_2H_5OSi(OC_2H_5)_2OP(S)(OC_2H_5)_2 \longrightarrow$$

$$\longrightarrow (C_2H_5O)_4Si + (C_2H_5O)_2P(S)OSi(OC_2H_5)_2OP(S)(OC_2H_5)_2 \qquad (1.115)$$

Feher and Blumcke [319] hoped to obtain bis(diethylthiophosphoxy)-diethoxysilane by this reaction. However, they were unable to isolate such a compound. The authors of [319] ascribed their failure to the impossibility of stopping the condensation at the first stage. The formation of exclusively macromolecular products by thermal decomposition of triethoxysilyl diethyl thiophosphate indicates a polycondensation of the type shown in (1.116).

$$\cdots + \overline{(C_2H_5O)_3Si}|OP(S)(OC_2H_5)|\overline{OC_2H_5 + (C_2H_5O)_3Si}|OP(S)(OC_2H_5)|\overline{OC_2H_5} + \cdots \longrightarrow$$
$$\longrightarrow \cdots -OP(S)(OC_2H_5)OP(S)(OC_2H_5)- \cdots \qquad (1.116)$$

Distillation of higher tris(trialkylsilyl) phosphates $(R_3SiO)_3PO$ ($R = C_3H_7$, C_4H_9) at a pressure of 1 mm is accompanied by their partial decomposition with the formation of luminescent volatile decomposition products.

Attempted distillation of bis(triorganosilyl) hydroxymethyl phosphates resulted in complete decomposition, even at 3 mm. Hexaethyldisiloxane was detected among the decomposition products [178].

On the other hand, bis(triorganosilyl) triorganosiloxymethylphosphonates are stable when distilled repeatedly in vacuum. Bis(trialkylsilyl)phosphonomethyl trialkylsilyl trialkylsiloxymethylphosphonates may be distilled in vacuum with partial decomposition. Crystalline cyclic compounds were isolated from the decomposition products [136, 180]:

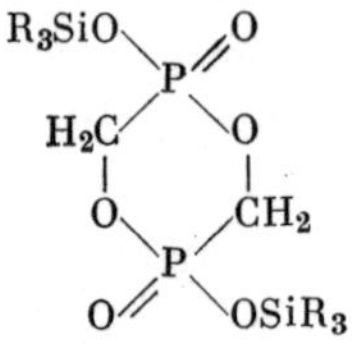

Bis(trialkylsilyl) acetoxymethylphosphonates are stable when distilled in vacuum and when heated to 170-190°C for at least 4 h. Trialkylsilyl fluorophosphates decompose partly when heated [101]:

$$3R_3SiOP(O)F_2 \longrightarrow 3R_3SiF + POF_3 + P_2O_5 \qquad (1.117)$$

Polymers of the type $(C_2H_5O)_2P(O)O[(CH_3)_2SiO]_nP(O)(OC_2H_5)_2$ lose triethyl phosphate to a greater or lesser extent when distilled; this indicates thermal condensation processes and not decomposition [525]:

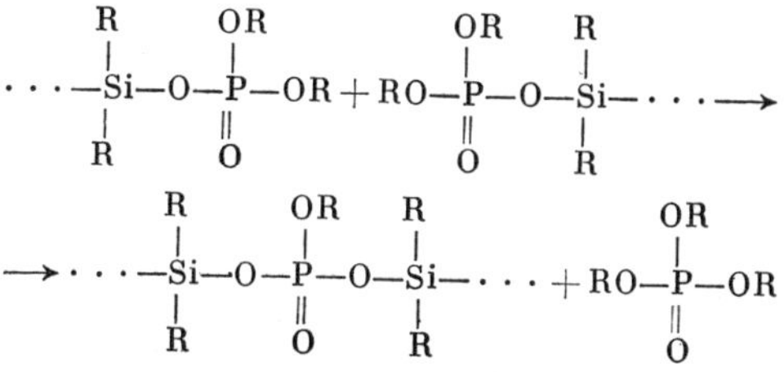

$$
\begin{array}{c}
\overset{\displaystyle R}{\underset{\displaystyle R}{\cdots-Si}}-O-\overset{\displaystyle OR}{\underset{\displaystyle O}{P}}-OR + RO-\overset{\displaystyle OR}{\underset{\displaystyle O}{P}}-O-\overset{\displaystyle R}{\underset{\displaystyle R}{Si}}-\cdots \longrightarrow \\[4mm]
\longrightarrow \overset{\displaystyle R}{\underset{\displaystyle R}{\cdots-Si}}-O-\overset{\displaystyle OR}{\underset{\displaystyle O}{P}}-O-\overset{\displaystyle R}{\underset{\displaystyle R}{Si}}-\cdots + RO-\overset{\displaystyle OR}{\underset{\displaystyle O}{P}}-OR \qquad (1.118)
\end{array}
$$

The polymer with $n = 2$ has the lowest thermal stability. When $n \geq 3$, cyclic substances, whose structures have not been reported, are formed in pyrolysis together with highly viscous linear polymers.

The thermal stability of triethylsilylalumophosphonoxanes [20] is quite high, polymers with a molecular weight of 2500 losing no more than 15% weight when heated to 300°C, and, in the presence of fillers, no more than 5%. The thermal stability of polymers of the type $R_3SiO[Al(OSiR_3)OP(O)R'O]_nH$ is higher when $R' = CH_3$ than when $R' = R_3SiO$. At 520°C polyphenylmethyltitanaphosphasiloxane loses almost half the weight of the corresponding polysiloxane [54].

The most characteristic property of compounds containing the group $Si-O-P$ is the comparative ease of their hydrolytic cleavage. The degree of hydrolysis and the composition of the products depend both on the nature of the phosphorus acid from which the organosilicon phosphorus compound is derived and the structure of its "silicon" part. Thus, for example, hydrolysis of trialkylsilyl phosphates and phosphonates may be represented in the general form in the following way [30, 63, 69, 136, 320, 424, 498]:

$$
R_xP(O)(OR')_y(OSiR_3'')_z \xrightarrow{\;H_2O\;} R_xP(O)(OH)_{3-x} + yR'OH + \frac{z}{2} R_3''SiOSiR_3'' \qquad (1.119)
$$

$$
(R = H, \text{ alkyl, aryl}; \quad R' = H, \text{ alkyl}; \quad R'' = \text{alkyl, aryl}; \quad x = 0-2;
$$
$$
y = 0-2; \quad z = 1-3; \quad x+y = 0-2; \quad y+z = 1-3)
$$

In all cases the first stage of the process is cleavage of the $Si-O-P$ group. In the case of trialkylsilyl dialkyl phosphates, for example, it is possible to titrate the free dialkylphosphoric acid formed in the hydrolysis [335].

With a molar ratio of tris(trimethylsilyl) phosphate:water = 1:1, it is possible to carry out its partial hydrolysis quantitatively

[335]:

$$[(CH_3)_3SiO]_3PO + H_2O \longrightarrow (CH_3)_3SiOP(O)(OH)_2 + (CH_3)_3SiOSi(CH_3)_3 \quad (1.120)$$

It is not possible to obtain the bis(trialkylsilyl) derivative analogously.

Hydrolysis of linear or cyclic phosphonasiloxanes containing R_2SiO units forms polydiorganosiloxanes. However, according to data in [15], by limiting the amount of water used it is possible to stop the hydrolytic cleavage of $\underline{(CH_3)_2 SiO[(CH_3)_2SiO]_n P(O)(CH_3)O}$ at ring opening only with the formation of linear phosphonasiloxanes $HO[(CH_3)_2SiO]_nP(O)(CH_3)OH$. Hydrolysis of phosphonasiloxanes containing the group $(RO)_2SiO$ and also trialkoxylsilyl esters of phosphoric acids leads to polymeric esters of silicic acid [320, 430]. In the case of solid polymers the reaction proceeds slowly from the surface into the depth of the substance, which is initially swollen. Then a highly viscous solution forms and silicic acid separates from this [430]. The hydrolysis of the polymer is much more rapid in 0.1 N NaOH solution [105].

The hydrolysis products of the compounds $[(CH_3)_3SiO]_3 \cdot SiOP(X)(OR)_2$, where X = O or S while R = alkyl, are tris(trimethylsiloxy) silanol and the corresponding acid. The process is catalized by hydrogen ions and in an acid meidum it proceeds as a first-order reaction. In the absence of acid, the hydrolysis is much slower and no longer corresponds to first order. The hydrolysis of thiophosphates is slower by a factor of 500-1000 than that of their oxygen analogs [322]. While the nature of the radical R in the alkoxyl groups has little effect on the hydrolysis rate, the replacement of methyl groups at the silicon atom by ethyl groups reduces it markedly.

The hydrolysis of silicophosphoric acid derivatives proceeds with the formation of orthophosphoric and silicic acids. The acid chloride of silicophosphoric acid $Cl_2SiP_2O_6$ is hydrolyzed most readily, the dimethyl ester $(CH_3O)_2SiP_2O_6$ to a lesser degree, and the diphenyl ester $(C_6H_5O)_2SiP_2O_6$ with greatest difficulty [390]. The action of cold water on bis(trialkylsilyl) acylphosphonates results in hydrolysis of the Si–O–P group without affecting the P–C bond, forming otherwise difficultly accessible acylphosphinic acids in quantitative yield [94]:

$$R'COP(O)(OSiR_3)_2 + 2H_2O \longrightarrow R'COP(O)(OH)_2 + 2R_3SiOH \quad (1.121)$$

Organosilicon esters of acids of trivalent phosphorus are also hydrolyzed readily [94, 417]:

$$(R_3SiO)_3P + 3H_2O \longrightarrow 3R_3SiOH + P(OH)_3 \quad (1.122)$$

$$R_3SiOPR_2 + H_2O \longrightarrow R_3SiOH + R_2P(O)H \quad (1.123)$$

Alcoholysis of tris(trialkylsilyl) phosphates is essentially the reverse of reaction (1.26) [320]:

$$(R_3SiO)_3PO + 3ROH \longrightarrow H_3PO_4 + 3R_3SiOR \quad (1.124)$$

From this it follows that to obtain tris(triorganosilyl) phosphates by scheme (1.26) in good yield it is essential to remove the alcohol formed from the reaction zone.

It is possible to put forward three schemes for the alcoholysis of bis(trialkylsilyl) methylphosphonate, namely, (1.125a) (attack by ROH on the silicon), (1.125b) (attack on the phosphorus), and (1.125c) (mixed attack):

$$CH_3P(O)(OSiR_3)_2 + 2ROH \longrightarrow CH_3P(O)(OH)_2 + 2R_3SiOR \quad (1.125a)$$

$$CH_3P(O)(OSiR_3)_2 + 2ROH \longrightarrow CH_3P(O)(OR)_2 + 2R_3SiOH \quad (1.125b)$$

$$CH_3P(O)(OSiR_3)_2 + 2ROH \longrightarrow CH_3P(O)(OR)OH + R_3SiOR \quad (1.125c)$$

In actual fact, n-alkanols, cyclohexanol, and phenol react with bis(trimethylsilyl) methylphosphonate in accordance with scheme (1.125a). Equilibrium is established instantaneously at room temperature [498]. Tertiary alcohols undergo reaction (1.125a) slowly at room temperature, but with heating to 80-110°C equilibrium is established in only a few minutes. If propylamine, which binds the methylphosphinic acid (or its acid ester), is added to the equilibrium mixture of (1.125a), the ratio of trimethylalkoxysilane:ROH is increased considerably.

In contrast to alcohols, amyl mercaptan and benzyl mercaptan do not react with bis(trimethylsilyl) methylphosphonate [498].

Of the monomeric cyclic compounds $[R_2SiOP(O)(X)O]_2$, the derivative with R = C_6H_5 and X = C_6H_5O reacts with alcohol most slowly (its alcoholysis occurs in butanol) [390].

In the reaction of triorganosilyl fluorophosphates with sodium methylate in methanol, there is gradual quantitative cleavage of all Si–O–P bonds [100, 102]:

$$(R_3SiO)_2P(O)F + CH_3ONa \longrightarrow R_3SiOP(O)F(ONa) + R_3SiOCH_3 \quad (1.126a)$$

$$R_3SiOP(O)F(ONa) + CH_3ONa \longrightarrow (NaO)_2P(O)F + R_3SiOCH_3 \quad (1.126b)$$

$$R_2Si[OP(O)F_2]_2 + 2CH_3ONa \longrightarrow R_2Si(OCH_3)_2 + 2NaOP(O)F_2 \quad (1.127)$$

$$R_2SiF[OP(O)F_2] + CH_3ONa \longrightarrow R_2SiF(OCH_3) + NaOP(O)F_2 \quad (1.128a)$$

$$R_2SiF(OCH_3) + CH_3ONa \longrightarrow R_2Si(OCH_3)_2 + NaF \quad (1.128b)$$

Tris (trialkylsilyl) phosphates react exothermally with sodium acetylide in benzene or xylene to form trialkylethynylsilanes in ~75% yield [95, 96]:

$$(R_3SiO)_3PO + 2NaC{\equiv}CH \longrightarrow 2R_3SiC{\equiv}CH + (R_3SiO)P(O)(ONa)_2 \quad (1.129)$$

Quantitative transsilylation of tris(trimethylsilyl) phosphate is possible by heating it with trialkylchlorosilanes with simultaneous distillation of the lower-boiling trimethylchlorosilane formed [58, 152, 153, 601]:

$$[(CH_3)_3SiO]_3PO + 3R_3SiCl \longrightarrow 3(CH_3)_3SiCl + (R_3SiO)_3PO \quad (1.130)$$

An analogous transsilylation may be achieved by heating tris-(trialkylsilyl) phosphates with higher trialkylsilanes in the presence of colloidal nickel [188, 204, 607]:

$$[(CH_3)_3SiO]_3PO + 3R_3SiH \xrightarrow{Ni} 3(CH_3)_3SiH + (R_3SiO)_3PO \quad (1.131)$$

The yield of trimethylsilane in this case is 89% and that of tris(triethylsilyl) phosphate, 84%. The transsilylation of bis-(triorganosilyl) alkylphosphonates proceeds analogously. It is considered [204] that this reaction proceeds with the formation of six-membered cyclic transition complexes:

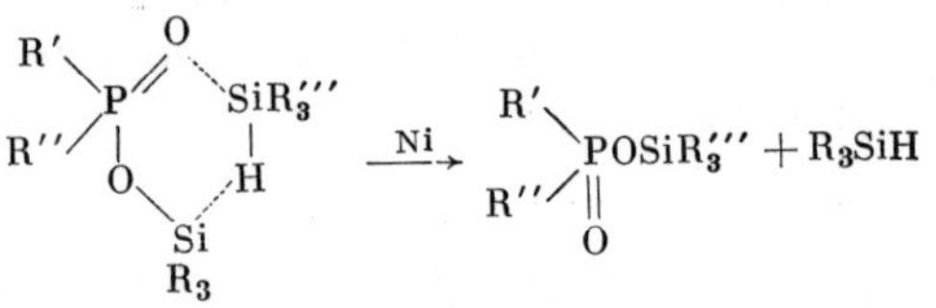

(1.132)

In the presence of alkyl halides, trialkylsilyl dialkyl thiophosphates isomerize in accordance with the scheme [52]

$$R_3SiOP(S)(OC_2H_5)_2 \xrightarrow{RX} R_3SiOP(O)(OC_2H_5)(SC_2H_5) \quad (1.133)$$

However, this reaction proceeds with much more difficulty than in the case of $(RO)_3PS$ and the yield of the isomeric product is only 55% when the reaction is carried out in sealed ampoules at 150°C for 70 h.

Bis(triorganosilyl) organophosphates containing functional groups in the organic radical undergo a series of reactions which do not affect the Si–O–P bonds. These reactions include:

a) reduction of bis(triorganosilyl) chloroalkylphosphonates to bis(triorganosilyl) alkylphosphonates by trialkylsilanes in the presence of colloidal nickel [188, 204]:

$$(R_3SiO)_2P(O)CH_2Cl + HSiR_3 \xrightarrow{Ni} (R_3SiO)_2P(O)CH_3 + R_3SiCl \qquad (1.134)$$

b) dehydrocondensation of bis(triorganosilyl) hydroxymethylphosphonates with trialkylsilanes in the presence of colloidal nickel [136, 178]:

$$(R_3SiO)_2P(O)CH_2OH + HSiR'_3 \xrightarrow{Ni} (R_3SiO)_2P(O)CH_2OSiR'_3 + H_2 \qquad (1.135)$$

c) acetylation of bis(triorganosilyl) hydroxymethylphosphonate by acetic anhydride [136, 178]:

$$(R_3SiO)_2P(O)CH_2OH + (CH_3CO)_2O \xrightarrow[-CH_3COOH]{(H_2SO_4)} (R_3SiO)_2P(O)CH_2OCOCH_3 \qquad (1.136)$$

A comparative study of the reaction of bis(trialkylsilyl) trialkylsiloxymethylphosphonates, bis(trialkylsilyl) acetoxymethylphosphonates, and dialkyl trialkylsiloxymethylphosphonates with acetic anhydride in the presence of a catalytic amount of sulfuric acid showed that in all cases there is elimination of trialkylsilyl groups in the form of trialkylacetoxysilanes and that the C–O–Si bond is broken much more readily than the P–O–Si bond, while the P–O–C group remains inert towards acetic anhydride [136, 176, 179]. Therefore it is possible to carry out the selective desilyation of bis(triethylsilyl) triethylsiloxymethylphosphonate with acetic anhydride to bis(triethylsilyl) acetoxymethylphosphonate (83% yield) [176, 179]:

$$(R_3SiO)_2P(O)CH_2OSiR_3 + (CH_3CO)_2O \xrightarrow[-CH_3COOH]{} (R_3SiO)_2P(O)CH_2OCOCH_3 \qquad (1.137)$$

Bis(trialkylsilyl) phosphites also undergo transsilylation with trialkylchlorosilanes in accordance with a scheme analogous to (1.130) [58], but in their reaction with trialkylsilanes in the presence of colloidal nickel further dehydrocondensation is observed with the formation of tris(trialkylsilyl) phosphites in accordance with scheme (1.103) [94, 171, 172, 204], with possible transsilylation in accordance with scheme (1.104) when there are unlike alkyl groups at the silicon atoms of the two reagents [204].

The reaction of bis(trimethylsilyl) phosphite with sulfur with the formation of the thiophosphate proceeds much more slowly than in the case of dialkyl phosphites [59]:

$$(R_3SiO)_2P(O)H + S \longrightarrow (R_3SiO)_2P(S)OH \qquad (1.138)$$

Bis(trialkylsilyl) phosphites add to acrylonitrile in the presence of sodium alcoholates to form bis(trialkylsilyl) β-cyanoethylphosphonates (20-30% yield) [59, 582]:

$$(R_3SiO)_2P(O)H + CH_2{=}CHCN \xrightarrow{RONa} (R_3SiO)_2P(O)CH_2CH_2CN \qquad (1.139)$$

Trialkylsilyl phosphites form complexes with cuprous chloride, are converted into phosphates by the action of atmospheric oxygen, add sulfur and halogens readily, and undergo the Arbuzov rearrangement [46, 52, 65, 205, 222]:

$$R_3SiOP(OC_2H_5)_2 + S \longrightarrow R_3SiOP(S)(OC_2H_5)_2 \qquad (1.140)$$

$$R_3SiOP(OC_2H_5)_2 + C_2H_5Br \longrightarrow C_2H_5P(O)(OC_2H_5)(OSiR_3) \qquad (1.141)$$

In the reaction trialklylchlorosilanes in the presence of sodium [581] or organic bases [612] they are converted into tris(trialkylsilyl) phosphites.

The synthesis of tris(triorganosilyl) phosphites in accordance with scheme (1.36) should also be carried out in the absence of air, as otherwise a by-product – or even the main reaction product – may be the corresponding tris(triorganosilyl) phosphate. Thus, for example, boiling a mixture of triethylmethoxysilane and phosphorous acid (3:1) under reflux in a stream of air for 8 h leads to the formation of tris(triethylsilyl) phosphate (58% yield) [63].

Tris(triorganosilyl) phosphites rearrange very actively under the effect of acyl halides* [94, 170, 604]:

$$(R_3SiO)_3P + R'COCl \longrightarrow R'COP(O)(OSiR_3)_2 + R_3SiCl \qquad (1.142)$$

In dry ether the yield of bis(trialkylsilyl) acylphosphonates reaches 70-90%. As a result of the rearrangement of bis(triethylsilyl) dimethylphenyl phosphite with acetyl chloride there is

*The reaction of tris(triorganosilyl) phosphites with alkyl halides proceeds analogously at 125-150° [596].

the elimination of a triethylsilyl group and the formation of triethylsilyl dimethylphenylsilyl acetylphosphonate. In the reaction of mixed esters of phosphorous acid with acetyl chloride there is the attack of the chlorine ion on the siloxy group, leading to the elimination of a trialkyl-chlorosilane, while the alkoxyl group remains unaffected [94, 170]:

$$(R'O)_nP(OSiR_3)_{3-n}+CH_3COCl \longrightarrow CH_3COP(O)(OR')_n(OSiR_3)_{2-n}+R_3SiCl \quad (1.143)$$

When triorganosilyl diorganophosphinates are treated with alkyl iodides there is complete cleavage of the P−O−Si bond [417]:

$$R_2POSiR'_3+R''I \longrightarrow R_2R''PO+R'_3SiI \quad (1.144)$$

Like tris(trialkylsilyl) phosphates, tris(trialkylsilyl) phosphites are transsilylated by trialkylsilanes in the presence of colloidal nickel [204]:

$$(R_3SiO)_3P+3R'_3SiH \xrightarrow{Ni} (R'_3SiO)_3P+3R_3SiH \quad (1.145)$$

and in the reaction with sulfur they form tris(trialkylsilyl) thiophosphates [597-599].

1.4. Analysis Methods

The analysis of silicon-containing organophosphorus compounds containing the grouping Si−O−P is based on their thermal, oxidative, or hydrolytic cleavage with subsequent determination of the phosphorus and the silicon in the cleavage products by gravimetric, titrimetric, or photocolorimetric methods. The spectrophotometric determination of phosphorus and silicon in silicon-containing organophosphorus compounds through their emission spectra has also been proposed [116].

The decomposition of silicon-containing organophosphorus compounds is carried out with a mixture of oleum or concentrated sulfuric acid with nitric acid [30, 97, 131, 157, 158, 333]; sulfuric acid with KIO_3 added [162], ammonium persulfate added [573], or potassium persulfate added [128-131]; a mixture of chromic and sulfuric acids in a stream of oxygen at 180°C [132]; perchloric acid [339, 390]; sodium peroxide at 1000°C [140, 141, 333, 424]; and potassium at 900°C [230].*

*Fusion with a mixture of sodium and potassium carbonates does not give satisfactory results in the determination of phosphorus in silicon-containing organophosphorus compounds [142].

After the decomposition, the phosphorus is determined gravimetrically as magnesium ammonium phosphate [30, 102, 157, 158, 202, 333, 389, 390], or photocolorimetrically in the form of phosphomolybdenum blue [31, 43, 102, 128-132, 141, 210, 211, 491, 572]. Silicon is determined gravimetrically as SiO_2 [30, 97, 157, 158, 333, 389], acidimetrically [132], or photocolorimetrically in the form of silicomolybdenum blue [77, 107].

The tritrimetric determination of phosphorus in organosilicon esters of phosphorus acids is based on their hydrolysis and the titration of the phosphorus acids thus formed with alkali in the presence of indicators, namely, methyl orange [61, 105-107, 114], bromothymol blue [62a, 63], or thymolphthalein [14, 31]. In these cases the silicon content may be determined by difference from the sum of the oxides formed by mineralization of the substances or the P_2O_5 content found by alkalimetry [62a].

Analysis of silicon-containing organophosphorus compounds in aqueous solutions has definite drawbacks, which are connected with further conversions of the primary hydrolysis products. Taking this into account, Kreshkov and his co-workers developed a method of analyzing alkylsilylphosphoric acids and alkylsilyl fluorophosphates by potentiometric [81, 99, 100, 102, 103, 212] and conductometric [82, 212] titration in nonaqueous solvents. As the titration media it is possible to use alcohols, ketones, and acetonitrile, and as the titrants, alkali metal methylates. Lithium methylate in methanol reacts in stages, which the authors explain this by the scheme

$$R_3SiOP(O)(OH)_2 + CH_3OLi \xrightarrow{-CH_3OH} R_3SiOP(O)(OH)Li \xrightarrow[-CH_3OH]{+2CH_3OLi} R_3SiOCH_3 + Li_3PO_4 \quad (1.146)$$

In other aliphatic alcohols the reaction of CH_3OLi with trimethylsilyl phosphate proceeds with the immediate consumption of three molecules of titrant. When potassium methylate and tetramethylammonium hydroxide are used, one potential jump is observed and there is evidence [103, 212] in favor of the reaction

$$(CH_3)_3SiOP(O)(OH)_2 + CH_3OK \longrightarrow (CH_3)_3SiOCH_3 + KH_2PO_4 \quad (1.147)$$

In titration in the presence of indicators, the best agreement between potentiometric and visual data is given by the use of methyl red.

It is recommended that silicon-containing organophosphorus compounds containing the group Si−O−P be analyzed for carbon and hydrogen by microcombustion of a sample in a stream of oxygen [166, 572], preferably in the presence of Cr_2O_3 catalyst [106, 113] (with simultaneous determination of the sum of the oxides $SiO_2 + P_2O_5$).

1.5. Practical Application

There are reports of the possibility of the use of phosphasiloxane polymers as hydraulic fluids [318], antifoaming additives for hydrocarbon oils [109, 318, 554, 555], plasticizers [109, 554], additives to improve the thermal and oil resistance of plastics [113], varnishes [109, 318], greases for ground joints which are resistant to aggressive gases and solvents at temperatures up to 200-250°C [110, 111], impregnating agents to fireproof cloth [488], and coatings to increase the resistance of glass [137].

According to patent data [467, 468] the introduction of alkyl diaryl or trialkyl phosphates (up to 40% by volume) into polydimethylsiloxane fluids makes the latter fire-proof and compatible with mineral oils. Such compositions are also high quality lubricants. Tin-containing polyphosphonasiloxanes also have improved lubricating power [72]. However, the addition of organophosphorus compounds increases the antiwear properties of polysiloxane fluids to a lesser degree than those of mineral oils [60, 147].

In many publications [91, 107, 111, 113, 117, 118, 218, 219, 229] there is an examination of the improvement in the properties of cement on the addition to the cement solutions of the polymeric products from the reaction of phosphorus pentoxide with alkoxysilanes. The maximum effect was obtained by adding 0.5% (on the dry cement) of $[Si_2P_2O_8(OC_2H_5)_2 \cdot H_2O]_n$. When phosphonasiloxanes are used there is some reduction in the water requirement (water: cement ratio), an increase in the compression strength of the cement (by an average of 20% for 7 days and 30% for 28 days "aging"), and improvement in both the frost resistance and water impermeability.

It is also recommended that sand used for making building mortar be impregnated with silicon-containing organophosphorus compounds [117].

Organophosphorous compounds containing nitrogen of the type $(-PX_2 = N-)_n$, $RR'NPX_2$, RR'NP = NR", $RR'NP(O)X_2$, and RR'NP(O) = NR" are agents which induce low-temperature conversions of polydiorganosiloxanes into elastomers [479, 557].

Andrianov [5, 6] recommends the use of alumaphosphonasiloxanes as intermediates for the preparation of heteroorganic graft and block polymers of high thermal stability.

The possibility of using phosphonasiloxanes as antistick coatings for tooth enamel has been investigated [241].

Elastomers obtained by using decomposing quaternary phosphonium bases as polymerization catalysts have an improved thermal stability. This is evidently connected with deactivation of the "living" polymer molecule on heating due to decomposition of the phosphonium ends of the macromolecular chain and also the stabilizing effect of the decomposition products. In any case, in the patents [377, 446-449] it is recommended that esters of phosphoric or phosphorous acids such as tritolyl phosphate or triphenyl phosphate be added to polymethylsiloxane rubbers to increase their thermal stability to 250-300°C. Triorgano phosphites (1-2 wt.%) also stabilize polysiloxane oils [372]. The addition of 2-5% of phenylphosphinic acid or its diphenyl ester to thermosetting polysiloxane resins increases their "lifetimes" at 260°C from 200 to 1200 h [314]. Tetrabutylphosphonium hydroxide has also been described as a stabilizer for resins [462].

A linear polymer with a viscosity of 2500 centipoises, obtained by alkaline polycondensation of polydimethylsiloxanes at 150°C, is completely depolymerized at 300°C. However, the addition of phosphoric esters, which deactivate the remaining traces of alkali, stabilize such polymers, making it possible to remove volatile components at 300°C without appreciable depolymerization. Vulcanizates from polydimethylsiloxane modified in this way have a reduced compression set [446-449].

The presence in the main chain of polysiloxane of P–O fragments, whose bond energy (145 kcal/mole)* is greater than the bond energy of Si–O (117 kcal/mole), should lead to an improvement in the resistance of the polymer to the action of heat [78].

* The P=O bond energy is 179 kcal/mole.

In actual fact, the thermal stability of polymethylphenylphosphonasiloxane (Si:P = 200 kcal/mole) at 350–400°C is higher than the thermal stability of polymethylphenylsiloxanes [89]. The introduction of 0.02% P into polydimethylsiloxane elastomer in the form of =P(O)O groups increases the thermooxidative stability of rubbers based on them at 300°C in comparison with rubbers from pure polydimethylsiloxane raw rubber [9, 15, 79]. Phosphorus also has the same effect on the strength and relative elongation [9, 15]. The thermal stability of rubbers from radiation vulcanizates of polyboraphosphonamethylphenylsiloxanes is higher than that of rubbers from the corresponding polysiloxanes which contain no phosphorus [1, 194, 195]. Rubbers obtained from polyalumaphosphonadimethylsiloxanes are also superior in mechanical strength and relative elongation to rubbers from polydimethylsiloxane raw rubber [28].

2. COMPOUNDS CONTAINING THE GROUPINGS Si − O − S − P AND Si − S − P

Compounds containing the grouping Si−O−S−P, which have been mentioned in only one paper [524], were obtained by the reaction of phosphines with the trimethylsilyl ester of chlorosulfonic acid at −78°C [524]:

$$(C_6H_5)_2PH + ClSO_2OSi(CH_3)_3 \xrightarrow[-HCl]{} (C_6H_5)_2PSO_2OSi(CH_3)_3 \quad (1.148)$$

The compounds $C_6H_5PHSO_2OSi(CH_3)_3$ and $(C_6H_5)_2PSO_2OSi(CH_3)_3$ are only stable at low temperatures and decompose above even 0°C.

Compounds containing the grouping Si−S−P were synthesized from salts of the corresponding thioacids of phosphorus according to the scheme [319, 408]

$$R_3SiCl + NH_4SP(S)(OR)_2 \longrightarrow R_3SiSP(S)(OR)_2 + NH_4Cl \quad (1.149)$$

TABLE 5. Organosilicon Compounds Containing the Grouping Si−S−P

Compound	B.p., °C (mm)	n_D^{20}	d_4^{20}	Literature
$(CH_3)_3SiSP(S)(OCH_3)_2$	44–46 (10^{-3})	1.5148	1.119	319
$(CH_3)_3SiSP(S)(OC_2H_5)_2$	55–56 (10^{-3})	1.5098	1.084	258, 319

The cleavage of Si−N bonds in trialkyl(alkylamino)silanes and hexaalkyldisilazans by derivatives of thiophosphoric acids may also be used for their preparation [258, 484]:

$$R_3SiNHR' + HSP(S)R''_2 \longrightarrow R_3SiSP(S)R''_2 + R'NH_2 \quad (1.150)$$

The physicochemical constants of compounds containing the grouping Si−S−P are given in Table 5.

3. COMPOUNDS CONTAINING THE GROUPING Si − N − P

3.1. Preparation Methods

Esters of trialkylsilylaminophosphinic acids were synthesized in accordance with the scheme [297, 319]

$$R_3SiCl + H_2NP(X)(OR)_2 \xrightarrow{R'_3N} R_3SiNHP(X)(OR)_2 + R'_3N \cdot HCl \quad (1.151)$$
$$(X = O \text{ or } S)$$

The reaction of amides [15, 55, 257, 430] and hydrazides [258] of dialkylphosphoric acids with trialkyl(alkylamino)silanes of hexaalkyldisilanes may be used conveniently for the preparation of dialkyl (trialkylsilylamino)phosphonates and dialkyl (trialkylsilylhydrazino)phosphonates:

$$R_3SiNHR' + H_2N(NH)_nP(O)(OR'')_2 \longrightarrow R_3SiNH(NH)_nP(O)(OR'')_2 + R'NH_2$$
$$(R' = \text{alkyl or } (CH_3)_3Si;\ n = 0 \text{ or } 1) \quad (1.152)$$

Hexachlorocyclotriphosphazene $(-PCl_2{=}N-)_3$ does not react with hexamethyldisilazan on heating to boiling for 37 h [457]. Silylation occurs only if one or two chlorine atoms in the hexachlorocyclotriphosphazene molecule are replaced by amino [457], methylamino [456], or ethylamino [456] groups. Then the reaction proceeds according to the scheme analogous to (1.152), with the replacement of a hydrogen atom in the amino group by a trimethylsilyl group.

The character and the direction of the reaction of compounds containing an Si−N bond with phosphorus compounds containing a P−X bond (where X is a halogen) depends to a considerable degree on the structure of the two reagents. In many cases there is simply cleavage of the Si – N bond with the formation of the corresponding halosilane:

$$\rangle Si-N\langle + X-P(O)\langle \longrightarrow \rangle Si-X + \rangle N-P(O)\langle \qquad (1.153)$$

This occurs in the reaction of PCl_3, $POCl_3$, $RPOCl_2$, and R_2POCl with organosilicon derivatives of heterocyclic amines containing the Si – N bond [261, 407, 572], the cleavage of trimethyl-(dimethylamino)silane by PF_5 [295], the reaction of N-ethylhexamethyldisilazan with PCl_3 [235], the cleavage of hexamethyldisilazan by diphenylchlorophosphine in boiling toluene [480, 481], and the reaction of triorganosilylazides with chlorophosphines [432]. In the reaction of hexamethyldisilazan with $POCl_3$ and $(C_6H_5O)_2P(O)Cl$ there is also cleavage of the Si – N bond with the formation of trimethylchlorosilane, but the second reaction product is not a phosphamide, formed in accordance with scheme (1.153), but either the acid chloride of trimethylsilylaminophosphinic acid or the corresponding trimethylsilylaminophosphonate [258]:

$$[(CH_3)_3Si]_2NH + ClP(O)Cl_2 \longrightarrow (CH_3)_3SiNHP(O)Cl_2 + (CH_3)_3SiCl \qquad (1.154)$$

$$[(CH_3)_3Si]_2NH + ClP(O)(OC_6H_5)_2 \longrightarrow (CH_3)_3SiNHP(O)(OC_6H_5)_2 + (CH_3)_3SiCl \qquad (1.155)$$

The reaction of hexamethyldisilazan with PCl_3 at –20°C proceeds analogously [256]:

$$[(CH_3)_3Si]_2NH + PCl_3 \longrightarrow (CH_3)_3SiNHPCl_2 + (CH_3)_3SiCl \qquad (1.156)$$

The reaction of trimethyl(methylamino)silane with the acid chloride of diphenylphosphoric acid forms not only diphenyl (methylamino)phosphonate and trimethylchlorosilane but also diphenyl [trimethylsilyl(methyl)amino]phosphonate, whose yield reaches 85% with a molar ratio of the agents of 2:1 [258]. The formation of the latter may be explained by the fact that the trimethylchlorosilane and diphenyl (methylamino)phosphonate obtained in accordance with scheme (1.153) interact in accordance with scheme (1.151) or there is the direct replacement of the hydrogen atom of the amino group in trimethyl(methylamino)silane with subsequent cleavage of the silylamide formed by the hydrogen chloride thus liberated:

$$2(CH_3)_3SiNHCH_3 + 2ClP(O)(OR)_2 \longrightarrow (CH_3)_3SiN(CH_3)P(O)(OR)_2 + HCl +$$
$$+ (CH_3)_3SiCl + CH_3NHP(O)(OR)_2 \qquad (1.157)$$

Sodium bis(trimethylsilyl)amide reacts with PCl_3 with the formation of a polymer which has the approximate composition $[(CH_3)_3SiNPCl]_n$ [436, 559]. However, its reaction with $POCl_3$ [560] or diphenylchlorophosphine [480] yields monomeric compounds ($R = CH_3$):

$$2(R_3Si)_2NNa + POCl_3 \longrightarrow [(R_3Si)_2N]_2POCl + 2NaCl \qquad (1.158)$$

$$(R_3Si)_2NNa + (C_6H_5)_2PCl \longrightarrow (R_3Si)_2NP(C_6H_5)_2 + NaCl \qquad (1.159)$$

The reaction of lithium (trimethylsilyl)methylamide with PCl_3 leads to a mixture of products containing the Si−N−P link [513], while the reaction of the dilithium derivative of octamethyltrisilazan with phenyldichlorophosphine forms a cyclic compound [337]:

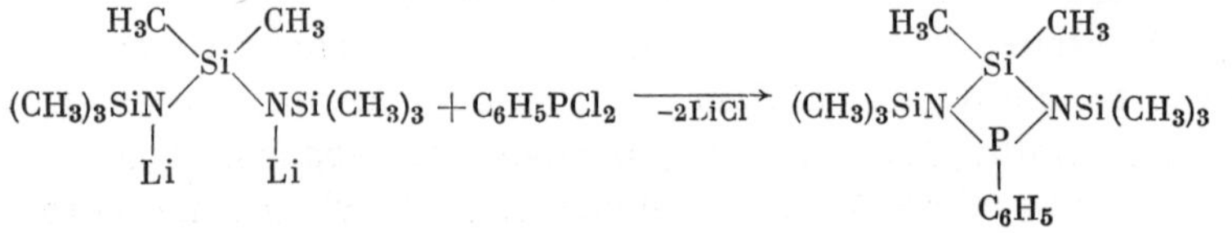
(1.160)

The reaction of triorganochlorosilanes with N-lithio-P,P,P-triorganophosphazenes forms N-triorganosilyl-P,P,P-triorganophosphazenes [515]:

$$R_3SiCl + LiN{=}PR'_3 \longrightarrow R_3SiN{=}PR'_3 + LiCl \qquad (1.161)$$

A more convenient general method of preparing N-silylphosphazenes is heating a mixture of organosilylazides with triorganophosphines at 100-150°C [262, 264, 500, 518-521, 552, 553, 562, 565-567, 571, 689]:

$$R_nSi(N_3)_{4-n} + (4-n)R'_3P \longrightarrow R_nSi(N{=}PR'_3)_{4-n} + (4-n)\,N_2 \qquad (1.162)$$

In the case of the reaction of dimethyldiazidosilane with trimethylphosphine, the replacement of the first azido group occurs even at room temperature (92% yield), while the replacement of the second requires heating in an autoclave at 110-130°C (96% yield) [519, 520].

The reaction of diphenylphosphine with triphenylazidosilane proceeds in accordance with the scheme [487]:

$$R_3SiN_3 + R_2PH \longrightarrow R_3SiNHPR_2 + R_3SiNHPR_2{=}NSiR_3 \quad (R = C_6H_5) \qquad (1.163)$$

Only the diadduct is obtained with excess triphenylazidosilane.

Heating tetraphenyldiphosphine with triphenylazidosilane in benzene in a sealed ampoule at 80-115°C for 14 days leads to the liberation of only 17.1% of the nitrogen. At 140°C in the absence of benzene, 93% of the nitrogen is liberated after 3 days and N,N'-bis(triphenylsilyl)-P,P,P',P'-tetraphenyldiphosphinodiimine is formed [487]:

$$2R_3SiN_3 + R_2PPR_2 \longrightarrow R_3SiN{=}PR_2PR_2{=}NSiR_3 + 2N_2 \qquad (1.164)$$

N-Trimethylsilyl-P,P,P-triphenylphosphazene may be obtained in accordance with the following scheme [481]:

$$[(CH_3)_3Si]_2NH + (C_6H_5)_3PBr_2 \xrightarrow[-(CH_3)_3SiBr]{} [(CH_3)_3SiNHP(C_6H_5)_3]\,Br \xrightarrow{+(C_2H_5)_3N}$$
$$\longrightarrow (CH_3)_3SiN{=}P(C_6H_5)_3 + (C_2H_5)_3N \cdot HBr$$

3.2. Physical Properties

Trialkylsilylamides of phosphoric acids dissolve readily in organic solvents. The greatest variation is shown by $(CH_3)_3SiNHP(O)Cl_2$ [an unusual compound which, like dialkyl (trimethylsilylamino)thiophosphonates, contains atoms of seven chemical elements], which dissolves readily in ether and acetonitrile, less readily in dioxane and nitro benzene, and is insoluble in benzene, cyclohexane, and chlorinated hydrocarbons [261]. Trimethylsilylaminodichlorophosphine is unstable and gradually decomposes even at room temperature. It is insoluble in most organic solvents but is slightly soluble in nitrobenzene [259].

N-Triphenylsilyl-P,P,P-triphenylphosphazene is soluble in tetrahydrofuran, but is insoluble in ether, benzene, and ligroin.

The infrared spectra of N-trimethylsilyl-P,P,P-triorganophosphazenes contain absorption bands in the region 1285-1315 cm^{-1}, which corresponds to P=N bond oscillations [521]. In the infrared spectra of their complexes with R_3M (M = Al, Ga, In) this band is strongly shifted toward the long-wave region and is observed at 1060-1120 cm^{-1} in the case of trialkyl derivatives [521] and 1015-1045 cm^{-1} in the case of the triphenyl derivative [522]. The greatest shift is observed in the spectra of complexes with AlR_3. In the spectra of trialkylsilyltrialkylphosphazenes the band in the region of 538-556 cm^{-1} is assigned to Si−N bond vibrations [521].

TABLE 6. NMR Spectra of N-Organosilyl-P,P,P-trimethylphosphazenes and Their Complexes. The spectra were plotted at 60 MHz. Negative values of δ are shifts to low fields relative to $(CH_3)_4Si$ (internal standard in CCl_4; tetramethylsilane was used as an external standard in solutions in CH_2Cl_2)

Compound	Chemical shifts δ, Hz			Spin-spin coupling constants J, Hz					Solvent; temp., °C	Literature
	CH_3–Si	CH_3–P	CH_3–M	H^1–C–Si^{29}	H^1–C^{13}–Si	H^1–C–P^{31}	H^1–C^{13}–P	H^1–C^{13}–M		
$(CH_3)_3SiN{=}P(CH_3)_3$	+3.8	−82.5	—	6.4	117	12.4	127	—	CH_2Cl_2; 30–35	523
$(CH_3)_2Si(N_3)N{=}P(CH_3)_3$	−4.4	−87.7	—	—	—	12.9	—	—	CCl_4; 30	520
$(CH_3)_2Si[N{=}P(CH_3)_3]_2$	+3.7	−86.4	—	—	—	12.75	—	—	CH_2Cl_2; 30	520
$(CH_3)_3SiN{=}P(CH_3)_3\cdot Al(CH_3)_3$	−13.2	−104.0	+61.0	6.6	119	12.8	130	109	CCl_4; 30	518, 521
$(CH_3)_3SiN{=}P(CH_3)_3\cdot AlCl_3$	−30.5	−122.5	—	6.8	120	13.15	131	—	CH_2Cl_2; 30–35	523
$(CH_3)_3SiN{=}P(CH_3)_3\cdot AlBr_3$	−36.8	−129.5	—	6.75	119.5	13.0	130.5	—	CH_2Cl_2; 30–35	523
$(CH_3)_3SiN{=}P(CH_3)_3\cdot AlI_3$	−43.4	−136.5	—	6.7	119	12.85	129.5	—	CH_2Cl_2; 30–35	523
$(CH_3)_3SiN{=}P(CH_3)_3\cdot Ga(CH_3)_3$	−9.0	−98.2	+36.5	6.55	118	12.7	129	117	CCl_4; 30	518, 521
$(CH_3)_3SiN{=}P(CH_3)_3\cdot GaCl_3$	−30.4	−123.5	—	6.9	119.5	13.05	—	—	CH_2Cl_2; 30–35	523
$(CH_3)_3SiN{=}P(CH_3)_3\cdot In(CH_3)_3$	−6.6	−95.6	+32.5	6.5	117.5	12.6	128.5	124.5	CCl_4; 30	518, 521
$(CH_3)_3SiN{=}P(CH_3)_3\cdot InCl_3$	−26.2	−119.4	—	6.8	—	13.15	—	—	CH_2Cl_2; 30–35	523
$(CH_3)_3SiN{=}P(CH_3)_3\cdot InBr_3$	−29.8	−122.6	—	6.75	119	12.9	—	—	CH_2Cl_2; 30–35	523
$(CH_3)_2Si[N{=}P(CH_3)_3]_2\cdot Al(CH_3)_3$	−10.2	−97.0	+58.5	—	—	13.0	—	—	CH_2Cl_2; 30	520
	—	−88; −108 *	—	—	—	13.3; 12.8 *	—	—	CH_2Cl_2; −60	520
$(CH_3)_2Si[N{=}P(CH_3)_3]_2\cdot Ga(CH_3)_3$	−5.0	−93.0	+37.5	—	—	12.9	—	—	CH_2Cl_2; 30	520
	—	−86; −102 *	—	—	—	13.0; 12.75 *	—	—	CH_2Cl_2; −60	520
$(CH_3)_3Si[N{=}P(CH_3)_3]_2\cdot In(CH_3)_3$	−5.0	−93.0	+31.3	—	—	12.8	—	—	CH_2Cl_2; 30	520
	—	−93.5	—	—	—	12.75	—	—	CH_2Cl_2; −60	520

*The first value refers to the signal of a free phosphinimino group and the second to the signal of a phosphinimino group bound in a complex.

The data on the NMR spectra of N-organosilyl-P,P,P-trimethylphosphazenes and their complexes with aluminum, gallium, and indium halides and trialkyl derivatives are given in Table 6.

It has been shown that in the NMR spectra of $R_3P = NSiR_2N = PR_3 \cdot AlR_3$ and $R_3P = NSiR_2N = PR_3 \cdot GaR_3$ ($R = CH_3$) at low temperatures (−60°C), two signals are observed for protons of the $(CH_3)_3P$ group. One of these may be assigned to a free $R_3P = N$ group, while the other (at lower fields) may be assigned to a $R_3P = N$ group bound in a complex with R_3M. With a rise in temperature these two signals merge into one due to the rapid transition:

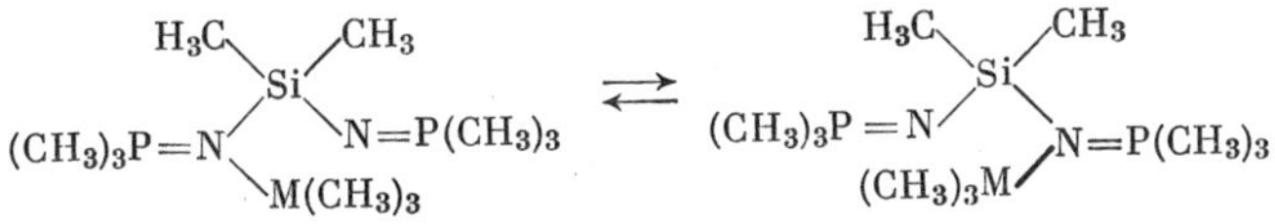

In an analogous complex with $(CH_3)_3In$ the signal of the $(CH_3)_3P$ group is not split, even with the temperature lowered to −70°C. This may be explained by extremely rapid exchange or the existence of a fixed structure with a pentacovalent indium atom [519, 520].

The physicochemical constants of compounds containing the grouping Si−N−P are given in Table 7.

3.3. Chemical Properties

Most compounds containing the grouping Si−N−P are readily hydrolyzed by water. Hydrolysis of $(CH_3)_3SiNHPOCl_2$ is caused even by moist air; cleavage occurs at the Si−N bond, since according to chromatography data the hydrolysis products are hexamethyldisiloxane and $H_2NP(O)(OH)_2$ [258]. Atmospheric moisture also effects cleavage of $(CH_3)_3SiNHNHP(O)(OC_6H_5)_2$. At the same time diorgano (trialkylsilylamino)phosphonates are quite resistant to the action of cold water and are hydrolyzed rapidly only by boiling water, also with the liberation of hexamethyldisiloxane. Hydrolysis of trimethylsilylaminodichlorophosphine proceeds with flashes, due to the formation of phosphorus and phosphine [256].

Triphenylsilyltriphenylphosphinimine is resistant to both hydrolysis and thermal decomposition and is cleaved only by strong acids [552].

TABLE 7. Compounds Containing the Grouping Si–N–P

Empirical formula	Compound	M.p., °C	B.p., °C (mm) * n_D^{20}	Literature
$C_3H_{10}Cl_2NOPSi$	$(CH_3)_3SiNHP(O)Cl_2$	94—95	—	258
$C_4H_{12}Cl_5N_4P_3Si$	$\overline{N(=PCl_2N=)_2P}ClN(CH_3)Si(CH_3)_3$	90.5—91	—	456
$C_5H_{14}Cl_5N_4P_3Si$	$\overline{N(=PCl_2N=)_2P}ClN(C_2H_5)Si(CH_3)_3$	—	135.5 (1.5) *1.5281[18]	456
$C_5H_{15}N_4PSi$	$(CH_3)_2Si(N_3)N=P(CH_3)_3$	—	40 (0.3)	520
$C_5H_{16}NO_2PSSi$	$(CH_3)_3SiNHP(S)(OCH_3)_2$	—	58—60 (10^{-3}) *1.4753	319
$C_6H_{18}NPSi$	$(CH_3)_3SiN=P(CH_3)_3$	3—4	169	518, 521
		—	57 (11)	515
$C_6H_{20}Cl_4N_5P_3Si_2$	$\overline{N=PCl_2N=P}(NHSi(CH_3)_3)ClN=\overline{P}ClNHSi(CH_3)_3$	173.5—174.2	—	457
$C_7H_{20}NO_2PSSi$	$(CH_3)_3SiNHP(S)(OC_2H_5)_2$	—	117—118 (10^{-3}) *1.4718	319
$C_8H_{24}N_2P_2Si$	$(CH_3)_2Si[N=P(CH_3)_3]_2$	24—25	77—78 (0.3)	520
$C_9H_{24}NPSi$	$(CH_3)_3SiN=P(C_2H_5)_3$	—	88—89 (11)	515
		—	89.5 (11)	262
$C_9H_{24}NPSi_2$	$(CH_3)_3SiN=P(CH_3)_2CH_2Si(CH_3)_3$	—	93—94 (11)	515
$C_9H_{27}AlNPSi$	$(CH_3)_3SiN=P(CH_3)_3 \cdot Al(CH_3)_3$	79—80	120 (1)	518, 521
$C_9H_{27}GaNPSi$	$(CH_3)_3SiN=P(CH_3)_3 \cdot Ga(CH_3)_3$	32—34	59 (1)	518, 521
$C_9H_{27}InNPSi$	$(CH_3)_3SiN=P(CH_3)_3 \cdot In(CH_3)_3$	43—44	67 (1)	518, 521
$C_{11}H_{33}AlN_2P_2Si$	$(CH_3)_2Si[N=P(CH_3)_3]_2 \cdot Al(CH_3)_3$	93—95	—	520
$C_{11}H_{33}GaN_2P_2Si$	$(CH_3)_2Si[N=P(CH_3)_3]_2 \cdot Ga(CH_3)_3$	80—82	—	520
$C_{11}H_{33}InN_2P_2Si$	$(CH_3)_2Si[N=P(CH_3)_3]_2 \cdot In(CH_3)_3$	74—76	—	520
$C_{12}H_{25}N_2O_2PSi_2$	$[(CH_3)_3SiNH]_2P(O)(OC_6H_5)$	119—120	—	258
$C_{12}H_{30}NPSi$	$(CH_3)_3SiN=P(C_3H_7)_3$	—	119 (11)	262
$C_{12}H_{33}AlNPSi$	$(CH_3)_3SiN=P(CH_3)_3 \cdot Al(C_2H_5)_3$	28—30	130 (1)	521
	$(CH_3)_3SiN=P(C_2H_5)_3 \cdot Al(CH_3)_3$	169	131 (1) sublim.	518, 521
$C_{12}H_{33}GaNPSi$	$(CH_3)_3SiN=P(C_2H_5)_3 \cdot Ga(CH_3)_3$	113—114	115 (1) sublim.	518, 521
$C_{12}H_{33}InNPSi$	$(CH_3)_3SiN=P(C_2H_5)_3 \cdot In(CH_3)_3$	107—110	125—128 (1)	521
$C_{12}H_{36}ClNOPSi$	$\{[(CH_3)_3Si]_2N\}_2P(O)Cl$	—	94 (0.1)	560

$C_{14}H_{29}N_2PSi_3$	$(CH_3)_3SiNSi(CH_3)_2N[Si(CH_3)_3]PC_6H_5$ (ring closed between first N and P)	—	107 (4) *1.5029	337
$C_{15}H_{20}NO_3PSi$	$(CH_3)_3SiNHP(O)(OC_6H_5)_2$	88	—	257
$C_{15}H_{21}N_2O_3PSi$	$(CH_3)_3SiNHNHP(O)(OC_6H_5)_2$	53—54	—	319
$C_{15}H_{36}NPSi$	$(CH_3)_3SiN{=}P(C_4H_9)_3$	—	94 (0.1)	262
			149 (11)	262
$C_{15}H_{39}AlNPSi$	$(CH_3)_3SiN{=}P(C_2H_5)_3 \cdot Al(C_2H_5)_3$	57—59	135 (1)	521
$C_{16}H_{22}NO_3PSi$	$(CH_3)_3Si(CH_3)NP(O)(OC_6H_5)_2$	38—40	—	258
$C_{17}H_{40}N_5P_3Si_2$	$(CH_3)_2Si\overset{C_2H_5}{N}Si(CH_3)_2\overset{C_2H_5}{N}-\left[P(O)(CH_3)\overset{C_2H_5}{N}-\right]_2P(O)(CH_3)_2\overset{C_2N_5}{N}$ (ring closed)	—	170 (1)	13, 55
$C_{18}H_{28}NPSi_2$	$[(CH_3)_3Si]_2NP(C_6H_5)_2$	49—50	157 (0.6)	480
$C_{21}H_{24}NPSi$	$(CH_3)_3SiN{=}P(C_6H_5)_3$	76—77	170—171 (0.06)	264
		75—76	162—164 (0.5)	515
$C_{24}H_{33}AlNPSi$	$(CH_3)_3SiN{=}P(C_6H_5)_3 \cdot Al(CH_3)_3$	126—130	—	521
$C_{24}H_{33}GaNPSi$	$(CH_3)_3SiN{=}P(C_6H_5)_3 \cdot Ga(CH_3)_3$	87—89	—	521
$C_{27}H_{29}BrNPSi$	$[(C_6H_5)_3P{=}N(C_6H_5)Si(CH_3)_3]Br$	201—202	—	452
$C_{30}H_{26}NPSi$	$(C_6H_5)_3SiNHP(C_6H_5)_2$	148—149	—	487
$C_{33}H_{33}AlNPSi$	$(C_6H_5)_2P{=}NSi(CH_3)_3$ 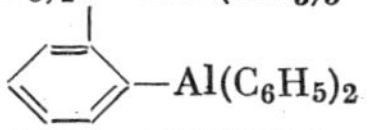	222—224	—	522
$C_{33}H_{33}GaNPSi$	$(C_6H_5)_2P{=}NSi(CH_3)_3$ 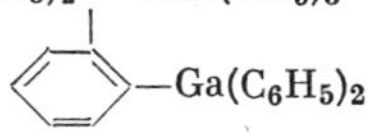	202—204	—	522
$C_{36}H_{30}NPSi$	$(C_6H_5)_3SiN{=}P(C_6H_5)_3$	213—215	—	565
		214	—	553
		215—216	—	566
		216—217	—	500
$C_{36}H_{37}N_4PSi$	$[2,4,6\text{-}(CH_3)_3C_6H_2]_2Si(N_3)N{=}P(C_6H_5)_3$	189	—	564
$C_{39}H_{39}AlNPSi$	$(CH_3)_3SiN{=}P(C_6H_5)_3 \cdot (C_6H_5)_3Al$	202—212 decomp.	—	522
$C_{39}H_{39}GaNPSi$	$(CH_3)_3SiN{=}P(C_6H_5)_3 \cdot (C_6H_5)_3Ga$	175—176	—	522
$C_{48}H_{40}N_2P_2Si$	$(C_6H_5)_2Si[N{=}P(C_6H_5)_3]_2$	194—195	—	500
		190—191	—	566
$C_{48}H_{41}N_2PSi$	$(C_6H_5)_3SiNHP(C_6H_5)_2{=}NSi(C_6H_5)_3$	161—162	—	487
$C_{60}H_{50}N_2P_2Si_2$	$(C_6H_5)_3SiN{=}P(C_6H_5)_2P(C_6H_5)_2{=}NSi(C_6H_5)_3$	236—238	—	487
$C_{60}H_{50}N_3P_3Si$	$C_6H_5Si[N{=}P(C_6H_5)_3]_3$	225—226	—	500

N-Trialkylsilyl-P,P,P-triorganophosphazenes are cleaved by alcohols in the presence of catalytic amounts of sulfuric acid [262, 263]:

$$R_3SiN{=}PR'_3 + R''OH \longrightarrow R_3SiOR'' + R'_3PNH \quad (1.165)$$

Diphenylchlorophosphine also cleaves the grouping Si–N–P at the Si–N bond with the formation of trialkylchlorosilanes [480, 481]:

$$[(CH_3)_3Si]_2NP(C_6H_5)_2 + 2(C_6H_5)_2PCl \longrightarrow 2(CH_3)_3SiCl + [(C_6H_5)_2P]_3N \quad (1.166)$$

$$(CH_3)_3SiN{=}P(C_6H_5)_3 + 2(C_6H_5)_2PCl \longrightarrow (CH_3)_3SiCl + \{(C_6H_5)_3P{=}N[P(C_6H_5)_2]_2\}^+Cl^- \quad (1.167)$$

In benzene or carbon tetrachloride, N-triorganosilyl-P,P,P-triorganophosphazenes add aluminum, gallium, and indium halides to form moisture-sensitive crystalline complexes which are soluble in benzene [523]:

$$R_3SiN{=}PR_3 + MX_3 \longrightarrow R_3SiN{=}PR_3 \cdot MX_3 \quad (1.168)$$

$$(M{=}Al,\ Ga,\ In;\ \ X{=}Cl,\ Br,\ I)$$

Heating to 215°C leads to the thermal decomposition of these complexes according to the scheme

$$2(CH_3)_3SiN{=}P(CH_3)_3 \cdot AlBr_3 \longrightarrow 2(CH_3)_3SiBr + [(CH_3)_3P{=}NAlBr_2]_2 \quad (1.169)$$

N-Silicon substituted phosphazenes add trialkyl and triphenyl derivatives of aluminum, gallium, and indium with the formation of complexes which are soluble in benzene and are sensitive to oxidation and hydrolysis [518-522]:

$$R_3SiN{=}PR_3 + R_3M \longrightarrow R_3SiN{=}PR_3 \cdot R_3M$$

$$(R{=}CH_3,\ C_2H_5,\ C_6H_5;\ \ M{=}Al,\ Ga,\ In) \quad (1.170)$$

These adducts are readily hydrolyzed by water, particularly in the presence of dilute acids [521, 522]:

$$2(CH_3)_3SiN{=}P(C_6H_5)_3 \cdot (C_6H_5)_3Al + 9H_2O \longrightarrow 2(C_6H_5)_3PO + 2Al(OH)_3 +$$

$$+ 2NH_3 + 6C_6H_6 + [(CH_3)_3Si]_2O \quad (1.171)$$

On heating above 170°C, complexes with trimethylalumane dissociate into the starting components, but in the case of complexes with triphenylalumane there is an intramolecular alumination with the elimination of a benzene molecule [522]:

$$(CH_3)_3SiN=P(C_6H_5)_3+(CH_3)_3Al \underset{175°C}{\overset{20°C}{\rightleftharpoons}} (CH_3)_3SiN=P(C_6H_5)_3\cdot(CH_3)_3Al \quad (1.172)$$

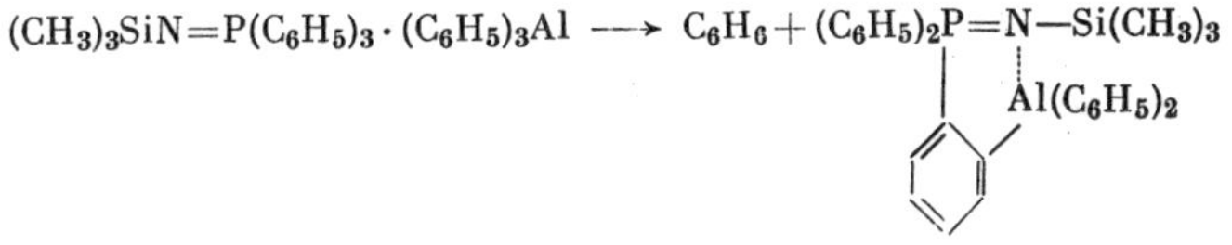

$$(CH_3)_3SiN=P(C_6H_5)_3\cdot(C_6H_5)_3Al \longrightarrow C_6H_6+(C_6H_5)_2P=N-Si(CH_3)_3 \quad (1.173)$$

$$\underset{}{\overset{}{\quad\quad\quad\quad\quad\quad\quad\quad\quad\quad\quad\quad\quad}}\quad \overset{|}{Al(C_6H_5)_2}$$

Methyllithium does not cleave the Si−N bond in trialkylsilyltrialkylphosphazenes, but metallates one of the P−CH$_3$ groups
in them [515]:

$$(CH_3)_3Si-N=P(CH_3)_3+LiCH_3 \longrightarrow CH_4+(CH_3)_3Si-N=P(CH_3)_2 \quad (1.174)$$

$$\overset{|}{CH_2Li}$$

$$+(CH_3)_3SiCl \Big\downarrow -LiCl$$

$$(CH_3)_3SiN=P(CH_3)_2CH_2Si(CH_3)_3$$

4. COMPOUNDS CONTAINING THE
BOND Si − P

4.1. Preparation Methods

The first known compound containing the Si−P bond was silicon phosphide, SiP (or Si$_3$P$_4$), which was obtained from the elements at a temperature above 1000°C as early as 1938 [260a]. It
was also reported [40a] that a phosphide of the composition Si$_5$P$_2$
may be formed.

In 1953 Fritz [345] reported the preparation of silylphosphine
H$_3$SiPH$_2$, the simplest monomeric compound containing the Si−P
bond. It is formed by passing a mixture of silane and phosphine
through a tube heated to 500°C and cooling the reaction products
with liquid air [345, 346]. The reaction of silane with phosphine
in a flow system at 450°C and 200–300 mm leads mainly to gaseous
and liquid compounds, from which it was possible to isolate H$_3$SiPH$_2$
much later [347, 348] by using gas-liquid chromatography. The
only solid product of this reaction was a blue–black amorphous
substance which was found to be phosphorus disilicide Si$_2$P, which
decomposes at 600°C to silicon and SiP, which was identical to
that obtained earlier from the elements [355]. The reaction in a

sealed tube at 440°C gives the same results as in a flow system at 500°C [445]. Taking into account the fact that silane loses hydrogen even at 400°C, while phosphine begins to decompose only at 550°C, it is possible to put forward the following radical mechanism for the formation of silylphosphine from silane and phosphine at 450°C [350]:

$$\begin{aligned} SiH_4 &\longrightarrow H\cdot + \cdot SiH_3 \\ H\cdot + PH_3 &\longrightarrow \cdot PH_2 + H_2 \\ H_3Si\cdot + \cdot PH_2 &\longrightarrow H_3SiPH_2 \end{aligned} \quad (1.175)$$

Silylphosphine is also formed from silane and phosphine in a silent electrical discharge. Disilylphosphine and disilanylphosphine are formed together with this [301, 387]. They are evidently the products of further conversions of silylphosphine and disilane formed initially, since these compounds were actually obtained by a silent electrical discharge on a mixture of silylphosphine with silane and disilane with phosphine [388]:

$$SiH_4 + H_3SiPH_2 \longrightarrow (H_3Si)_2PH + H_2 \quad (1.176)$$

$$H_3SiSiH_3 + PH_3 \longrightarrow H_3SiSiH_2PH_2 + H_2 \quad (1.177)$$

In 1953, compounds containing Si–P bonds were also obtained by another method. The action of iodosilane on white phosphorus at 20–100°C forms silyliodophosphines $(H_3Si)_n PI_{3-n}$ with n = 1 or 2, which are unstable and spontaneously flammable in air, and trisilylphosphine [245, 246, 317].* Trisilylphosphine is evidently the product of disproportionation:

$$\begin{array}{c} 6H_3SiI + 4P \longrightarrow 2H_3SiPI_2 + 2(H_3Si)_2PI \\ \downarrow \\ H_3SiPI_2 + (H_3Si)_3P \end{array} \quad (1.178)$$

It was possible to isolate only silyldiiodophosphine in a sufficiently pure form to characterize it. The explanation is that disilyliodophosphine is practically inseparable from the H_2SiI_2 formed by disproportionation of iodosilane under the reaction conditions, while trisilylphosphine does not differ in volatility from iodosilane.

* Triethyliodosilane was unchanged after boiling with red phosphorus for 24 h [67].

With excess iodosilane the products of reaction (1.178) include the quaternary compound $[(H_3Si)_4P]I$. It is evidently obtained through the secondary reaction of the addition of iodosilane to trisilylphosphine. It should be noted that salt-like compounds of this type are also formed by the reaction of bromosilane and iodosilane with methylphosphines $(CH_3)_nPH_{3-n}$ (n = 1-3) and triethylphosphine [244, 246]. The ionic (phosphonium) character of these adducts is confirmed by the results of measuring the electrical conductance of their solutions in acetonitrile. Thus, while the specific conductance of acetonitrile itself is $2.43 \cdot 10^{-6}\,\Omega^{-1} \cdot cm^{-1}$, for a solution of 226 μmole of $[H_3SiP(C_2H_5)_3]I$ in 25 ml of acetonitrile it is it is raised to $4.95 \cdot 10^{-3}\,\Omega^{-1} \cdot cm^{-1}$ [246].

A general reaction for the preparation of compounds containing the Si−P bond, including organic substituted silylphosphines, is the reaction of halosilanes with metal phosphides:

$$\geq Si-X + M-P\langle \longrightarrow \geq Si-P\langle + MX \qquad (1.179)$$

The reaction of bromosilane with potassium dihydrophosphide in dimethyl ether begins even at −120°C and, despite an equimolar ratio of the reagents, ends in the formation of trisilylphosphine [239, 240]*:

$$3H_3SiBr + 3KPH_2 \longrightarrow (H_3Si)_3P + 2PH_3 + 3KBr \qquad (1.180)$$

The reaction of potassium dihydrophosphide with chlorosilane [293, 294, 312] and chlorodeuterosilane D_3SiCl [294] proceeds analogously.

The formation of secondary and tertiary silylphosphines by the reaction of trimethylfluorosilane with potassium dihydrophosphide is treated as a series of successive condensation and metallation processes [44]:

$$R_3SiF \xrightarrow[-KF]{+KPH_2} R_3SiPH_2 \xrightarrow{+KPH_2} R_3SiPHK + PH_3$$
$$R_3SiPHK \xrightarrow[-KF]{+R_3SiF} (R_3Si)_2PH \xrightarrow{+KPH_2} (R_3Si)_2PK, \text{ etc.} \qquad (1.181)$$

*The reactions of lithium dihydrophosphide with iodosilane and bromosilane did not lead to the expected silylphosphines, but to sublimable products of undetermined structure [350].

The weak point of this scheme is the hypothesis that KPH_2 reacts with the intermediate products more readily than with R_3SiF. According to data in [44], trialkylbromosilanes and trialkylchlorosilanes are less active in this reaction than the corresponding fluoro derivatives, and when the latter are used the yields do not exceed 40%. At the same time, tris(trimethylsilyl)phosphine is formed in 40% yield by the reaction of trimethylchlorosilane with sodium dihydrophosphide [441, 442] by a scheme analogous to (1.180).

In contrast to this, methylbis(diethylamino)chlorosilane reacts with lithium dihydrophosphide in ether at −20°C with the formation of a derivative of a primary silyl phosphine (70% yield) [37]:

$$CH_3[(C_2H_5)_2N]_2SiCl + LiPH_2 \longrightarrow CH_3[(C_2H_5)_2N]_2SiPH_2 + LiCl \qquad (1.182)$$

The reaction of dimethyl(diethylamino)chlorosilane with lithium methylhydrophosphide again forms a mixture of derivatives of secondary and tertiary silylphosphines [37]:

$$3(CH_3)_2[(C_2H_5)_2N]SiCl + 3LiPHCH_3 \longrightarrow (CH_3)_2[(C_2H_5)_2N]SiPHCH_3 + CH_3PH_2 +$$
$$+ \{(CH_3)_2[(C_2H_5)_2N]Si\}_2PCH_3 + 3LiCl \qquad (1.183)$$

The reaction of trimethylchlorosilane with lithium tetraphosphinoaluminate $LiAl(PH_2)_4$ leads to the formation of a mixture of phosphine, trimethylsilylphosphine, and bis(trimethylsilyl)phosphine [336]. In this reaction bromosilane forms tris(phosphino)silane, while dibromosilane gives bis(phosphino)silane [666, 667].

The reaction of trimethylchlorosilane with a mixture of di- and trilithium phosphides formed by rapidly passing phosphine into a solution of butyllithium in ether gives largely bis- and tris-(trimethylsilyl)phosphines. It is possible to increase the yield of tris(trimethylsilyl)phosphine from 24 to 45% by adding trimethylchlorosilane to a suspension of trilithium phosphide obtained by passing PH_3 over the surface of a solution of butyllithium [490]. A mixture of Li_2PH and Li_3P reacts with diethyldichlorosilane with the formation of cyclic products ($R = C_2H_5$) [489, 490]:

$$2R_2SiCl_2 + 2Li_2PH \xrightarrow[-4LiCl]{} R_2Si\begin{array}{c} PH \\ \diagup \quad \diagdown \\ \diagdown \quad \diagup \\ PH \end{array}SiR_2 \qquad (1.184)$$

$$3R_3SiCl_2 + 2Li_3P \xrightarrow[-6LiCl]{} R_2Si\begin{array}{c} P \\ \diagup | \diagdown \\ SiR_2 \diagdown SiR_2 \\ | \\ P \end{array} \qquad (1.185)$$

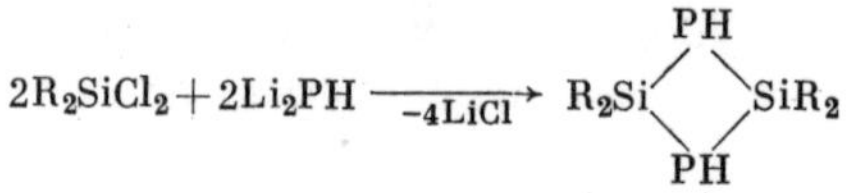

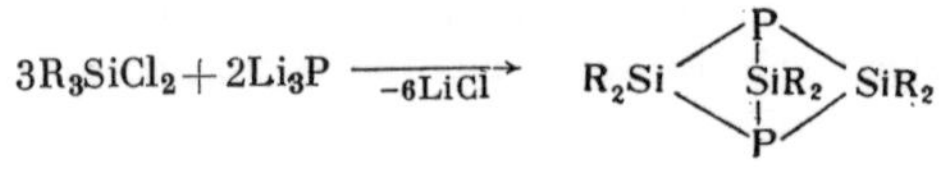

The reaction of halosilanes with alkali metal diorganophosphides proceeds according to the following scheme [315, 349-353, 357, 359, 361, 434, 435, 483, 570, 621]:

$$R_{4-n}SiX_n + nMPR'_2 \longrightarrow R_{4-n}Si(PR'_2)_n + nMX \quad (1.186)$$

(R=H, alkyl, C_6H_5; R″=alkyl, C_6H_5; X=Cl, Br; M=Na, Li; n=1–4)

Tetrakis(diethylphosphino)silane is formed with great difficulty in this reaction [361]. When R = H the reaction (1.186) is complicated by a series of side reactions. Thus, for example, in the reaction of excess bromosilane with lithium diethylphosphide, bis(diethylphosphino)silane and tris(diethylphosphino)silane are formed in addition to silyldiethylphosphine [351, 352, 642]:

$$H_3SiP(C_2H_5)_2 + LiP(C_2H_5)_2 \longrightarrow H_2Si[P(C_2H_5)_2]_2 + LiH \quad (1.187)$$

$$H_2SiP(C_2H_5)_2 + LiP(C_2H_5)_2 \longrightarrow HSi[P(C_2H_5)_2]_3 + LiH \quad (1.188)$$

The reaction of trichlorosilane with lithium diethylphosphide forms a whole series of products of total and partial substitution at the Si−Cl bonds, disproportionation, metallation, and further conversion of the products of a series of side reactions, and this may be represented very approximately by the following scheme ($R = C_2H_5$) [352, 353]:

$$\begin{gathered}
HSiCl_3 \xrightarrow[-LiCl]{+LiPR_2} HSiCl_2PR_2 \xrightarrow[-LiCl]{+LiPR_2} HSiCl(PR_2)_2 \xrightarrow[-LiCl]{+LiPR_2} HSi(PR_2)_3 \\
2HSiCl(PR_2)_2 \longrightarrow Cl_2Si(PR_2)_2 + H_2Si(PR_2)_2 \\
HSi(PR_2)_3 + LiPR_2 \longrightarrow (R_2P)_3SiLi + R_2PH \\
(R_2P)_3SiLi + ClSiH(PR_2)_2 \longrightarrow (R_2P)_3SiCl + LiHSi(PR_2)_2 \\
(R_2P)_3SiCl + LiPR_2 \longrightarrow (R_2P)_3SiLi + ClPR_2 \\
(R_2P)_3SiLi + ClSi(PR_2)_3 \xrightarrow[-LiCl]{} (R_2P)_3SiSi(PR_2)_3 \\
R_2PCl + LiPR_2 \xrightarrow[-LiCl]{} R_2PPR_2
\end{gathered} \quad (1.189)$$

In any case, the reaction products have been shown to include $H_2Si(PR_2)_2$, $ClSiH(PR_2)_2$, $HSi(PR_2)_3$, $(R_2P)_3SiLi$, R_2PPR_2, R_2PH, and polysilanes.

In the reaction of R_2PLi with alkylhydrochlorosilanes there is also replacement of the hydrogen atom by a dialkylphosphino group, but no formation of a dialkylphosphine and polysilanes was observed [350, 352]:

$$(CH_3)_2SiHCl + LiP(C_2H_5)_2 \longrightarrow (CH_3)_2SiHP(C_2H_5)_2 + LiCl \quad (1.190)$$

$$(CH_3)_2SiHP(C_2H_5)_2 + LiP(C_2H_5)_2 \longrightarrow (CH_3)_2Si[P(C_2H_5)_2]_2 + LiH \quad (1.191)$$

The problem of the impossibility of forming compounds containing the Si−P bond in reactions of halosilanes with sodium dialkylphosphites by scheme (1.20) was discussed in Section 1.1.1. Nonetheless, in the patents [323, 327] the products of the reaction of dimethyl(chloromethyl)chlorosilane with compounds of the type $NaP(O)(OR)_2$ or NaP(O)(OR)R' have been assigned a structure with the grouping P−C−Si−P without any proof. It is logical to assume that the reaction products in this case also are compounds with a skeleton of the type P−C−Si−O−P.

A general method of preparing triarylsilyldiorganophosphines may be the reaction of triarylsilyllithium with organohalophosphines [405, 406, 451]:

$$n(C_6H_5)_3SiLi + X_nPR_{3-n} \longrightarrow [(C_6H_5)_3Si]_nPR_{3-n} + nLiX \qquad (1.192)$$

(X = Cl, Br; R = organic; n = 1 or 2).

When n = 3 (PCl_3, PBr_3) no Si−P bond is formed and the reaction product is hexaphenyldisilane [369].

Newlands [470], a supporter of scheme (1.15), proposed a confirmatory synthesis of compounds containing the grouping Si−P(O):

$$(C_6H_5)_3SiLi + ClP(O)(OCH_2C_6H_5)_2 \xrightarrow[-LiCl]{} (C_6H_5)_3SiP(O)(OCH_2C_6H_5)_2 \qquad (1.193)$$

This scheme is more probable than scheme (1.15). However, the possibility of the formation of the grouping Si−O−P cannot be excluded completely in this case also, since Gilman and his coworkers [382-384, 569], who studied the reaction of triphenylsilyllithium with esters of phosphorus acids, showed that while $(C_6H_5)_3SiP(O)(OR)_2$ is formed initially in these reactions, it subsequently undergoes a series of conversions, since the final reaction products are a triphenylorganosilane or hexaphenyldisilane, depending on the reaction conditions.

In the presence of $AlCl_3$, phosphorus trichloride cleaves the Si−C bond in phenyltrichlorosilane, forming phenyldichlorophosphine and $SiCl_4$. In all probability this cleavage does not include the formation of a Si−P bond [243]. The formation of silicon-containing organophosphorus compounds was not observed in the cleavage of the Si−Si bond in methylchlorodisilanes by quaternary phosphonium salts [366, 376].

In conclusion we should note that from the point of view of Haiduk [216] cyclic or chain polymers consisting of alternating atoms of any two elements A and B must be stable if $x_{AB} = (x_A + x_B)/2 = 2.5 \pm 0.35$ (x_A and x_B are the electronegativities of the elements A and B, respectively; 2.5 is the electronegativity of carbon). This condition does not hold for the Si−P bond since in this case x = 2.0. Naturally, this hypothesis requires substantiation.

4.2. Physical Properties

The vapor pressure of disilylphosphine at 0°C equals 28 ± 1 mm, while that of disilanylphosphine is 31 ± 1 mm at 0°C and 81 ± 1 mm at 20°C [388]. The vapor pressure of trisilylphosphine obeys the equation $\log p = -1901.8/T + 7.792$ [240]. The dissociation energy of silylphosphine D (H_3Si-PH_2) = 88.2 kcal/mole, and the standard enthalpy of formation $\Delta H_f^\circ = 1.9$ kcal/mole [508].

The Raman spectrum of trisilylphosphine contains an intense polarized line at 414 cm^{-1}, corresponding to symmetrical valence vibrations of the PSi_3 skeleton. The absence of this band from the IR spectrum of $P(SiH_3)_3$ vapor may indicate its planar rather than pyramidal structure [293, 294]. A less intense line at 455 cm^{-1} in the Raman spectrum and an intense doublet at 456 and 463 cm^{-1} in the IR spectrum corresponds to antisymmetric bond vibrations of the PSi_3 skeleton [241, 293, 294, 445]. A weak line at 134 cm^{-1} corresponds to planar deformational vibrations of PSi_3. The character of the spectra does not make it possible to establish completely whether the molecule of trisilylphosphine is actually planar, but the molecule does obey the spectral rules of planar structures [296]. The deviation from the pyramidal structure might be explained by the formation of $(p \rightarrow d)\pi$-bonds in the trisilylphosphine molecule [293], but according to calculations of the effective force constant of the Si−P bond, given in [294], the degree of π-bonding in this case is small, while on the other hand, the planarity of the molecule is not sufficient demonstration of the formation of $(p \rightarrow d)\pi$-bonds in trisilylphosphine [311].

An electron diffraction investigation of trisilylphosphine established that its molecule is pyramidal. The length of the Si−P bond found (2.247 ± 0.005 Å) is close to the sum of the covalent atomic radii of silicon and phosphorus (2.27 Å). The Si−P−Si and H−Si−H valence angles equal 95 ± 2° and 110 ± 7°, respectively [251a].

A line at 2152 cm^{-1} in the Raman spectrum of trisilylphosphine corresponds to the vibrations of the Si–H bonds, while bands at 905 cm^{-1} [301], 885, and 939 cm^{-1}[294] in the IR spectrum correspond to deformation vibrations, and bands at 932 and 942 cm^{-1}[387] correspond to deformation vibrations in the IR spectrum of disilylphosphine.

In the NMR spectra of silylphosphine a signal from the protons of the SiH_3 group is observed at lower values (τ = 6.27 ppm) than in the spectrum of disilane (τ = 6.68 ppm). In the spectrum of disilylphosphine τ_{SiH_3} = 6.30 ppm, in the spectrum of disilanylphosphine τ_{SiH_3} = 6.81, τ_{SiH_2} = 6.43 ppm [387], in the spectrum of trisilylphosphine τ_{SiH_3} = 6.040 ± 0.03 ppm [312], and in the spectrum of the complex of silylphosphine with borane τ_{SiH_3} = 6.0 ppm [303]. In the P^{31} magnetic resonance spectrum of trisilylphosphine there is a signal with δ = 485 ± 2 ppm on the high field side of P_4O_6, which was used as an external standard [312]. See [302, 303, 312, 387] for more details on the NMR spectra of these compounds.

The physicochemical constants of compounds containing the Si–P bond are given in Table 8.

4.3. Chemical Properties

The chemical properties of compounds containing the Si–P bond are determined primarily by two factors: the low stability of this bond toward the action of various reagents and the trivalent state of phosphorus in these compounds. The "secondary properties," which are due, for example, to the formation, on hydrolysis, of extremely oxidation-sensitive substances containing the P–H bond, are connected with the first of these factors. It may be due to this that most compounds containing silyl groups at a phosphorus atom ignite spontaneously in air [37, 239, 246, 357, 490]. At the same time, tris(trimethylsilyl)phosphine hardly changes on heating for 8 h at 243°C in an inert gas atmosphere, while the bicyclic compound $[(C_2H_5)_2Si]_3P_2$ is stable up to 280°C [490].

Salts of the type $[H_3SiPR_3]I$ decompose slowly even at room temperature with the evolution of SiH_4 [246]. By slow oxidation by air or NO_2 it is possible to convert tris(trimethylsilyl)phosphine into tris(trimethylsilyl) phosphate [441, 442, 490]. This has no analogy in the chemistry of organophosphorus compounds (R_3P can be converted only into R_3PO by means of nitrogen dioxide).

Silicon phosphide, SiP, is hardly oxidized by dry air, this being explained by the formation of a protective film of silicophosphate on its surface [260a]. SiP is hydrolyzed appreciably by boiling water with the formation of PH_3 [260a].

Alkaline hydrolysis of silylphosphine proceeds readily even at low temperatures and corresponds mainly to the equation

$$H_3SiPH_2 + 4H_2O \xrightarrow{OH^-} Si(OH)_4 + PH_3 + 3H_2 \quad (1.194)$$

A small amount of SiH_4 is formed at the same time, and in the opinion of the authors of [354] this indicates the initial cleavage of the Si–P bond of the silylphosphine with the formation of phosphine and H_3SiOH. The latter then both partly disproportionates with the evolution of silane and is broken down at the Si–H bonds by the action of alkali. In an acid medium the disproportionation of the silanol formed initially becomes more marked. In the presence of methanol, hydrolysis proceeds readily even at –50°C and is accelerated with a rise in temperature. Ethanol saturated with HCl acts similarly on silylphosphine, i.e., with the initial formation of ethoxysilane and PH_3:

$$H_3SiPH_2 + C_2H_5OH \longrightarrow H_3SiOC_2H_5 + PH_3 \quad (1.195)$$

Subsequent conversions of ethoxysilane give hydrogen, silane, and polysilicic esters containing Si–H bonds.

Disilylphosphine [388], trisilylphosphine [240], and disilanylphosphine [387, 388] are hydrolyzed according to the following schemes:

$$(H_3Si)_2PH + H_2O \longrightarrow (H_3Si)_2O + PH_3 \quad (1.196)$$

$$(H_3Si)_3P + 6NaOH + 3H_2O \longrightarrow 3Na_2SiO_3 + PH_3 + 9H_2 \quad (1.197)$$

$$2H_3SiSiH_2PH_2 + H_2O \longrightarrow (Si_2H_5)_2O + 2PH_3 \quad (1.198)$$

The hydrolyses of tris(trimethylsilyl)phosphine by water or aqueous alkali is reflected by the scheme [44, 442]

$$2[(CH_3)_3Si]_3P + 6H_2O \longrightarrow 3[(CH_3)_3Si]_2O + 2PH_3 \quad (1.199)$$

The hydrolyses of bis(trimethylsilyl)phosphine [442], trimethylsilyldiethylphosphine [360], and tetrakis(diethylphosphino)silane [359] proceed analogously.

TABLE 8. Compounds Containing the Si–P Bond

Empirical formula	Compound	B.p., °C (mm)	n_D^{20}	d_4^{20}	Literature
H_3I_2PSi	H_3SiPI_2	190±5 (extrapol.) m.p. —1.8	—	2.9±0.2	246
H_9PSi_3	$(H_3Si)_3P$	114 (extrapol.)	—	—	239, 240
$C_3H_{11}PSi$	$(CH_3)_3SiPH_2$	77.5	1.4368^{25}	—	490
$C_4H_{10}Cl_3PSi$	$Cl_3SiP(C_2H_5)_2$	50 (25)	—	—	357
$C_4H_{13}PSi$	$H_3SiP(C_2H_5)_2$	53.5 (79)	—	—	351
	$(CH_3)_3SiPHCH_3$	54—55 (153)	—	0.7969	44
$C_6H_{16}ClPSi$	$(CH_3)_2ClSiP(C_2H_5)_2$	53—54 (5)	—	—	357
$C_6H_{17}PSi$	$(CH_3)_2SiHP(C_2H_5)_2$	58.5 (25)	—	—	351
$C_6H_{19}PSi_2$	$[(CH_3)_3Si]_2PH$	60—61 (20)	—	0.8188	44
		170—172	1.4637^{25}	—	489, 490
$C_7H_{19}PSi$	$(CH_3)_3SiP(C_2H_5)_2$	56—57 (11)	—	—	483
		71—72 (20)	—	—	357
$C_7H_{19}PS_2Si$	$(CH_3)_3SiP(SC_2H_5)_2$	69—72 (3)	1.5357	1.0263	45
$C_7H_{20}NPSi$	$(CH_3)_2[(C_2H_5)_2N]SiPHCH_3$	75—76 (19)	1.4702	0.8711	37
$C_7H_{21}PSi_2$	$[(CH_3)_3Si]_2PCH_3$	75—76 (20)	—	0.8433	44
$C_7H_{21}O_2PSi_2$	$[(CH_3)_2(CH_3O)Si]_2PCH_3$	85—86 (10)	1.4670	0.9600	37
$C_7H_{22}BPSi$	$(CH_3)_3SiP(C_2H_5)_2 \cdot BH_3$	m.p. 12	1.4656	—	483
$C_8H_{20}Cl_2P_2Si$	$Cl_2Si[P(C_2H_5)_2]_2$	~47 (5)	—	—	357
$C_8H_{22}PSi$	$H_2Si[P(C_2H_5)_2]_2$	~110 (13)	—	—	351
$C_8H_{22}P_2Si_2$	$(C_2H_5)_2SiPHSi(C_2H_5)_2PH$ (ring: first Si bonded to last PH)	107—110 (0.06)	1.5829^{25}	—	490
$C_9H_{24}IPSi$	$[(CH_3)_3SiP(C_2H_5)_3]I$	m.p. 123 (decomp.)	—	—	358
$C_9H_{25}N_2PSi$	$CH_3[(C_2H_5)_2N]_2SiPH_2$	80 (5)	1.4606	0.8756	37
$C_9H_{27}PSi_3$	$[(CH_3)_3Si]_3P$	49—55 (1)	1.5007^{25}	—	489
		66.8—67 (2)	—	0.8670	44
		95 (11) 102 (16)	—	~0.87	441, 442

		237—240	1.5068[25]	—	489
		242—243	1.5027[25]	—	490
$C_{10}H_{17}PSi$	$C_6H_5SiH_2P(C_2H_5)_2$	60—61 (1)	—	—	361
$C_{10}H_{26}P_2Si$	$(CH_3)_2Si[P(C_2H_5)_2]_2$	83 (3)	—	—	351
		60—65 (1)	—	—	357
$C_{10}H_{27}N_2PSi$	$CH_3[(C_2H_5)_2N]_2SiPHCH_3$	82—83 (2.5)	1.4800	0.8962	37
$C_{11}H_{27}PSi$	$(CH_3)_3SiP(C_4H_9)_2$	106 (10)	1.4683	—	482, 483
$C_{12}H_{30}P_2Si_3$	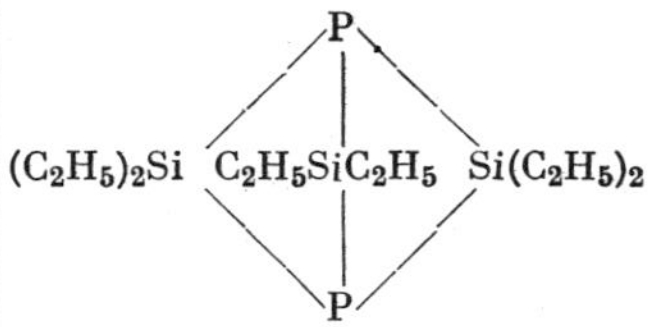	145—155 (0.6)	1.5900[25]	—	489
$C_{12}H_{30}ClP_3Si$	$ClSi[P(C_2H_5)_2]_3$	138—142 (1)	—	—	357, 361
$C_{13}H_{33}O_2PSi_2$	$[(CH_3)_2$ iso-$C_4H_9OSi]_2PCH_3$	117—118 (4)	1.4603	0.8991	37
$C_{13}H_{35}N_2PSi_2$	$\{(CH_3)_2[(C_2H_5)_2N]Si\}_2PCH_3$	106—108 (1)	1.5020	0.9184	37
$C_{14}H_{38}P_2Si_4$	$(C_2H_5)_2SiP[Si(CH_3)_3]Si(C_2H_5)_2PSi(CH_3)_3$ (ring)	96—98 (0.2)	1.5522[25]	—	490
$C_{15}H_{19}PSi$	$(CH_3)_3SiP(C_6H_5)_2$	126—127 (1)	1.6000	—	435
$C_{16}H_{40}P_4Si$	$Si[P(C_2H_5)_2]_4$	~160 (1)	—	—	359
$C_{20}H_{30}P_2Si_2$	$(C_2H_5)_2SiP(C_6H_5)Si(C_2H_5)_2PC_6H_5$ (ring)	151—153 (0.02)	—	—	490
		m.p. 43—47	—	—	
$C_{37}H_{33}PSi_2$	$[(C_6H_5)_3Si]_2PCH_3$	m.p. 80	—	—	451
$C_{53}H_{48}P_4Si_2$	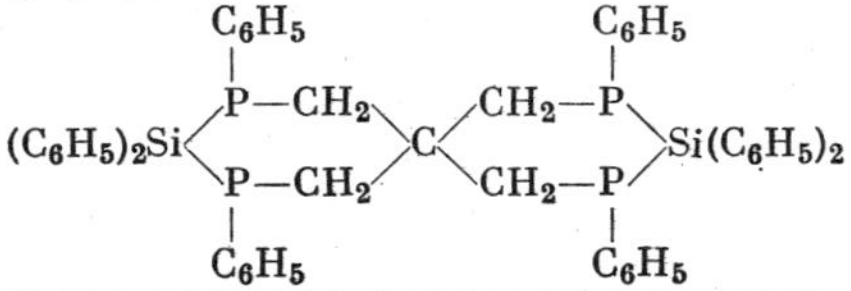	m.p. 126	—	—	315
$C_{54}H_{45}PSi_4$	$(C_6H_5)_2SiSi(C_6H_5)_2Si(C_6H_5)_2Si(C_6H_5)_2PC_6H_5$ (ring)	m.p. 198—201	—	—	406
		m.p. 203—206	—	—	405

The reaction of silylphosphine with liquid ammonia leads to the formation of phosphine, silane, and solid products containing nitrogen and phosphorus. The following scheme has been proposed to explain this reaction [354]:

$$2H_3SiPH_2 + 2NH_3 \longrightarrow 2H_3SiNH_2 + PH_3 \quad (1.200)$$
$$\downarrow -NH_3$$
$$(H_3Si)_2NH \xrightarrow[-SiH_4]{} [-H_2SiNH-]_x$$

Disilylphosphine and disilanylphosphine are cleaved by hydrogen chloride with the formation of phosphine and the corresponding chlorosilanes [388]:

$$(H_3Si)_2PH + HCl \longrightarrow 2H_3SiCl + PH_3 \quad (1.201)$$

$$Si_2H_5PH_2 + HCl \longrightarrow Si_2H_5Cl + PH_3 \quad (1.202)$$

Trimethylsilyldiethylphosphine reacts at −78°C with an equimolar amount of HI to form the colorless crystalline salt $(CH_3)_3SiP(C_2H_5)_2 \cdot HI$, which is stable only at −78°C [358]. A stable adduct of trimethylsilyldiethylphosphine and methyl iodide is formed at the same temperature; a solution of this compound in formamide has electrical conductivity [358]:

$$(CH_3)_3SiP(C_2H_5)_2 + C_2H_5I \longrightarrow [(CH_3)_3SiP(C_2H_5)_3]I \quad (1.203)$$

Excess ethyl iodide or ethyl bromide produce cleavage of the Si–P bond [358]:

$$(CH_3)_3SiP(C_2H_5)_2 + 2C_2H_5X \longrightarrow (CH_3)_3SiX + [(C_2H_5)_4P]X \quad (1.204)$$

The cleavage of tris(trimethylsilyl)phosphine by methyl iodide proceeds analogously.

Iodine and bromine also cleave tris(trimethylsilyl)phosphine with the liberation of the corresponding trimethylhalosilane. An orange polymer $[(CH_3)_3SiP]_n$ is also formed, and this is insoluble in all proton-active solvents (HCl produces cleavage of the Si–P bond).

Trimethylsilylalkylphosphines are cleaved readily by boron halides, in many cases with the intermediate formation of 1:1 adducts with a phosphonium character. When heated, they decompose with the formation of a trimethylhalosilane and a boron- and phosphorus-containing compound [482, 483]. The reaction of trimethylsilyldiethylphosphine with boron trifluoride forms an adduct which decomposes at 100°C:

$$(CH_3)_3SiP(C_2H_5)_2 + BF_3 \longrightarrow (CH_3)_3SiP(C_2H_5)_2 \cdot BF_3 \longrightarrow (CH_3)_3SiF + (C_2H_5)_2PBF_2 \quad (1.205)$$

Adducts with boron trichloride and tribromide decompose quantitatively at 120 and 150-160°C, respectively. The analogous adduct of chlorodibutoxyborine decomposes at 180°C, while the decomposition of the adduct of trimethylsilyldiethylphosphine with bis(dimethylamino)chloroborine begins only at 300°C [483]. The product from the addition of phenyldichloroborine to tris(trimethylsilyl)phosphine decomposes at 150°C with the elimination of two molecules of trimethylchlorosilane. However, the Si – P bond is preserved in the polymer obtained $[(CH_3)_3SiPB(C_6H_5)]_n$. The reaction of equimolar amounts of tris(trimethylsilyl)phosphine and BF_3 at 120°C forms trimethylfluorosilane and boron phosphide, i.e., cleavage proceeds to completion. The adduct of silylphosphine and boron trifluoride forms at –134°C, but decomposes even at –96°C in two directions [506, 507]:

$$H_3SiPH_2 \cdot BF_3 - \begin{cases} \longrightarrow (H_3Si)_3P + PH_3 + BF_3 \\ \longrightarrow H_3SiF + (-H_2PBF_2-)_x \end{cases} \quad (1.206)$$

No reaction is observed between the trisilylphosphine and BF_3 formed. The adduct of silylphosphine and BCl_3, which is obtained at –78°, decomposes at –23°C [637].

Phosphorus pentachloride cleaves tris(trimethylsilyl)phosphine but is not converted into PCl_3 [442]. In the reaction with transition metal halides such as cobalt chloride, trimethylchlorosilane is liberated and finely divided pyrophoric powders of phosphides of the type M_3P_2 are formed. The action of diphenyldichlorosilane on tris(trimethylsilyl)phosphine also splits out trimethylchlorosilane, but the main product of the reaction is a mixture of low molecular weight substances with the ratio Si : P = 5 : 2 [442]. The reaction of trisilylphosphine with a small excess of bromogermane at room temperature leads to the formation of trigermylphosphine [288]:

$$(H_3Si)_3P + 3H_3GeBr \longrightarrow (H_3Ge)_3P + 3H_3SiBr \quad (1.207)$$

Tris(trimethylsilyl)phosphine adds diborane with the formation of an adduct with BH_3 (1:1), which decomposes slowly at 20°C and rapidly at 100-135°C [441, 442, 482, 490]:

$$(R_3Si)_3P \cdot BH_3 \xrightarrow[-R_3SiH]{} (R_3Si)_2PBH_2 \xrightarrow[-R_3SiH]{} \frac{1}{n}(R_3SiPBH)_n \quad (1.208)$$

Diborane also adds to trimethylsilyldiethylphosphine. The adduct formed may be distilled in high vacuum. Its decomposition with the elimination of trimethylsilane begins at 80°C, but proceeds quantitatively only at 250-300°C with the formation of trimethylsilane and $[-P(C_2H_5)_2BH-]_3$ [482, 483]. With silylphosphine, diborane forms a liquid adduct $H_3SiPH_2 \cdot BH_3$ [303]. The reaction of tris(trimethylsilyl)phopshine with pentaborane leads to an adduct, which decomposes in 9 h at 90°C with the liberation of trimethylsilane, hydrogen, and CH_4 [441, 442]. The adduct of tris-(trimethylsilyl)phosphine with trimethylborine is stable at −78°C, but dissociates readily even at room temperature [482].

Tris(trimethylsilyl)phosphine readily adds elemental sulfur with the formation of $[(CH_3)_3Si]_3PS$ [44], but is cleaved by diethyl disulfide [45]:

$$[(CH_3)_3Si]_3P + C_2H_5SSC_2H_5 \longrightarrow (CH_3)_3SiP(SC_2H_5)_2 + 2(CH_3)_3SiSC_2H_5 \qquad (1.209)$$

Among the other cleavage reactions of compounds containing the Si−P bond we should mention the reaction of dimethyl(diethylamino)silylmethylphosphine with acetone, which in all probability reacts in the enol form [37]:

$$(CH_3)_2[(C_2H_5)_2N]SiPHCH_3 + CH_2{=}\underset{\displaystyle OH}{\underset{|}{C}}{-}CH_3 \rightarrow CH_3PH_2 + CH_2{=}\underset{\displaystyle OSi[N(C_2H_5)_2](CH_3)_2}{\underset{|}{C}}{-}CH_3 \qquad (1.210)$$

Insertion reactions at the Si−P bond in trimethylsilyldiphenylphosphine have been carried out with oxygen, carbon dioxide, SO_2, organic isothiocyanates, and ketene [622].

Tris(trimethylsilyl)phosphine forms complexes with nickel, cobalt, and chromium carbonyls [682, 683].

5. COMPOUNDS CONTAINING THE GROUPING Si – $(C)_n$ – P

5.1. Preparation Methods

5.1.1. Reactions of Organosilicon Grignard Reagents with Halogen-Containing Phosphorus Compounds

Trimethylsilylmethylmagnesium chloride reacts with phosphorus trichloride and tribromide according to the scheme

$$nR_3SiCH_2MgX + PX_3 \xrightarrow[-n\,MgX_2]{} (R_3SiCH_2)_nPX_{3-n} \quad (n=1-3) \tag{1.211}$$

The reaction is carried out in tetrahydrofuran [534] or a mixture of it with ether [533]. The yield of tris(trimethylsilylmethyl)-phosphine reaches 66%. When the reaction is carried out in ether at −10°C, trimethylsilylmethyldichlorophosphine is obtained in 45% yield. In this case it is essential to flush the reaction mixture with dry HCl as otherwise the yield of $(CH_3)_3SiCH_2PCl_2$ is low. The treatment with HCl is evidently necessary to decompose the complex of trimethylsilyldichlorophosphine with $MgCl_2$ [535].

When the reaction with a triorganosilylmethylmagnesium halide is carried out with organohalophosphines, the number of silicon-containing groups in the compound obtained is determined primarily by the number of halogen atoms at the phosphorus [47, 51, 534]:

$$nR_3Si(CH_2)_mMgX + R'_{3-n}PX_n \xrightarrow[-nMgX_2]{} [R_3Si(CH_2)_m]_nPR'_{3-n} \quad (n=1 \text{ or } 2;\ m=1 \text{ or } 3) \tag{1.212}$$

When PX_3 is replaced by phosphorus oxychloride, the reaction with R_3SiCH_2MgX leads to the formation of the corresponding tris(trialkylsilyl)phosphine oxide [280, 391, 392, 415]:

$$3R'R_2Si(CH_2)_nMgX + POCl_3 \xrightarrow[-3MgXCl]{} [R'R_2Si(CH_2)_n]_3PO \quad (n=1 \text{ or } 3) \tag{1.213}$$

However, good yields are achieved only if R' = CH_3 or p-$CH_3C_6H_4$, while when R' = p-ClC_6H_4 and n = 1 the triorganosilylphosphine oxide is formed in 13% yield [281], while the main reaction product is [p-$ClC_6H_4(CH_3)_2SiCH_2]_2P(O)OH$. In this connection we should point out that tris(trimethylsilylmethyl)phosphine and tris(p-chlorophenyldimethylsilylmethyl)phosphine are obtained quite readily by scheme (1.211) [534].

Canavan and Eaborn [280] were unable to identify a substance with m.p. 140°C, which they obtained by the reaction of $POCl_3$ with a Grignard reagent from pentamethylchloromethyldisiloxane. However, under the same conditions $(C_6H_5O)_2P(O)Cl$ reacts with $(CH_3)_3SiOSi(CH_3)_2CH_2MgBr$ with the formation of the corresponding phosphonate [365, 366].

Silicon-substituted arylmagnesium halides react with PCl_3 [343], $POCl_3$ [280, 343], and $(RO)_2POCl$ [269, 270, 272, 365, 366] like the alkylmagnesium halides examined above. For example:

$$3(CH_3)_3SiC_6H_4MgBr + PCl_3 \xrightarrow[-3MgClBr]{} [(CH_3)_3SiC_6H_4]_3P \tag{1.214}$$

This compound is also obtained by the action of p-trimethylsilylphenylmagnesium bromide with phosphorus pentachloride, but it is accompanied by a second reaction product which is believed to be $[(CH_3)_3SiC_6H_4]_2PCl$.* The reaction of triphenylsilylphenylmagnesium bromide with $POCl_3$ (by a scheme analogous to 1.213) yielded bis(p-trimethylsilylphenyl)phosphinic acid in addition to tris(trimethylsilylphenyl)phosphine oxide. This indicates the difficulty of replacing all the chlorine atoms in $POCl_3$ by trimethylsilylphenyl groups [280, 343].

5.1.2. Reactions of Haloalkylsilanes with Salts and Esters of Phosphorus Acids

In 1955 Keeber and Post [424] obtained dibutyl trimethylsilylmethylphosphonate in accordance with the scheme

$$R_3SiCH_2X + NaP(Y)(OR')_2 \longrightarrow R_3SiCH_2P(Y)(OR')_2 + NaX \qquad Y = O \text{ or } S \tag{1.215}$$

This reaction was subsequently extended [280, 365, 366] to compounds with very different substituents R and R'. The replace-

*The opinion in [424] that free chlorine is liberated in this case is evidently based on a misunderstanding.

ment of R_3SiCH_2X by compounds of the type $XZ\overset{|}{\underset{|}{Si}}O\overset{|}{\underset{|}{Si}}ZX$ (Z = alkylene or phenylene) leads to linear polymers with terminal phosphonate groups [365, 366]. The following reaction is analogous in sense, but opposite in character [193]:

$$R_3SiCl + 2Na + ClCH_2P(O)(OCH_3)_2 \longrightarrow R_3SiCH_2P(O)(OCH_3)_2 + 2NaCl \qquad (1.216)$$

Carrying out reaction (1.215) with bifunctional organosilicon compounds such as $ClCH_2Si(CH_3)_2Cl$ or $BrC_6H_4Si(C_6H_5)_2Cl$ makes it possible to obtain compounds in which the phosphorus atoms are attached to silicon both through carbon atoms and oxygen atoms [323] (see Section 4.1, p. 84).

By using metal phosphides in this reaction it is possible to synthesize triorganosilylalkyldiorganophosphines [327, 328]:

$$R_3Si(CH_2)_nCl + NaPR'_2 \xrightarrow[-NaCl]{} R_3Si(CH_2)_nPR'_2 \qquad (1.217)$$

The reaction of haloalkylsilanes with triethyl phosphite is a particular case of the Arbuzov rearrangement [46, 47, 50, 51, 75, 224-226, 280, 281, 286, 373, 379]:

$$\rangle\!-Si(CH_2)_nX + (C_2H_5O)_3P \longrightarrow \rangle\!-Si(CH_2)_nP(O)(OC_2H_5)_2 + C_2H_5X \qquad (1.218)$$

When n = 1 or 3 and X = Cl, it is necessary to heat the reaction mixture for 6-15 h at 150-200°C to obtain a satisfactory yield of the diethyl silylalkylphosphonate. However, these conditions are considerably milder than those required to carry out the reaction of chloromethylphosphines with triethyl phosphite.

The replacement of ethoxyl groups at the phosphorus atom by phenyl groups considerably facilitates the reaction. Thus, heating trimethyl(chloromethyl)silane with triethyl phosphite for 6 h at 170°C leads to $(CH_3)_3SiCH_2P(O)(OC_2H_5)_2$ in 32% yield [75]; 62% yield is achieved only by boiling the reaction mixture for 3 days after having increased the temperature of the reaction mixture from 100 to 185°C [373]. However, $(CH_3)_3SiCh_2P(O)(OC_2H_5)C_6H_5$ was synthesized from $C_6H_5P(OC_2H_5)_2$ in 80% yield by heating the mixture for 10 h at 110-130°C [225], while in the case of $(C_6H_5)_2P(O)C_2H_5$ the yield of $(CH_3)_3SiCH_2P(O)(C_6H_5)_2$ was 95% under the same conditions (in 2.5 h) [51].

Trimethylchloromethylsilane has the lowest activity in reaction (1.218). The replacement of one of the methyl radicals by an ethyl radical in it has an insignificant effect, but the presence of aromatic substituents or alkoxyl groups at the silicon atom appreciably reduces the duration of the reaction of (haloalkyl)silanes with triethyl phosphite or phenyl phosphinates. Carrying out the synthesis under strictly identical conditions [140 ± 2°C, molar ratio of (haloalkyl)silane:phosphite = 1.5:1] made it possible to establish, through the duration of reaction (1.218), the following series of activities [46]:

$$(C_2H_5O)_3SiCH_2Cl > CH_3(C_2H_5O)_2SiCH_2Cl > (CH_3)_2(C_2H_5O)SiCH_2Cl >$$
$$> C_6H_5(CH_3)_2SiCH_2Cl > CH_3(C_2H_5O)_2Si(CH_2)_3Cl > (CH_3)_3SiCH_2Cl$$

When β-haloalkylsilanes are used in reaction (1.218), there is β-decomposition together with the Arbuzov reaction [51]:

$$(C_2H_5O)_3SiCH_2CH_2Cl + (C_2H_5O)_3P - \begin{cases} \rightarrow (C_2H_5O)_3SiCH_2CH_2P(O)(OC_2H_5)_2 + C_2H_5Cl \\ \rightarrow (C_2H_5O)_3SiCl + CH_2{=}CH_2 \end{cases} \quad (1.219)$$

The halosilanes formed in the decomposition naturally react with triethyl phosphite (see Section 1.1.1.), and this complicates the process still further. As a result of this, diethyl β-triethoxysilylethylphosphate is obtained in low yield (~35%). In the case of trimethyl(α,β-dibromoethyl)silane the first direction is suppressed and the yield of $(CH_3)_3SiCHBrCH_2P(O)(OC_2H_5)_2$ is only ~8%. At the same time, triethoxy(β-chloropropyl)silane does not react with triethyl phosphite in 11 h at ~175°C [51], i.e., it behaves analogously to secondary alkyl halides.

In the reaction of $P(OC_2H_5)_3$ with $ClCH_2Si(CH_3)_2OSi(CH_3)_2CH_2Cl$, even with a molar ratio of 1:2, almost three times as much of the monosubstituted product is obtained as of the disubstituted product [50]. Their total yield is low since a considerable amount of polymer is also formed.

5.1.3. Hydrophosphorylation and Hydrosilylation of Unsaturated Organic Compounds of Silicon and Phosphorus

A method of preparing silicon-containing organophosphorus compounds with a carbon bridge between the Si and P atoms, based on the addition of phosphorus compounds containing a P—H bond to alkenylsilanes, was patented in 1958 [368, 450] for application to vinylsilanes and dialkyl phosphonates. The authors of the first journal article on this subject [250, 292] assessed the patent data critically and reported that they were not sufficiently reproducible. Nonetheless, the addition of $(RO)_2P(O)H$ to alkenylsilanes was rapidly introduced into synthesis practice. In the general form it is described well by the scheme

$$R_3Si(CH_2)_nCH{=}CH_2 + HP(O)(OR')_2 \longrightarrow R_3Si(CH_2)_{n+2}P(O)(OR')_2 \qquad (1.220)$$

In all the cases investigated the phosphonate residue is added to the terminal carbon atom of the unsaturated compound.

The use of metallic potassium as a catalyst was proposed in the patent [450]. At the same time, it was reported that it is expedient to use free radical initiators (azobisalkanenitriles and peroxides). Tert-butyl peroxide (3-5 mol.%) was used in subsequent work [46, 50, 226, 250, 326, 455, 496]. The reagents are taken in an equimolar ratio, since it was shown [250] that a two- to three-fold excess of dialkyl phosphite does not give any advantage. The yield of products from addition to vinyltriethoxysilane is less than in the case of triethylvinylsilane, indicating [46, 50] the greater tendency of the former to polymerize. The yields of products from the addition of $(RO)_2P(O)H$ to allyltriethoxysilane are higher than to vinyltriethoxysilane [373], but in addition to the triethyl derivative the reverse relation is observed [50, 224, 226].

When reaction (1.220) is carried out with pentamethylvinyldisiloxane its conversion reaches 73% [250]. However, it is not possible to separate the mixture of reaction products formed, and this is explained by the possibility of cleavage of the disiloxane group under the reaction conditions with the formation of various by-products.

The presence of ethoxyl groups at the silicon atom does not complicate reaction (1.220) by any side processes, but the addition of dibutylphosphine to alkenylchlorosilanes is accompanied by the elimination of butyl groups with the formation of butyl chloride [250].

In addition to $(RO)_2P(O)H$, reaction (1.220) may be carried out with esters of other acids of phosphorus such as $R_2P(S)H$ [474,

475], $R_2P(O)H$ [475], $R(RO)P(O)H$ [324, 455], and also alkyl- and dialkylphosphines [328, 396, 464, 472-474, 485, 486, 496]:

$$R_xSi[(CH_2)_yCH{=}CH_2]_{4-x} + (4-x)\,HPR'R'' \longrightarrow R_xSi[(CH_2)_{y+2}PR'R'']_{4-x} \qquad (1.221)$$

In scheme (1.221) R may be a hydrocarbon radical and also a functional group (C_2H_5O, Cl), x = 0-3, y = 0-2, R' = H or an organic radical, while R" may be an organic radical. When R = Cl a side reaction is observed [475]:

$$(CH_2{=}CH)_2SiCl_2 + HP(C_2H_5)_2 \longrightarrow (CH_2{=}CH)_2SiClP(C_2H_5)_2 + HCl \qquad (1.222)$$

Reaction (1.221) is initiated by ultraviolet radiation and azobisisobutyronitrile or tert-butyl peroxide, but it may also be carried out by heating to ~160°C in the absence of initiators. The process is carried out in ether or hydrocarbons in an atmosphere of nitrogen or carbon dioxide.

When reaction (1.221) is carried out between dialkenylsilanes or dialkenylsiloxanes with terminal alkenyl groups and organophosphines RPH_2, it is possible to obtain polymeric silylalkylphosphines with a consistency varying from an oil to a solid [90, 396, 397, 472, 475]:

$$CH_2{=}CH(CH_2)_m(SiR'_2O)_nSiR_2(CH_2)_mCH{=}CH_2 + H_2PR'' \longrightarrow$$
$$\longrightarrow [-(R'_2SiO)_nSiR_2(CH_2)_{m+2}-PR''-(CH_2)_{m+2}-]_x \qquad (1.223)$$

In 1961-1962 it was shown that in the presence of $H_2PtCl_6 \cdot 6H_2O$ or peroxides, compounds containing an Si−H bond may add to unsaturated derivatives of phosphorus (hydrosilylation) [46, 50, 221, 224, 325, 586, 618]:

$$R_3SiH + CH_2{=}CH(CH_2)_nP(O)R'_2 \longrightarrow R_3Si(CH_2)_{n+2}P(O)R'_2 \qquad (1.224)^*$$

The substituents R at the silicon atom may be very varied (hydrocarbon radical, alkoxyl group, and halogen). The same applies to the substituent R' at the phosphorus. However, attempts to add triethylsilane to the acid chloride of allylphosphinic acid and alkylhydrochlorosilanes to the acid chloride of vinylphosphinic acid were unsuccessful [50].

Hydrosilylation of the C≡C triple bond in diethyl propargylphosphonates has also been described [201, 585].

* Alkenylphosphines may also be used in the reaction [325, 328].

5.1.4. Reactions of Tetraorganosilanes, Organochlorosilanes, and (Chloroalkyl)silanes with Phosphorus Halides

Acid chlorides of silicon-substituted alkylphosphinic acids may be obtained by the Clayton–Soborovskii reaction, i.e., by the reaction of alkyl-substituted silanes with PCl_3 and oxygen [220, 227, 379, 380]:

$$R_4Si + 2PCl_3 + O_2 \longrightarrow R_3SiR'P(O)Cl_2 + POCl_3 + HCl \qquad (1.225)$$

The reaction begins in the cold and proceeds with heat evolution. In principle the presence of only one alkyl radical at the silicon atom is sufficient for the process to occur. However, a gradual increase of the number of chlorine atoms in molecules of the type R_nSiCl_{4-n} reduces the reactivity. On the example of ethylchlorosilanes ($R = C_2H_5$) it was shown [227] that with a change from $n = 0$ to $n = 3$ there is a decrease in both the degree of conversion of starting substituted silane (from 14 to 6%) and in the yield of acid chloride, calculated on the silane reacting (from 62 to 40%). In the case of ethyltrichlorosilane the β-derivative is formed exclusively. An increase in the number of ethyl groups at the silicon atom promotes the formation of β- and α-isomers. At the same time, α- and β-chloroethyltrichlorosilanes give such unstable acid chlorides, which decompose with the formation of $SiCl_4$ even when stored in vacuum, that their yield is only 1-2%. When $R = CH_3$ the oxidation of PCl_3 to $POCl_3$ (which, incidently, does not occur in the absence of organic compounds) proceeds more markedly. Therefore, in [380] for example, in the reaction of $(CH_3)_4Si$ it is recommended that a molar ratio of 1:4 be used instead of 1:2. However, with this excess of phosphorus trichloride the yield of acid chloride of trimethylsilylmethylphosphinic acid is only $\sim$23%, calculated on the starting tetramethylsilane. Dimethyldichlorosilane and methyltrichlorosilane are recovered from the reaction unchanged almost in quantitative yield [220].

Benzyltrichlorosilane and (β-phenylethyl)trichlorosilane are unchanged under the conditions of reaction (1.225), though PCl_3 is converted into phosphorus oxychloride to a considerable extent. Neither PBr_3 nor $C_6H_5PCl_2$ undergoes the Clayton–Soborovskii reaction with alkylsilanes. Phosphorus tribromide is practically unchanged in this case, while phenyldichlorophosphine is converted into $C_6H_5P(O)Cl_2$ [220].

It has been suggested [380] that reaction (1.225) may be used to introduce phosphorus into cyclic and linear polymethylsiloxanes. However, according to data in [297] an attempt to apply this reaction to hexaethyldisiloxane lead to a 56% yield of triethylchlorosilane.

The reaction of PCl_3 and oxygen with triethylvinylsilane proceeds according to a different scheme [220]:

$$2R_3SiCH{=}CH_2 + 2PCl_3 + O_2 \longrightarrow 2R_3SiCHClCH_2P(O)Cl_2 \qquad (1.226)$$

Formally it may be assumed that in this case there is addition of $POCl_3$ to the alkenylsilane.

The addition of trimethylallylsilane to a suspension of PCl_5 in benzene at room temperature is accompanied by the evolution of heat. After decomposition of the crystalline complex formed with dry sulfur dioxide it is possible to isolate from the reaction mixture the acid chloride of allylphosphinic acid, whose formation is explained by the decomposition of the intermediate acid chloride of γ-trimethylsilyl-β-chloropropylphosphinic acid [198, 199]. Carrying out the reaction with PCl_5 and the decomposition in diethyl ether at -20°C makes it possible to isolate the acid chloride of γ-trimethylsilyl-β-chloropropylphosphinic acid in 84% yield:

$$(CH_3)_3SiCH_2CH{=}CH_2 + PCl_5 \xrightarrow{(SO_2)} (CH_3)_3SiCH_2CHClCH_2P(O)Cl_2 \qquad (1.227)$$

Dimethylallylchlorosilane, methylallyldichlorosilane, and allyltrichlorosilane do not add PCl_5 at -20°C. The reaction of PCl_5 with dimethylallylchlorosilane begins at 5°C and proceeds with the elimination of HCl and the formation of the acid chloride of 3-dimethylchlorosilyl-1-propenylphosphinic acid:

$$(CH_3)_2SiClCH_2CH{=}CH_2 + PCl_5 \xrightarrow{(SO_2)} (CH_3)_2SiClCH_2CH{=}CHP(O)Cl_2 + HCl \qquad (1.228)$$

Allyltrichlorosilane does not add phosphorus pentachloride at room temperature even in 5 days. Chlorination occurs instead of addition [198]:

$$Cl_3SiCH_2CH{=}CH_2 + PCl_5 \longrightarrow Cl_3SiCH_2CHClCH_2Cl + PCl_3 \qquad (1.229)$$

p-Trimethylsilylstyrene reacts with PCl_5 at room temperature in accordance with the scheme [197, 198]

$$(CH_3)_3SiC_6H_4CH{=}CH_2 + PCl_5 \xrightarrow{(SO_2)} (CH_3)_3SiC_6H_4CH{=}CHPOCl_2 + HCl \qquad (1.230)$$

Trimethylvinylsilane, dimethylvinylchlorosilane, methylvinyldichlorosilane, and vinyltrichlorosilane do not add PCl_5 even on prolonged contact of the reagents at room temperature [198]. In contrast to this, trimethyl(2-alkoxyvinyl)silanes react with PCl_5 to form complexes [589].

In the reaction of aralkylsilanes with phosphorus trichloride at 60°C in the presence of $AlCl_3$, there is phosphorylation of the benzene nucleus [196, 198]:

$$(CH_3)_nSiCl_{3-n}(CH_2)_mC_6H_5 + PCl_3 \xrightarrow[-HCl]{(AlCl_3)} (CH_3)_nSiCl_{3-n}(CH_2)_mC_6H_4PCl_2 \quad (n=0-3;\ m=1-3) \tag{1.231}$$

The rate of the reaction is increased with an increase in the number of methyl groups at the silicon atom. However, when n = 3 the silicon-containing organophosphorus compound could not be isolated, and this was associated with decomposition of the reaction product formed by aluminum chloride on heating. In the case of phenyltrichlorosilane (n = m = 0) this cleavage proceeds quantitatively and the reaction products are phenyldichlorophosphine and $SiCl_4$ [243]:

$$C_6H_5SiCl_3 + PCl_3 \xrightarrow{(AlCl_3)} C_6H_5PCl_2 + SiCl_4 \tag{1.231a}$$

Cleavage of the Si – C bond in silacyclobutanes by phosphorus trichloride in the presence of $AlCl_3$ leads to the formation of 3-(dialkylchlorosilyl)propyldichlorophosphines [577, 578]:

$$R_2Si\langle{}^{CH_2}_{CH_2}\rangle CH_2 + PCl_3 \xrightarrow{(AlCl_3)} ClR_2SiCH_2CH_2CH_2PCl_2 \tag{1.232}$$

The reaction of chloroalkylsilanes with PCl_3 in the presence of $AlCl_3$ proceeds according to the scheme [48, 49, 399]

$$R_nSiCl_{3-n}(CH_2)_mCl + PCl_3 + AlCl_3 \longrightarrow [R_nSiCl_{3-n}(CH_2)_mPCl_3]^+[AlCl_4]^- \xrightarrow{H_2O}$$
$$\longrightarrow HCl + AlCl_3\cdot 6H_2O + [R_nSi(O)_{3-n}(CH_2)_mP(O)Cl_2]_x \quad (n=0-3;\ m=1-3). \tag{1.233}$$

Trimethyl(chloromethyl)silane reacts with PCl_3 and $AlCl_3$ even at room temperature, but in the case where R = CH_3, n = 1 and 2, heating at 70-80°C for 6-8 h is necessary to carry out reaction (1.233) successfully. The yields of acid chlorides of silicon-containing alkylphosphinic acids do not exceed 30-35%, since simultaneously with the substitution reaction there are side processes of disproportionation, dehydrochlorination, cleavage of the Si–C bond, and hydrolysis of the P–Cl bond [48, 49]. Thus, for example, in the case of trimethyl(chloromethyl)silane it was possible to isolate 31.8% of the acid chloride of (trimethylsilylmethyl)phosphinic acid, 7.2% of (trimethylsilylmethyl)phosphinic acid, 52.1% of trimethylchlorosilane, 5.4% of ethyldimethylchlorosilane, and traces of dimethyldichlorosilane [48].

5.1.5. Synthesis of Organosilicon Ylides of Phosphorus

The synthesis of organosilicon ylides of phosphorus is based on dehydrohalogenation of organosilicon phosphonium halides. The latter are obtained by two methods, namely, by the reaction of triorgano(halomethyl)silanes with triorganophosphines [304, 394, 465, 517, 534, 536, 538]:

$$R_3SiCH_2X + R'_3P \longrightarrow [R_3SiCH_2PR'_3]X \qquad (1.234)$$

and the addition of triorganohalosilanes to ylides of phosphorus (for example, triphenylmethylenephosphorane) [385, 394, 517, 536, 538, 542]:

$$R_3SiX + (C_6H_5)_3P{=}CH_2 \longrightarrow [R_3SiCH_2P(C_6H_5)_3]X \qquad (1.235)$$

These quaternary phosphonium salts are then treated with butyllithium [466, 517] or phenyllithium [385, 538] or decomposed thermally in vacuum [517], and organosilicon ylides of phosphorus are obtained:

$$[R_3SiCH_2PR''_3]X + R''Li \longrightarrow R_3SiCH{=}PR''_3 + LiCl + R''H \qquad (1.236)$$

$$2[(CH_3)_3SiCH_2P(CH_3)_3]Cl \longrightarrow (CH_3)_3SiCH{=}P(CH_3)_3 + (CH_3)_3SiCl + [(CH_3)_4P]Cl \qquad (1.237)$$

The reaction (1.236) is more convenient and the yields of ylides in this case reach 80-90%.

It should be noted that the salt formed in reaction (1.235) may react with excess triphenylmethylenephosphorane in accordance with the scheme [517]

$$[(C_6H_5)_3PCH_2Si(CH_3)_3]Cl + (C_6H_5)_3P{=}CH_2 \rightleftarrows$$
$$\rightleftarrows (C_6H_5)_3P{=}CHSi(CH_3)_3 + [(C_6H_5)_3PCH_3]\,Cl \qquad (1.238)$$

5.1.6. Other Preparation Methods

A compound in which the silicon and phosphorus atoms were separated by a fluorinated ethylene bridge was obtained in accordance with the following scheme [541]:

$$(C_2H_5)_3SiCF{=}CF_2 + LiP(C_6H_5)_2 \xrightarrow[-LiF]{} (C_2H_5)_3SiCF{=}CFP(C_6H_5)_2 \qquad (1.239)$$

(Trimethylsilylmethyl)dichlorophosphine is formed in a small amount by the action of PCl_3 on tetrakis(trimethylsilylmethyl)-plumbane [535]:

$$[(CH_3)_3SiCH_2]_4Pb + 4PCl_3 \longrightarrow 4(CH_3)_3SiCH_2PCl_2 + (PbCl_4) \qquad (1.240)$$

However, bis(trimethylsilylmethyl)mercurane does not react with phosphorus trichloride [535].

Acid-resistant and alkali-resistant polymers containing silicon and phosphorus atoms in the main or side chains are obtained by copolymerization of alkenyl derivatives of silicon and phosphorus initiated by peroxides [251, 476] and Lewis acids ($AlCl_3$, BF_3) [251].

5.2. Physical Properties

The NMR spectra of compounds containing the grouping $Si-(C)_n-P$ have been investigated in a series of studies [220, 465, 466, 517, 538]. Data on the NMR spectra of organosilicon ylides of phosphorus are given in Table 9.

In the infrared spectra of $(C_6H_5)_3P=CHSi(CH_3)_3$ the bond vibrations $\nu P{=}C$ are represented by a very intense band at 1060 cm^{-1} and in the spectrum of $[(CH_3)_3Si]_2C=P(C_6H_5)_3$ at 1142 cm^{-1} [517].

The physicochemical constants of compounds containing carbon bridges between silicon and phosphorus atoms are given in Table 10.

5.3. Chemical Properties

Investigation of the thermal stability of alkyl esters of (triorganosilylalkyl)phosphinic acids at 200-280°C showed that esters of the type $(CH_3)_3SiCH_2P(O)(OR)_2$ with $R = C_4H_9$ or C_5H_{11} decompose with the formation of hexamethyldisiloxane and olefins (butene and

TABLE 9. NMR Spectra of Organosilicon Ylides of Phosphorus [517]
(in C_6D_6 at 60 MHz; negative values of δ indicate shifts toward low fields relative to an external standard − $(CH_3)_4Si$)

Compound	Chemical shift δ, Hz				Spin-spin coupling constant J, Hz						
	CH_3—Si	Si—CH=P	C_6H_5—P CH_3—P	CH_3—Ge CH_3—Sn	H_3C—P^{31}	H^1—C—P^{31}	H_3C—Si—C—P^{31}	H_3C—Si^{29}	H^1—C—Si^{29}	H^1—C^{13}—Si	H_3C—$Sn^{117,119}$
$(CH_3)_3SiCH=P(C_6H_5)_3$	+21.0	+16.0	−389 to −449	—	—	9.5	0.45	6.5	—	—	—
$(CH_3)_3SiCH=P(CH_3)_3$	+14.0	+62.0	−36.0	—	12.0	7.9	0.30	6.5	4.5	117	—
$[(CH_3)_3Si]_2C=P(CH_3)_3$	+13.5	—	−32.5	—	12.0	—	—	6.5	—	121	—
$(CH_3)_3Si[(CH_3)_3Ge]C=P(CH_3)_3$	+11.5	—	−39.0	+3.5	12.0	—	—	6.6	—	118	—
$(CH_3)_3Si[(CH_3)_3Sn]C=P(CH_3)_3$	+12.5	—	−34.0	+13.3	12.0	—	—	6.7	—	119	49.5 51.5

TABLE 10. Compounds Containing the Grouping Si−(C)$_n$−P

Empirical formula	Compound	B.p., °C (mm)	n_D^{20}	d_4^{20}	Literature
$C_2H_3Cl_6OPSi$	$Cl_3SiCHClCH_2P(O)Cl_2$	135—140 (4.5)	1.5068	1.6328	220
	$Cl_3SiCH_2CHClP(O)Cl_2$	141—146 (4)	1.5125	1.6812	220
$C_2H_4Cl_5OPSi$	$Cl_3SiCH_2CH_2P(O)Cl_2$	130 (8) m.p. 44	—	—	227
$C_3H_6Cl_5OPSi$	$Cl_3Si(CH_2)_3P(O)Cl_2$	134—142 (4)	1.5003	1.5202	220
$C_4H_9Cl_4OPSi$	$Cl_2(C_2H_5)SiCH_2CH_2P(O)Cl_2$	140—142.5 (6)	1.4985	1.5160	50, 325
		142—145.5 (8)	1.4960	1.4483	227
	$Cl_2(CH_3)Si(CH_2)_3P(O)Cl_2$	137—139 (6)	1.4955	1.4040	50, 221
		137—145 (4)	1.4978	1.4118	220
$C_4H_{11}Cl_2PSi$	$(CH_3)_3SiCH_2PCl_2$	50 (1.5)	—	—	535
$C_4H_{11}Cl_2OPSi$	$(CH_3)_3SiCH_2P(O)Cl_2$	64 (1.4)	1.4710^{25}	—	399

		75—76 (3)	1·4745	1·2494	49
		102.8—103.5 (51)	—	—	380
$C_4H_{13}O_3PSi$	$(CH_3)_3SiCH_2P(O)(OH)_2$	m.p.119—121	—	—	424
		m.p.120—121	—	—	49
		m.p. 121	—	—	399
$C_5H_{10}Cl_3OPSi$	$Cl(CH_3)_2SiCH_2CH{=}CHP(O)Cl_2$	102—106 (3)	1·5012	—	198
$C_{15}H_{11}Cl_4OPSi$	$Cl_2(C_2H_5)Si(CH_2)_3P(O)Cl_2$	125—126 (3)	1·4963	1·3580	50, 221
$C_5H_{15}O_3PSi$	$(CH_3)_3SiCH_2CH_2P(O)(OH)_2$	m.p. 147	—	—	250
$C_6H_{14}Cl_3OPSi$	$Cl(C_2H_5)_2SiCH_2CH_2P(O)Cl_2$	143—148 (8)	1·4910	1·2499	227
	$(CH_3)_3SiCH_2CHClCH_2P(O)Cl_2$	108—110 (2)	—	—	198
		108—110 (3)	—	—	199
		m.p. 32	—	—	198
		m.p. 38	—	—	199
$C_6H_{17}O_3PSi$	$(CH_3)_3SiCH_2P(O)(OCH_3)_2$	93 (7)	1.4363^{25}	—	399
		111—114 (21.5)	1·4350	—	281
$C_7H_6Cl_5PSi$	$Cl_3SiCH_2C_6H_4PCl_2$ *	128—132 (1)	1·5860	1·4735	198
$C_7H_{19}PSi$	$(CH_3)_3SiCH{=}P(CH_3)_3$	66 (11)	—	—	517
		70—75 (14)	—	—	465
		m.p.—36	—	—	465, 517
$C_7H_{19}O_3PSi$	$(C_2H_5)_3SiCH_2P(O)(OH)_2$	m.p. 98.5—99.5	—	—	226
$C_8H_8Cl_5PSi$	$Cl_3SiCH_2CH_2C_6H_4PCl_2$ *	142 (2)	1·5740	1·4344	196
		160—162 (2)	1·5748	1·4344	198
$C_8H_9Cl_4PSi$	$Cl_2(CH_3)SiCH_2C_6H_4PCl_2$ *	105—108 (0.5)	1·5800	1·3675	198
			1·5790	1·3675	196
$C_8H_{18}Cl_3OPSi$	$(C_2H_5)_3SiCHClCH_2P(O)Cl_2$	130—131 (5)	1·4948	1·1844	220
$C_8H_{19}Cl_2OPSi$	$(C_2H_5)_3SiCH_2CH_2P(O)Cl_2$	140—146 (8)	1·4895	1·1552	227
$C_8H_{21}OPSi$	$(CH_3)_3Si(CH_2)_3P(O)(CH_3)_2$	130—140 (0.1)	—	—	391
$C_8H_{21}O_3PSi$	$(CH_3)_3SiCH_2P(O)(OC_2H_5)_2$	79—80 (3)	1·4280	0.9878	75
		93—94 (4)	1·4325	0.9911	193
		96—98 (5.5)	1.4302^{25}	—	399
		101—103 (13)	1·4348	0.9992	49
		118—121 (22)	1·4321	—	280, 373
	$(CH_3)_2(C_2H_5)SiCH_2CH_2P(O)(OCH_3)_2$	119 (0.5)	1.4341^{25}	—	455
	$(C_2H_5)_3SiCH_2CH_2P(O)(OH)_2$	m.p.112—113	—	—	226
$C_8H_{21}O_4PSi$	$(CH_3)_2(C_2H_5O)SiCH_2CH_2P(O)(OCH_3)_2$	119 (0.5)	1.4341^{25}	—	496
$C_8H_{23}O_4PSi_2$	$(CH_3)_3SiOSi(CH_3)_2CH_2P(O)(OCH_3)_2$	130—132.5 (20—21)	1·4258	—	281
$C_9H_9Cl_4OPSi$	$Cl_2(CH_3)SiC_6H_4CH{=}CHP(O)Cl_2$-*p*	146—150 (0.3)	—	—	197, 198
$C_9H_{10}Cl_5PSi$	$Cl_3Si(CH_2)_3C_6H_4PCl_2$ *	m.p. 37	—	—	198
		130—133 (0.5)	1·5655	1·4054	196, 198

*Neither o-, m-, nor p- was indicated in the original.

TABLE 10 (Continued)

Empirical formula	Compound	B.p., °C (mm)	n_D^{20}	d_4^{20}	Literature
$C_9H_{11}Cl_4PSi$	$Cl_2(CH_3)Si(CH_2)_2C_6H_4PCl_2$ *	105—109 (0.1)	1.5702	1.3344	198
$C_9H_{12}Cl_3PSi$	$Cl(CH_3)_2SiCH_2C_6H_4PCl_2$ *	111—113 (1)	1.5720	1.2556	198
$C_9H_{13}Cl_2OPSi$	p-$(CH_3)_3SiC_6H_4P(O)Cl_2$	124—128 (1.12)	—	—	272
		m.p. 44—45	—	—	272
	m-$(CH_3)_3SiC_6H_4P(O)Cl_2$	131—132 (2.5)	1.5352^{25}	—	272
$C_9H_{15}O_3PSi$	p-$(CH_3)_3SiC_6H_4P(O)(OH)_2$	m.p. 174—175	—	—	269, 272
	m-$(CH_3)_3SiC_6H_4P(O)(OH)_2$	m.p. 96—98	—	—	269, 272
$C_9H_{22}BrO_3PSi$	$(CH_3)_3SiCHBrCH_2P(O)(OC_2H_5)_2$	95—98 (1.5)	1.4562	1.1433	51
$C_9H_{23}O_3PSi$	$(C_2H_5)_3Si(CH_2)_3P(O)(OH)_2$	m.p. 72.5—73	—	—	226
	$C_2H_5(CH_3)_2SiCH_2P(O)(OC_2H_5)_2$	128.5—131.5 (13)	1.4375	0.9690	50
$C_9H_{23}O_4PSi$	$(CH_3)_2(C_2H_5O)SiCH_2P(O)(OC_2H_5)_2$	79—81 (3)	1.4302	1.0060	50
$C_9H_{23}O_5PSi$	$CH_3(C_2H_5O)_2SiCH_2CH_2P(O)(OCH_3)_2$	125.5—129 (3)	1.4288^{25}	—	455, 496
		133 (5)	1.4300^{25}	—	455, 496
		146—154 (1—2)	1.4400	—	450
$C_{10}H_{12}Cl_3OPSi$	$Cl(CH_3)_2SiC_6H_4CH{=}CHP(O)Cl_2$ *	170—173 (1)	—	—	198
		170—175 (0.6)	—	—	197
		m.p. 36	—	—	198
$C_{10}H_{15}Cl_2OPSi$	p-$(CH_3)_3SiCH_2C_6H_4P(O)Cl_2$	136—138 (0.5)	—	—	269
	m-$(CH_3)_3SiCH_2C_6H_4P(O)Cl_2$	119—122 (0.7)	—	—	270
$C_{10}H_{17}O_3PSi$	p-$(CH_3)_3SiCH_2C_6H_4P(O)(OH)_2$	m.p. 178	—	—	269, 270
	m-$(CH_3)_3SiCH_2C_6H_4P(O)(OH)_2$	m.p. 133	—	—	269, 270
	p-$(CH_3)_3SiC_6H_4CH_2P(O)(OH)_2$	167—168	—	—	269, 272
	m-$(CH_3)_3SiC_6H_4CH_2P(O)(OH)_2$	m.p. 144	—	—	269, 272
$C_{10}H_{19}BPSi$	$(CH_3)_3SiCH{=}P(CH_3)_3 \cdot B(CH_3)_3$	m.p. 28—29.5	—	—	466
$C_{10}H_{23}O_3PSi$	$(CH_3)_2(CH_2{=}CHCH_2)SiCH_2P(O)(OC_2H_5)_2$	73—75 (1.5)	1.4490	0.9854	51
$C_{10}H_{25}OPSi$	$(CH_3)_3Si(CH_2)_3P(O)(C_2H_5)_2$	110—135 (0.05)	—	—	391, 392
$C_{10}H_{25}O_3PSi$	$(CH_3)_3SiCH_2P(O)(OC_3H_7)_2$	101—102 (2.3)	1.4350^{25}	—	399
		125—128 (10)	1.4331	—	281
	$(C_2H_5)_3SiCH_2CH_2P(O)(OCH_3)_2$	150—153 (10)	1.4527	—	281
	$(C_2H_5)_3Si(CH_2)_4P(O)(OH)_2$	m.p. 53—56	—	—	226
$C_{10}H_{25}O_4PSi$	$(CH_3)_2(C_2H_5O)SiCH_2CH_2P(O)(OC_2H_5)_2$	115 (1)	1.4299^{25}	0.999	250
$C_{10}H_{25}O_5PSi$	$CH_3(C_2H_5O)_2SiCH_2P(O)(OC_2H_5)_2$	99—99.5 (1)	1.4277	1.0481	51
$C_{10}H_{25}O_6PSi$	$(C_2H_5O)_3SiCH_2CH_2P(O)(OCH_3)_2$	142.5—143.5 (5)	1.4230^{25}	—	455, 496
		143 (5)	1.4237^{25}	—	455, 496
		158—161 (10)	1.4272	—	281

$C_{10}H_{26}ClO_4PSi_2$	$ClCH_2(CH_3)_2SiOSi(CH_3)_2CH_2P(O)(OC_2H_5)_2$	79—81 (3)	1.4335	1.0478	50
$C_{10}H_{27}PSiSn$	$(CH_3)_3Si[(CH_3)_3Sn]C{=}P(CH_3)_3$	51—53 (1)	—	—	517
		m.p. 11—13	—	—	517
$C_{10}H_{27}PSi_2$	$[(CH_3)_3Si]_2C{=}P(CH_3)_3$	60—62 (1)	—	—	517
		m.p. 14—18	—	—	466, 517
$C_{10}H_{27}GePSi$	$(CH_3)_3Si[(CH_3)_3Ge]C{=}P(CH_3)_3$	60—65 (1)	—	—	517
		m.p. 14—15	—	—	517
$C_{10}H_{27}O_4PSi_2$	$(CH_3)_3SiOSi(CH_3)_2CH_2P(O)(OC_2H_5)_2$	88—90 (0.05)	—	—	373
		116—124 (9)	—	—	365
		154—156 (36)	—	—	280
		154—157 (36)	1.4240	—	373
$C_{11}H_{15}Cl_2OPSi$	$(CH_3)_3SiC_6H_4CH{=}CHP(O)Cl_2$ *	138—140 (0.2)	—	—	197, 198
		m.p. 38	—	—	197, 198
$C_{11}H_{19}OPSi$	p-$(CH_3)_3SiC_6H_4P(O)(CH_3)_2$	m.p. 114	—	—	269, 272
$C_{11}H_{27}O_3PSi$	$(C_2H_5)_3SiCH_2P(O)(OC_2H_5)_2$	93—95 (2.5)	1.4475	0.9773	226
$C_{11}H_{27}O_4PSi$	$(CH_3)_2(C_2H_5O)Si(CH_2)_3P(O)(OC_2H_5)_2$	120 (1)	1.4320^{25}	1.986^{25}	250
$C_{11}H_{27}O_5PSi$	$CH_3(C_2H_5O)_2SiCH_2CH_2P(O)(OC_2H_5)_2$	124 (2)	1.4270^{25}	1.019^{25}	250
		170—176 (3—4)	1.4365	—	450
$C_{11}H_{27}O_6PSi$	$(C_6H_5O)_3SiCH_2P(O)(OC_2H_5)_2$	96.5—97 (2)	1.4234	1.0547	51
		133—135 (3)	—	0.995	75
	$(C_2H_5O)_3Si(CH_2)_3P(O)(OCH_3)_2$	150 (5)	1.4273^{25}	—	496
$C_{12}H_{19}Cl_2PSi$	$Cl_2(C_6H_5)SiCH_2CH_2P(C_2H_5)_2$	126—127.5 (2)	—	—	474, 475
$C_{12}H_{21}OPSi$	p-$(CH_3)_3SiCH_2C_6H_4P(O)(CH_3)_2$	m.p. 134	—	—	269, 270
	m-$(CH_3)_3SiCH_2C_6H_4P(O)(CH_3)_2$	m.p. 36—38	—	—	269, 270
$C_{12}H_{21}O_2PSi$	$(CH_3)_3SiCH_2P(O)(OC_2H_5)C_6H_5$	93—94 (1)	1.5039	1.0288	225
$C_{12}H_{22}IPSi$	[p-$(CH_3)_3SiC_6H_4P(CH_3)_3$]I	m.p. 183—184	—	—	269, 272
	[m-$(CH_3)_3SiC_6H_4P(CH_3)_3$]I	m.p. 173	—	—	269, 272
$C_{12}H_{25}O_3PSi$	$CH_2(CH_2)_3Si(CH_3)CH{=}CHCH_2P(O)(OC_2H_5)_2$ (ring: CH_2 bonded to Si)	130—132 (1)	1.4708	1.0189	201
$C_{12}H_{27}O_5PSi$	$CH_3(C_2H_5O)_2SiCH{=}CHCH_2P(O)(OC_2H_5)_2$	145—146 (4—5)	1.4365	1.2067	201
$C_{12}H_{28}Cl_2P_2Si$	$Cl_2Si[CH_2CH_2P(C_2H_5)_2]_2$	139—140.5 (2)	—	—	474, 475
		165—170 (3)	—	—	473
$C_{12}H_{29}O_3PSi$	$(CH_3)_3SiCH_2P(O)(OC_4H_9)_2$	119 (2)	1.4372	—	399
		154—158 (17)	1.4360^{25}	0.9425^{25}	280, 424
	$(C_2H_5)_3SiCH_2CH_2P(O)(OC_2H_5)_2$	116—118 (2)	1.4493	0.9726	226
		135—138 (4)	1.4470^{25}	0.9718^{25}	50
	$(C_2H_5O)_3SiCH_2CH_2P(C_2H_5)_2$	123—124 (10)	—	—	474, 475

*See bottom of p. 100.

TABLE 10 (Continued)

Empirical formula	Compound	B.p., °C (mm)	n_D^{20}	d_4^{20}	Literature
$C_{12}H_{29}O_3PSSi$	$(C_2H_5O)_3SiCH_2CH_2P(S)(C_2H_5)_2$	137—140 (2)	—	—	475
		145—148 (2)	—	—	474
$C_{12}H_{29}O_5PSi$	$CH_3(C_2H_5O)_2Si(CH_2)_3P(O)(OC_2H_5)_2$	124—127 (2)	1.4325	1.0118	224
		125—126 (2)	1.4320	1.0113	224
$C_{12}H_{29}O_6PSi$	$(C_2H_5O)_3SiCH_2CH_2P(O)(OC_2H_5)_2$	135—140 (2)	1.4253	1.0356	51
		138—141 (2)	1.4250	1.0337	50
		141 (2)	1.4216^{25}	1.031^{25}	250
	$(C_2H_5O)_3Si(CH_2)_4P(O)(OCH_3)_2$	148 (5)	1.4229^{25}	—	455, 496
$C_{12}H_{31}O_4PSi_2$	$(CH_3)_3SiOSi(CH_3)_2CH_2P(O)(OC_3H_7)_2$	145—147 (10)	1.4272	—	281
$C_{12}H_{33}PSi_3$	$[(CH_3)_3SiCH_2]_3P$	m.p. 66—69 (sealed tube)	—	—	533
		m.p. 110—113	—	—	534
$C_{12}H_{33}Cl_2HgPSi_3$	$[(CH_3)_3SiCH_2]_3P \cdot HgCl_2$	m.p. 175—177	—	—	533
$C_{12}H_{33}OPSi_3$	$[(CH_3)_3SiCH_2]_3PO$	m.p. 168—172	—	—	533
		m.p. 182	—	—	280
$C_{12}H_{33}O_7PSi_4$	$O[(CH_3)_2SiO]_3Si(CH_3)CH_2P(O)(OC_2H_5)_2$ (cyclic)	161 (11)	1.4212	—	373
$C_{13}H_{23}O_3PSi$	$(CH_3)_2(C_2H_5O)SiCH_2P(O)(OC_2H_5)C_6H_5$	121—123 (2)	1.4962	1.0553	225
	p-$(CH_3)_3SiC_6H_4P(O)(OC_2H_5)_2$	154—156 (2.5)	1.4933^{25}	—	269, 272
	m-$(CH_3)_3SiC_6H_4P(O)(OC_2H_5)_2$	142—143 (2.2)	1.4915^{25}	—	269, 272
$C_{13}H_{24}IPSi$	[p-$(CH_3)_3SiCH_2C_6H_4P(CH_3)_3$]I	m.p. 186	—	—	269, 270
	[m-$(CH_3)_3SiCH_2C_6H_4P(CH_3)_3$]I	m.p. 160—162	—	—	269, 270
$C_{13}H_{29}O_6PSi$	$(C_2H_5O)_3SiCH{=}CHCH_2P(O)(OC_2H_5)_2$	153—155 (1)	1.4325	1.0493	201
$C_{13}H_{31}O_3PSi$	$(CH_3)_3SiCH_2CH_2P(O)(OC_4H_9)_2$	128 (1)	1.4853^{25}	0.936^{25}	250
	$(C_2H_5)_3Si(CH_2)_3P(O)(OC_2H_5)_2$	118—120 (1.5)	1.4498	0.9640	226
		118—120 (2)	1.4500	0.9639	224
		153—156 (8)	1.4501	0.9646	224
$C_{13}H_{31}O_5PSi$	$C_2H_5(C_2H_5O)_2Si(CH_2)_3P(O)(OC_2H_5)_2$	136—139 (3)	1.4370	1.0250	224
		138—140 (3)	1.4372	1.0256	224
$C_{13}H_{31}O_6PSi$	$(C_2H_5O)_3Si(CH_2)_3P(O)(OC_2H_5)_2$	134—136 (1.1)	1.4257^{25}	—	496
		143—145 (3)	1.4303	1.0332	224
		143—146 (3)	1.4305	1.0339	224
$C_{13}H_{32}OPSi_2$	$[(CH_3)_3Si(CH_2)_3]_2P(O)C_6H_5$	130—140 (0.1)	—	—	415
$C_{13}H_{36}IPSi_3$	$\{[(CH_3)_3SiCH_2]_3PCH_3\}I$	m.p. 175—177	—	—	533
$C_{14}H_{14}Cl_3PSi$	$Cl_3SiCH_2CH_2P(C_6H_5)_2$	156—160 (0.45)	—	—	485

$C_{14}H_{25}Cl_2PSi_2$	$[Cl(CH_3)_2SiCH_2CH_2]_2PC_6H_5$	134—156 (0.3—0.9)	—	—	464, 486
$C_{14}H_{25}O_2PSi$	$CH_3(C_2H_5)_2SiCH_2P(O)(OC_2H_5)C_6H_5$	126—129 (2)	1.4868	1.0743	225
	$C_3H_7(CH_3)_2SiCH_2P(O)(OC_2H_5)C_6H_5$	114—117 (1.5)	1.5053	1.0227	225
$C_{14}H_{25}O_3PSi$	$(C_2H_5O)_3SiCH_2CH_2PHC_6H_5$	157—213 (2.5)	1.4844^{25}	—	328
		157—161 (2.5)	1.4844^{25}	—	496
	p-$(CH_3)_3SiCH_2C_6H_4P(O)(OC_2H_5)_2$	134 (0.2)	1.4955^{25}	—	269, 270
	m-$(CH_3)_3SiCH_2C_6H_4P(O)(OC_2H_5)_2$	160—161 (3)	1.4942^{25}	—	269, 270
	p-$(CH_3)_3SiC_6H_4CH_2P(O)(OC_2H_5)_2$	151—153 (1)	1.4897^{25}	—	269, 270
	m-$(CH_3)_3SiC_6H_4CH_2P(O)(OC_2H_5)_2$	149—151 (2.3)	1.4885^{25}	—	269, 270
$C_{14}H_{25}O_5PSi$	$C_6H_5(C_2H_5O)_2Si(CH_2)_2P(O)(OCH_3)_2$	186 (5)	1.4762^{25}	—	496
$C_{14}H_{33}O_3PSi$	$(C_2H_5)_3Si(CH_2)_4P(O)(OC_2H_5)_2$	142—143 (2)	1.4489	0.9722	226
	$(CH_3)_3SiCH_2P(O)(OC_5H_{11})_2$	186—188 (18)	1.4418	—	280
$C_{14}H_{34}P_2Si$	$(CH_3)_2Si[CH_2CH_2P(C_2H_5)_2]_2$	155—160 (3)	—	—	474
		164—165 (9)	—	—	475
$C_{14}H_{35}OPSi_2$	$[(CH_3)_3Si(CH_2)_3]_2P(O)C_2H_5$	110—153 (0.5)	—	—	415
$C_{14}H_{35}O_4PSi_2$	$(CH_3)_3SiOSi(CH_3)_2CH_2P(O)(OC_4H_9)_2$	104—108 (0.01)	—	—	365
		150—152 (10)	—	—	280
$C_{14}H_{36}O_7P_2Si_2$	$[(C_2H_5O)_2P(O)CH_2Si(CH_3)_2]_2O$	157—160 (3)	1.4425	1.0805	50
$C_{14}H_{38}IPSi_3$	$\{[(CH_3)_3SiCH_2]_3PC_2H_5\}I$	m.p. 125.5—127	—	—	533
$C_{15}H_{17}Cl_2PSi$	$Cl_2(CH_3)SiCH_2CH_2P(C_6H_5)_2$	140—153 (0.08—0.35)	—	—	485
$C_{15}H_{17}Cl_2PSSi$	$Cl_2(CH_3)SiCH_2CH_2P(S)(C_6H_5)_2$	m.p. 97—100	—	—	485
$C_{15}H_{27}O_5PSi$	$(C_2H_5O)_3SiCH_2P(O)(OC_2H_5)C_6H_5$	151—153 (3)	1.4796	1.0875	225
$C_{16}H_{20}ClPSSi$	$Cl(CH_3)_2SiCH_2CH_2P(S)(C_6H_5)_2$	m.p. 95—96	—	—	485
$C_{16}H_{21}PSi$	$(CH_3)_3SiCH_2P(C_6H_5)_2$	124—125 (1.5)	1.5810	1.0046	535
$C_{16}H_{21}PSSi$	$(CH_3)_3SiCH_2P(S)(C_6H_5)_2$	m.p. 89.5	—	—	51
$C_{16}H_{21}PSeSi$	$(CH_3)_3SiCH_2P(Se)(C_6H_5)_2$	m.p. 116	—	—	51
$C_{16}H_{21}OPSi$	$(CH_3)_3SiCH_2P(O)(C_6H_5)_2$	m.p. 107.5—108.5	—	—	51
$C_{16}H_{29}O_4PSi$	$CH_3(C_2H_5O)_2Si(CH_2)_3P(O)(OC_2H_5)C_6H_5$	167—171 (2.5)	1.4863	1.0534	225
$C_{16}H_{29}O_5PSi$	$(C_2H_5O)_3SiCH_2CH_2P(O)(OC_2H_5)C_6H_5$	163—167 (3)	1.4740	1.0783	225
$C_{16}H_{37}O_3PSi$	$(CH_3)_3SiCH_2P(O)(OC_6H_{13})_2$	145—149 (2)	1.4422	—	399
$C_{16}H_{37}O_4PSi$	$(CH_3)_2(C_4H_9O)SiCH_2CH_2P(O)(OC_4H_9)_2$	157 (1)	1.4388^{25}	0.954^{25}	250
$C_{16}H_{37}O_6PSi$	$(C_2H_5O)_3SiCH_2CH_2P(O)(C_4H_9)_2$	149—159 (1—2)	1.4320	—	450
$C_{16}H_{38}P_2Si$	$(CH_3)_2Si[(CH_2)_3P(C_2H_5)_2]_2$	170—171 (4)	—	—	474, 475
$C_{16}H_{38}O_2P_2Si$	$(CH_3O)_2Si[(CH_2)_3P(C_2H_5)_2]_2$	182—184 (3)	—	—	474, 475
$C_{16}H_{39}O_4PSi_2$	$(CH_3)_3SiOSi(CH_3)_2CH_2P(O)(OC_5H_{11})_2$	183.5—184 (10)	1.4300	—	281
$C_{16}H_{44}IPSi_4$	$[(CH_3)_3SiCH_2]_2PI$	m.p. 136.5—138	—	—	533

TABLE 10 (Continued)

Empirical formula	Compound	B.p., °C (mm)	n_D^{20}	d_4^{20}	Literature
$C_{17}H_{23}O_2PSi$	$C_6H_5(CH_3)_2SiCH_2P(O)(OC_2H_5)C_6H_5$	152—154 (3)	1·5508	1·0888	225
$C_{17}H_{30}ClO_3PSi$	p-$ClC_6H_4(CH_3)_2SiCH_2P(O)(OC_4H_9)_2$	186·5—188 (2)	1·4942	—	280
$C_{17}H_{31}O_3PSi$	$C_6H_5(CH_3)_2SiCH_2P(O)(OC_4H_9)_2$	200—202 (10)	1·4864	—	280
	$(C_2H_5O)_3Si(CH_2)_3P(C_2H_5)C_6H_5$	90—120 (0·2)	—	—	325
		129—130 (0·55)	$1·4840^{25}$	—	327, 328
$C_{17}H_{31}O_4PSi$	$C_2H_5(C_2H_5O)_2Si(CH_2)_3P(O)(OC_2H_5)C_6H_5$	184—188 (3)	1·4890	1·0775	225
$C_{17}H_{31}O_5PSi$	$(C_2H_5O)_3Si(CH_2)_2P(O)(OC_2H_5)C_6H_5$	184—186 (2·5)	1·4792	1·0730	225
	$CH_3(C_2H_5O)_2SiCH_2CH_2—C_6H_4—P(O)(OC_2H_5)_2$	198 (2)	$1·4515^{25}$	1·071	250
$C_{18}H_{25}OPSi$	$(CH_3)_3Si(CH_2)_3P(O)(C_6H_5)_2$	152—154 (0·1)	—	—	391, 392
		m.p. 41—45	—	—	391, 392
$C_{18}H_{25}Cl_2O_2PSi_2$	[p-$ClC_6H_4(CH_3)_2SiCH_2]_2P(O)OH$	m.p. 95	—	—	280
$C_{18}H_{25}O_3PSi$	$(CH_3)_3SiCH_2P(O)(OC_6H_4CH_3$-n$)_2$	193—195 (2)	1·5262	—	399
$C_{18}H_{26}ClPSi_2$	[p-$(CH_3)_3SiC_6H_4]_2PCl$	72 (43)	—	—	343
$C_{18}H_{27}O_2PSi_2$	[p-$(CH_3)_3SiC_6H_4]_2P(O)OH$	m.p 213—214·5	—	—	343
$C_{18}H_{27}O_4PSi_2$	$(CH_3)_3SiOSi(CH_3)_2CH_2P(O)(OC_6H_5)_2$	158—162 (0·1)	—	—	365
$C_{18}H_{33}O_3PSi$	p-$CH_3C_6H_4(CH_3)_2SiCH_2P(O)(OC_4H_9)_2$	187—189 (2·6)	1·4879	—	280
	$(C_2H_5O)_3SiCH_2CH_2P(C_4H_9)C_6H_5$	126 (0·25)	$1·4911^{25}$	—	328, 496
$C_{18}H_{35}OPSi_2$	$[(CH_3)_3Si(CH_2)_3]_2P(O)C_6H_5$	152—154 (<0·1)	—	—	415
$C_{18}H_{45}OPSi_3$	$[(CH_3)_3Si(CH_2)_3]_3PO$	154—156 (0·1)	—	—	391, 415
$C_{19}H_{27}O_2PSi$	$CH_3(C_2H_5O)_2SiCH_2CH_2P(C_6H_5)_2$	137—151 (0·05)	$1·5496^{25}$	—	328, 496
$C_{20}H_{25}F_2PSi$	$(C_2H_5)_3SiCF{=}CFP(C_6H_5)_2$	153—155 (0·35)	$1·5619^{25}$	—	541
$C_{20}H_{29}O_3PSi$	$CH_3(C_2H_5O)_2Si(CH_2)_3P(O)(C_6H_5)_2$	200—205 (2)	1·5395	1·0943	51
	$(C_2H_5O)_3SiCH_2CH_2P(C_6H_5)_2$	178—179 (2)	—	—	474, 475
$C_{20}H_{31}O_2PSi_2$	$[(CH_3)_2$(p-$CH_3C_6H_4)SiCH_2]_2P(O)H$	m.p. 122	—	—	281
$C_{20}H_{32}OP_2Si_2$	$[(CH_3)_2SiCH_2CH_2PHC_6H_5]_2O$	223—228 (3)	—	—	90
$C_{20}H_{45}O_3PSi$	$(CH_3)_3SiCH_2P(O)[OCH_2CH(C_2H_5)C_4H_9]_2$	157—158 (1·5)	1·4462	—	399
$C_{21}H_{23}OPSi$	$C_6H_5(CH_3)_2SiCH_2P(O)(C_6H_5)_2$	m.p. 95—96	—	—	51
	p-$(CH_3)_3SiC_6H_4P(O)(C_6H_5)_2$	m.p. 93	—	—	272
	m-$(CH_3)_3SiC_6H_4P(O)(C_6H_5)_2$	212—215 (0·1)	—	—	272
$C_{21}H_{31}O_3PSi$	$C_2H_5(C_2H_5O)_2Si(CH_2)_3P(O)(C_6H_5)_2$	205—207 (2)	1·5362	1·0873	51
	$(C_2H_5O)_3Si(CH_2)_3P(O)(C_6H_5)_2$	206—208 (1)	—	—	51
		m.p 53—56	—	—	51

$C_{22}H_{25}PSi$	$(CH_3)_3SiCH{=}P(C_6H_5)_3$	150—153 (1)	—	—	517
		m.p. 76—77	—	—	517
$C_{22}H_{25}OPSi$	p-$(CH_3)_3SiCH_2C_6H_4P(O)(C_6H_5)_2$	m.p. 151—152	—	—	270
	m-$(CH_3)_3SiCH_2C_6H_4P(O)(C_6H_5)_2$	215—218 (0.1)	—	—	270
$C_{22}H_{26}BrPSi$	$[(CH_3)_3SiCH_2(C_6H_5)_3P]Br$	m.p. 175 (decomp.)	—	—	394, 536
$C_{22}H_{26}Br_3HgPSi$	$[(CH_3)_3SiCH_2(C_6H_5)_3P][HgBr_3]$	m.p. 134—135	—	—	394, 536
$C_{22}H_{26}IPSi$	$[(C_6H_5)_3PCH_2Si(CH_3)_3]I$	m.p. 168—169	—	—	538
$C_{22}H_{41}O_2PSi$	$(C_2H_5O)_3SiCH_2CH_2P(O)(OC_8H_{17})C_6H_5$	211—215 (1)	1.4770	—	455
		211—215 (1)	1.4802^{25}	—	324
$C_{23}H_{28}IPSi$	$[(C_6H_5)_3PCH(CH_3)Si(CH_3)_3]I$	m.p. 176—178	—	—	538
$C_{23}H_{51}O_5PSi$	$CH_3(C_2H_5O)_2SiCH_2CH_2P(O)[OCH_2CH(C_2H_5)C_4H_9]_2$	—	1.4468^{25}	—	496
$C_{24}H_{56}P_4Si$	$Si[CH_2CH_2P(C_6H_5)_2]_4$	224.5—228	—	—	474, 475
$C_{26}H_{32}CrN_6PS_4Si$	$[(CH_3)_3SiCH_2(C_6H_5)_3P][Cr(NH_3)_2(SCN)_4]$	142—144	—	—	292, 536
$C_{27}H_{36}Cl_3OPSi_3$	[(p-$ClC_6H_4)(CH_3)_2SiCH_2]_3PO$	m.p. 121.5	—	—	281
$C_{27}H_{39}PSi_3$	[p-$(CH_3)_3SiC_6H_4]_3P$	127 (31)	—	—	343
		m.p. 95—96	—	—	343
$C_{27}H_{39}OPSi_3$	[p-$(CH_3)_3SiC_6H_4]_3PO$	m.p. 259	—	—	280, 343
$C_{28}H_{28}N_3O_7PSi$	$[(CH_3)_3SiCH_2(C_6H_5)_3P][OC_6H_2(NO_2)_3$-2,4,6]	110—122 (decomp.)	—	—	536
$C_{30}H_{45}OPSi_3$	[p-$CH_3C_6H_4(CH_3)_2SiCH_2]_3PO$	m.p. 111.5	—	—	280
$C_{32}H_{38}P_2Si$	$(CH_3)_2Si[(CH_2)_3P(C_6H_5)_2]_2$	280—285 (2)	—	—	475
$C_{46}H_{46}BPSi$	$[(CH_3)_3SiCH_2P(C_6H_5)_3][B(C_6H_5)_4]$	m.p. 195—197	—	—	538
$C_{47}H_{48}BPSi$	$[(CH_3)_3SiCH(CH_3)P(C_6H_5)_3][B(C_6H_5)_4]$	m.p. 179—180	—	—	538
		m.p. 182—184	—	—	517
$C_{50}H_{60}N_8O_{15}P_2Si_2$	$[(CH_3)_3CNHP(C_6H_5)_2CH_2Si(CH_3)_2]_2O[OC_6H_2(NO_2)_3$-2,4,6$]_2$	m.p. 164 (decomp.)	—	—	543
$C_{56}H_{56}P_4Si$	$Si[CH_2CH_2P(C_6H_5)_2]_4$	m.p. 208—211	—	—	474, 475
$C_{100}H_{84}Br_4P_4Si$	Si{[p-$C_6H_4CH_2P(C_6H_5)_3]Br\}_4$	m.p. 255—260	—	—	304

pentene). When R = CH_3 or C_3H_7 the volatile products are $(CH_3)_3SiOR$ and $CH_3P(O)(OR)_2$ and hardly any olefins are formed [280, 281]. Siloxanes of the type $(CH_3)_3SiOSi(CH_3)_2CH_2P(O)(OR)_2$ with R = CH_3, C_2H_5, C_3H_7, C_4H_9 are less stable: the products from their thermal decomposition are $(CH_3)_3SiOSi(CH_3)_2OR$ and $CH_3P(O)(OR)_2$. Compounds containing the group $SiCH_2CH_2P$ are more stable than the corresponding esters with the group $SiCH_2P$. Thus, for example, when dimethyl β-triethylsilylethylphosphonate is heated at 222-284°C for 26 h, 73% of it remains unchanged [281]. When tris-(trimethylsilylmethyl)phosphine oxide is heated at 268-285°C for 38 h hexamethyldisiloxane is formed, but 42.5% of the starting substance remains unchanged [281].

When the acid chloride of γ-trimethylsilyl-β-chloropropylphosphinic acid is heated at 150-170°C for half an hour, it decomposes completely in accordance with the scheme:

$$(CH_3)_3SiCH_2CHClCH_2P(O)Cl_2 \longrightarrow (CH_3)_3SiCl + CH_2{=}CHCH_2P(O)Cl_2 \qquad (1.241)$$

The decomposition of organosilicon phosphinic acid salts at 200°C proceeds in accordance with the scheme (1.237) with the intermediate formation of trimethylchlorosilane and a triorganomethylenephosphorane, which reacts with a second molecule of the starting salt in accordance with the scheme (1.238) with the formation of a trimethylsilylmethylenetriorganophosphorane and a tetraorganophosphonium chloride [517].

The introduction of a silicon atom into the organic radical of alkylphosphoric acid reduces its acidity more, the smaller the number of carbon atoms which separate the Si and P atoms (Table 11). The steric factor also has some effect – (triethylsilylalkyl)-phosphinic acids have a lower acidity than their trimethylsilyl analogs. However, the inductive effect of the R_3Si group has a greater effect on the pK value, and this is extinguished only by a bridge of three or more methylene groups separating the silicon and phosphorus atoms. The elimination of one methyl radical from the silicon atom in $(CH_3)_3SiCH_2CH_2$ with a change to the structure $-OSi(CH_3)_2CH_2CH_2-$ has no effect on the acidity. Further replacement of $Si-CH_3$ bonds by siloxane bonds produces an increase in acidity.

Prolonged boiling of dialkyl trialkylsilylalkylphosphonates with concentrated hydrochloric acid does not lead to cleavage of

TABLE 11. Ionization Constants of (Trialkylsilylalkyl)phosphinic Acids Determined by Potentiometric Titration [46, 226, 250]

Acid	In water		In 50% ethanol	
	pK_1	pK_2	pK_1	pK_2
$(CH_3)_3SiCH_2P(O)(OH)_2$	3.20	8.70	—	—
$C_2H_5(CH_3)_2SiCH_2P(O)(OH)_2$	3.50	8.80	4.10	9.55
$(C_2H_5)_3SiCH_2P(O)(OH)_2$	3.60	9.10	4.30	9.60
$CH_3P(O)(OH)_2$	—	—	3.65	8.90
$(CH_3)_3SiCH_2CH_2P(O)(OH)_2$	3.40	8.40	—	—
$(C_2H_5)_3SiCH_2CH_2P(O)(OH)_2$	3.45	8.25	3.90	9.40
$C_2H_5P(O)(OH)_2$	2.45	7.85	3.75	9.10
$[(HO)_2(O)PCH_2CH_2Si(CH_3)_2]_2O$	3.40	8.50	—	—
$[(HO)_2(O)PCH_2CH_2Si(CH_3)O]_x$	2.4	7.7	—	—
$[(HO)_2(O)PCH_2CH_2SiO_{1,5}]_x$	2.5	7.6	—	—
$(CH_3)_3SiCH_2CH_2CH_2P(O)(OH)_2$	3.45	8.40	3.80	9.40
$(C_2H_5)_3SiCH_2CH_2CH_2P(O)(OH)_2$	3.50	8.50	3.85	9.40
$C_3H_7P(O)(OH)_2$	—	—	3.85	9.20
$[(HO)_2(O)PCH_2CH_2CH_2Si(CH_3)_2]_2O$	2.5	8.3	—	—
$(C_2H_5)_3SiCH_2CH_2CH_2CH_2P(O)(OH)_2$	3.50	8.40	3.90	9.35
$C_4H_9P(O)(OH)_2$	—	—	3.95	9.35
$[(HO)_2(O)PC_6H_{10}CH_2CH_2Si(CH_3)O]_x$	3.1	8.40	—	—

the Si$-CH_2$ bond [46, 221, 251, 414]:

$$R_3Si(CH_2)_nP(O)(OR')_2+2H_2O \xrightarrow{(H+)} R_3Si(CH_2)_nP(O)(OH)_2+2R'OH \quad (1.242)$$
$$(n=1-4; \ R'=C_2H_5, C_4H_9)$$

The yield of silylalkylphosphinic acids varies from 85% to quantitative. If one of the ethoxyl groups in $R(CH_3)_2SiCH_2P(O)\cdot(OC_2H_5)_2$ is replaced by a phenyl radical with R = CH_3, C_2H_5 or C_6H_5, the Si$-CH_2$ bond is ruptured if the compound is boiled with 38% HCl for 25 h. The replacement of the second ethoxyl group by a phenyl group facilitates the cleavage which occurs even with brief boiling (2 h) with distilled water. In either case the hydrolysis products are hexamethyldisiloxane and the corresponding mono- and diorganophosphinic acid [46, 51].

In the acid hydrolysis of silylalkylphosphonates containing alkoxyl groups at the silicon atom, concentrated hydrochloric acid hydrolyzes both the P$-$OR and the Si$-$OR groups. Siloxane polymers are formed containing terminal or side groups $-(CH_2)_nP(O)(OH)_2$ [250, 292]. Under milder conditions (in an aqueous ether medium) the hydrolysis may be limited to cleavage of the Si – OR groups with the formation of a disiloxane containing the group $-(CH_2)_nP(O)(OR)_2$ at the silicon atom [50].

Hydrolysis of the acid chlorides $Cl_3SiCH_2CH_2P(O)Cl_2$ and $Cl_2(C_2H_5)SiCH_2CH_2P(O)Cl_2$ forms resins which are unexpectedly readily soluble in water [220, 227]. However, cohydrolysis of the second of these compounds with methyltrichlorosilane gives insoluble resins.

After prolonged boiling of trialkylsilylethylphosphonates with 0.9 N NaOH it was not possible to isolate pure hydrolysis products [220]. However, the $Si-CH_2$ bond was not cleaved. Trimethylsilylmethyltriphenylphosphonium bromide is cleaved by 0.1 N NaOH with the formation of hexamethyldisiloxane and methyltriphenylphosphonium derivatives [536].

Trimethylsilylmethylenetrimethylphosphorane reacts vigorously with water. The products of its controlled hydrolysis in dilute hydrochloric acid are trimethylsilylmethyltriphenylphosphonium chloride, tetramethylphosphonium chloride, and hexamethyldisiloxane [465, 466]. On alkaline hydrolysis, there is quantitative cleavage of the Si – C bond with the formation of hexamethyldisiloxane:

$$(CH_3)_3SiCH{=}P(CH_3)_3 \xrightarrow{H_3O^+} [(CH_3)_3SiCH_2P(CH_3)_3]^+ + H_2O \xrightarrow[H_2O]{OH^-} [(CH_3)_3Si]_2O + [(CH_3)_4P]^+ \quad (1.243)$$

In the desilylation of compounds of the type $XC_6H_4Si(CH_3)_3$ by sulfuric acid the reaction rates increase in the following order, depending on the structure and position of the substituent X [269, 271]:

$$\text{p-}P(O)(C_6H_5)_2 < \text{p-}P(O)(CH_3)_2 < \text{p-}[P(CH_3)_3]^+ < \text{p-}NO_2 < \text{p-}P(O)(OC_2H_5)_2 <$$
$$< \text{p-}P(O)(OH)_2 \text{ and p-}[N(CH_3)_3]^+ < \text{m-}P(O)(C_6H_5)_2 < \text{m-}P(O)(OC_2H_5)_2 <$$
$$< \text{m-}P(O)(OH)_2 < \text{m-}CF_3 < \text{m-}OP(O)(OC_2H_5)_2 < \text{m-}Cl < \text{m-}CH_2P(O)(OC_2H_5)_2 <$$
$$< \text{m-}CH_2P(O)(OH)_2 < \text{p-}OP(O)(OC_2H_5)_2 < \text{p-}CH_2P(O)(OC_2H_5)_2 <$$
$$< \text{p-}CH_2P(O)(OH)_2 < H$$

The rate of desilylation of compounds of the type $XC_6H_4CH_2Si \cdot (CH_3)_3$ by 39% aqueous methanolic NaOH solution falls in the following series, depending on the structure and the position of the substituent X [269, 270]:

$$p\text{-}[P(CH_3)_3]^+ > p\text{-}P(O)(C_6H_5)_2 > p\text{-}P(O)(OC_2H_5)_2 > m\text{-}[P(CH_3)_3]^+ >$$
$$> p\text{-}P(O)(CH_3)_2 > m\text{-}P(O)(C_6H_5)_2 > m\text{-}P(O)(CH_3)_2$$

Compounds of the type $Cl_2(R)Si(CH_2)_3POCl_2$ are completely ethoxylated by treatment with alcohol in the presence of pyridine [224].

The series of conversions of diethyl(trimethylsilylphenyl)-phosphonate and diethyl(triethylsilylmethylphenyl)phosphonate is illustrated by the following scheme [269, 270, 272]:

$$\begin{array}{c} RP(O)(OH)_2 \\ \uparrow H_2O \\ RP(O)(OC_2H_5)_2 \xrightarrow{PCl_5} RP(O)Cl_2 \xrightarrow{R'MgX} RP(O)R'_2 \xrightarrow{LiAlH_4} RPR'_2 \xrightarrow{CH_3I} \\ \longrightarrow [RR'_2PCH_3]I \xrightarrow{AgOH} [RR'_2PCH_3]OH \\ [R = m\text{- or } p\text{-}(CH_3)_3Si(CH_2)_nC_6H_4;\ n = 0 \text{ or } 1] \end{array} \quad (1.244)$$

Heating a mixture of equal amounts of $CH_3(C_2H_5O)_2SiCH_2CH_2 \cdot P(O)(OC_4H_9)_2$, tetramethyldisiloxanediol, and hexamethyldisiloxane at 80°C and a pressure of 30 mm in the presence of phosphorus oxychloride is accompanied by the liberation of water and ethanol and the formation of a polymer which does not dissolve in a mixture of toluene and isooctane even after several days [380].

The possibility of the "equilibration" of mixtures of linear methylphenylsiloxanes with compounds of the type $[(RO)_2P(O)XSi \cdot (CH_3)_2]_2O$ [$X = (-CH_2-)_n$ or p- $-C_6H_4-$] under the action of sulfuric acid is reported in patents [365, 366]. In this case the phosphorus-containing group is not eliminated, and siloxanes are formed with a degree of polymerization up to 20 and terminal groups $(RO)_2P(O)-$.

The unsaturation of the phosphorus atoms in silylorganophosphines makes possible their variety of conversions. However, the character of the carbon bridge between the Si and P atoms also has a considerable effect on the properties of these compounds. Thus, for example, tris(p-trimethylsilylphenyl)-phosphine is resistant to oxidation. Boiling it with an aqueous or pyridine solution of potassium permanganate does not lead to the formation of compounds with a P=O bond [343]. At the same time, tris(trimethylsilylmethyl)phosphine (which has a marked unpleasant "phosphine" odor) fumes in air and is oxidized and melted as a result of the heat of reaction with atmospheric oxygen (m.p. 66-69°C). If there is contact with combustible substances (for example, with ether) the latter bursts into flame [533]. Boiling trimethylsilylmethyldiphenylphosphine in ether with MnO_2 in an atmosphere of dry nitrogen for 6 h gives 77% of the corresponding oxide [51]. Hydrogen peroxide may also be used for oxidation [475]. The addition of sulfur and selenium to $(CH_3)_3SiCH_2P(C_6H_5)_2$ proceeds even more readily [51]. β-Triethoxysilylethyl(diethyl)-phosphine reacts analogously with sulfur [475]:

$$R_3Si(CH_2)_nPR'_2 + S \longrightarrow R_3Si(CH_2)_nP(S)R'_2 \qquad (1.245)$$

Bis[β-(diethylphosphinyl)ethyl]dichlorosilane $[(C_2H_5)_2PCH_2 \cdot CH_2]_2SiCl_2$ is hydrolyzed by an aqueous solution of $NaHCO_3$ or NaOH with the formation of a colorless, viscous oil, treatment of which with hydrogen peroxide or sulfur gives polymers containing the groupings $>P(O)CH_2-$ and $>P(S)CH_2-$ respectively [473].

The gradual addition of an equivalent amount of bromine and iodine to a solution of tris(trimethylsilylmethyl)phosphine in ligroin and, correspondingly, methylene chloride produces the exothermal formation of $[(CH_3)_3SiCH_2]_3PX_2$ (X = Br, I), which, however, are so hygroscopic that it is impossible to isolate them in a pure form [533]. At the same time, the thermal reaction of trimethylsilylmethyldichlorophosphine with chlorine in tetrachloroethane at −20°C leads to cleavage of the $Si-CH_2$ bond. This is confirmed by the distillation of trimethylchlorosilane from the reaction mixture and the detection of derivatives of chloromethylphosphinic acid in the hydrolysis products [535].

The presence of three "siliconeopentyl" groups at a phosphorus atom does not create steric hindrance to the formation of quaternary phosphonium derivatives [533]:

$$[(CH_3)_3SiCH_2]_3P + RI \longrightarrow \{[(CH_3)_3SiCH_2]_3PR\}\,I \qquad (1.246)$$
$$[R = CH_3,\ C_2H_5 \text{ and even } (CH_3)_3SiCH_2]$$

Tris(trimethylsilylmethyl)phosphine also forms an adduct with $HgCl_2$ readily. Interestingly, the reaction of 1-(triethylsilyl)-2-(diphenylphosphinyl)-1,2-difluoroethylene $(C_2H_5)_3SiCF =$

$CFP(C_6H_5)_2$ with methyl iodide leads unexpectedly to the formation of dimethylphenylphosphonium iodide [541].

Organosilicon ylides of phosphorus readily add hydrogen halides [465, 517, 538, 539], alkyl halides [466, 517, 538, 539, 662, 673], trialkylchlorosilanes [466, 517, 662], trialkylchlorogermanes, trialkylchlorostannanes [517], and trialkylborines [466]:

$$R_3SiCH{=}PR'_3 \xrightarrow{HX} [R_3SiCH_2PR'_3]X \quad (1.247a)$$

$$R_3SiCH{=}PR'_3 \xrightarrow{R''I} [R_3SiCHR''PR'_3]I \quad (1.247b)$$

$$R_3SiCH{=}PR'_3 \xrightarrow{R_3SiCl} [(R_3Si)_2CHPR'_3]Cl \quad (1.247c)$$

$$R_3SiCH{=}PR'_3 \xrightarrow[-[R_3SiCH_2PR'_3]Cl]{R_3MCl\ (M=Ge,\ Sn)} R_3Si(R_3M)C{=}PR'_3 \quad (1.247d)$$

$$R_3SiCH{=}PR'_3 \xrightarrow{R''_3B} R_3SiCH{=}PR'_3 \cdot BR''_3 \quad (1.247e)$$

By treatment with NH_4PF_6 and $NaB(C_6H_5)_4$, phosphonium salts formed by scheme (1.247a) may be converted into the corresponding hexafluorophosphates [466] and tetraphenylborates [538, 539].

Phosphonium salts formed by schemes (1.247a) and (1.247c) undergo transylidization [466, 517, 538, 539]:

$$[R_3SiCH_2PR'_3]X + CH_2{=}PR'_3 \longrightarrow R_3SiCH{=}PR'_3 + [R'_3PCH_3]X \quad (1.248a)$$

$$[(R_3Si)_2CHPR'_3]Cl + R_3SiCH{=}PR'_3 \longrightarrow (R_3Si)_2C{=}PR'_3 + [R_3SiCH_2PR'_3]X \quad (1.248b)$$

Therefore, reaction (1.247c) may be continued by scheme (1.248b) to the final formation of a bis(trialkylsilyl)methylenetrialkylphosphorane [466, 517]. This compound adds methyl iodide in accordance with scheme (1.247b), but does not form a complex with trimethylborine in accordance with scheme (1.247e).

In all probability the reaction (1.247d) also proceeds through an addition stage in analogy with (1.247c).

Of the other reactions of organosilicon ylides of phosphorus investigated, we should note their cleavage by trimethylsilanol, which is quantitative even at 0°C [516, 674]:

$$(CH_3)_3SiCH{=}P(CH_3)_3 + (CH_3)_3SiOH \longrightarrow (CH_3)_3P{=}CH_2 + [(CH_3)_3Si]_2O \quad (1.249)$$

The disproportionation in accordance with the following scheme is also interesting [516]:

$$[(CH_3)_3Si]_2C{=}P(CH_3)_3 + (CH_3)_3P{=}CH_2 \xrightarrow{100^\circ C} 2(CH_3)_3SiCH{=}P(CH_3)_3 \quad (1.250)$$

The reaction of benzaldehyde with trimethylsilylmethylene-triphenylphosphorane is reminiscent of the Wittig reaction [517]:

$$(CH_3)_3SiCH{=}P(C_6H_5)_3 + C_6H_5CHO \longrightarrow (CH_3)_3SiCH{=}CHC_6H_5 + (C_6H_5)_3PO \quad (1.251)$$

In the case of benzophenone, tetraphenylallene and hexamethyldisiloxane were isolated from the reaction mixture in addition to triphenylphosphine oxide [385].

5.4. Practical Application

In most publications concerned with the application of compounds containing the grouping $Si-(C)_n-P$, the main recommendations are for their use as additives for siloxane oils and greases, which considerably improves their lubricating properties and flame resistance [46, 197, 324, 343, 344, 365, 366, 373, 379, 396, 450]. According to data in [46], a phosphorus content of the composition of 0.1-0.2% is sufficient for a substantial improvement in the lubricating properties of polysiloxanes over the range of 20-300°C. In addition to this, compounds containing silylorganic groups at phosphorus atoms are recommended for use as plasticizers for synthetic resins [193, 343, 344, 373, 396], antifoaming agents [343, 344], hydraulic fluids [396], insecticides [373], and protective varnishes for metals [324, 396]. According to data in [220], resins formed by the hydrolysis of compounds of the type $Cl_2(R)Si(CH_2)_3P(O)Cl_2$ and mixtures of them with methyltrichlorosilane do not burn in the flame of a gas burner.

Silyl alkyl phosphites (phosphonates) are also mentioned as intermediates for the production of siloxane polymers, particularly rubbery polymers [193, 380, 450], with an improved resistance to frost and solvents [380]. It is reported in a patent [554] that the product from the treatment of a linear polymethylsiloxane oil with phosphorus trichloride and oxygen is converted on brief contact with air into an elastomer; the latter in a rubber form has a tensile strength of 39 kg/cm^2 with a relative elongation of 225%.

6. COMPOUNDS CONTAINING THE GROUPINGS $Si-(C)_n-O-P$, $Si-(C)_n-S-P$, $Si-(C)_n-N-P$, AND $Si-O-(C)_n-P$

6.1. Preparation Methods

The simplest method of synthesizing compounds containing the grouping $Si-(C)_n-O-P$ is the reaction of organosilicon alcoholates with phosphorus halides [269, 280, 281]:

$$nR_3SiXONa + Cl_nP(O)R'_{3-n} \xrightarrow[-nNaCl]{} [R_3SiXO]_nP(O)R'_{3-n} \qquad (1.252)$$

$$(X = CH_2,\ C_6H_4;\ n = 1-3)$$

Instead of alcoholates it is also possible to use the organosilicon alcohols or phenols themselves by carrying out the reaction in the presence of tertiary amines [272, 426]. However, this modification of the reaction (1.252) has not been used widely for the preparation of compounds containing the grouping $Si-(C)_n-O-P$. A similar reaction was used successfully for the synthesis of compounds containing the grouping $Si-(C)_n-N-P$ [329-332, 439]:

$$nR_3Si(CH_2)_3NH_2 + Cl_nP(O)R'_{3-n} \xrightarrow[-nR''_3N\cdot HCl]{+nR''_3N} [R_3Si(CH_2)_3NH]_nP(O)R'_{3-n} \qquad (1.253)$$

$$(n = 1-3)$$

Organodichlorophosphines and diorganochlorophosphines have also been used in this reaction [331].

A second general method for synthesizing compounds containing the groupings $Si-(C)_n-O-P$, $Si-(C)_n-S-P$, and also $Si-(C)_n-O-(C)_n-P$ is the reaction of chloroalkylsilanes and chloroalkylsiloxanes with salts of phosphoric and thiophosphoric acids [21, 23, 25, 70, 71, 73, 74, 136, 174, 175, 290]:

$$\rangle SiCH_2Cl + MXP(Y)R_2 \xrightarrow[-MCl]{} \rangle SiCH_2XP(Y)R_2 \qquad (1.254)$$

$$(X = O,\ S,\ OCH_2;\ Y = O,\ S;\ M = Na,\ K)$$

In the presence of a catalyst, namely, diethylamine, the yield of the products of the reaction (1.254) reaches 80-90% [23].

In the patents [463, 497] it is suggested that compounds containing the grouping Si $-(C)_n-$ O$-$P may also be synthesized by opening an epoxide ring in organosilicon oxides with orthophosphoric or phenylphosphinic acid:

$$\rangle Si-(C)_n-\underset{\diagdown O \diagup}{CH-CH_2} + H_3PO_4 \longrightarrow \rangle Si-(C)_n-\underset{OH}{\underset{|}{C}}HCH_2OP(O)(OH)_2 \quad (1.255)$$

Dialkyldithiophosphoric acids add smoothly to alkenylsilanes in the absence of any catalysts with the evolution of heat. In the case of vinylsilanes the addition proceeds in accordance with Farmer's rule, but in the case of allylsilanes, in accordance with Markovnikov's rule [139, 190-192]:

$$\rangle SiCH=CH_2 + HSP(S)(OR)_2 \longrightarrow \rangle SiCH_2CH_2SP(S)(OR)_2 \quad (1.256)$$

$$\rangle SiCH_2CH=CH_2 + HSP(S)(OR)_2 \longrightarrow \rangle SiCH_2\underset{CH_3}{\underset{|}{C}}HSP(S)(OR)_2 \quad (1.257)$$

Dipropargyl (alkyl)phosphonates are hydrolyzed by triethylsilane in the presence of $H_2PtCl_6 \cdot 6H_2O$ at 85-90°C in 8 h to form the product from addition at both C$\equiv$C bonds [2]:

$$2(C_2H_5)_3SiH + (CH\equiv CCH_2O)_2P(O)R \longrightarrow [(C_2H_5)_3SiCH=CHCH_2O]_2P(O)R \quad (1.258)$$

The reaction of chlorosilanes with esters of hydroxylalkylphosphinic acids (most often in the form of their sodium derivatives) is used for the synthesis of compounds containing the grouping Si$-$O$-(C)_n-$P [32, 136, 181, 182, 232, 265]:

$$\rangle Si-Cl + MOCH_2P(O)(OR)_2 \xrightarrow[-MCl]{} \rangle SiOCH_2P(O)(OR)_2 \quad (1.259)$$

(M=H, Na).

The reaction of esters of chloromethylphosphinic acid with sodium triethylsilanolate in benzene does not lead to compounds containing the grouping Si$-$O$-$(C) $-$P. Instead, a salt of an ester of chloromethylphosphinic acid is formed in accordance with the scheme [136]

$$(R_3SiO)_2P(O)CH_2Cl + NaOSiR_3 \longrightarrow (R_3SiO)(NaO)P(O)CH_2Cl + (R_3Si)_2O \qquad (1.260)$$

The reaction of dialkyl(glycidoxymethyl)phosphonates with triorganochlorosilanes and triorganobromosilanes proceeds with opening of the oxide ring on the side of the most hydrogenated carbon atom [57, 155-158]:

$$R_3SiX + (R'O)_2P(O)CH_2OCH_2\underset{\diagdown O \diagup}{CH-CH_2} \longrightarrow (R'O)_2P(O)CH_2OCH_2\underset{\displaystyle OSiR_3}{CH}CH_2X \qquad (1.261)$$

$$(X = Cl,\ Br)$$

The structure of the products obtained was demonstrated by confirmatory synthesis:

$$R_3SiOC_2H_5 + (R'O)_2P(O)CH_2OCH_2\underset{\displaystyle OH}{CH}CH_2X \xrightarrow[-C_2H_5OH]{} (R'O)_2P(O)CH_2OCH_2\underset{\displaystyle OSiR_3}{CH}CH_2X$$

$$(1.262)$$

The reactions of hydroxymethylphosphinic acid with triorganosilanes, alkoxysilanes, and acetoxysilanes, which proceed with the formation of compounds containing both Si−O−P and $Si-O-(C)_n-P$ groups, have already been described in Section 1.1.

The reaction of tetraphenoxysilane with P_4S_{10} with the formation of $(C_6H_5O)_3SiOC_6H_4PS_2$ was examined in Section 1.1.2.

6.2. Physical and Chemical Properties

The physicochemical constants of compounds containing the groupings $Si-(C)_n-O-P$, $Si-(C)_n-S-P$, $Si-(C)_n-N-P$, and $Si-(C)_n-O-(C)_n-P$ are given in Table 12, and those of compounds containing the grouping $Si-O-(C)_n-P-$, in Table 13.

Organosilicon esters of dialkyldithiophosphoric acids decompose at elevated temperatures with the formation of pyrolysis products, which have an unpleasant odor. Adducts of $HSP(S)(OR)_2$ with diallyl- and triallylsilanes will not distill without decomposition. Compounds of the type $R_{4-n}Si[OCH_2P(O)(OR')_2]_n$ with n = 1 ($R = CH_3$, C_2H_5, $CH = CH_2$; $R' = C_2H_5$, iso-C_3H_7) distill in vacuum without decomposition. Compounds with n = 2 and particularly

TABLE 12. Compounds Containing the Groupings $Si-(C)_n-O-P$, $Si-(C)_n-S-P$, $Si-(C)_n-N-P$, and $Si-(C)_n-O-(C)_n-P$

Empirical formula	Compound	B.p., °C (mm)	n_D^{20}	d_4^{20}	Literature
$C_6H_{17}O_4PSi$	$(CH_3)_3SiCH_2OP(O)(OCH_3)_2$	61—62 (0.4)	1.4158	—	281
$C_7H_{19}O_3PSi$	$C_2H_5O(CH_3)_2SiCH_2OP(O)(CH_3)_2$	106—108 (4)	1.4355	1.0040	21
$C_8H_{21}O_2PS_2Si$	$(CH_3)_3SiCH_2CH(CH_3)SP(S)(OCH_3)_2$	115—117 (3)	1.5045	1.0696	191
$C_8H_{21}O_4PSi$	$(C_2H_5O)_2(CH_3)SiCH_2OP(O)(CH_3)_2$	114—115 (2)	1.4325	1.0390	21
$C_9H_{23}O_3PS_2Si$	$C_2H_5O(CH_3)_2SiCH_2SP(S)(OC_2H_5)_2$	126 (1)	1.4863	1.0715	23
$C_9H_{23}O_4PSSi$	$C_2H_5O(CH_3)_2SiCH_2OP(S)(OC_2H_5)_2$	89—94 (15)	1.4450	1.0581	70, 71
$C_9H_{23}O_5PSi$	$C_2H_5O(CH_3)_2SiCH_2OP(O)(OC_2H_5)_2$	105—107 (1)	1.4220	1.0400	21
$C_{10}H_{25}O_2PSSi$	$(C_2H_5)_3SiCH_2CH_2SP(S)(OCH_3)_2$	121 (0.5)	1.5102	1.0577	191
$C_{10}H_{25}O_2PS_2Si$	$(CH_3)_3SiCH_2CH(CH_3)SP(S)(OC_2H_5)_2$	94 (1)	1.4945	1.0261	191
$C_{10}H_{25}O_4PS_2Si$	$(C_2H_5O)_2(CH_3)SiCH_2SP(S)(OC_2H_5)_2$	159—160 (6)	1.4815	1.0919	23
$C_{10}H_{25}O_6PSi$	$(C_2H_5O)_2(CH_3)SiCH_2OP(O)(OC_2H_5)_2$	113—115 (1)	1.4182	1.0590	21
$C_{10}H_{28}O_5P_2Si_2$	$[(CH_3)_2P(O)OCH_2Si(CH_3)_2]_2O$	208—210 (5)	1.4550	1.0869	21
$C_{11}H_{19}O_4PSi$	p-$(CH_3)_3SiC_6H_4OP(O)(OCH_3)_2$	152—154 (4)	1.4862	—	280
$C_{11}H_{27}O_2PS_2Si$	$(C_2H_5)_3SiCH_2CH(CH_3)SP(S)(OCH_3)_2$	139.5 (5.5)	1.5090	1.0510	191
$C_{11}H_{27}O_4PSi$	$CH_3(C_2H_5)_2SiCH_2OCH_2P(O)(OC_2H_5)_2$	140 (6)	1.4401	0.9917	174
$C_{11}H_{27}O_6PSi$	$CH_3(C_2H_5O)_2SiCH_2OCH_2P(O)(OC_2H_5)_2$	120 (2)	1.4282	1.0595	174
$C_{11}H_{28}O_4P_2S_4Si$	$CH_3SiH[CH_2CH(CH_3)SP(S)(OCH_3)_2]_2$	136 (1)	1.5360	1.1796	191
$C_{11}H_{29}O_3PSi_2$	$(C_2H_5)_3SiOSi(CH_3)_2CH_2OP(O)(CH_3)_2$	117—118 (2)	1.4409	0.9609	22
$C_{12}H_{27}O_5PSi_2$	$CH_2{=}C(CH_3)COOCH_2Si(CH_3)_2OSi(CH_3)_2CH_2OP(O)(CH_3)_2$	154—155 (2)	1.4460	1.0285	16
$C_{12}H_{29}O_2PS_2Si$	$(CH_3)_3SiCH_2CH(CH_3)SP(S)(OC_3H_7)_2$	124 (2.5)	1.4912	1.0100	191
	$(CH_3)_3SiCH_2CH(CH_3)SP(S)(OC_3H_7\text{-iso})_2$	119—121 (3.5)	1.4820	0.9944	191
	$(C_2H_5)_3SiCH_2CH_2SP(S)(OC_2H_5)_2$	133 (0.5)	1.5021	1.0258	191
$C_{12}H_{29}O_4PSi$	$(C_2H_5)_3SiCH_2OCH_2P(O)(OC_2H_5)_2$	121 (3)	1.4450	0.9883	174
$C_{12}H_{30}O_4P_2S_4Si$	$(CH_3)_2Si[CH_2CH(CH_3)SP(S)(OCH_3)_2]_2$	—	1.5405	1.1882	191
$C_{12}H_{33}O_4PSi$	$[(CH_3)_3SiCH_2O]_3PO$	134—136 (5)	1.4331	—	280
$C_{13}H_{23}O_2PSi$	$(CH_3)_3Si(CH_2)_4OP(O)(H)C_6H_5$	142—148 (1)	—	—	426
$C_{13}H_{23}O_2PS_2Si$	$(CH_3)_2(C_6H_5)SiCH_2CH(CH_3)SP(S)(OCH_3)_2$	136.5 (0.5)	1.5490	1.1029	191
$C_{13}H_{23}O_4PSi$	p-$(CH_3)_3SiC_6H_4OP(O)(OC_2H_5)_2$	145—150 (2—2.5)	1.4840^{25}	—	269, 272
	m-$(CH_3)_3SiC_6H_4OP(O)(OC_2H_5)_2$	148—150 (1)	1.4900	—	269, 272
$C_{13}H_{25}O_3PSi_2$	$(CH_3)_2(C_6H_5)SiOSi(CH_3)_2CH_2OP(O)(CH_3)_2$	155—157 (2)	1.4820	1.0329	22

$C_{13}H_{26}NO_4PSi$	$CH_3(C_2H_5)_2SiCH_2OCH_2P(O)(OH)_2 \cdot C_6H_5NH_2$	m.p. 139.5—140.5	—	—	174
$C_{13}H_{31}O_2PS_2Si$	$(C_2H_5)_3SiCH_2CH(CH_3)SP(S)(OC_2H_5)_2$	134 (1)	1.4990	1.0131	191
$C_{13}H_{31}O_3PS_2Si$	$C_2H_5O(CH_3)_2SiCH_2SP(S)(OC_4H_9)_2$	166—168 (2—3)	1.4821	1.0267	23
$C_{13}H_{31}O_4PSi$	$CH_3(C_2H_5)_2SiCH_2OCH_2P(O)(OC_3H_7\text{-iso})_2$	135 (5)	1.4355	0.8404	174
$C_{13}H_{33}O_3PS_2Si_2$	$(C_2H_5)_3SiOSi(CH_3)_2CH_2SP(S)(OC_2H_5)_2$	153 (2)	1.4811	1.026	25
		168—170 (3—4)	1.4823	1.028	25
$C_{14}H_{28}NO_4PSi$	$(C_2H_5)_3SiCH_2OCH_2P(O)(OH)_2$	m.p. 145—146	—	—	174
$C_{14}H_{33}O_2PS_2Si$	$(C_3H_7)_3SiCH_2CH(CH_3)SP(S)(OCH_3)_2$	132.5 (1)	1.5028	1.0158	191
$C_{14}H_{33}O_4PS_2Si$	$(C_2H_5O)_2(CH_3)SiCH_2SP(S)(OC_4H_9)_2$	170—171 (2)	1.4770	1.0514	23
$C_{14}H_{36}OP_2S_4Si_2$	$[(C_2H_5O)_2P(S)SCH_2Si(CH_3)_2]_2O$	100 (0.001)	1.4915	1.1134	23
$C_{15}H_{35}O_2PS_2Si$	$(C_2H_5)_3SiCH_2CH(CH_3)SP(S)(OC_3H_7)_2$	130 (1)	1.4960	1.0014	191
$C_{16}H_{38}O_4P_2S_4Si$	$(CH_3)_2Si[CH_2CH(CH_3)SP(S)(OC_2H_5)_2]_2$	—	1.5252	1.1115	191
$C_{16}H_{38}O_8P_2Si_3$	$[(CH_3)_2P(O)OCH_2Si(CH_3)_2O]_2Si(CH_3)CH_2OCOC(CH_3){=}CH_2$	—	1.4485	1.0860	16
$C_{16}H_{39}O_6P_3S_6Si$	$CH_3Si[CH_2CH(CH_3)SP(S)(OCH_3)_2]_3$	—	1.5568	1.2232	191
$C_{16}H_{41}O_4PSi_3$	$[(C_2H_5)_3SiO]_2Si(CH_3)CH_2OP(O)(CH_3)_2$	153—156 (1—1.5)	1.4442	0.9634	22
$C_{17}H_{23}OPS_2Si$	$C_2H_5O(CH_3)_2SiCH_2SP(S)(C_6H_5)_2$	210—220 (0.001)	1.6056	1.1481	23
$C_{17}H_{32}O_4P_2S_4Si$	$CH_3(C_6H_5)Si[CH_2CH(CH_3)SP(S)(OCH_3)_2]_2$	—	1.5550	1.1799	192
$C_{17}H_{41}O_3PS_2Si_2$	$(C_2H_5)_3SiOSi(CH_3)_2CH_2SP(S)(OC_4H_9)_2$	100—103 (0.001)	1.4780	0.9971	25
		150 (0.1)	1.4769	0.9947	25
$C_{17}H_{41}NO_9P_2Si$	$(C_2H_5O_3)Si(CH_2)_3N[P(O)(OC_2H_5)_2]_2$	—	1.4303^{25}	—	332
$C_{18}H_{25}O_2PS_2Si$	$CH_3(C_2H_5O)_2SiCH_2SP(S)(C_6H_5)_2$	230—235 (0.001)	1.5848	1.1545	23
$C_{18}H_{27}O_3PSi_2$	$CH_3(C_6H_5)_2SiOSi(CH_3)_2CH_2OP(O)(CH_3)_2$	172—174 (0.15)	1.5275	1.0780	22
$C_{18}H_{37}O_8PSi_3$	$[CH_2{=}C(CH_3)COOCH_2Si(CH_3)_2O]_2Si(CH_3)CH_2OP(O)(CH_3)_2$	—	1.4488	1.0453	16
$C_{19}H_{46}ClN_2O_7PSi_2$	$ClCH_2P(O)[NH(CH_2)_3Si(OC_2H_5)_3]_2$	—	1.4560^{25}	—	330
$C_{20}H_{33}O_4PSi_3$	$[C_6H_5(CH_3)_2SiO]_2Si(CH_3)CH_2OP(O)(CH_3)_2$	147—150 (0.15)	1.5034	1.0615	22
$C_{22}H_{51}O_6P_3S_6Si$	$CH_3Si[CH_2CH(CH_3)SP(S)(OC_2H_5)_2]_3$	—	1.5410	1.1655	191
$C_{22}H_{52}O_5P_2S_4Si_2$	$[(C_4H_9O)_2P(S)SCH_2Si(CH_3)_2]_2O$	200—204 (0.001)	1.4996	1.0753	23
$C_{25}H_{51}O_3PSi_2$	$CH_3C_6H_{10}P(O)[OCH_2CH{=}CHSi(C_2H_5)_3]_2$	145—147 (0.4)	1.4650	0.9400	2
$C_{27}H_{39}O_4PSi_3$	$[p\text{-}(CH_3)_3SiC_6H_4O]_3PO$	m.p. 99	—	—	280
$C_{30}H_{36}OP_2S_4Si_2$	$[(C_6H_5)_2P(S)SCH_2Si(CH_3)_2]_2O$	m.p. 126—127	—	—	23
$C_{30}H_{37}O_4PSi_3$	$[(C_6H_5)_2(CH_3)SiO]_2Si(CH_3)CH_2OP(O)(CH_3)_2$	250 ($1 \cdot 10^{-4}$)	1.5580	1.1178	22

TABLE 13. Compounds Containing the Grouping $Si-O-(C)_n-P$

Empirical formula	Compound	B.p., °C (mm)	n_D^{20}	d_4^{20}	Literature
$C_8H_{21}O_4PSi$	$(CH_3)_3SiOCH_2P(O)(OC_2H_5)_2$	113 (9)	1.4205	1.0116	181, 182
$C_9H_{23}O_4PSi$	$(CH_3)_2(C_2H_5)SiOCH_2P(O)(OC_2H_5)_2$	104 (5)	1.4292	1.0145	181, 142
	$(CH_3)_3SiOCH(CH_3)OP(OC_2H_5)_2$	55—56 (0.1—0.2)	1.4204^{25}	—	265
$C_{10}H_{23}Cl_2O_4PSi$	$(CH_3)_2(CHCl_2)SiOCH(C_2H_5)OP(OC_2H_5)_2$	114—118 (0.15—0.2)	—	—	265
$C_{10}H_{23}Cl_2O_5PSi$	$(C_2H_5O)_2P(O)CH_2OCH_2CH(CH_2Cl)OSiCl(CH_3)_2$	168—170 (1)	1.4675	1.2400	154
$C_{11}H_{26}ClO_5PSi$	$(C_2H_5O)_2P(O)CH_2OCH_2CH(CH_2Cl)OSi(CH_3)_3$	145—146 (1)	1.4452	1.1030	151, 152
$C_{11}H_{27}O_4PSi$	$(C_2H_5)_3SiOCH_2P(O)(OC_2H_5)_2$	115 (2.5)	1.4360	1.0086	181, 182
$C_{11}H_{27}O_5PSi$	$(CH_3)_2(C_2H_5O)SiOCH(C_2H_5)OP(OC_2H_5)_2$	92—93 (0.4)	—	—	265
	$(CH_3)_2(C_2H_5O)SiOC(CH_3)_2OP(OC_2H_5)_2$	80 (0.4)	—	—	265
$C_{11}H_{27}O_6PSi$	$C_2H_5(C_2H_5O)_2SiOCH_2P(O)(OC_2H_5)_2$	124—125 (3)	1.4309	1.0283	182
$C_{12}H_{27}Cl_2O_5PSi$	$(C_2H_5O)_2P(O)CH_2OCH_2CH(CH_2Cl)OSiCl(C_2H_5)_2$	173—175 (1)	1.4715	1.2097	154
$C_{12}H_{29}O_4PSi$	$(C_2H_5)_3SiOCH(CH_3)OP(OC_2H_5)_2$	101—103 (0.2—0.25)	—	—	265
	$CH_3(C_2H_5O)_2SiOCH(C_2H_5)OP(OC_2H_5)_2$	102—104 (0.33—0.4)	—	—	265
$C_{12}H_{30}O_8P_2Si$	$(CH_3)_2Si[OCH_2P(O)(OC_2H_5)_2]_2$	164—166 (1.5)	1.4387	1.1371	181, 182
$C_{13}H_{23}O_4PSi$	$(CH_3)_3SiOCH(C_6H_4CH_3\text{-p})OP(OCH_3)_2$	126 (0.44)	—	—	265
$C_{13}H_{29}O_6PSi$	$(CH_3)_3SiOCH(CH_2CH_2COOC_2H_5)OP(OC_2H_5)_2$	125—126 (0.45)	—	—	265
$C_{13}H_{30}ClO_5PSi$	$(C_3H_7O)_2P(O)CH_2OCH_2CH(CH_2Cl)OSi(CH_3)_3$	163—164 (2)	1.4460	1.0666	153
$C_{13}H_{31}O_4PSi$	$(C_2H_5)_3SiOCH_2P(O)(OC_3H_7\text{-iso})_2$	121 (2)	1.4340	0.9712	181, 182
$C_{13}H_{31}O_4PS_3Si$	$(C_2H_5S)_3SiOCH(C_2H_5)OP(OC_2H_5)_2$	172—173 (0.6—0.7)	—	—	265
$C_{14}H_{24}ClO_4PSi$	$(CH_3)_3SiOCH(C_6H_4Cl\text{-}o)OP(OC_2H_5)_2$	132—134 (0.6)	—	—	265
$C_{14}H_{25}O_3PSi$	$(CH_3)_3SiOCH(C_2H_5)OP(OC_2H_5)C_6H_5$	127—128 (0.4)	—	—	265
$C_{14}H_{30}NO_4PSi$	$(CH_3)_3SiOCH[C(CH_3)_2CH_2CH_2CN]OP(OC_2H_5)_2$	143—145 (0.7)	—	—	265

$C_{14}H_{31}O_6PSi$	$OCH(CH_3)CH_2C(CH_3)_2OSi(CH_3)OCH(C_2H_5)OP(OC_2H_5)_2$	115—116 (0.6)	—	—	265
$C_{14}H_{32}ClO_5PSi$	$(C_2H_5)_3SiOCH(CH_2Cl)CH_2OCH_2P(O)(OC_2H_5)_2$	170—171 (1)	1.4550	1.0893	153
		175—176 (2)	—	—	151
$C_{14}H_{32}BrO_5PSi$	$(C_2H_5)_3SiOCH(CH_2Br)CH_2OCH_2P(O)(OC_2H_5)_2$	182—183 (1)	1.4640	1.1896	153
$C_{14}H_{33}O_7PSi$	$(C_2H_5O)_3SiOCH(C_3H_7)OP(OC_2H_5)_2$	114.5—117 (0.23—0.25)	—	—	265
$C_{14}H_{34}O_8P_2Si$	$(CH_3)_2Si[OCH(CH_3)OP(OC_2H_5)_2]_2$	148—154 (0.2—0.4)	1.4355^{25}	—	265
	$(C_2H_5)_2Si[OCH_2P(O)P(OC_2H_5)_2]_2$	179—180 (2.5)	1.4446	1.1210	181, 182
$C_{15}H_{27}O_4PSi$	$(CH_3)_3SiOC(CH_3)(C_6H_5)OP(OC_2H_5)_2$	129—130 (0.4—0.5)	—	—	265
$C_{15}H_{27}O_5PSi$	$(CH_3)_3SiOCH(C_6H_4OCH_3\text{-}p)OP(OC_2H_5)_2$	114—145 (0.6)	—	—	265
$C_{15}H_{31}O_4PSi$	$(CH_3)_3SiOCHOP(OC_2H_5)_2$ 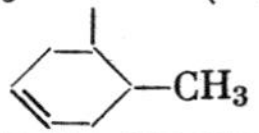	124—125 (0.6)	—	—	265
$C_{15}H_{34}ClO_5PSi$	$C_3H_7(C_2H_5)_2SiOCH(CH_2Cl)CH_2OCH_2P(O)(OC_2H_5)_2$	175—176 (1)	1.4592	1.0806	151
$C_{16}H_{28}ClO_5PSi$	$C_6H_5(CH_3)_2SiOCH(CH_2Cl)CH_2OCH_2P(O)(OC_2H_5)_2$	190—191 (1)	1.4922	1.1498	151, 153
$C_{16}H_{29}O_{12}P_3Si$	$CH_3Si[OCH_2P(O)(OC_2H_5)_2]_3$	—	1.4408	1.1732	182
$C_{16}H_{36}ClO_5PSi$	$(C_2H_5)_3SiOCH(CH_2Cl)CH_2OCH_2P(O)(OC_3H_7\text{-iso})_2$	184—185 (1)	1.4539	1.0631	151
	$(C_2H_5)_3SiOCH(CH_2Cl)CH_2OCH_2P(O)(OC_3H_7)_2$	184—185 (1)	1.4560	1.0591	153
$C_{17}H_{29}O_{12}P_3Si$	$CH_2{=}CHSi[OCH_2P(O)(OC_2H_5)_2]_3$	—	1.4455	1.1731	182
$C_{18}H_{32}ClO_5PSi$	$C_6H_5(CH_3)_2SiOCH(CH_2Cl)CH_2OCH_2P(O)(OC_3H_7)_2$	212—213 (1)	1.4926	1.1301	153
$C_{18}H_{40}Cl_2O_{10}P_2Si$	$(CH_3)_2Si[OCH(CH_2Cl)CH_2OCH_2P(O)(OC_2H_5)_2]_2$	212—215 (1)	1.4560	1.1967	153
$C_{19}H_{32}O_8P_2Si$	$CH_2(CH_2)_4Si[OCH(C_2H_5)OP(OC_2H_5)_2]_2$	186—188 (0.65)	—	—	265
$C_{20}H_{39}Cl_6O_{12}PSi$	$CH_2{=}CHSi[OCH(CH_3)OP(OCH_2CH_2Cl)_2]_3$	150 (0.5)	1.4860^{25}	—	265
$C_{20}H_{44}Cl_2O_{10}P_2Si$	$(C_2H_5)_2Si[OCH(CH_2Cl)CH_2OCH_2P(O)(OC_2H_5)_2]_2$	248—250 (1)	1.4600	1.1802	153
$C_{21}H_{30}ClO_5PSi$	$CH_3(C_6H_5)_2SiOCH(CH_2Cl)CH_2OCH_2P(O)(OC_2H_5)_2$	238 (1)	1.5265	1.1725	153
$C_{21}H_{47}O_7PSi$	$(iso\text{-}C_3H_7O)_3SiO(CH_2)_4P(O)(OC_4H_9)_2$	141—144 (0.5)	1.4128	—	232
$C_{22}H_{47}O_6PSi$	$OCH(CH_3)CH_2C(CH_3)_2OSi(CH_3)OCH(C_{10}H_{21})OP(OC_2H_5)_2$	178—179 (0.35—0.4)	—	—	265
$C_{22}H_{49}O_7PSi$	$(sec\text{-}C_4H_9O)_3SiOCH_2CH_2P(O)(OC_4H_9)_2$	154—158 (1)	1.4251	0.9675	232
$C_{46}H_{73}O_4PSi$	$[C_6H_5(CH_2)_3]_3SiOCH(C_2H_5)OP[OCH_2CH(C_2H_5)C_4H_9]_2$	180 (0.04)	—	—	265

n = 3 partly decompose when vacuum distilled [136, 182]. The thermal decomposition process proceeds through the gradual replacement of phosphorus-containing groups at the silicon atom by alkoxyl radicals, which were previously attached to the phosphorus. Thus, for example, when bis[(diethylphosphono)methoxy]-diethylsilane is heated at 245-255°C and a pressure of 50 mm, the decomposition proceeds according to the scheme

$$2[(C_2H_5O)_2P(O)CH_2O]_2Si(C_2H_5)_2 \longrightarrow 2(C_2H_5O)_2P(O)CH_2OSiOC_2H_5(C_2H_5)_2$$

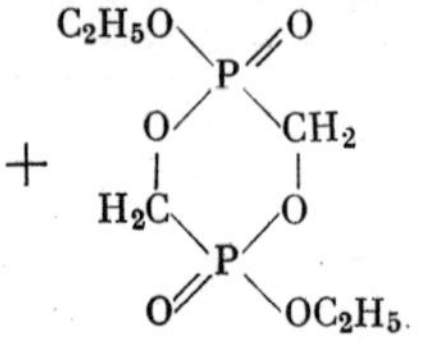

(1.263)

With continuation of the heating (260-280°C, 760 mm) there is further decomposition with the formation of diethyldiethoxysilane (31.4%), and then there begins vigorous evolution of gases which contain unsaturated compounds [136, 182]. Triethyl(diethylphosphonomethoxy)silane decomposes completely when heated to 260-288°C (760 mm) for 2 h, forming triethylethoxysilane (45%). The thermal decomposition of tris[(diethylphosphono)methoxy]-methylsilane (230-250°C, 55 mm) leads to a complex mixture in which methyltriethoxysilane has been detected [136].

Hydrolysis of dialkyl (trialkylsilylmethoxy)methylphosphonates by 18% HCl for 20 h leads to the formation of the corresponding (trialkylsilylmethoxy)methylphosphinic acids (yield 85-90%), which were characterized in the form of the aniline salts [136, 174]:

$$R_3SiCH_2OCH_2P(O)(OC_2H_5)_2 + 2H_2O \xrightarrow[-2C_2H_5OH]{(HCl)} R_3SiCH_2OCH_2P(O)(OH)_2 \quad (1.264)$$

If $R_3 = CH_3(C_2H_5O)_2$, then under the reaction conditions there is almost complete decomposition of the grouping Si–C–O–C–P with the formation of cross-linked siloxane polymer.

Compounds of the type $\rangle SiCH_2YP(Y)R_2$, which contain ethoxyl groups at the silicon atom (Y = O or S), readily undergo further conversions. Thus, by using triorganosilanols it is possible to grow the silicon part of the molecule [22, 25]:

$$R_3'SiOH + C_2H_5O(CH_3)_2SiCH_2YP(Y)R_2 \xrightarrow[-C_2H_5OH]{} R_3'SiOSi(CH_3)_2CH_2YP(Y)R_2 \quad (1.265)$$

The liberation of alcohol at 140-150°C is complete in a few hours. In the reaction with $CH_3(C_2H_5O)_2SiCH_2OP(O)(CH_3)_2$, triethylsilanol replaces both ethoxyl groups at the silicon [22], but in the case of $CH_3(C_2H_5O)_2SiCH_2SP(S)(OR'')_2$ the reaction does not proceed even with a reagent ratio from 1 : 4 ($R'' = C_2H_5$) to 1:6 ($R'' = C_4H_9$). No explanation has been found for this difference as yet. At the same time, this offers the possibility of using silicon-containing organophosphorus compounds of a similar type for the build-up of polymers, which are obtained, for example, by heating $CH_3(C_2H_5O)_2SiCH_2OP(O)(CH_3)_2$ with α,ω-siloxanediols [24, 27].

Analogous to the reactions of silanols with alkoxyl derivatives of silicon-containing organophosphorus compounds actually is the following [16]:

$$R(CH_3)_2SiOC_2H_5 + C_2H_5O(CH_3)_2SiCH_2OP(O)(CH_3)_2 \xrightarrow{(H_2O)}$$
$$\longrightarrow R(CH_3)_2SiOSi(CH_3)_2CH_2OP(O)(CH_3)_2$$
$$[R{=}CH_2{=}C(CH_3)COO] \qquad (1.266)$$

It is also possible to obtain siloxane polymers with terminal phosphorus-containing groups by a condensation of a different type [71]:

$$2(C_2H_5O)_2P(S)OCH_2Si(CH_3)_2OC_2H_5 + Cl[(CH_3)_2SiO]_nSi(CH_3)_2Cl \xrightarrow{0.5\%\,FeCl_3}$$
$$\longrightarrow 2C_2H_5Cl + (C_2H_5O)_2P(S)OCH_2[(CH_3)_2SiO]_{n+2}Si(CH_3)_2CH_2OP(S)(OC_2H_5)_2$$
$$(n = 3-7) \qquad (1.267)$$

In the presence of sulfuric acid, acetic anhydride cleaves the Si–O–C bond in organosilicon esters of hydroxymethylphosphinic acid [136, 179]:

$$(RO)_2P(O)CH_2OSiR'_3 + (CH_3CO)_2O \xrightarrow{H_2SO_4} (RO)_2P(O)CH_2OCOCH_3 + R'_3SiOCOCH_3 \qquad (1.268)$$

When organosilicon esters of hydroxyalkylhaloalkylphosphinic acids are heated with excess of a triorganohalosilane there is the possibility of both transsilylation and the exchange of halogen atoms in accordance with the scheme [57, 153]

$$(R_3SiO)_2P(O)CH_2OCH_2CH(CH_2X)OSiR_3 + 3YSiR'_3 \longrightarrow$$
$$\longrightarrow 2YSiR_3 + XSiR_3 + (R'_3SiO)_2P(O)CH_2OCH_2CH(CH_2Y)OSiR'_3 \qquad (1.269)$$
$$(X{=}Cl;\ Y{=}Cl \text{ or } Br)$$

6.3. Practical Application

Liquid monomeric and polymeric compounds containing the groupings $Si-(C)_n-O-P$, $Si-(C)_n-S-P$, and $Si-(C)_n-N-P$ are recommended as additives for lubricants to increase the wear resistance [70, 136, 290, 332, 426], and as hydraulic fluids [232, 426]. Thus, for example, it is reported that the addition to liquid polydimethylsiloxane and polymethylphenylsiloxanes of 0.5-3% of polymers containing one $-YP(S)(OR)_2$ (Y = O or S) per 2-22 silicon atoms reduces the wear coefficient of steel on steel by a factor of 2-2.5 at temperatures from room temperature to 350°C. The use of siloxanes with terminal phosphonoxy or thiophosphonoxy groups is particularly effective [70]. Polysiloxanes obtained from $(CH_3)_2(C_2H_5O)Si(CH_2)_3N(CH_3)P(OC_2H_5)_2$ have been suggested for producing coatings which are resistant to the action of flames and solvents [331]. The products from cleavage of epoxysilanes by phosphorus acids [463] are proposed as base coatings for steel surfaces to improve the adhesion of siloxane rubbers to metals and also as catalysts for hardening epoxy, melamine and urea resins. According to data in [16], compounds containing methacrylate groups may be used to obtain phosphorus- and silicon-containing organic polymers.

7. COMPLEX COMPOUNDS CONTAINING Si AND P ATOMS

Compressing a mixture of SiF_4 with PH_3 to 50 atm (excess) and cooling to −22°C gave white crystals to which was assigned the composition $3SiF_4 \cdot 2PH_3$. The adduct $SiF_4 \cdot 2PH_3$ has also been obtained [259]. However, subsequent radiospectroscopic investigations did not confirm the existence of this adduct [313]. At −50°C $SiCl_4$ absorbs phosphine with the formation of a liquid adduct, for which the composition $SiCl_4 \cdot 2PH_3$ has been proposed [260]. The reaction of SiX_4 with PR_3 (R = C_2H_5, C_3H_7, C_6H_5, or cyclohexyl) in ether in a nitrogen atmosphere forms precipitates − instantaneous when X = Br and gradually when X = Cl. Then, regardless of the starting ratio of the reagents, the composition of the adducts was found to be the same, namely $SiX_4 \cdot 2PR_3$. The only exception is the adduct $SiBr_4 \cdot 4P(C_6H_5)_3$ [416]. However, when this work was repeated the formation of the adduct $SiCl_4 \cdot$

$2P(C_6H_{11})_3$ was not confirmed [3, 252, 254]. In anhydrous solvents in the absence of oxygen, $SiCl_4$ does not react with tricyclohexylphosphine. The determination of the molecular weight of these components (in benzene) with a molar ratio of 1:2 indicates the complete absence of interaction between them [254].

Iodosilane forms 1:1 adducts with methylphosphine and dimethylphosphine [550]. Diethylphosphine gives 2:1 adducts with trichlorosilane and $SiCl_4$ [643].

With SiX_4, triphenylphosphine oxide forms adducts with the composition $SiX_4 \cdot 2(C_6H_5)_3PO$ (X = F, Cl) and $SiBr_4 \cdot 4(C_6H_5)_3PO$ [255, 340, 416]. Conductimetric investigations show that the compounds $SiBr_4 \cdot 4(C_6H_5)_3PO$, $SiBr_2(OClO_3)_2 \cdot 4(C_6H_5)_3PO$ and $Si(OClO_3)_4 \cdot 4(C_6H_5)_3PO$ contain the ion $\{[(C_6H_5)_3PO]_4Si\}^{4+}$, while the compound $SiCl_4 \cdot 4(CH_3)_3PO$ contains the ion cis-$\{[(CH_3)_3PO]_4SiCl_2\}^{2+}$ [255]. Trimethyliodosilane and trimethylsilyl perchlorate react with trimethylphosphine oxide to form the adducts $(CH_3)_3SiI \cdot (CH_3)_3PO$ and $(CH_3)_3SiOClO_3 \cdot (CH_3)_3PO$ [253].

When the ethyl ester of diethylphosphinous acid is mixed with trialkylchlorosilanes at 0°C, the temperature rises to 5-27°C and a voluminous white precipitate is formed. The authors of [144, 200] explain this by the formation of complexes through a donor-acceptor interaction of the unshared pair of electrons of phosphorus or oxygen with vacant 3d-orbitals of the silicon atom.

No appreciable interaction between the components in solution was observed in the system $SiCl_4-PCl_3$, $SiCl_4-POCl_3$, and $SiCl_4-PCl_5$ [38, 134, 146, 395, 512].

In a discussion of the mechanism of interaction of tetraethoxysilane with phosphorus tribromide, the hypothesis was put forward [494, 495, 499] that the first stage of the reaction is the formation of the complex $(C_2H_5O)_4Si \cdot PBr_3$.

In the reaction of methyltriphenylphosphonium iodide with the complex compound $Li[(C_6H_5)_3BSi(C_6H_5)_3]$ there is formed a salt-like substance $[CH_3(C_6H_5)_3P][(C_6H_5)_3BSi(C_6H_5)_3]$, which decomposes above 200°C [537].

Phosphorus-containing complexes, in which the silicon is attached to an atom of a Group VIII metal, were obtained by the schemes [283, 287, 386, 623, 631, 632, 649, 650, 687]

$$\geq SiH + [(C_6H_5)_3P]_2Ir(CO)Cl \longrightarrow [(C_6H_5)_3P]_2IrClH(CO)Si \leq \qquad (1.270)$$

$$(C_6H_5)_2CH_3SiLi + [(CH_3)_2C_6H_5]_2PtCl_2 \longrightarrow [(CH_3)_2C_6H_5P]_2Pt[SiCH_3(C_6H_5)_2]_2 \qquad (1.271a)$$

$$R_3SiLi + (R'_3P)_2PtCl_2 \rightarrow (R'_3P)_2Pt(SiR)_2 \qquad (1.271b)$$

$$[(CH_3)_3Si]_2Hg + [(C_2H_5)_3P]_2PtCl_2 \longrightarrow (CH_3)_3SiPtCl[P(C_2H_5)_3]_2 + Hg + (CH_3)_3SiCl \qquad (1.272)$$

Hydrosilanes contaning electronegative substituents [$HSiCl_3$, $C_2H_5SiHCl_2$, $(C_2H_5O)_3SiH$] will undergo reaction (1.270). Trialkylsilanes and triarylsilanes do not react analogously. Octahedral complexes of Ir(III) are formed at room temperature and do not decompose up to 150-200°C [231].

Complexes of organosilicon phosphazenes with triorganoaluminanes [518-522] and aluminum halides [523] have already been examined in Sections 3.2 and 3.3; complexes of silylphosphines with boranes [303, 441, 442, 482, 483, 490] and boron halides [482, 483, 506, 507] have been examined in Section 4.3; and complexes of organosilicon ylides of phosphorus with trialkylborines [466] were examined in Section 5.3. See Sections 4.1, 4.3, 5.1, and 5.3 for information on organosilicon phosphonium salts [244-246, 269, 270, 272, 304, 385, 394, 465, 466, 517, 533, 534, 536, 538, 539, 542].

LITERATURE CITED IN CHAPTER ONE

1. Publications of Russian Authors

1. S. V. Aver'yanov, I. Ya. Poddubnyi, L. A. Aver'yanova, and Yu. V. Trenko, Kauchuk i Rezina, 22:1 (1963).
2. M. I. Aliev, I. A. Shikhiev, S. A. Balezin, S. Z. Israfilova, and N. I. Podobaev, Azerb. Khim. Zh., No. 5, p. 44 (1965).
3. N. M. Alpat'eva and Yu. M. Kessler, Zh. Strukt. Khim., 5:332 (1964).
4. K. A. Andrianov, Usp. Khim., 26:895 (1957).
5. K. A. Andrianov, Usp. Khim., 27:1257 (1958).
6. K. A. Andrianov, in: The Chemistry and Practical Use of Organosilicon Compounds, No. 2, TsBTI, Leningrad (1958), p. 3.
7. K. A. Andrianov, Organosilicon Compounds, Goskhimizdat (1956).
8. K. A. Andrianov, Polymers with Inorganic Main Molecular Chains, Izd. AN SSSR (1962).
9. K. A. Andrianov, Dokl. Akad. Nauk SSSR, 151:1093 (1963).
10. K. A. Andrianov, V. V. Astakhin, and I. V. Sukhanova, Izv. Akad. Nauk SSSR, Otdel Khim. Nauk, p. 1478 (1962).

11. K. A. Andrianov, N. V. Varlamova, M. F. Borisov, A. G. Kolchina, and G. V. Grebenshchikova, Plast. Massy, No. 3, p. 33 (1966).
12. K. A. Andrianov, T. V. Vasil'eva, and L. V. Kozlova, Izv. Akad. Nauk SSSR, Ser. Khim., p. 381 (1965).
13. K. A. Andrianov, T. V. Vasil'eva, and A. A. Minaeva, Izv. Akad. Nauk SSSR, Ser. Khim., p. 2227 (1963).
14. K. A. Andrianov, T. V. Vasil'eva, and L. M. Khananashvili, Izv. Akad. Nauk SSSR, Otdel Khim. Nauk, p. 1030 (1961).
15. K. A. Andrianov, T. V. Vasil'eva, and L. M. Khananashvili, Vysokomol. Soed., 4:708 (1962).
16. K. A. Andrianov, A. K. Dabagova, and I. K. Kuznetsova, Izv. Akad. Nauk SSSR, Otdel. Khim. Nauk, p. 1664 (1962).
17. K. A. Andrianov and A. A. Zhdanov, Abstracts of Reports to the International Symposium on High Molecular Compounds, Prague, p. 13 (1957).
18. K. A. Andrianov, A. A. Zhdanov, and A. A. Kazakova, Izv. Akad. Nauk SSSR, Otdel Khim. Nauk, p. 466 (1959).
19. K. A. Andrianov, A. A. Zhdanov, and A. A. Kazakova, Zh. Obshch. Khim., 29:1281 (1959).
20. K. A. Andrianov and A. A. Kazakova, Plast. Massy, No. 3, p. 24 (1963).
21. K. A. Andrianov and I. K. Kuznetsova, Izv. Akad. Nauk SSSR, Otdel Khim. Nauk, p. 1454 (1961).
22. K. A. Andrianov and I. K. Kuznetsova, Izv. Akad. Nauk SSSR, Otdel Khim. Nauk, p. 1792 (1961).
23. K. A. Andrianov and I. K. Kuznetsova, Izv. Akad. Nauk SSSR, Otdel Khim. Nauk, p. 456 (1962).
24. K. A. Andrianov, I. K. Kuznetsova, and M. N. Ermakova, Izv. Akad. Nauk SSSR, Ser. Khim., p. 454 (1964).
25. K. A. Andrianov, I. K. Kuznetsova, and I. Pakhomova, Izv. Akad. Nauk SSSR, Otdel Khim. Nauk, p. 500 (1963).
26. K. A. Andrianov, I. K. Kuznetsova, and Yu. N. Smirnov, Izv. Akad. Nauk SSSR, Neorg. Mater., 1:301 (1965).
27. K. A. Andrianov, N. A. Kurasheva, I. K. Kuznetsova, and É. I. Gerkhardt, Dokl. Akad. Nauk SSSR, 140:365 (1961).
28. K. A. Andrianov, G. I. Marfenkova, L. M. Khananashvili, and A. S. Shapatin, Vysokomol. Soed., 5:1552 (1963).
29. K. A. Andrianov, B. N. Rutovskii, and A. A. Kazakova, Author's Cert. 99821, 1955; Byull. Izobr., No. 1, p. 9 (1955).
30. K. A. Andrianov, B. N. Rutovskii, and A. A. Kazakova, Zh. Obshch. Khim., 26:267 (1956).
31. K. A. Andrianov, L. M. Khananashvili, and T. V. Vasil'eva, Zh. Analit. Khim., 16:739 (1961).
32. E. S. Andronov, N. F. Orlov, and V. P. Mileshkevich, Abstracts of the Twenty-Second Conference of the Students' Scientific Society of the Leningrad Institute of Textile and Light Industries, Chemical Fibers Section (1965).
33. B. A. Arbuzov and N. P. Bogonostseva, Izv. Akad. Nauk SSSR, Otdel Khim. Nauk, p. 484 (1953).

34. B. A. Arbuzov and N. P. Grechkin, Izv. Akad. Nauk SSSR, Otdel Khim. Nauk, p. 440 (1956).
35. B. A. Arbuzov and A. N. Pudovik, Zh. Obshch. Khim., 17:2158 (1947).
36. B. A. Arbuzov and A. N. Pudovik, Dokl. Akad. Nauk SSSR, 59:1433 (1948).
37. L. D. Balashova, A. B. Bruker, and L. Z. Soborovskii, Zh. Obshch. Khim., 36:73 (1966).
38. A. S. Barabanova and B. A. Voitovich, Ukr. Khim. Zh., 31:352 (1964).
39. G. F. Bebikh, Phosphorylation of High Molecular Compounds and Phenol Ethers with Phosphorus Pentasulfide, Author's abstract of Cand. Dissertation, MITKhT, Moscow (1962).
40. G. F. Bebikh, R. A. Pentin, and T. V. Ershova, Zh. Obshch. Khim., 33:3544 (1963).
40a. A. S. Berezhnoi, Silicon and Its Binary, Izd. AN USSR, Kiev (1958).
41. S. N. Borisov, Usp. Khim., 28:63 (1959).
42. S. N. Borisov, M. G. Voronkov, and É. Ya. Lukevits, Heteroorganic Silicon Compounds [in Russian], Izd. "Khimiya," (1966) [English translation: S. N. Borisov, M. G. Voronkov, and É. Ya. Lukevits, Organosilicon Heteropolymers and Heterocompounds, Plenum Press, New York (1969)].
43. V. A. Bork and L. A. Shvyrkova, Abstracts of Reports to the Scientific and Technical Conference of MKhTI, p. 40 (1958).
44. A. B. Bruker, M. D. Balashova, and L. Z. Soborovskii, Dokl. Akad. Nauk SSSR, 135:843 (1960).
45. A. B. Bruker, L. D. Balashova, and L. Z. Soborovskii, Zh. Obshch. Khim., 36:75 (1966).
46. E. F. Bugerenko, Synthesis and Conversions of Silicon-Containing Organophosphorus Compounds, Author's abstract of Cand. Dissertation, IOKh, Moscow (1963).
47. E. F. Bugerenko, in: Synthesis and Properties of Monomers, Izd. "Nauka" (1964), p. 145.
48. E. F. Bugerenko, Proceedings of the Conference on Organosilicon Compounds, No. 1, NIITÉKh, p. 78 (1966).
49. E. F. Bugerenko, Author's cert. 184269 (1966); Izobr., Prom. Obr., Tov. Zn., No. 15, 31 (1966); C. A., 66:11049 (1967).
50. E. F. Bugerenko, E. A. Chernyshev, and A. D. Petrov, Dokl. Akad. Nauk SSSR, 143:840 (1962).
51. E. F. Bugerenko, E. A. Chernyshev, and A. D. Petrov, Izv. Akad. Nauk SSSR, Ser. Khim., p. 286 (1965).
52. E. F. Bugerenko, E. A. Chernyshev, and E. M. Popov, Izv. Akad. Nauk SSSR, Ser. Khim., p. 1391 (1966).
53. V. M. Vainburg, N. F. Orlov, and V. P. Mileshkevich, Abstracts of Reports to the Twenty-Second Conference of the Students' Scientific Society of the Leningrad Institute of Textile and Light Industries, Section on the Chemistry and Technology of Fibrous Materials (1965).
54. N. V. Varlamova, K. A. Andrianov, V. V. Severnyi, A. G. Kolchina, and F. N. Vishnevskii, Proceedings of the Conference on Organosilicon Compounds, No. 4, NIITÉKh, p. 86 (1966).

55. T. V. Vasil'eva, The Synthesis of Cyclic Silicon-Containing Organophosphorus and Some Other Compounds and Their Properties, Author's abstract of Cand. Dissertation, MITKhT, Moscow (1967).
56. T. V. Vasil'eva and K. A. Andrianov, Abstracts of Reports to the Scientific and Technical Conference of the Lomonosov MITKhT, p. 54 (1964).
57. O. F. Viktorov, Reaction of Organic and Organophosphorus α-Oxides with Organohalosilanes, Author's abstract of Cand. dissertation, LITLP, Leningrad (1967).
58. L. N. Volodina, Zh. Obshch. Khim., 37:513 (1967).
59. L. N. Volodina, Zh. Obshch. Khim., 37:1842 (1967).
60. N. I. Volchinskaya, A. I. Dintses, S. S. Khasinevich, and Z. M. Stolyar, Proceedings of the Conference on Organosilicon Compounds, No. 4, NIITÉKh, p. 29 (1966).
61. M. G. Voronkov, Zh. Obshch. Khim., 25:469 (1955).
62. M. G. Voronkov, Heterolytic Cleavage Reactions of the Siloxane Bond, Report of Scientific Works, presented instead of doctor's dissertation, INKhS, Moscow (1961).
62a. M. G. Voronkov and V. N. Zgonnik, Zh. Obshch. Khim., 27:1483 (1957).
63. M. G. Voronkov, V. A. Kolesova, and V. N. Zgonnik, Izv. Akad. Nauk SSSR, Otdel Khim. Nauk, p. 1363 (1957).
64. M. G. Voronkov, N. F. Orlov, B. L. Kaufman, and V. A. Pestunovich, Zh. Obshch. Khim., 37:2065 (1967).
65. M. G. Voronkov and Yu. I. Skorik, Izv. Akad. Nauk SSSR, Otdel Khim. Nauk, p. 119 (1958).
66. M. G. Voronkov and Yu. I. Skorik, Zh. Obshch. Khim., 35:106 (1965).
67. M. G. Voronkov and Yu. I. Khudobin, Zh. Obshch. Khim., 26:584 (1956).
68. G. F. Gavrilin and B. A. Vovsi, Authors' cert. 168443, 1965; Byull. Izobr., No. 4, p. 62 (1965); C. A., 63:706 (1965).
69. M. L. Galashina, M. V. Sobolevskii, and T. P. Alekseeva, Plast. Massy, No. 1, p. 18 (1965).
70. M. L. Galashina, M. V. Sobolevskii, and K. A. Andrianov, Authors' cert. 140061, 1960; Byull. Izobr., No. 15, p. 19 (1961).
71. M. L. Galashina, M. V. Sobolevskii, K. A. Andrianov, and T. P. Alekseeva, Plast. Massy, No. 4, p. 16 (1962).
72. M. L. Galashina, M. V. Sobolevskii, G. V. Kaznina, and T. P. Alekseeva, Authors' cert. 178988, 1966; Izobr., Prom. Obr., Tov. Zn., No. 4, p. 70 (1966); C. A., 65:2374 (1966).
73. M. L. Galashina, M. V. Sobolevskii, D. Z. Levina, and T. P. Alekseeva, Plast. Massy, No. 8, p. 16 (1964).
74. M. L. Galashina, M. V. Sobolevskii, L. A. Chistyakova, D. Z. Levina-Shteinblat, and T. P. Alekseeva, Authors' cert. 148477, 1962; Byull. Izobr., No. 13, p. 25 (1962); C. A., 58:3558 (1963).
75. V. A. Ginsburg and A. Ya. Yakubovich, Zh. Obshch. Khim., 28:728 (1958).
76. B. M. Gladshtein, I. P. Kulyulin, and L. Z. Soborovskii, Zh. Obshch. Khim., 36:488 (1966).

77. N. I. Gludina, Photocolorimetric Determination of Silicon in Silicates and Organosilicon Compounds, Author's abstract of Cand. dissertation, MKhTI, Moscow (1955).
78. V. N. Gruber, A. L. Klebanskii, T. G. Degteva, A. S. Kuz'minskii, T. A. Mikhailova, and E. V. Kuz'mina, Vysokomol. Soed., 7:462 (1965).
79. T. G. Degteva, V. N. Gruber, and A. S. Kuz'minskii, Kauchuk i Rezina, 24(5):1 (1965).
80. V. A. Drozdov, G. S. Karetnikov, and I. Yu. Orlova, Zh. Fiz. Khim., 40:695 (1966).
81. V. A. Drozdov and R. R. Tarasyants, Trudy MKhTI, 44:143 (1963).
82. V. A. Drozdov, R. R. Tarasyants, E. G. Vlasova, and Z. A. Kubyak, Izv. Vyzov. Khim. i Khim. Tekhnol., 6:960 (1963).
83. S. M. Zhivukhin, V. V. Kireev, and V. B. Tolstoguzov, Authors' cert. 158415, 1962; Byull. Izobr., No. 21, p. 53 (1963); C. A., 60:10821 (1964).
84. S. M. Zhivukhin and V. B. Tolstoguzov, Authors' cert. 141298, 1964; Byull. Izobr., No. 18, p. 42 (1961).
85. S. M. Zhivukhin and V. B. Tolstoguzov, Plast. Massy, No. 5, p. 24 (1963).
86. S. M. Zhivukhin, V. B. Tolstoguzov, and S. I. Belykh, Zh. Neorgan. Khim., 9:134 (1964).
87. S. M. Zhivukhin, V. B. Tolstoguzov, and A. I. Ivanov, Authors' cert. 149568, 1962; Byull. Izobr., No. 16, p. 47 (1962).
88. S. M. Zhivukhin, V. B. Tolstoguzov, and A. I. Ivanov, Zh. Neorgan. Khim., 7:2192 (1962).
89. L. N. Zhinkina, V. V. Severnyi, and T. F. Altukhova, Proceedings of the Conference on Organosilicon Compounds, No. 3, NIITÉKh, p. 77 (1967).
90. V. A. Zamyatina, A. I. Solomatina, and V. V. Korshak, Proceedings of the Conference on Organosilicon Compounds, No. 3, NIITÉKh, p. 110 (1967).
91. D. A. Karateev, Abstracts of Reports to the Scientific and Technical Conference of MKhTI, p. 41 (1957).
91a. D. A. Karateev, Abstracts of Reports to the Scientific and Technical Conference of MKhTI, p. 29 (1958).
92. D. A. Karateev, Trudy MKhTI, 29:114 (1959).
93. D. A. Karateev, Reactions of Some Organosilicon Compounds with Inorganic Phosphorus Compounds, Author's abstract of Cand. dissertation, MITKhT, Moscow (1960).
94. B. L. Kaufman, Investigation of the Synthesis of Compounds with Phosphosiloxane Bonds from Derivatives of Trivalent Phosphorus, Author's abstract of Cand. dissertation, LITLP, Leningrad (1966).
95. N. V. Komarov and O. G. Yarosh, Authors' cert. 166337, 1963; Byull. Izobr., No. 22, p. 20 (1964); C. A., 62:10460 (1965).
96. N. V. Komarov and O. G. Yarosh, Zh. Obshch. Khim., 36:101 (1966).
97. A. P. Kreshkov, V. A. Bork, E. A. Bondarevskaya, L. V. Myshlyaeva, S. V. Syavtsillo, and V. T. Shemyatenkova, Practical Handbook on the Analysis of Organosilicon Compounds, Goskhimizdat (1962).
98. A. P. Kreshkov, V. A. Drozdov, and I. Yu. Orlova, Authors' cert. 173228, 1965; Byull. Izobr., No. 15, p. 31 (1965); C. A., 64:755 (1966).

99. A. P. Kreshkov, V. A. Drozdov, and I. Yu. Orlova, Zh. Analit. Khim., 21:214 (1966).
100. A. P. Kreshkov, V. A. Drozdov, and I. Yu. Orlova, Zh. Obshch. Khim., 36:307 (1966).
101. A. P. Kreshkov, V. A. Drozdov, and I. Yu. Orlova, Zh. Obshch. Khim., 36:525 (1966).
102. A. P. Kreshkov, V. A. Drozdov, and I. Yu. Orlova, Zh. Obshch. Khim., 36:2014 (1966).
103. A. P. Kreshkov, V. A. Drozdov, and R. R. Tarasyants, Plast. Massy, No. 4, p. 57 (1963).
104. A. P. Kreshkov and D. A. Karateev, Trudy MKhTI, 25:116 (1957).
105. A. P. Kreshkov and D. A. Karateev, Zh. Priklad. Khim., 30:1416 (1957).
106. A. P. Kreshkov and D. A. Karateev, Zh. Obshch. Khim., 27:2715 (1957).
107. A. P. Kreshkov and D. A. Karateev, Zh. Priklad. Khim., 32:369 (1959).
108. A. P. Kreshkov and D. A. Karateev, Zh. Obshch. Khim., 29:4082 (1959).
109. A. P. Kreshkov and D. A. Karateev, Authors' cert. 127262, 1960; Byull. Izobr., No. 7, p. 17 (1960); C. A., 54:20309 (1960).
110. A. P. Kreshkov and D. A. Karateev, Zh. Priklad. Khim., 33:413 (1960).
111. A. P. Kreshkov and D. A. Karateev, in: Chemistry and Use of Organophosphorus Compounds, Proceedings of Second Conference, Izd. AN SSSR (1962), p. 324.
112. A. P. Kreshkov, D. A. Karateev, and V. Fyurst, Authors' cert. 140799 1961; Byull. Izobr., No. 17, p. 18 (1961).
113. A. P. Kreshkov, D. A. Karateev, and V. Fyurst, Zh. Priklad. Khim., 34:2711 (1961).
114. A. P. Kreshkov, D. A. Karateev, and V. Fyurst, Plast. Massy, No. 3, p. 63 (1962).
115. A. P. Kreshkov, D. A. Karateev, V. Fyurst, and É. N. Pavlova, Zh. Obshch. Khim., 33:261 (1963).
116. A. P. Kreshkov, Yu. Ya. Mikhailenko, and E. A. Kuchkarev, Vestn. Tekhn. Ékon. Inform., No. 12, p. 28 (1964); C. A., 64:5746 (1966).
117. A. P. Kreshkov, V. I. Khramova, and D. A. Karateev, Authors' cert. 113143, 1958; Byull. Izobr., No. 5, p. 167 (1958).
118. A. P. Kreshkov, V. I. Khramova, and D. A. Karateev, Trudy MKhTI, 27:306 (1959).
119. S. P. Krukovskii, "Organosilicon Polymers," Ch. 5 in Scientific Results. The Chemical Sciences, The Chemistry and Technology of Synthetic High-Molecular Compounds, Heterochain Polymers, Vol. 8, Edit. V. V. Korshak, Izd. "Nauka" (1966), p. 542.
120. V. K. Kuskov and G. F. Bebikh, Dokl. Akad. Nauk SSSR, 136:354 (1961).
121. N. S. Leznov, L. A. Sabun, and K. A. Andrianov, Zh. Obshch. Khim., 29:1276 (1959).
122. V. D. Lobkov, A. L. Klebanskii, and É. V. Kogan, Vysokomol. Soed., 7:163 (1965).
123. V. D. Lobkov, A. L. Klebanskii, and É. V. Kogan, Vysokomol. Soed., 7:290 (1965).

124. V. D. Lobkov, A. L. Klebanskii, and É. V. Kogan, Vysokomol. Soed., 7:1535 (1965).
125. V. D. Lobkov, A. L. Klebanskii, and É. V. Kogan, Kauchuk i Rezina, No. 2, p. 1 (1965).
126. V. D. Lobkov, A. L. Klebanskii, and É. V. Kogan, Scientific Communications of the International Symposium on Organosilicon Chemistry, A/36-178, Prague (1965).
127. V. D. Lobkov, A. L. Klebanskii, and É. V. Kogan, Vysokomol. Soed., 9A:1099 (1967).
128. B. M. Luskina, A. P. Terent'ev, and N. A. Gradskova, Zh. Analit. Khim., 19:1251 (1964).
129. B. M. Luskina, A. P. Terent'ev, and N. A. Gradskova, Zh. Analit. Khim., 20:990 (1965).
130. B. M. Luskina, A. P. Terent'ev, and N. A. Gradskova, Zh. Analit. Khim., 20:2743 (1966).
131. B. M. Luskina, A. P. Terent'ev, and S. V. Syavtsillo, Zh. Analit. Khim., 17:639 (1962).
132. B. M. Luskina, A. P. Terent'ev, and S. V. Syavtsillo, Trudy komissii po analiticheskoi khimii, 13:3 (1963).
133. S. D. Lyukas, N. P. Smetankina, and V. P. Kuznetsova, Zh. Obshch. Khim., 36:2003 (1966).
134. B. F. Markov, B. A. Voitovich, and A. S. Barabanova, Zh. Neorgan. Khim., 6:1204 (1961).
135. D. I. Mendeleev, Khim. Zh., 4:65 (1860); Papers, Vol. 5, Izd. AN SSSR, p. 40 (1947).
136. V. P. Mileshkevich, Synthesis and Properties of Organosilicon Derivatives of α-Hydroxy- and α-Chloroalkylphosphinic Acids, Author's abstract of Cand. dissertation, LITLP; Leningrad (1965).
137. V. P. Mileshkevich, V. I. Troenko, and N. V. Suikovskaya, Author's cert. 159614, 1963; Byull. Izobr., No. 1, p. 35 (1964).
138. E. I. Minsker, R. R. Tarasyants, N. F. Orlov, and V. V. Severnyi, Proceedings of the Conference on Organosilicon Compounds, No. 4, NITÉKh, p. 39 (1966).
139. V. F. Mironov and N. A. Pogonkina, Izv. Akad. Nauk SSSR, Otdel Khim. Nauk, p. 85 (1959).
140. R. A. Mogilevskaya and N. P. Guseva, VNIISK Report 2103 (1964).
141. R. A. Mogilevskaya and G. S. Roslavtseva, VNIISK Report 2148 (1965).
142. R. A. Mogilevskaya and V. F. Smirnova, VNIISK Report 1730 (1961).
143. N. N. Motovilova and E. I. Starovoitova, Kauchuk i Rezina, 25(5):4 (1966).
144. A. A. Muratova, Reactions of Esters of Some Phosphorus Acids with Organic and Heteroorganic Halogen Derivatives, Author's abstract of Cand. dissertation, KGU, Kazan' (1964).
145. N. S. Nametkin, A. V. Topchiev, and L. I. Kartasheva, Dokl. Akad. Nauk SSSR, 93:667 (1953).
146. L. A. Nisen'son and G. V. Ceryakov, Zh. Neorgan. Khim., 5:1139 (1960).
147. M. I. Nosov and G. V. Vinogradov, Khim. i Tekh. Topl. i Masel, 9(8):50 (1964).

148. N. F. Orlov, Synthesis of Triorganosiloxy Derivatives of Group III, IV, and V Elements from Triorganosilanols and Hexaalkyldisiloxanes, Author's abstract of Cand. dissertation, LGU, Leningrad (1960).
149. N. F. Orlov, Proceedings of the Conference on the Chemistry and Practical Use of Organosilicon Compounds, Leningrad, p. 10 (1966).
150. N. F. Orlov, 3 Internationales Symposium über Metallorganische Chemie, München (1967), S. 20.
151. N. F. Orlov and O. F. Viktorov, Authors' cert. 178802, 1966; Byull. Izobr., No. 4, p. 17 (1966); C. A., 65:2296 (1966).
152. N. F. Orlov and O. F. Viktorov, Material of the Conference on the Chemistry and Practical Use of Organosilicon Compounds, Leningrad, p. 10 (1966).
153. N. F. Orlov and O. F. Viktorov, Proceedings of the Conference on Organosilicon Compounds, No. 1, NIITÉKh, p. 70 (1966).
154. N. F. Orlov and O. F. Viktorov, Authors' cert. 192207, 1967; Izobr., Prom. Obr., Tov. Zn., No. 5, p. 38 (1967).
155. N. F. Orlov and L. N. Volodina, Authors' cert. 172327, 1965; Byull. Izobr., No. 13, p. 20 (1965); C. A., 63:16383 (1965).
156. N. F. Orlov and L. N. Volodina, Authors' cert. 172787, 1965; Byull. Izobr., No. 14, p. 24 (1965); C. A., 64:755 (1966).
157. N. F. Orlov and L. N. Volodina, Zh. Obshch. Khim., 36:920 (1966).
158. N. F. Orlov and M. G. Voronkov, Zh. Obshch. Khim., 30:2223 (1960).
159. N. F. Orlov and M. G. Voronkov, Authors' cert., 136373, 1961; Byull. Izobr., No. 5, p. 25 (1961).
160. N. F. Orlov and M. G. Voronkov, Authors' cert. 140800, 1961; Byull. Izobr., No. 17, p. 18 (1961); C. A., 56:12946 (1962).
161. N. F. Orlov and M. G. Voronkov, in: The Chemistry and Use of Organophosphorus Compounds, Proceedings of the Second Conference, Izd. AN SSSR (1962), p. 212.
162. N. F. Orlov and M. G. Voronkov, Zh. Obshch. Khim., 32:608 (1962).
163. N. F. Orlov, B. N. Dolgov, and M. G. Voronkov, Proceedings of the Conference on the Chemistry and Practical Use of Organosilicon Compounds, No. 1, TsBTI LSNKh, p. 172 (1958).
164. N. F. Orlov and B. L. Kaufman, Authors' cert. 176585, 1965; Byull. Izobr., No. 23, p. 18 (1965); C. A., 64:9765 (1966).
165. N. F. Orlov and B. L. Kaufman, Authors' cert. 182146, 1966; Izobr., Prom. Obr., Tov. Zn., No. 11, p. 19 (1966); C. A., 65:17002 (1966).
166. N. F. Orlov and B. L. Kaufman, Izv. Akad. Nauk SSSR, Neorgan. Mater., 2:946 (1966).
167. N. F. Orlov and B. L. Kaufman, Zh. Obshch. Khim., 36:1155 (1966).
168. N. F. Orlov and B. L. Kaufman, Proceedings of the Conference on Organosilicon Compounds, No. 1, NIITÉKh, p. 74 (1966).
169. N. F. Orlov and B. L. Kaufman, Material of the Conference on the Chemistry and Practical Use of Organosilicon Compounds, Leningrad (1966), p. 10.
170. N. F. Orlov and B. L. Kaufman, Authors' cert. 193505, 1967; Izobr., Prom. Obr., Tov. Zn., No. 7, p. 36 (1967).
171. N. F. Orlov and B. L. Kaufman, Authors' cert. 193509, 1967; Izobr., Prom. Obr., Tov. Zn., No. 7, p. 37 (1967).

172. N. F. Orlov and B. L. Kaufman, Authors' cert., 196834, 1967; Izobr., Prom. Obr., Tov. Zn., No. 12, p. 34 (1967).
173. N. F. Orlov and V. P. Mileshkevich, Scientific Communications of the International Symposium on Organosilicon Chemistry, A/5-22, Prague (1965).
174. N. F. Orlov and V. P. Mileshkevich, Dokl. Akad. Nauk SSSR, 164:344 (1965).
175. N. F. Orlov and V. P. Mileshkevich, Authors' cert. 172789, 1965; Byull. Izobr., No. 14, p. 25 (1965); C. A., 64:2128 (1966).
176. N. F. Orlov and V. P. Mileshkevich, Authors' cert. 178376, 1966; Izobr., Prom. Obr., Tov. Zn., No. 3, p. 24 (1966); C. A., 64:19680 (1966).
177. N. F. Orlov and V. P. Mileshkevich, Authors' cert. 183207, 1966; Izobr., Prom. Obr., Tov. Zn., No. 13, p. 20 (1966); C. A., 65:18618 (1966).
178. N. F. Orlov and V. P. Mileshkevich, Zh. Obshch. Khim., 36:518 (1966).
179. N. F. Orlov and V. P. Mileshkevich, Zh. Obshch. Khim., 36:699 (1966).
180. N. F. Orlov and V. P. Mileshkevich, Zh. Obshch. Khim., 36:892 (1966).
181. N. F. Orlov, V. P. Mileshkevich, and E. S. Andronov, Authors' cert. 173761, 1965; Byull. Izobr., No. 16, p. 33 (1965); C. A., 64:3603 (1966).
182. N. F. Orlov, V. P. Mileshkevich, and E. S. Andronov, Zh. Obshch. Khim., 35:2193 (1965).
183. N. F. Orlov, V. P. Mileshkevich, and V. M. Vainburg, Authors' cert. 172788, 1965; Byull. Izobr., No. 14, p. 24 (1965); C. A., 64:757 (1966).
184. N. F. Orlov, V. P. Mileshkevich, and V. M. Vainburg, Zh. Obshch. Khim., 36:1075 (1966).
185. N. F. Orlov, V. P. Mileshkevich, and E. L. Gefter, Authors' cert. 166682, 1964; Byull. Izobr., No. 23, p. 21 (1964); C. A., 62:10460 (1965).
186. N. F. Orlov, V. P. Mileshkevich, and E. L. Gefter, Zh. Obshch. Khim., 35:590 (1965).
187. N. F. Orlov, V. P. Mileshkevich, and E. L. Gefter, Zh. Obshch. Khim., 35:1312 (1965).
188. N. F. Orlov and L. N. Slesar', Authors' cert. 186479 (1966); Izobr., Prom. Obr., Tov. Zn., No. 19, p. 33 (1966); C. A., 66:95182k (1967).
189. N. F. Orlov and L. N. Slesar', Material of the Conference on the Chemistry and Practical Use of Organosilicon Compounds, Leningrad (1966), p. 4.
190. A. D. Petrov, Bull. Soc. Chim. Fr., p. 1098 (1956).
191. A. D. Petrov, V. F. Mironov, V. G. Glukhovtsev, Dokl. Akad. Nauk SSSR, 93:499 (1953).
192. A. D. Petrov, V. F. Mironov, and V. G. Glukhovtsev, Zh. Obshch. Khim., 27:1535 (1957).
193. K. A. Petrov, F. L. Maklyaev, and N. K. Bliznyuk, Authors' cert. 124439, 1959; Byull. Izobr., No. 23, p. 16 (1959); C. A., 54:10861 (1960).
194. I. Ya. Poddubnyi and S. V. Aver'yanov, Abstracts of Reports presented at the Twentieth International Congress on Theoretical and Applied Chemistry, B56-184, Izd. "Nauka" (1965); Abbreviated English Translations of Scientific Papers Presented in Russian.
195. I. Ya. Poddubnyi and S. V. Aver'yanov, Vysokomol. Soed., 8:1549 (1966).

196. V. V. Ponomarev, A. S. Shapatin, and S. A. Golubtsov, Authors' cert. 184856, 1965; Izobr., Prom. Obr., Tov. Zn., No. 16, p. 33 (1966); C. A., 66:85855t (1967).
197. V. V. Ponomarev, A. S. Shapatin, and S. A. Golubtsov, Authors' cert. 186477, 1965; Izobr., Prom. Obr., Tov. Zn., No. 19, p. 32 (1966).
198. V. V. Ponomarev, A. S. Shapatin, and S. A. Golubtsov, Proceedings of the Conference on Organosilicon Compounds, No. 1, NIITÉKh, p. 63 (1966).
199. V. V. Ponomarev, A. S. Shapatin, and S. A. Golubtsov, Zh. Obshch. Khim., 36:364 (1966).
200. A. N. Pudovik and A. A. Muratova, Dokl. Akad. Nauk SSSR, 158:419 (1964).
201. S. I. Sadykh-zade, R. A. Sultanov, B. A. Mamedov, and D. A. Ashurov, Authors' cert. 172320, 1964; Byull. Izobr., No. 13, p. 19 (1965); C. A., 63:16385 (1965).
202. V. N. Sveshnikova and E. P. Danilova, Zh. Neorgan. Khim., 2:928 (1957).
203. G. B. Seifer, Zh. Neorgan. Khim., 7:2806 (1962).
204. A. S. Sergeev, Investigations of Polydimethylsiloxanes and Polyorganoheterodimethyl(methylphenyl)siloxanes and Rubbers Based on Them, Authors' abstract of Cand. dissertation, MITKhT, Moscow (1966).
205. V. I. Skorik, Study of the Reactions of Phosphorus Trihalides with Trialkylalkoxysilanes and Hexaalkyldisiloxanes. Synthesis of Tris(trialkylsilyl) Phosphites and Bis(trialkylsilyl) Alkylphosphonates, Diploma work, LGU (1957).
206. L. N. Slesar', Investigation of the Reaction of Triorganosilanes with Carboxylic and Phosphorus-Containing Acids and Their Organosilicon Derivatives, Author's abstract of Cand. dissertation, LITLP (1967).
207. L. Z. Soborovskii, B. M. Gladshtein, and I. P. Kulyulin, Authors' cert. 168694, 1965; Byull. Izobr., No. 5, p. 23 (1965).
208. I. K. Stavitskii, B. E. Neimark, and A. G. Zatekina, VNIISK Report (1948).
209. E. I. Starovoitova and N. N. Motovilova, Kauchuk i Rezina, 25(10):11 (1966).
210. F. P. Sudakov, V. I. Klitina, and N. T. Maslova, Vestn. MGU, Khim. Fak., No. 1, p. 98 (1966).
211. F. P. Sudakov, V. I. Klitina, and N. T. Maslova, Zh. Analit. Khim., 21:1089 (1966).
212. R. R. Tarasyants, Titration of Organosilicon Compounds in Nonaqueous Media, Author's abstract of Cand. dissertation, MKhTI, Moscow (1965).
213. V. B. Tolstoguzov, S. M. Zhivukhin, and L. M. Samorodova, Authors' cert. 140206, 1960; Byull. Izobr., No. 15, p. 40 (1961).
214. V. Fyurst, Reactions of Organosilicon Compounds with Inorganic Compounds of Boron and Phosphorus, Author's abstract of Cand. dissertation, MKhTI, Moscow (1963).
215. L. M. Khananashvili, A. N. Chivikova, A. P. Kreshkov, and M. L. Darashkevich, in: The Chemistry and Practical Use of Organosilicon Compounds, No. 6, Izd. AN SSSR (1961), p. 159.
216. I. Khaiduk, Zh. Obshch. Khim., 30:1395 (1960).
217. I. Khaiduk and K. A. Andrianov, Izv. Akad. Nauk SSSR, Otdel. Khim. Nauk, 1963:1537.

218. V. I. Khramova and A. P. Kreshkov, in: The Chemistry and Practical Use of Organosilicon Compounds, No. 4, TsBTI LSNKh (1958), p. 101.
219. V. I. Khramova and A. P. Kreshkov, in: The Chemistry and Practical Use of Organosilicon Compounds, No. 6, Izd. AN SSSR (1961), p. 342.
220. E. A. Chernyshev, Izv. Akad. Nauk SSSR, Otdel. Khim. Nauk, 1958:96.
221. E. A. Chernyshev and E. F. Bugerenko, Authors' cert. 148049, 1961; Byull. Izobr., No. 12, p. 26 (1962).
222. E. A. Chernyshev and E. F. Bugerenko, Izv. Akad. Nauk SSSR, Otdel. Khim. Nauk, 1963:769.
223. E. A. Chernyshev and E. F. Bugerenko, Organomet. Chem. Rev., 3A:469 (1968).
224. E. A. Chernyshev, E. F. Bugerenko, E. D. Lubuzh, and A. D. Petrov, Izv. Akad. Nauk SSSR, Otdel. Khim. Nauk, 1962:1001.
225. E. A. Chernyshev, E. F. Bugerenko, N. A. Nikolaeva, and A. D. Petrov, Dokl. Akad. Nauk SSSR, 147:117 (1962).
226. E. A. Chernyshev, E. F. Bugerenko, and A. D. Petrov, Dokl. Akad. Nauk SSSR, 148:875 (1963).
227. E. A. Chernyshev and A. D. Petrov, Dokl. Akad. Nauk SSSR, 105:282 (1955).
228. A. N. Shivikova, A. P. Kreshkov, M. D. Darashkevich, L. V. Mishlyaeva, D. A. Karateev, and L. M. Khananashvili, in: The Chemistry and Practical Use of Organosilicon Compounds, No. 1, TsBTI LSNKh (1958), p. 178.
229. M. N. Chumachenko and V. P. Burlaka, Izv. Akad. Nauk SSSR, Otdel. Khim. Nauk, 1963:5.
230. A. S. Shapatin, Some Reactions of Aluminum Alcoholates, Synthesis and Properties of Aluminum Derivatives of Phosphinic Acids and Polyorganophosphinatoaluminodimethylsiloxanes, Author's abstract of Cand. dissertation, MITKhT, Moscow (1965).
231. B. I. Yakovlev and N. V. Vinogradova, Zh. Obshch. Khim., 29:695 (1959).

2. Publications of Foreign Authors

232. A. D. Abbott, J. R. Wright, A. Goldschmidt, W. T. Stewart, and R. O. Bolt, J. Chem. Eng. Data, 6:437 (1961).
233. E. W. Abel, D. A. Armitage, and R. P. Bush, J. Chem. Soc., Suppl., 5584 (1964).
234. E. W. Abel, D. A. Armitage, and R. P. Bush, J. Chem. Soc., 1965:7098.
235. E. W. Abel and G. Willey, Proc. Chem. Soc., 1962:308.
236. M. C. Agens (General Electric Co.), USA Patent 2842515, 1958; C. A., 52:16780 (1958).
237. T. Alfrey, F. J. Hohn, and H. Mark, J. Polymer Sci., 1:102 (1946).
238. E. Amberger, Angew. Chem., 75:579 (1963).
239. E. Amberger and H. D. Boeters, Angew. Chem., 74:32 (1962).
240. E. Amberger and H. D. Boeters, Chem. Ber., 97:1999 (1964).
241. Anonymous, Ind. Eng. Chem., 51(9):I:37a (1959).
242. B. A. Ashby (General Electric Co.), British Patent 990657, 1965; C. A., 63:3076 (1965).

243. J. D. Austin, C. Eaborn, and J. D. Smith, J. Chem. Soc., 1963:4744.
244. B. J. Aylett, J. Inorg. Nucl. Chem., 15:87 (1960).
245. B. J. Aylett, H. J. Emelèus, and A. G. Maddock, Research, 6:30 (1953).
246. B. J. Aylett, H. J. Emelèus, and A. G. Maddock, J. Inorg. Nucl. Chem., 1:187 (1955).
247. B. J. Aylett and R. A. Sinclair, Proceedings of the Eighth International Conference on Coordination Chemistry, Vienna (1964); Springer Verlag, Vienna, New York, p. 166 (1964).
248. W. Bamford (Imperial Chemical Industires Ltd.), British Patent 933191, 1963; C. A., 59:14175 (1963).
249. W. Bamford (Imperial Chemical Industries Ltd.), British Patent 940547, 1963; C. A., 60:684 (1964).
250. G. N. Barnes and M. P. David, J. Org. Chem., 25:1191 (1960).
251. Bataafse Petroleum Maatschappij N. V., Federal Germ. Rep. 1084027, 1960; C. A., 55:18176 (1961).
251a. B. Beagley, A. G. Robiette, and G. M. Sheldrick, Chem. Comm., 1967:601.
252. I. R. Beattie, Quart. Rev., 17:382 (1963).
253. I. R. Beattie and F. W. Parrett, J. Chem. Soc., 1966(A):1784.
254. I. R. Beattie and M. Webster, J. Chem. Soc., 1963:4285.
255. I. R. Beattie and M. Webster, J. Chem. Soc., 1965:3672.
256. M. Becke-Goehring and H. Krill, Chem. Ber., 94:1059 (1961).
257. M. Becke-Goehring and G. Wunsch, Ann., 618:43 (1958).
258. M. Becke-Goehring and G. Wunsch, Chem. Ber., 93:326 (1960).
259. M. Besson, C. R., 110:80 (1890).
260. M. Besson, C. R., 110:240 (1890).
260a. W. Biltz, H. Hartmann, F. W. Wrigge, and F. W. Wiechman, Sitzber. Preuss. Akad. Wiss., Phys.-Mat. Kl., 7:99 (1938); C. A., 32:8976 (1938).
261. L. Birkofer, W. Gildenberg, and A. Ritter, Angew. Chem., 73:143 (1961).
262. L. Birkofer and S. M. Kim, Chem. Ber., 97:2100 (1964).
263. L. Birkofer, A. Ritter, and S. M. Kim, Chem. Ber., 96:3099 (1963).
264. L. Birkofer, A. Ritter, and R. Richter, Chem. Ber., 96:2750 (1963).
265. G. H. Birum and G. A. Richardson (Monsanto Chemical Co.), USA Patent 3113139, 1963; C. A., 60:5551 (1964).
266. B. A. Bluestein (Compagnie Francaise Thomson-Houston), French Patent 1377915, 1964; C. A., 63:5783 (1965).
267. B. A. Bluestein (Compagnie Francaise Thomson-Houston), French Patent 1379261, 1964; C. A., 62:12001 (1965).
268. B. A. Bluestein (Compagnie Francaise Thomson-Houston), French Patent 1379262, 1964; C. A., 62:12001 (1965).
269. R. W. Bott, B. F. Dowden, and C. Eaborn, Scientific Communications of the International Symposium on Organosilicon Chemistry, B/19-290, Prague (1965).
270. R. W. Bott, B. F. Dowden, and C. Eaborn, J. Chem. Soc., 1965:4994.
271. R. W. Bott, B. F. Dowden, and C. Eaborn, J. Chem. Soc., 1965:6306.
272. R. W. Bott, B. F. Dowden, and C. Eaborn, J. Organomet. Chem., 4:291 (1965).

273. A. Boulle and R. Jary, C. R., 237:161 (1953).
274. A. Boulle and R. Jary, C. R., 237:328 (1953).
275. British Thomson-Houston Co., Ltd., British Patent 629642, 1949; C. A., 44:2271 (1950).
276. J. L. Brooks and D. Williams (Imperial Chemical Industries Ltd.), British Patent 981675, 1965; C. A., 62:12026 (1965).
277. J. L. Brooks and D. Williams (Imperial Chemical Industries Ltd.), British Patent 986905, 1965; C. A., 63:1905 (1965).
278. A. H. Campbell, British Patent 609324, 1948; C. A., 43:2221 (1949).
279. A. E. Canavan, B. F. Dowden, and C. Eaborn, J. Chem. Soc., 1962:331.
280. A. E. Canavan and C. Eaborn, J. Chem. Soc., 1959:3751.
281. A. E. Canavan and C. Eaborn, J. Chem. Soc., 1962:592.
282. G. Chainani, W. Gerrard, J. K. Patel, and R. Twaits, J. Appl. Chem., 13:365 (1963).
283. A. J. Chalk and J. F. Harrod, J. Am. Chem. Soc., 87:16 (1965).
284. M. M. Chamberlain, AD 601 415, 1964; C. A., 62:1311 (1965).
285. M. M. Chamberlain, G. Kern, G. A. Jabs, D. Germanes, A. Greene, K. Brain, and B. Wayland, US Dept. Comm., Office Tech. Services, PB Rept. 152086 (1960); C. A., 58:2508 (1963).
286. E. Chang, Synthesis and Properties of Organic Compounds Containing Phosphorus and Silicon, Diss. Abstr., 24:2265 (1963).
287. J. Chatt, C. Eaborn, and S. Ibekwe, Chem. Comm., 1966:700.
288. S. Cradock, G. Davidson, E. A. V. Ebsworth, and L. A. Woodward, Chem. Comm., 1965:515.
289. K. O. Christe, Dissertation, Tech. Hochschule Stuttgart (1961); cited from [390].
290. G. D. Cooper (General Electric Co.), USA Patent 2811540, 1957; C. A., 52:2395 (1958).
291. B. R. Currel, M. J. Frazer, and W. Gerrard, J. Chem. Soc., 1960:2776.
292. M. P. David and G. N. Barnes, Abstracts of Papers to the 136th Meeting Am. Chem. Soc., Atlantic City, p. 104, P-167 (1959).
293. G. Davidson, E. A. V. Ebsworth, G. M. Sheldrick, and L. A. Woodward, Chem. Comm., 1965:122.
294. G. Davidson, E. A. V. Ebsworth, G. M. Sheldrick, and L. A. Woodward, Spectrochim. Acta, 22:67 (1966).
295. G. C. Demitras, R. A. Hent, and A. G. MacDiarmid, Chem. Ind., 1964:1712.
296. F. G. A. DeMonterey (General Electric Co.), USA Patent 3153006, 1964; C. A., 61:16260 (1964).
297. R. T. Dickerson (Dow Chemical Co.), USA Patent 3304270, 1967; C. A., 67:3388b (1967).
298. Dow Corning Corp., French Patent 1036694, 1953; R. Zh. Khim., 47500P (1955).
299. Dow Corning Corp., Netherlands Applic. 6401175, 1964; C. A., 62:10647 (1965).
300. Dow Corning Ltd., British Patent 687759, 1953; C. A., 1954:6365; Rubb. Abstr., 31(6):265 (1953).
301. J. E. Drake and W. L. Jolly, Chem. Ind., 1962:1470.

302. J. E. Drake and W. L. Jolly, J. Chem. Phys., 38:1033 (1963).
303. J. E. Drake and J. Simpson, Chem. Comm., 1967:249.
304. G. Drefahl and D. Lorenz, J. Prakt. Chem., 24:312 (1964).
305. N. Duffaut, R. Calas, and J. Dunogues, Bull. Soc. Chim. Fr., 1960:597.
306. N. Duffaut, R. Calas, and J. Dunogues, Bull. Soc. Chim. Fr., 1963:512.
307. E. Dumont, Federal Germ. Rep. 1042896, 1958; C. A., 55:2204 (1961).
308. E. Dumont, Gummi u. Asb., 11:764 (1958).
309. E. Dumont, Kunstst.-Plast., 6:30 (1959).
310. E. A. V. Ebsworth, Volatile Silicon Compounds, Pergamon Press, Oxford, London, New York, Paris (1963).
311. E. A. V. Ebsworth, Chem. Comm., 1966:530.
312. E. A. V. Ebsworth and G. M. Sheldrick, Trans. Far. Soc., 62:3282 (1966).
313. E. A. V. Ebsworth and R. M. Truscott, cited from [312].
314. H. Ellenhorst, USA Patent 2553643, 1951; C. A., 45:7817 (1951).
315. J. Ellerman and F. Poersch, Angew. Chem., 79:380 (1967); Inter. Ed., 6:355 (1967).
316. H. G. Emblem, C. E. Oxley, and S. A. Trow (Unilever Ltd.), British Patent 1052388, 1966.
317. H. J. Emelèus, Angew. Chem., 66:714 (1954).
318. Farbenfabriken Bayer, British Patent 706781, 1954; C., 1956:575; J. Appl. Chem., 4:538 (1954).
319. F. Feher and A. Blümcke, Chem. Ber., 90:1934 (1957).
320. F. Feher, G. Kühlbörsch, A. Blümcke, H. Keller, and K. Lippert, Chem. Ber., 90:134 (1957).
321. F. Feher and K. Lippert, Chem. Ber., 92:2998 (1959).
322. F. Feher and K. Lippert, Chem. Ber., 94:2437 (1961).
323. F. Fekete (Union Carbide Corp.), USA Patent 2920094, 1960; C. A., 54:24399 (1960).
324. F. Fekete (Union Carbide Corp.), USA Patent 2978471, 1961; C. A., 55:16482 (1961).
325. F. Fekete (Union Carbide Corp.), USA Patent 2995594, 1961; C. A., 56:3516 (1962).
326. F. Fekete (Union Carbide Corp.), USA Patent 2996530, 1961; C. A., 55:25806 (1961).
327. F. Fekete (Union Carbide Corp.), USA Patent 3019248, 1962; C. A., 57:11238 (1963).
328. F. Fekete (Union Carbide Corp.), USA Patent 3067229, 1962; C. A., 58:10239 (1963).
329. F. Fekete (Union Carbide Corp.), USA Patent 3202633, 1965; C. A., 63:18156 (1965).
330. F. Fekete (Union Carbide Corp.), USA Patent 3203923, 1965; C. A., 63:18157 (1965).
331. F. Fekete (Union Carbide Corp.), USA Patent 3203924, 1965; C. A., 64:3597 (1966).
332. F. Fekete (Union Carbide Corp.), USA Patent 3203925, 1965; C. A., 63:16384 (1965).
333. T. R. Fennel and J. R. Webb, Talanta, 2:389 (1959).

334. J. Fertig and W. Gerrard, Chem. Ind., 1956:83.
335. J. Fertig, W. Gerrard, and H. Herbst, J. Chem. Soc., 1957:1488.
336. A. E. Finholt, C. Helling, V. Imhof, L. Nielsen, and E. Jacboson, Inorg. Chem., 2:504 (1963).
337. W. Fink, Chem. Ber., 96:1071 (1963).
338. E. Fluck, Angew. Chem., 72:752 (1960).
339. M. J. Frazer, W. Gerrard, and A. P. Singh, J. Chem. Soc., 1961:4680.
340. M. J. Frazer, W. Gerrard, and R. Tvaits, J. Inorg. Nucl. Chem., 25:637 (1963).
341. C. Friedel and M. Crafts, Lieb. Ann., 127:28 (1863).
342. C. Friedel and A. Ladenburg, Ber., 3:15 (1870).
343. K. C. Frisch and H. Lyons, J. Am. Chem. Soc., 75:4078 (1953).
344. K. C. Frisch and H. Lyons (General Electric Co.), USA Patent 2673210, 1954; C. A., 49:4018 (1955).
345. G. Fritz, Z. Naturf., 8b:776 (1953).
346. G. Fritz, Z. Anorg. Allg. Chem., 280:332 (1955).
347. G. Fritz, Angew. Chem., 72:566 (1960).
348. G. Fritz, Fortschr. Chem. Forsch., 4:459 (1963).
349. G. Fritz, Angew. Chem., 76:994 (1964).
350. G. Fritz, Angew. Chem., 78:80 (1966).
351. G. Fritz and G. Becker, Z. Anorg. Allg. Chem., cited from [350].
352 G. Fritz and G. Becker, 3 Internationales Symposium über Metallorganische Chimie, München (1967).
353. G. Fritz, G. Becker, and D. Kummer, Z. Anorg. Allg. Chem., cited from [350].
354. G. Fritz and H. O. Berkenhoff, Z. Anorg. Allg. Chem., 289:250 (1957).
355. G. Fritz and H. O. Berkenhoff, Z. Anorg. Allg. Chem., 300:205 (1959).
356. G. Fritz and J. Grobe, Angew. Chem., 75:579 (1963).
357. G. Fritz and G. Poppenburg, Angew. Chem., 72:208 (1960).
358. G. Fritz and G. Poppenburg, Naturwiss., 49:449 (1962).
359. G. Fritz and G. Poppenburg, Angew. Chem., 75:297 (1963).
360. G. Fritz and G. Poppenburg, Z. Anorg. Allg. Chem., 331:147 (1964).
361. G. Fritz, G. Poppenburg, and M. G. Rocholl, Naturwiss., 49:255 (1962).
362. H. Fritzsche, U. Hasserodt, F. Korte, G. Friese, and K. Adrian, Chem. Ber., 98:171 (1965).
363. H. Fritzsche, U. Hasserodt, F. Korte, G. Friese, K. Adrian, and H. J. Arenz, Chem. Ber., 97:1988 (1964).
364. H. Fritzsche, U. Hasserodt, F. Korte, G. Friese, K. Adrian, and H. J. Arenz, Chem. Ber., 98:1681 (1965).
365. W. D. Garden and J. M. C. Thompson, USA Patent 2889349, 1959; C. A., 54:418 (1960).
366. W. D. Garden and J. M. C. Thompson, British Patent 815231, 1959; C. A., 54: 7134 (1960).
367. General Electric Co., British Patent 794119, 1958; C. A., 52:21227 (1958).
368. General Electric Co., British Patent 836241, 1960; C. A., 54:24398 (1960).
369. M. V. George, B. J. Gaj, and H. Gilman, J. Org. Chem., 24:624 (1959).

370. W. Gerrard, Abstracts of Reports to the Twentieth International Congress on Theoretical and Applied Chemistry, D 27 [Russian translation], Izd. "Nauka" (1965).
371. W. Gerrard and G. J. Jeacocke, Chem. Ind., 1959:704.
372. A. R. Gilbert (General Electric Co.,), USA Patent 2717902, 1956; C. A., 50:1299 (1956).
373. A. R. Gilbert (General Electric Co.), USA Patent 2768193, 1956; C. A., 51:5816 (1957).
374. A. R. Gilbert (General Electric Co.), USA Patent 2867606, 1959; C. A., 53:12738 (1959).
375. A. R. Gilbert and G. D. Cooper (General Electric Co.), USA Patent 2842580, 1958; C. A., 52:18216 (1958).
376. A. R. Gilbert and G. R. Cooper, Federal Germ. Rep. Patent 1026314, 1958; C. A., 54:8634 (1960).
377. A. R. Gilbert and S. W. Kantor (General Electric Co.), USA Patent 2837494, 1958; C. A., 52:19226 (1958).
378. A. R. Gilbert and S. W. Kantor, J. Polymer Sci., 40(136):35 (1959).
379. A. R. Gilbert and F. M. Precopio, Abstracts of Papers to the 125th Meeting of the Am. Chem. Soc., p. 16 (1954).
380. A. R. Gilbert and F. M. Precopio (General Electric Co.), USA Patent 2835651, 1958; C. A., 52:15125 (1958).
381. W. F. Gilliam, R. N. Meals, and R. O. Sauer, J. Am. Chem. Soc., 68:1161 (1946).
382. H. Gilman and B. J. Gaj, J. Am. Chem. Soc., 82:6326 (1960).
383. H. Gilman and B. J. Gaj, J. Org. Chem., 26:1305 (1961).
384. H. Gilman and B. J. Gaj, J. Org. Chem., 26:2471 (1961).
385. H. Gilman and R. A. Tomasi, J. Org. Chem., 27:3647 (1962).
386. F. Glockling and K. A. Hooton, Chem. Comm., 1966:218.
387. S. D. Gokhale and W. L. Jolly, Inorg. Chem., 3:1141 (1964).
388. S. D. Gokhale and W. L. Jolly, Inorg. Chem., 4:596 (1965).
388a. J. Goubeau, Organosilicon Chemistry (Special Lectures Presented at the International Symposium on Organosilicon Chemistry Held in Prague, Czechoslovakia, Sept. 6-9, 1965), Butterworths, London, p. 61 (1966).
389. J. Goubeau, K. O. Christe, W. Teske, and W. Wilborn, Z. Anorg. Allg. Chem. , 325:26 (1963).
390. J. Goubeau, W. Wilborn, and K. Kerger, Mitteilungsblatt Chem. Ges. DDR, 10(3):47 (1963).
391. T. E. Graham and J. M. Thompson (Imperial Chemical Industries Ltd.), British Patent 1002129, 1965; R. Zh. Khim., 13H107 (1966).
392. T. E. Graham and J. M. Thompson (Imperial Chemical Industries Ltd.), French Patent 1318080, 1963; C. A., 59:10123 (1963).
393. G. L. Greenfield, US Dept. Comm., Office Tech. Serv., AD 285370 (1961); US Govt. Res. Rept., 38:52 (1963); C. A., 60:5647 (1964).
394. S. O. Grim and D. Seyferth, Chem. Ind., 1959:849.
395. V. Gutman, Z. Anorg. Allg. Chem., 269:279 (1952).
396. K. Gutweiler and H. Niebergall, Federal Germ. Rep. Patent 1103590, 1961; C. A., 55:24102 (1961).

397. K. Gutweiler and H. Niebergall (Shell Oil Co.), USA Patent 3070582, 1962; C. A., 58:12600 (1963).
398. R. L. Hancock, J. Gas Chromatogr., 4:363 (1966); C. A., 66:8716s (1967).
399. R. J. Hartle, J. Org. Chem., 31:4288 (1966).
400. T. Hashizume and Y. Sasaki, Anal. Biochem., 15:199 (1966).
401. T. Hashizume and Y. Sasaki, Anal. Biochem., 15:346 (1966).
402. P. Hautefeuille and J. Margottet, Compt. Rend., 96:1052 (1883).
403. P. Hautefeuille and J. Margottet, Compt. Rend., 99:789 (1884).
404. P. Hautefeuille and J. Margottet, Compt. Rend., 104:56 (1887).
405. E. Hengge and U. Brychcy, Monatsh. Chem., 97:1309 (1966).
406. E. Hengge, R. Petzold, and U. Brychcy, Z. Naturf., 20b:397 (1965).
407. H. G. Henning and G. Haack, Z. Chem., 6:261 (1966).
408. O. A. Homberg and I. Hechenbleikner (Carlisle Chemical Works, Inc.), French Patent 1365375, 1964; C. A., 61:14711 (1964).
409. D. T. Hurd, J. Am. Chem. Soc., 77:2998 (1955).
410. J. F. Hyde, USA Patent 2571039, 1951; C. A., 46:2837 (1952).
411. J. F. Hyde, Federal Germ. Rep. Patent 889046, 1953; C., 1954:2280.
412. J. F. Hyde (Dow Corning Corp.), USA Patent 3160601, 1964; C. A., 62:11932 (1965).
413. J. F. Hyde, R. R. McGregor, and E. L. Warrick, Federal Germ. Rep. Patent 880487, 1953; C., 1954:8232.
414. J. F. Hyde and E. M. Schultz (Dow Corning Corp.), French Patent 1352529, 1964; C. A., 61:9653 (1964).
415. Imperial Chemical Industries Ltd., French Patent 1318080, 1963; C. A., 59:10123 (1963).
416. K. Issleib and H. Reinhold, Z. Anorg. Allg. Chem., 314:113 (1962).
417. K. Issleib and B. Walther, Angew Chem., 79:59 (1967).
418. J. Jack, USA Patent 2934550, 1960; C. A., 54:14777 (1960).
419. J. Jack, British Patent 857343, 1960; C. A., 55:15999 (1961).
419a. V. B. Jex and R. Y. Mixer, Federal Germ. Rep. 1091567, 1956; C. A., 56:1481 (1962).
420. J. I. Jones, "Polymetallosiloxanes," Part I, in Development in Inorganic Polymer Chemistry, Ed. M. F. Lappert, G. J. Leigh, Amsterdam, London, New York (1962), p. 162.
421. J. I. Jones, "Polymetallosiloxanes," Part II, in: Development in Inorganic Polymer Chemistry, Ed. M. F. Lappert, G. J. Leigh, Amsterdam, London, New York (1962), p. 200.
422. S. W. Kantor (General Electric Co.), USA Patent 2785147, 1957; C. A., 51:7756 (1957).
423. S. W. Kantor and R. C. Osthoff (General Electric Co.), USA Patent 2793222, 1957; C. A., 51:12533 (1957).
424. W. H. Keeber and H. W. Post, J. Org. Chem., 21:509 (1956).
425. K. Kerger, Dissertation, Tech. Hochschule Stuttgart, cited from [390].
426. P. M. Kerschner and B. W. Greenwald, USA Patent 2864845, 1958; C. A., 53:12239 (1959).
427. N. Kirk, Ind. Eng. Chem., 51:515 (1959).

428. N. Kirk, USA Patent 2883272, 1959; C. A., 53:12739 (1959).
429. H. W. Kohlschütter, Angew. Chem., 71:82 (1959).
430. H. W. Kohlschütter and H. Simoleit, Kunstst.-Plast., 6:9 (1959).
431. G. M. Kosolapoff, Organophosphorus Compounds, New York (1950), p. 196.
432. R. H. Kratzer and K. L. Paciorek, Inorg. Chem., 4:1767 (1965).
433. C. Kruger, E. G. Rochow, and U. Wannagat, Chem. Ber., 96:2132 (1963).
434. W. Kuchen and H. Buchwald, Angew. Chem., 69:307 (1957).
435. W. Kuchen and H. Buchwald, Chem. Ber., 92:227 (1959).
436. H. Kuckertz, Dissertation, Tech. Hochschule Aachen (1962); cited from [433].
437. A. Ladenburg, Lieb. Ann., 164:300 (1872).
438. Z. Laita, P. Hlozek, B. Buček, and M. Jelinek, Abstracts of Scientific Communications of the International Symposium on Macromolecular Chemistry, A-90/49, Prague (1965); Reprints of Scientific Comm., P49.
439. W. M. Lanham and P. L. Smith (Union Carbide Corp.), USA Patent 3197431, 1965; C. A., 63:9988 (1965).
440. E. Larsson and B. Smith, Svensk. Kem. Tidskr., 62:87 (1950).
441. A. J. Leffler and E. G. Teach, Abstracts of Papers of the 136th Meeting of the American Chemical Society, Atlantic City, p. 31 (1959).
442. A. J. Leffier and E. G. Teach, J. Am. Chem. Soc., 82:2710 (1961).
443. J. H. Letcher and J. R. van Wazer, J. Chem. Phys., 44:815 (1966).
444. G. R. Levi and G. Peyronel, Z. Krist., 92:191 (1935).
445. H. R. Linton and E. R. Nixon, Spectrochim. Acta, 15:146 (1959).
446. R. G. Linville (General Electric Co.), USA Patent 2739952, 1956; C. A., 50:9784 (1956).
447. R. G. Linville (General Electric Co.), French Patent 1108764, 1956; C., 1957:14215.
448. R. G. Linville (General Electric Co.), British Patent 772088, 1957; C. A., 51:10121 (1957).
449. R. G. Linville (General Electric Co.), Federal Germ. Rep. Patent 1018221, 1957; C. A., 53:17568 (1959).
450. R. G. Linville (General Electric Co.), USA Patent 2843615, 1958; C. A., 53:1147 (1959).
451. L. Maier, Helv. Chim. Acta, 46:2667 (1963).
452. L. Maier (Monsanto Co.), USA Patent 3188294, 1965; C. A., 63:13318 (1965).
453. L. Malatesta, Gazz., 80:527 (1950).
454. L. Malatesta and A. Sacco, Gazz., 80:658 (1950).
455. J. G. Marsden, USA Patent 2963503, 1960; C. A., 55:8292 (1961).
456. G. Mattogno, A. Monaci, and F. Tarli, Ann. Chim. (Rome), 55:599 (1965).
457. G. Mattogno and F. Tarli, Ric. Sc. Rend., Ser., A4(4):487 (1964).
458. P. A. McCusker and E. L. Reilly, J. Am. Chem. Soc., 75:1583 (1953).
459. R. R. McGregor and E. L. Warrick (Corning Glass Works), USA Patent 2386488, 1945; C. A., 40:592 (1946).
460. R. R. McGregor and E. L. Warrick, USA Patent 2435147, 1948; C. A., 42:2819 (1948).

461. R. R. McGregor and E. L. Warrick, USA Patent 2459387, 1949; C. A., 43:2815 (1949).
462. J. K. McLoughlin, USA Patent 2986548, 1961; C. A., 55:19328 (1961).
463. Midland Silicones Ltd., British Patent 844421, 1960; C. A., 55:4362 (1961).
464. Midland Silicones Ltd., French Patent 1340233, 1963; C. A., 60:6954 (1964).
465. N. E. Miller, J. Am. Chem. Soc., 87:390 (1965).
466. N. E. Miller, Inorg. Chem., 4:1458 (1965).
467. D. H. Moreton (Douglas Aircraft Co. Inc.), USA Patent 2618600, 1952; C. A., 47:1928 (1953).
468. D. H. Moreton (Douglas Aircraft Co. Inc.), USA Patent 2618601, 1952; C. A., 47:1928 (1953).
469. W. Moschel, H. Jonas, and W. Noll, Federal Germ. Rep. Patent 832499, 1952; C., 1952:6774.
470. M. J. Newlands, Proc. Chem. Soc., 1960:123.
471. M. J. Newlands (Philadelphia Quartz Co.), USA Patent 3164622, 1965; C. A., 62:16297 (1965).
472. H. Niebergall, Federal Germ. Rep. Patent 1086897, 1960; C. A., 55:14977 (1961).
473. H. Niebergall, Federal Germ. Rep. Patent 1113827, 1961; C. A., 56:14475 (1962).
474. H. Niebergall, Federal Germ. Rep. Patent, 1118781, 1961; C. A., 56:11622 (1962).
475. H. Niebergall, Makromol. Chem., 52:218 (1962).
476. H. Niebergall (Shell Oil Co.), USA Patent 3142663, 1964; C. A., 61:8436 (1964).
477. S. Nitzsche, Makromol. Chem., 34:231 (1959).
478. S. Nitzsche, E. Schmidt, and M. Wick, USA Patent 2990419, 1961; C. A., 55:24103 (1961).
479. S. Nitzsche and M. Wick, Federal Germ. Rep. Patent 930481, 1955; C. A., 52:6830 (1958).
480. H. Nöth and L. Meinel, Z. Anorg. Allg. Chem., 349:225 (1967).
481. H. Nöth, L. Meinel, and H. Madersteig, Angew. Chem., 77:734 (1965).
482. H. Nöth and W. Schrägle, Z. Naturf., 16b:473 (1961).
483. H. Nöth and W. Schrägle, Chem. Ber., 98:352 (1965).
484. G. Oertel, H. Holtschmidt, and H. Malz (Farbenfabriken Bayer A. G.), Federal Germ. Rep. Patent 1157226, 1963; C. A., 60:6858 (1964).
485. W. J. Owen and F. C. Saunders (Midland Silicones Ltd.), British Patent 1007333, 1965; C. A., 63:18154 (1965).
486. W. J. Owen and F. C. Saunders (Midland Silicones Ltd.), British Patent 1014051, 1965; R. Zh. Khim., 22C339 (1966).
487. K. L. Paciorek and R. H. Kratzer, J. Org. Chem., 31:2426 (1966).
488. W. G. Parks, J. G. Erhardt, and D. R. Roberts, Am. Dyestuff Rept. 39, Proc. Am. Assoc. Textile Chem. Colorists, 1950, p. 294; C. A., 44:6619 (1950).
489. G. W. Parshall, USA Patent 2907785, 1959; C. A., 54:4389 (1960).
490. G. W. Parshall and R. V. Linsey, J. Am. Chem. Soc., 81:6273 (1959).
491. J. Paul, Mikrochim. Ichnoanal. Acta, 1965:836.

492. R. Piekoš and A. Radecki, Roczn. Chem., 36:1303 (1962).
493. R. Piekoš and A. Radecki, Roczn. Chem., 36:1329 (1962).
494. R. Piekoš and A. Radecki, Roczn. Chem., 39:969 (1965).
495. R. Piekoš and A. Radecki, J. Inorg. Nucl. Chem., 27:2589 (1965).
496. R. A. Pike (Union Carbide Co.), USA Patent 3122581, 1964; C. A., 60:12053 (1964).
497. E. P. Plueddemann, USA Patent 2951860, 1960; C. A., 55:11917 (1961).
498. R. Rabinowitz, J. Org. Chem., 28:2975 (1963).
499. A. Radecki and R. Piekoš, Roczn. Chem., 36:1329 (1962).
500. W. T. Reichle, Inorg. Chem., 3:402 (1964).
501. E. G. Rochow, British Patent 549081, 1942; C. A., 38:552 (1944); Paint Techn., 11:277 (1946).
502. E. G. Rochow, USA Patent 2371068, 1945; C. A., 39:4889 (1945).
503. E. G. Rochow, Federal Germ. Rep. Patent 926811, 1955; C., 1955:8982.
504. F. Runge and K. Helke, Germ. Dem. Rep. Patent 16966, 1959; C. A., 54:16915 (1960).
505. F. Runge and K. Helke, Federal Germ. Rep. Patent 1099175, 1961; C. A., 55:24102 (1961).
506. C. R. Russ, An Investigation of the Silicon-Nitrogen, Silicon-Phosphorus and Silicon-Arsenic Bond, Diss. Abstr., 26:5027 (1966).
507. C. R. Russ and A. G. MacDiarmid, Angew. Chem., 78:391 (1966).
508. F. E. Saafeld and H. J. Svec, Inorg. Chem., 3:1442 (1964).
509. R. O. Sauer, J. Am. Chem. Soc., 66:1707 (1944).
510. R. M. Savage (General Electric Co.), USA Patent 2671069, 1954; C. A., 48:7332 (1954).
511. R. M. Savage, Federal Germ. Rep. Patent 949903, 1956; C. A., 53:2673 (1959).
512. J. C. Scheldon and S. Y. Teree, J. Am. Chem. Soc., 81:2290 (1959).
513. O. Scherer and M. Schmidt, Scientific Communications of the International Symposium on Organosilicon Chemistry, Prague (1965), p. 315.
514. H. Schmidbaur, Z. Anorg. Allg. Chem., 326:272 (1964).
515. H. Schmidbaur, and G. Jonas, Chem. Ber., 100:1120 (1967).
516. H. Schmidbaur, and W. Tronich, Angew. Chem., 79:412 (1967).
517. H. Schmidbaur, and W. Tronich, Chem. Ber., 100:1032 (1967).
518. H. Schmidbaur, and W. Wolfsberger, Angew. Chem., 78:306 (1966).
519. H. Schmidbaur, and W. Wolfsberger, 3, Internationales Symposium über Metallorganische Chemie, München (1967), p. 52.
520. H. Schmidbaur, and W. Wolfsberger, Angew. Chem., 79:411 (1967).
521. H. Schmidbaur, and W. Wolfsberger, Chem. Ber., 100:1000 (1967).
522. H. Schmidbaur, and W. Wolfsberger, Chem. Ber., 100:1016 (1967).
523. H. Schmidbaur, W. Wolfsberger, and H. Kröner, Chem. Ber., 100:1023 (1967).
524. M. Schmidt and H. Bipp, Sitzber. Ges. Beförder, Ges. Naturwiss. Marburg 83/84:523 (1961/1962); C. A., 59:6436 (1963).
525. M. Schmidt and H. Schmidbaur, Angew. Chem., 71:553 (1959).
526. M. Schmidt, H. Schmidbaur, and A. Binger, Chem. Ber., 93:872 (1960).

527. M. Schmidt, H. Schmidbaur, and I. Ruidisch, Angew. Chem., 73:408 (1961).
528. W. Schrägle, Dissertation, Univ. Münden (1963), cited from [483].
529. R. Schwarz and K. Schoeller, Angew. Chem., 69:93 (1957).
530. R. Schwarz and K. Schoeller, Chem. Ber., 91:2103 (1958).
531. M. P. Seidel, J. Swiss, and C. E. Arntzen, Swedish Patent 128667, 1950; C., 1951:I, 2970.
532. M. P. Seidel, J. Swiss, and C. E. Arntzen, Canadian Patent 496623, 1953; C., 1958:14514.
533. D. Seyferth, J. Am. Chem. Soc., 80:1336 (1958).
534. D. Seyferth, USA Patent 2964550, 1960; C. A., 55:6439 (1961).
535. D. Seyferth and W. Freyer, J. Org. Chem., 26:2604 (1961).
536. D. Seyferth and S. O. Grim, J. Am. Chem. Soc., 83:1610 (1961).
537. D. Seyferth, G. Raab, and S. O. Grim, J. Org. Chem., 26:3034 (1961).
538. D. Seyferth and G. Singh, J. Am. Chem. Soc., 87:4156 (1965).
539. D. Seyferth and G. Singh, Organosilicon Chemistry (special lectures presented at the International Symposium on Organosilicon Chemistry held in Prague, Czechoslovakia, 6-9 Sept. 1965), Butterworths, London (1966), p. 159.
540. D. Seyferth, G. Singh, and R. Suzuki, Pure Appl. Chem., 13:159 (1966).
541. D. Seyferth and T. Wada, Inorg. Chem., 1:79 (1962).
542. D. Seyferth, M. A. Weiner, S. O. Grim, and N. Kahlen, Abstracts of Papers to the 135th Meeting of the American Chemical Society, Boston, pp. 20-31 (1959).
543. N. L. Smith and H. H. Sister, J. Org. Chem., 28:272 (1963).
544. L. H. Sommer, H. D. Blankman, and P. C. Miller, J. Am. Chem. Soc., 73:3542 (1951).
545. M. M. Sprung, USA Patent 2492129, 1949; C. A., 44:2271 (1950).
546. M. M. Sprung and C. A. Burkhard, Canadian Patent 504114, 1954; R. Zh. Khim., 56167 (1956).
547. H. N. Stokes, Am. Chem. J., 13:244 (1891).
548. H. N. Stokes, Ber., 24:933 (1891).
549. H. N. Stokes, Am. Chem. J., 14:545 (1892).
549a. F. G. A. Stone, Hydrogen Compounds of the Group IV Elements, London (1962).
550. S. Sujishi and S. Witz, Office of Ordnance Research Control, Da-11-011 Ord. 1264, Rep. 1 (1954); cited from [312].
551. R. A. Sutton (Distillers Co. Ltd.), British Patent 991284, 1965; C. A., 63:4489 (1965).
552. J. S. Thayer, Preparation and Properties of Group IV A Organometallic Azides, Wisconsin Univ., Dissertation Abstr., 25:2222 (1964).
553. J. S. Thayer and R. West, Inorg. Chem., 3:406 (1964).
554. C. E. Trautmen, USA Patent 2488449, 1949; C. A., 44:2287 (1950).
555. C. E. Trautmen, USA Patent 2515024, 1950; C. A., 44:9144 (1950).
556. W. Tronich, Diplomarb., Univ. Würzburg. (1966); cited from [517].
557. Wacker-Chemie G. m. b. H., British Patent 765744, 1957; C. A., 51:9202 (1957).

558. Wacker-Chemie G. m. b. H., French Patent 1376923, 1964; C. A., 62:5416 (1965).
559. U. Wannagat, Angew. Chem., 75:173 (1963).
560. U. Wannagat, Angew. Chem., 76:234 (1964).
561. U. Wannagat, Adv. Inorg. Radiochem., 6:225 (1964).
562. R. M. Waschburn and R. A. Baldwin (American Potash and Chemical Corp.), USA Patent 3112331, 1963; C. A., 60:5554 (1964).
563. W. W. Wells, T. Katagi, R. Bently, and C. C. Sweeley, Bioch. Bioph. Acta, 82:408 (1964).
564. N. Wiberg and B. Neruda, Chem. Ber., 99:740 (1966).
565. N. Wiberg, F. Raschig, and R. Sustmann, Angew. Chem., 74:388 (1962).
566. N. Wiberg, F. Raschig, and R. Sustmann, Angew. Chem., 74:716 (1962).
567. N. Wiberg, K. H. Schmid, and Wan-Chul Joo, Angew Chem., 77:1042 (1965).
568. W. Wilborn, K. Christe, J. Goubeau, and W. Teske, Mitteilungsblatt Chem. Ges. DDR, 10(3):59 (1963).
569. H. J. S. Winkler, A. W.-P. Jarvie, D. J. Peterson, and H. Gilman, J. Am. Chem. Soc., 83:4089 (1961).
570. D. Wittenberg and H. Gilman, J. Org. Chem., 23:1063 (1958).
571. W. Wolfsberger, Diplomarb. Univ. Würzburg (1966), cited from [521].
572. C. H. Yoder and J. J. Zuckermann, J. Am. Chem. Soc., 88:2170 (1966).
573. H. Y. Yu and I. H. Sha, Hua Hsuech Pao, 1965:557; C. A., 64:16630 (1966).

SUPPLEMENT TO LITERATURE REFERENCES

574. N. F. Baina, O. I. Kochkina, and L. Z. Soborovskii, Authors' cert. 226607, 1967; Izobr., Prom. Obr., Tov. Zn., No. 29, p. 21 (1968).
575. S. I. Belykh, S. M. Zhivukhin, V. V. Kireev, and G. S. Kolesnikov, Vysokomol. Soed., 11A:625 (1969).
576. S. I. Belykh, S. M. Zhivukhin, V. V. Kireev, and G. S. Kolesnikov, Zh. Obshch. Khim., 39:799 (1969).
577. E. F. Bugerenko, A. S. Petukhova, and E. A. Chernyshev, Abstracts of Reports to the Fourth Conference on the Chemistry and Use of Organosilicon Compounds, NIITÉKhIM, Moscow, p. 31 (1968).
578. E. F. Bugerenko, E. A. Chernyshev, L. I. Shul'gina, and A. S. Petukhova, Authors' cert. 229516, 1967; Izobr., Prom. Obr., Tov. Zn., No. 33, p. 40 (1968).
579. N. V. Varlamova, A. I. Sidnev, V. V. Severnyi, K. A. Andrianov, and I. A. Zubkov, Abstracts of Reports of the Fourth Conference on the Chemistry and Use of Organosilicon Compounds, NIITÉKhIM, Moscow, p. 89 (1968).
580. T. V. Vasil'eva, K. A. Andrianov, and V. D. Nedorosol, Abstracts of Reports to the Fourth Conference on the Chemistry and Use of Organosilicon Compounds, NIITÉKhIM, Moscow, p. 89 (1968).
581. L. N. Volodina, Zh. Obshch. Khim., 38:200 (1968).

582. L. N. Volodina, Synthesis and Properties of Triorganosilyl Derivatives of Phosphorus Acid, Author's abstract of Cand. dissertation, Tashkent (1969).
583. V. F. Gridina, A. L. Klebanskii, L. P. Dorofeenko, L. E. Krupnova, and N. V. Kozlova, Vysokomol. Soed., 11A:426 (1969).
584. V. A. Zamyatina, A. I. Solomatina, and V. V. Korshak, Izv. Akad. Nauk SSSR, Neorg. Mater., 3:891 (1967).
585. S. Z. Ivin, I. D. Shelakova, and V. K. Promonenkov, Authors' cert. 199874, 1967; Izobr., Prom. Obr., Tov. Zn., No. 16, p. 22 (1967).
586. S. Z. Ivin, I. D. Shelakova, V. K. Promonenkov, V. V. Levin, and I. N. Fetin, Authors' cert. 199873, 1967; Izobr., Prom. Obr., Tov. Zn., No. 16, p. 22 (1967).
587. G. S. Kolesnikov, S. M. Zhivukhin, V. V. Kireev, S. I. Belykh, and I. M. Raigorodskii, Abstracts of Reports to the Fourth Conference on the Chemistry and Use of Organosilicon Compounds, NIITÉKhIM, Moscow, p. 43 (1968).
588. S. I. Kol'tsov, A. N. Volkova, and V. B. Aleskovskii, Zh. Prikl. Khim., 42:73 (1969).
589. N. V. Komarov, V. G. Rozinov, L. P. Vakhrushev, E. F. Grechkin, and N. F. Chernov, Izv. Akad. Nauk SSSR, Ser. Khim., 1969:729.
590. L. A. Mai, Izv. Akad. Nauk Latv.SSR, Ser. Khim., 1967:504.
591. L. A. Mitrofanov and A. V. Karlin, The Chemistry and Practical Application of Organosilicon Compounds, Leningrad (1968), p. 146.
592. T.N. Naumova, Zh. Fiz. Khim., 42:227 (1968).
593. N. F. Orlov, The Chemistry and Practical Application of Organosilicon Compounds, Leningrad (1968), p. 94.
594. N. F. Orlov, The Synthesis and Properties of Silicon-Containing Organophosphorus Compounds with a Phosphosiloxane Grouping, Author's abstract of Doctor's dissertation, Leningrad (1969).
595. N. F. Orlov and M. A. Belokrinitskii, Authors' cert. 217393, 1967; Izobr., Prom. Obr., Tov. Zn., No. 16, p. 20 (1968).
596. N. F. Orlov and M. A. Belokrinitskii, Authors' cert. 225878, 1967; Izobr., Prom. Obr., Tov. Zn., No. 28, p. 16 (1968).
597. N. F. Orlov, M. A. Belokrinitskii, and B. L. Kaufman, Authors' cert. 226604 (1967); Izobr., Prom. Obr., Tov. Zn., No. 29, p. 21 (1968).
598. N. F. Orlov, M. A. Belokrinitskii, B. L. Kaufman, and É. V. Sudakova, The Chemistry and Practical Application of Organosilicon Compounds, Leningrad (1968), p. 117.
599. N. F. Orlov, M. A. Belokrinitskii, É. V. Sudakova, and B. L. Kaufman, Zh. Obshch. Khim., 38:1656 (1968).
600. N. F. Orlov and O. F. Viktorov, The Chemistry and Practical Application of Organosilicon Compounds, Leningrad (1968), p. 129.
601. N. F. Orlov and L. N. Volodina, Authors' cert. 203682, 1965; Izobr., Prom. Obr., Tov. Zn., No. 21, p. 29 (1967).
602. N. F. Orlov and O. N. Dolgov, Abstracts of Reports to the Fourth Conference on the Chemistry and Use of Organosilicon Compounds, NIITÉKhIM, Moscow, p. 122 (1968).
603. N. F. Orlov and B. L. Kaufman, Zh. Obshch. Khim., 37:2320 (1967).

604. N. F. Orlov and B. L. Kaufman, Zh. Obshch. Khim., 38:1842 (1968).
605. N. F. Orlov, B. L. Kaufman, L. Sukhi, L. N. Slesap', and É. V. Sudakova, The Chemistry and Practical Application of Organosilicon Compounds, Leningrad (1968), p. 111.
606. N. F. Orlov, G. L. Korichev, and I. T. Gumenyuk, Abstracts of Reports to the Fourth Conference on the Chemistry and Use of Organosilicon Compounds, NIITÉKhIM, Moscow (1968), p. 54.
607. N. F. Orlov and L. N. Slesar', Dokl. Akad. Nauk SSSR, 179:108 (1968).
608. N. F. Orlov and L. N. Slesar', Abstracts of Reports to the Fourth Conference on the Chemistry and Use of Organosilicon Compounds, NIITÉKhIM, Moscow (1968), p. 55.
609. N. F. Orlov, L. N. Slesar', V. P. Mileshkevich, and M. A. Belokrinitskii, The Chemistry and Practical Application of Organosilicon Compounds, Leningrad (1968), p. 41.
610. N. F. Orlov and É. V. Sudakova, Authors' cert. 229511, 1967; Izobr., Prom. Obr., Tov. Zn., No. 33, p. 39 (1968).
611. N. F. Orlov and É. V. Sudakova, The Chemistry and Practical Use of Organosilicon Compounds, Leningrad (1968), p. 123.
612. N. F. Orlov and É. V. Sudakova, Zh. Obshch. Khim., 39:222 (1969).
613. N. F. Orlov, É. V. Sudakova, and M. A. Belokrinitskii, Abstracts of Reports to the Fourth Conference on the Chemistry and Use of Organosilicon Compounds, NIITÉKhIM, Moscow (1968), p. 55.
614. N. F. Orlov, É. V. Sudakova, and L. N. Slesar', II Symposium International sur la Chimie des Composés Organiques du Silicium, Résumés des Communications, Bordeaux (1968), p. 145.
615. N. F. Orlov, N. G. Shelkunov, and I. V. Klimenko, Abstracts of Reports to the Fourth Conference on the Chemistry and Use of Organosilicon Compounds, NIITÉKhIM, Moscow (1968), p. 56.
616. V. V. Ponomarev, Investigation of the Reactions of Phosphorus Trichloride and Pentachloride with Alkenyl- and Arylsilanes, Author's abstract of Cand. dissertation, Moscow (1969).
617. V. V. Ponomarev, S. A. Golubtsov, and K. A. Andrianov, Abstracts of Reports to the Fourth Conference on the Chemistry and Use of Organosilicon Compounds, NIITÉKhIM, Moscow (1968), p. 18.
618. E. A. Chernyshev, E. F. Bugerenko, A. S. Petukhova, V. K. Promonenkov, and C. Z. Ivin, Authors' cert. 215983, 1967; Izobr., Prom. Obr., Tov. Zn., No. 12, p. 28 (1968).
619. A. S. Shapatin, I. G. Kirsanova, K. A. Andrianov, and L. D. Prigozhina, Abstracts of Reports to the Fourth Conference on the Chemistry and Use of Organosilicon Compounds, NIITÉKhIM, Moscow (1968), p. 64.
620. E. W. Abel, R. A. N. McLean, and I. H. Sabherwal, J. Chem. Soc., 1968A:2371.
621. E. W. Abel and I. H. Sabherwal, J. Organomet. Chem., 10:491 (1967).
622. E. W. Abel and I. H. Sabherwal, J. Chem. Soc., 1968(A):1105.
623. M. C. Baird, J. Inorg. Nucl. Chem., 29:367 (1967).
624. A. Bazouin and M. Lefort, French Patent 1466546 (1967); C. A., 67:108 747 w (1967).

625. A. G. Brook and D. G. Anderson, Can. J. Chem., 46:2115 (1968).
626. A. G. Brook and S. A. Fieldhouse, J. Organomet. Chem., 10:235 (1967).
627. A. B. Burg and J. S. Basi, J. Am. Chem. Soc., 90:3361 (1968).
628. H. Bürger, II Symposium International sur la Chimie des Composés Organiques du Silicium, Résumés des Communications, Bordeaux (1968), p. 28.
629. H. Bürger, Organomet. Chem. Rev., 3A:425 (1968).
630. H. Bürger and U. Goetze, J. Organomet. Chem., 12:451 (1968).
631. F. de Charentenay, J. A. Osborn, and G. Wilkinson, J. Chem. Soc., 1968B:787.
632. J. Chatt, C. Eaborn, and P. N. Kapoor, J. Organomet. Chem., 13P:21 (1968).
633. H. C. Clark, P. W. R. Corfield, K. R. Dixon, and J. A. Ibers, J. Am. Chem. Soc., 89:3360 (1967).
634. H. C. Clark and K. R. Dixon, Chem. Comm., 1967:717.
635. A. F. Clemmit and F. Glockeing, II Symposium International sur la Chimie des Composés Organiques du Silicium, Résumés des Communications; Bordeaux (1968), p. 45.
636. S. Cradock, E. A. V. Ebsworth, G. Davidson, and L. A. Woodward, J. Chem. Soc., 1967A:1229.
637. J. E. Drake and J. Simpson, Inorg. Chem., 6:1984 (1967).
638. J. E. Drake and J. Simpson, J. Chem. Soc., 1968A:1039.
639. J. E. Drake and N. Goddard, J. Chem. Soc., 1969A:662.
640. E. A. V. Ebsworth, C. Glidewell, and G. M. Sheldrick, J. Chem. Soc., 1969A:352.
641. E. Fluck, H. Bürger, and U. Goetze, Z. Naturf., 22b:912 (1968).
642. G. Fritz and G. Becker, Angew. Chem., 79:1068 (1967).
643. G. Fritz, R. Wiermers, and U. Protzer, Z. Anorg. Allg. Chem., 363:225 (1968).
644. R. Gahm, H. Nöth, R. Schroen, N. Sprague, and H. Vahrenkamp, II Symposium International sur la Chimie des Composés Organiques du Silicium, Résumés des Communications, Bordeaux (1968), p. 74.
645. P. P. Gaspar, S. A. Bock, and W. C. Eckelman, J. Am. Chem. Soc., 90:6914 (1968).
646. C. W. Gehrke, D. L. Stalling, and C. D. Ruyle, Biochem. Biophys. Res. Comm., 28:869 (1967).
647. O. Glemser and E. Niecke, Z. Naturf., 23b:743 (1968).
648. C. Glidewell and G. M. Sheldrick, J. Chem. Soc., 1969A:350.
649. F. Glockeing and K. A. Hooton, J. Chem. Soc., 1967A:1066.
650. F. Glockeing and K. A. Hooton, J. Chem. Soc., 1968A:826.
651. R. E. Goldsberry, D. E. Lewis, and K. Cohn, J. Organomet. Chem., 15:491 (1968).
652. B. T. Heaton and A. Pidcock, J. Organomet. Chem., 14:235 (1968).
653. M. G. Horning, E. A. Boucher, and A. M. Moss, J. Gas Chromat., 5:297 (1967).
654. W. H. Keeber, The Synthesis and Reactions of Some Organosilicon Phosphonates, Diss. Abstr., 28B:1854 (1967).
655. K. Kerger and R. Kohlhaas, Z. Anorg. Allg. Chem., 354:44 (1967).
656. R. B. King and K. H. Pannell, Inorg. Chem., 7:1510 (1968).
657. Z. Laita, P. Hložek, B. Buček, and M. Jelinek, J. Polym. Sc., C(16):669 (1967); C. A., 67:64780 v (1967).

658. K. R. Lundberg, The Chemistry of Dimethylphosphinomethyl Substitution on Silicon, Nitrogen, and Sulfur, Diss. Abstr., 29B:1596 (1968).
659. G. Märkl, F. Lieb, and A. Merz, Angew. Chem., 79:475 (1967).
660. G. Märkl, and H. Olbrich, Tetrahedron Letters, 1968:3813.
661. D. R. Mathiason, Trimethylsilyl Substituted, Ylides, Diss. Abstr., 27B:4294 (1967).
662. D. R. Mathiason and N. E. Miller, Inorg. Chem., 7:709 (1968).
663. D. C. McKean, Spectrochim. Acta, 24A:1253 (1968).
664. M. Murray and R. Schmutzler, Chem. Ind., 1968:1730.
665. E. Niecke and J. Stenzel, Z. Naturf., 22b:785 (1967).
666. A. D. Norman, Chem. Comm., 1968:812.
667. A. D. Norman, J. Am. Chem. Soc., 90:6556 (1968).
668. G. A. Ozin, J. Chem. Soc., D, Chem. Comm., 1969:104.
669. D. J. Peterson, J. Org. Chem., 33:780 (1968).
670. O. J. Scherer and G. Schieder, Angew. Chem., 80:83 (1968).
671. H. Schmidbauer, Allg. Prakt. Chem., 18(5):138 (1967).
672. H. Schmidbauer, K. Schwirten, and H. H. Pickel, Chem. Ber., 102:564 (1969).
673. H. Schmidbaur, and W. Tronich, Angew. Chem., 80:239 (1968).
674. H. Schmidbaur, and W. Tronich, Chem. Ber., 101:595 (1968).
675. H. Schmidbaur, and W. Tronich, Chem. Ber., 101:604 (1968).
676. H. Schmidbaur, and W. Tronich, Chem. Ber., 101:3545 (1968).
677. H. Schmidbaur, and W. Tronich, Chem. Ber., 101:3556 (1968).
678. H. Schmidbaur, W. Tronich, and W. Malisch, II Symposium International sur la Chimie des Composés Organiques du Silicium, Résumés des Communications, Bordeaux (1968), p. 167.
679. H. Schmidbaur, W. Wolfsberger, and K. Schwirten, Chem. Ber., 102:556 (1969).
680. A. Schmidpeter and K. Stoll, Angew. Chem., 80:558 (1968).
681. O. Schmitz-Du Mont and W. Jansen, Angew. Chem., 80:399 (1968).
682. H. Schumann and O. Stelzer, Angew. Chem., 79:692 (1967).
683. H. Schumann and O. Stelzer, Angew. Chem., 80:318 (1968).
684. H. Schumann and O. Stelzer, II Symposium International sur la Chimie des Composés Organiques du Silicium, Résumés des Communications, Bordeaux (1968), p. 171.
685. H. Siebert, J. Eints, and E. Fluck, Z. Naturf., 23b:1006 (1968).
686. W. Siebert, W. E. Davidson, and M. C. Henry, J. Organomet. Chem., 15:69 (1968).
687. L. H. Sommer and J. E. Lyons, J. Am. Chem. Soc., 90:4197 (1968).
688. Unilever Ltd., British Patent 1052388, 1966; Brit. Pat. Abstr., 7(3):1, 6 (1967).
689. R. M. Washburn and R. A. Baldwin (American Potash and Chemical Corp.), USA Patent 3341477, 1967; R. Zh. Khim., 3C427 II (1969).
690. I. M. White, M. A. H. Hewins, and S. M. Somerville, II Symposium International sur la Chimie des Composés Organiques du Silicium, Résumés des Communications Suppl., Bordeaux (1968).
691. J. K. Wittle, Some Studies of Metal to Metal Bonding of Group IV B, Diss. Abstr. 29B:926 (1968).

Chapter II

Organosilicon Derivatives of Sulfur

1. COMPOUNDS CONTAINING THE GROUPING Si – S – H (SILANETHIOLS)

1.1. Preparation Methods

The first compound of silicon containing the grouping Si – S – H was obtained in 1847 by passing a mixture of hydrogen sulfide and silicon tetrachloride through a red-hot porcelain tube. However, the substance formed was then taken to be Cl_2SiS [554-561]. Only much later was it established that this reaction proceeds with the formation of trichlorosilanethiol in accordance with the scheme [328-331]

$$SiCl_4 + H_2S \longrightarrow Cl_3SiSH + HCl \quad (2.1)$$

At the present time the high-temperature sulfhydrolysis of silicon tetrachloride is the main method of synthesizing trichlorosilanethiol [369, 545, 546, 634, 718]. The yield when the reaction is carried out at 650°C varies from 21-25 [369] to 40% (with excess hydrogen sulfide); with excess $SiCl_4$ the yield is only 9% [634]. Below 500°C $SiCl_4$ does not react with H_2S [634], but at 550°C the pyrolysis of trichlorosilanethiol begins and proceeds at a high rate at 700°C [545]. The reaction of hydrogen sulfide with trichlorobromosilane leads to the formation of Cl_3SiSH in yields from 17 (at 400°C) to 30% (at 650°C) [369].

Trichlorosilanethiol is also formed by passing a mixture of hydrogen chloride and silicon tetrachloride over silicon disulfide heated to 1000°C [545]. This process evidently proceeds in two stages:

$$SiS_2 + 4HCl \longrightarrow SiCl_4 + 2H_2S \quad (2.2)$$

$$2SiCl_4 + 2H_2S \longrightarrow 2Cl_3SiSH + 2HCl \quad (2.3)$$

It may be represented by the general equation

$$SiS_2 + SiCl_4 + 2HCl \longrightarrow 2Cl_3SiSH \quad (2.4)$$

In the reaction of hydrogen chloride with thiosilicates of heavy metals (Ag, Cu, Pb, Fe) heated to ~500°C, the formation of trichlorosilanethiol is also observed [511]. This also occurs when the black product from the reaction of ferrosilicon with sulfur is heated to 290-300°C in a stream of hydrogen chloride [239]. It may be surmised that the compound close in composition to ClSiSSH, which was isolated from the reaction products of hydrogen sulfide and silicon tetrachloride [719], is sym-dichlorocyclodisilthianedithiol:

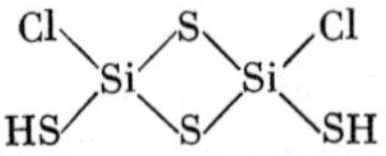

Silicon tetrabromide does not react with hydrogen sulfide in the absence of catalysts. The statement in [151] that $SiBr_4$ reacts very readily with H_2S to give a high yield of tribromosilanethiol is based on a misunderstanding. Nonetheless, when the reaction of $SiBr_4$ with H_2S is carried out in the presence of aluminum bromide, it is possible to isolate tribromosilanethiol [705]. Analogously, the reaction of hydrogen sulfide with hexabromodisilane forms bromine-substituted disilanethiols such as $Br_3SiSiBr_2SH$ [705].

Like $SiCl_4$ and H_3SiCl [674],* organochlorosilanes do not react with hydrogen sulfide at low temperatures. This may be explained by the fact that the equilibrium of the reaction

$$\geq Si{-}X + H_2S \rightleftarrows \geq Si{-}SH + HX \quad (2.5)$$

*It is possible that H_3SiSH is one of the products from the reaction of hydrogen sulfide with chlorosilane in the presence of aluminum chloride at 150°C [674].

is displaced completely to the left due to the great ease of cleavage of the Si – S bond by polar reagents. However, by removing the hydrogen halide formed from the sphere of the reaction, e.g., by binding it with a tertiary amine, the equilibrium of the reaction of halosilanes with hydrogen sulfides may be shifted to the right. Precisely this is used in the simplest method of synthesizing triorganosilanethiols, which is based on the reaction of hydrogen sulfide with triorganohalosilanes in the presence of a tertiary amine. For example, by this method it was possible to obtain smoothly tribenzylsilanethiol [867], triphenylsilanethiol [301]

$$(C_6H_5)_3SiCl + H_2S \xrightarrow[-R_3N \cdot HCl]{R_3N} (C_6H_5)_3SiSH, \quad (2.6)$$

and diethyl(triethylsilylmethyl)silanethiol (80% yield) [89]

$$(C_2H_5)_3SiCH_2Si(C_2H_5)_2Br + H_2S \xrightarrow[-R_3N \cdot HCl]{R_3N} (C_2H_5)_3SiCH_2Si(C_2H_5)_2SH \quad (2.7)$$

Dimethyl(trimethylsilylmethyl)silanethiol was obtained analogously but in a much lower yield (36%) [89]. Trialkylfluorosilanes do not react with hydrogen sulfide in the presence of tertiary amines [246].

It has not been possible to obtain lower trialkylsilanethiols by the reaction of trialkylhalosilanes with hydrogen sulfide, as under the reaction conditions the lower trialkylsilanethiols are converted extremely readily into hexaalkyldisilthianes:

$$R_3SiSH + R_3SiX \xrightarrow[-HX]{} R_3SiSSiR_3 \quad (2.8)$$

$$2R_3SiSH \rightleftarrows R_3SiSSiR_3 + H_2S \quad (2.9)$$

Trialkylsilanethiols, R_3SiSH with R = C_2H_5 or C_3H_7, may be obtained in about 19% yield by the action of hydrogen sulfide on trialkyl(amino)silanes [448, 449]:

$$R_3SiNH_2 + H_2S \longrightarrow R_3SiSH + NH_3 \quad (2.10)$$

Trimethylsilanethiol is formed in 14% yield by the reaction of hydrogen sulfide with trimethyl(phenylamino)silane [449]:

$$(CH_3)_3SiNHC_6H_5 + H_2S \longrightarrow (CH_3)_3SiSH + C_6H_5NH_2 \quad (2.11)$$

The most general method of synthesizing trialkylsilanethiols is based on the reaction of trialkylhalosilanes with alkali and alkaline earth metal hydrosulfides [246]:

$$nR_3SiX + M(SH)_n \longrightarrow nR_3SiSH + MX_n \quad (M = Li, Na, Mg;\ n = \text{valence of M}) \tag{2.12}$$

This method was also used for the synthesis of trichlorosilanethiol, which was obtained in 30-35% yield by heating silicon tetrachloride with sodium hydrosulfide in a sealed tube at 200°C [369]. However, an attempt to obtain silanethiol analogously was unsuccessful; the only product from the reaction of sodium hydrosulfide and iodosilane was disilthiane [294].

The equilibrium reaction between disilthiane and hydrogen sulfide

$$H_3SiSSiH_3 + H_2S \rightleftarrows 2H_3SiSH \tag{2.13}$$

is still the only convenient method of synthesizing silanethiol [296, 466], since the reaction of iodosilane with sulfur forms a whole series of other substances in addition to silanethiol [164].

Triethylsilane reacts with hydrogen sulfide in the presence of 5% Pd/Al_2O_3 and is converted quantitatively into triethylsilanethiol [853, 962]:

$$(C_2H_5)_3SiH + H_2S \xrightarrow{Pd/Al_2O_3} (C_2H_5)_3SiSH + H_2 \tag{2.14}$$

The latter is also formed by heating triethylsilane with sulfur at 240°C [36, 747].

Triphenylsilane does not react with H_2S in the absence of a catalyst even at 200-210°C. On the other hand, its reaction with sulfur in tetralin proceeds even at 180°C, and the yield of triphenylsilanethiol after the reaction mixture has been heated for 15 h is 80% [196, 237, 744, 751, 867, 903]:

$$R_3SiH + S_n \longrightarrow R_3SiS_nH \tag{2.15a}$$

$$R_3SiS_nH + (n-1)R_3SiH \rightarrow nR_3SiSH \tag{2.15b}$$

The reaction of sulfur with triphenylsilyllithium forms lithium triphenylsilanethiolate [354]. It must be assumed that this reaction is of a general nature and is a convenient method for synthesizing triarylsilanethiolates of alkali metals:

$$R_3SiM + S \longrightarrow R_3SiSM \quad (M = Li, Na, K) \tag{2.16}$$

The corresponding trialkyl derivatives are formed by cleavage of dialkylcyclopolysilthianes by an alkyllithium [617, 946], for example:

$$[(CH_3)_2SiS]_3 + 3LiCH_3 \longrightarrow 3(CH_3)_3SiSLi \qquad (2.17)$$

The formation of triethoxysilanethiol has been observed in the reaction of trichlorosilanethiol with ethyl alcohol [331]. However, it was not possible to isolate the triethoxysilanethiol in a pure form, since this compound has a boiling point close to the boiling point of tetraethoxysilane, which is formed at the same time.

A very convenient method of synthesizing trialkoxysilanethiols is based on the reaction of silicon disulfide with alcohols [553, 722-725]:

$$SiS_2 + 3ROH \longrightarrow (RO)_3SiSH + H_2S \qquad (2.18)$$

An attempt to obtain triisocyanatosilanethiol $HSSi(NCO)_3$ by the reaction of tetraisocyanatosilane $Si(NCO)_4$ with hydrogen sulfide at 180°C even in the presence of aluminum chloride was unsuccessful [148, 151]. It was impossible to synthesize dimethylsilanedithiol by the reaction of dimethylbis(methylamino)silane with hydrogen sulfide [449]. The formation of the product from the replacement of only one methylamino group was observed in this case:

$$(CH_3)_2Si(NHCH_3)_2 + H_2S \longrightarrow (CH_3)_2Si(NHCH_3)SH + CH_3NH_2 \qquad (2.19)$$

Compounds containing more than one sulfhydryl group at the silicon atom are unknown.* All attempts to obtain them have been unsuccessful and have lead to the formation of the corresponding cyclosilthianes.

1.2. Physical Properties

Silanethiol is a gas at room temperature. All known trialkylsilanethiols are liquids, while triphenylsilanethiol and α-naphthylphenylmethylsilanethiol are crystalline substances. The physical properties of compounds containing the grouping Si–S–H are given in Table 14. The Si–S and Si–Cl interatomic distances in the Cl_3SiSH molecule equal 2.14 and 2.02 ± 0.02 Å, respec-

*Diphenylsilanedithiol was mentioned in a patent [322] without confirmation of the structure.

TABLE 14. Compounds Containing the Grouping Si—S—H (Silanethiols)

Empirical formula	Compound	B.p., °C (mm)	n_D^{20}	d_4^{20}	Literature
Br_3HSSi	Br_3SiSH	m.p. 96	—	—	705
Cl_3HSSi	Cl_3SiSH	95 (768)	—	—	634
		95.5—95.6 (754)	—	—	545
		m.p. —57			
		95.6—95.7 (759)	—	—	718
		95.8	*	—	369
H_4SSi	H_3SiSH	14.2 (760)	**	—	337
		m.p. —34			
$C_3H_{10}SSi$	$(CH_3)_3SiSH$	75—76	—	—	448
		77—78	—	—	246, 369
$C_3H_{10}O_3SSi$	$(CH_3O)_3SiSH$	132—134	—	—	724
$C_6H_{16}SSi$	$(C_2H_5)_3SiSH$	158	—	—	448
		160—167	—	—	449
$C_6H_{16}O_3SSi$	$(C_2H_5O)_3SiSH$	164—167	—	—	331
$C_6H_{18}SSi$	$(CH_3)_3SiCH_2Si(CH_3)_2SH$	58—60 (14)	1.4604	0.8632	89, 99, 111
$C_9H_{22}SSi$	$(C_3H_7)_3SiSH$	83—84 (7)	—	—	448
$C_9H_{22}O_3SSi$	(iso-$C_3H_7O)_3SiSH$	188	1.4092	0.930	725
		105 (50)	—	—	724
$C_{11}H_{28}SSi_2$	$(C_2H_5)_3SiCH_2Si(C_2H_5)_2SH$	113—114 (3.5)	1.4852	0.8989	89, 99
					111
$C_{12}H_{28}O_3SSi$	(sec-$C_4H_9O)_3SiSH$	231—233	1.4200	0.916	724, 725
		138 (35)			
	(tert-$C_4H_9O)_3SiSH$	115 (35)	1.4231	0.924	553
		m.p. $<$—20			
$C_{15}H_{34}O_3SSi$	$(2\text{-}C_5H_{11}O)_3SiSH$	166.5 (16)	1.4305	0.913	722
	(tert-$C_5H_{11}O)_3SiSH$	109—110 (10)	1.4382	0.930	553
$C_{17}H_{16}SSi$	$\alpha\text{-}C_{10}H_7(C_6H_5)CH_3SiSH$	m.p. 63—64	—	—	664
$C_{18}H_{16}SSi$	$(C_6H_5)_3SiSH$	152—158 (0.06)	—	—	196, 301
		m.p. 103	—	—	301
$C_{21}H_{27}NSSi$	$\alpha\text{-}C_{10}H_7(C_6H_5)CH_3SiSH \cdot HN(C_2H_5)_2$	m.p. 105—130 (decomp.)	—	—	664

*$\log p = 7.5927 - 1738.6/T$; molar heat of evaporation 7950 cal/mole; Trouton's constant 21.55.

**$\log p = 7.596 - 1355/T$ (from −75 to −35°C); molar heat of evaporation 6201 cal/mole; Trouton's constant 21.9.

tively [718]. Data on the lengths of the Si–S and Si–Br bonds in Br_3SiSH (2.25 and 2.28 Å, respectively) are evidently less reliable [705]. The ν S–H bond vibrations in the Raman spectra of $(CH_3)_3SiSH$ and Cl_3SiSH [369, 424, 427] correspond to the frequencies 2575 cm^{-1} and 2565 cm^{-1}, respectively, which are also characteristic of the sulfhydryl group in alkanethiols and hydrogen sulfide. However, the intensity of the Raman line of ν S–H in silanethiols is appreciably lower than in alkanethiols. The frequency ν_s Si–S is represented by quite an intense line in the region of 450-465 cm^{-1} [454 cm^{-1} for $(CH_3)_3SiSH$ and 462 cm^{-1} for Cl_3SiSH]. An absorption band at 467 cm^{-1} in the infrared spectrum of $(C_2H_5)_3SiSH$ corresponds to this [759]. The force constant of the Si–SH bond calculated from spectroscopic data equals 2.20 mdyn/Å, which corresponds to the value f(Si–S) = 2.22 mdyn/Å calculated for a pure single Si–S bond [369].

Spectroscopic determination of the acidity of triorganosilanethiols show that they are stronger proton donors than their carbon analogs:

$$(C_6H_5)_3SiSH \gg C_6H_5SH > (CH_3)_3SiSH > (C_6H_5)_3CSH > (CH_3)_3CSH$$

An analogous result was obtained by potentiometric titration of triorganosilanethiols and alkanethiols with tetrabutylammonium hydroxide in pyridine. This increase in acidity of silanethiols in comparison with organic thiols (as in the case of analogous silanols and tertiary alkanols) may be explained by delocalization of the free electron pair of the sulfur atom in the free 3d-orbital of the silicon atom [938]. The presence of a $p_\pi - d_\pi$ interaction between the silicon and sulfur atoms in triorganosilanethiols is also confirmed by a study of their NMR spectra [756].

On the whole, the degree of $p_\pi - d_\pi$ interaction between a silicon atom and Group VI elements falls with an increase in their atomic number, i.e., in the series O > S > Se > Te [47a, 611, 614, 756].

1.3. Chemical Properties

As a result of the decreased degree of $p_\pi - d_\pi$ interaction between the silicon and sulfur atoms, the Si–S bond is more reactive in heterolytic reactions than the Si–O bond which has a considerable degree of double bonding.

The high reactivity of the Si−S bond is responsible for the considerable instability of silanethiols, which is the reason for the comparative difficulty of obtaining them. The intermolecular condensation with the liberation of H_2S is even more characteristic of silanethiols than the analogous dehydration reaction of silanols:

$$2R_3SiSH \longrightarrow R_3SiSSiR_3 + H_2S \qquad (2.20)$$

In the case of silanethiol itself, this reaction, which leads to the formation of disilthiane, proceeds spontaneously even at −78°C, and above −30°C it proceeds at a high rate [296]. However, if at the silicon atom attached to the sulfhydryl group there are substituents which create steric hindrance or have a high electronegativity, the intermolecular liberation of H_2S is inhibited. The conversion of such quite stable substituted silanethiols to the corresponding disilthianes may be achieved by the scheme [89, 99, 111]

$$R_3SiSH + XSiR_3 \xrightarrow[-R_3N\cdot HCl]{R_3N} R_3SiSSiR_3 \qquad (2.21)$$

The condensation of trichlorosilanethiol proceeds with particular difficulty, the conversion to hexachlorodisilthiane occurring at an appreciable rate only at 650–700°C [545]:

$$2Cl_3SiSH \longrightarrow Cl_3SiSSiCl_3 + H_2S \qquad (2.22)$$

At the same time there is intermolecular condensation with the elimination of hydrogen chloride and also the condensation of trichlorosilanethiol with $SiCl_4$:

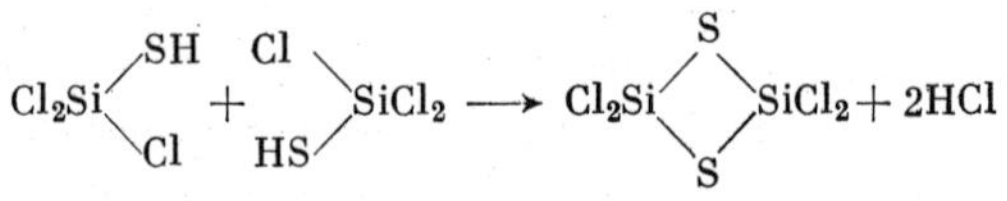
(2.23)

$$Cl_3SiSH + SiCl_4 \longrightarrow Cl_3SiSSiCl_3 + HCl \qquad (2.24)$$

The difficulty of condensing trichlorosilanethiol to hexachlorodisilthiane should be ascribed to the accumulation of electronegative chlorine atoms at the silicon atom, which leads to a fall in the polarity of the Si−S bond.

The condensation of tribromosilanethiol to tetrabromocyclodisilthiane proceeds in the presence of aluminum bromide without difficulty [305].

A sulfhydryl group attached to the silicon atom very readily undergoes many exchange and condensation reactions. In particular, we should point out the following well-known conversions of silanethiols:

$$\equiv SiSH + H_2O \longrightarrow \equiv SiOH + H_2S \qquad (2.25)$$

$$\equiv SiSH + HOR \longrightarrow \equiv SiOR + H_2S \quad [327, 329-331] \qquad (2.26)$$

$$2 \equiv SiSH + 3Br_2 \longrightarrow 2 \equiv SiBr + S_2Br_2 + 2HBr \quad [327, 329-331] \qquad (2.27)$$

$$\equiv SiSH + AgNCO \longrightarrow \equiv SiNCO + AgSH \quad [148] \qquad (2.28)$$

$$\equiv SiSH + (RCO)_2O \longrightarrow \equiv SiOCOR + RCOSH \quad [282] \qquad (2.29)$$

Silanethiols are readily hydrolyzed by water, which accounts for the odor of hydrogen sulfide that is characteristic of many of them, and also they are readily oxidized by atmospheric oxygen with the formation of free sulfur [725]. Silanethiols are readily oxidized by iodine and react with lead, copper, nickel, and mercury carbonates and acetates; this may be used for their quantitative and qualitative determination [725]. The reduction of silanethiols with $LiAlH_4$ forms hydrosilanes [664]. N-Phenyl-α-chlorobenzalimine cleaves the Si−S bond in triphenylsilanethiol on heating in xylene. Triphenylchlorosilane and N-phenyl(thiobenzamide) are then formed in 65% yield [475]:

$$C_6H_5CCl{=}NC_6H_5 + (C_6H_5)_3SiSH \longrightarrow (C_6H_5)_3SiCl + C_6H_5CSNHC_6H_5 \qquad (2.30)$$

The reactivity of the Si−SH group in silanethiols containing branched substituents at the silicon atom is considerably reduced. Thus, for example, tri(sec-alkoxy)- and tri(tert-alkoxy)silanethiols are resistant to the action of atmospheric moisture (in particular they have no odor of hydrogen sulfide) and also to the action of secondary and tertiary alcohols [553, 725].

Tribromosilanethiol reacts readily with diazomethane [705]. Concentrated benzene solutions of Br_3SiSH are electrically conducting. When tribromosilanethiol is electrolyzed, hydrogen is liberated at the cathode and hexabromodisildithiane is formed at the anode:

$$2Br_3SiSH \longrightarrow H_2 + Br_3SiSSSiBr_3 \qquad (2.31)$$

The latter compound is very unstable, and with prolonged electrolysis it decomposes with the formation of tetrabromocyclodisilthiane and other substances [705].

Of the reactions of silanethiols which proceed with retention of the Si – S bond, that with alkali metals has been studied in greatest detail [196, 750]. When triphenylsilanethiol is heated with sodium or potassium in benzene for 4 h the corresponding triphenylsilanethiolate is formed in 87–88% yield:

$$2\,(C_6H_5)_3SiSH + 2M \longrightarrow 2\,(C_6H_5)_3SiSM + H_2 \qquad (2.32)$$
$$(M = Na,\ K)$$

Lithium does not react with triphenylsilanethiol under analogous conditions. However lithium triphenylsilanethiolate may be obtained by the reaction of triphenylsilanethiol with phenyllithium at room temperature (81% yield) [196]:

$$(C_6H_5)_3SiSH + C_6H_5Li \longrightarrow (C_6H_5)_3SiSLi + C_6H_6 \qquad (2.33)$$

The diethylamine salt of α-naphthyl(phenyl)methylsilanethiol, which was formed by the hydrolysis of the corresponding disilthiane with an aqueous solution of diethylamine, has also been characterized [664].

Heating triphenylsilanethiol with benzoyl chloride in xylene forms O-triphenylsilyl thiobenzoate (75% yield) and not the expected S-triphenylsilyl derivative [282]:

$$(C_6H_5)_3SiSH + ClCOC_6H_5 \xrightarrow{-HCl} C_6H_5\underset{\underset{S}{\|}}{C}OSi(C_6H_5)_3 \qquad (2.34)$$

2. COMPOUNDS CONTAINING THE GROUPING Si – S – C (ORGANOTHIOSILANES)

2.1. Preparation Methods

Like hydrogen sulfide, alkanethiols hardly react with chloro-silanes at room or moderate temperatures [269]. Thiophenols (arenethiols) behave similarly. Only on prolonged heating of thiophenol with silicon tetrachloride is it possible to isolate (phenylthio)trichlorosilane [398]:

$$C_6H_5SH + SiCl_4 \longrightarrow C_6H_5SSiCl_3 + HCl \qquad (2.35)$$

However, by carrying out the reaction of halosilanes with thiols in the presence of hydrogen halide acceptors (tertiary amines) or by using alkali metal thiolates, it is possible to obtain the corresponding alkylthiosilanes in good yields. Thus, for example, tetra(alkylthio)silanes were obtained by the reaction of silicon tetrachloride with sodium alkanethiolates [167, 168, 170, 416, 417]:

$$4RSNa + SiCl_4 \longrightarrow Si(SR)_4 + 4NaCl \qquad (2.36)$$

The reaction of silicon tetrachloride with the disodium de-rivative of ethanedithiol forms a spirocyclic compound [165]:

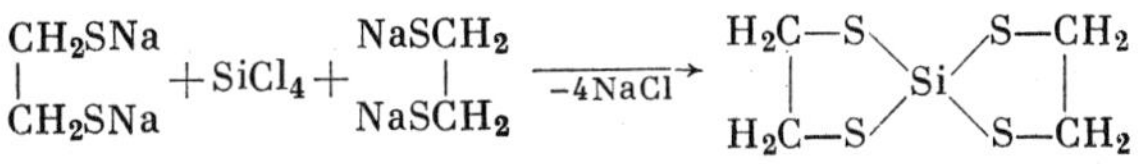

$$
\begin{array}{l}
CH_2SNa \\
| \qquad\quad + SiCl_4 + \\
CH_2SNa
\end{array}
\begin{array}{l}
NaSCH_2 \\
| \\
NaSCH_2
\end{array}
\xrightarrow[-4NaCl]{}
\begin{array}{l}
H_2C-S \diagdown \qquad \diagup S-CH_2 \\
\qquad\quad | \qquad Si \qquad | \\
H_2C-S \diagup \qquad \diagdown S-CH_2
\end{array}
\qquad (2.37)
$$

The reaction of sodium alkanethiolates with silicon tetra-chloride may also be used to obtain alkylthiochlorosilanes [166, 168]:

$$nRSNa + SiCl_4 \longrightarrow (RS)_nSiCl_{4-n} + nNaCl \qquad (2.38)$$
$$(n = 1 - 3)$$

Further reaction of alkylthiochlorosilanes obtained in this way with sodium thiolates containing an alkyl radical of a different structure makes it possible to synthesize tetra(alkylthio)silanes with different substituents [166, 169, 415, 416]:

$$(RS)_nSiCl_{4-n} + (4-n)NaSR' \longrightarrow (RS)_nSi(SR')_{4-n} + (4-n)NaCl \quad (n=1-3) \tag{2.39}$$

The reaction of difluorodichlorosilane with sodium ethanethiolate leads to the simultaneous formation of $(C_2H_5S)_2SiF_2$ and $(C_2H_5S)_3SiF$ [454, 632]. The reaction of SiF_2Cl_2 with sodium 1,6-hexanedithiolate gave a yellow oily polymer [632], which evidently had the structure $[-S(CH_2)_6SSiF_2-]_n$.

The reaction of trichlorosilane with sodium alkanethiolate makes it possible to obtain tri(alkylthio)silanes in low yield (3-49%) [727]:

$$3RSNa + HSiCl_3 \longrightarrow HSi(SR)_3 + 3NaCl \tag{2.40}$$

Tetra(alkylthio)silanes are formed simultaneously with these, and their yield increases with a rise in the reaction temperature.

Organochlorosilanes react with alkanethiolates and thiophenolates of alkali metals such as $SiCl_4$ to form the corresponding alkyl- or arylthio derivatives [69, 199, 359, 442, 508, 775-778, 843, 859]:

$$nRSNa + Cl_nSiR'_{4-n} \longrightarrow R'_{4-n}Si(SR)_n + nNaCl \quad (n=1-3) \tag{2.41}$$

In the case of sodium thiophenolates the addition of copper powder accelerates the reaction.

In addition to sodium (and other alkali metals) alkanethiolates, for the synthesis of alkylthiosilanes it is also possible to use magnesium [385], lead [129, 262, 372, 462, 608, 609], and mercury [276] alkanethiolates. Thus, for example, when trimethylchlorosilane is added to a suspension of CH_3SMgI in ether an exothermic reaction begins and leads to the formation of trimethyl(methylthio)silane (~80% yield). Dimethyl(methylthio)silane and methyltri(methylthio)silane were obtained analogously [385]:

$$(CH_3)_nSiCl_{4-n} + (4-n)CH_3SMgI \xrightarrow[-(4-n)MgICl]{} (CH_3)_nSi(SCH_3)_{4-n} \quad (n=1-3) \tag{2.42}$$

It is remarkable that $SiCl_4$ does not form tetra(methylthio)silane under the same conditions. Likewise, triphenylchlorosilane does not react with CH_3SMgBr, while trimethylchlorosilane reacts with C_6H_6SMgBr to give trimethyl(phenylthio)silane in good yield [385].

The reaction of lead alkanethiolates and thiophenolates with chlorosilanes proceeds in accordance with the equation [608, 609, 944]

$$\frac{m}{2}(RS)_2Pb + R'_nSiCl_{4-n} \longrightarrow R'_n(RS)_mSiCl_{4-(m+n)} + \frac{m}{2}PbCl_2 \qquad (2.43)$$

(R = alkyl, aryl; R′ = alkyl; $n = 0-3$; $m = 1-4$)

The use of lead thiolates, which have a yellow color, is particularly convenient in that the course of the reaction may be followed through the change in color of the reaction mixture (white lead chloride is formed). However, it should be noted that trimethyl(methylthio)silane cannot be obtained by this method since no reaction with lead methanethiolate is observed at the boiling point of trimethylchlorosilane.* The lead methanethiolate decomposes when the reaction is carried out in a sealed tube at 120°C [385].

The reaction of iodosilane with mercury trifluoromethanethiolate proceeds smoothly at room temperature with the formation of (trifluoromethylthio)silane (90% yield) [276]:

$$2H_3SiI + Hg(SCF_3)_2 \longrightarrow 2H_3SiSCF_3 + HgI_2 \qquad (2.44)$$

The complex of silicon tetrachloride with pyridine reacts smoothly with methanethiol in the presence of pyridine to form tetra(methylthio)silane [385]:

$$SiCl_4 \cdot 2C_5H_5N + 4CH_3SH + 2C_5H_5N \longrightarrow (CH_3S)_4Si + 4C_5H_5N \cdot HCl \qquad (2.45)$$

This method also gives more satisfactory (in comparison with thiolates) yields (45-65%) in the synthesis of tri(alkylthio)-silanes [727]:

$$HSiCl_3 + 3HSR \xrightarrow[-3C_5H_5N \cdot HCl]{+3C_5H_5N} (RS)_3SiH \qquad (2.46)$$

Methylthiosilane has been synthesized analogously [673]. However, it is not possible to obtain tri(tert-butylthio)silane by this method.

*It was shown recently that trimethyl(methylthio)silane is formed in 74% yield by carrying out this reaction in diethyl ether for 4 days [133].

In the presence of tertiary amines, alkanethils also react with organochlorosilanes [69, 380, 504, 541], alkoxychlorosilanes [539], and hexachlorodisiloxane [540] to form the corresponding alkylthio derivatives:

$$R_{4-n}SiCl_n + nHSR' \xrightarrow[-nR_3''N \cdot HCl]{+nR_3''N} R_{4-n}Si(SR')_n \qquad (2.47)$$
$$(n = 1-3)$$

It should be noted that according to data in [385] methanethiol does not react with trimethylchlorosilane on heating to 70°C in the presence of pyridine, while dimethyldi(methylthio)silane is formed smoothly in the reaction with dimethyldichlorosilane in the presence of triethylamine [504].

Alkylchlorosilanes also react with mercaptocarboxylic acids and their esters according to Eq. (2.47) [577].

Cyclic compounds containing one or two Si$-$S$-$C groupings are formed by using, in reaction (2.47), diorganodichlorosilanes or dialkylchloromethylchlorosilanes and bifunctional mercapto compounds such as alkanedithiols [575, 710, 712, 715], dithiophenols (arenedithiols) [710, 717], o-aminothiophenol [711, 716], and o-mercaptobenzoic acid [713]. Thus, for example, the reaction of dimethyldichlorosilane with 1,2-ethanedithiol in the presence of triethylamine gives an 80% yield of 2,2-dimethyl-2-sila-1,3-dithiacyclopentane [710, 715]:

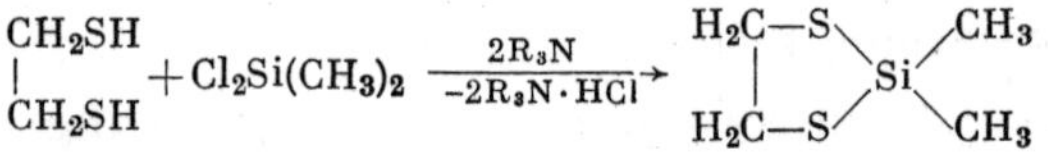

$$\begin{array}{c} CH_2SH \\ | \\ CH_2SH \end{array} + Cl_2Si(CH_3)_2 \xrightarrow[-2R_3N \cdot HCl]{2R_3N} \begin{array}{c} H_2C-S \\ | \\ H_2C-S \end{array} Si \begin{array}{c} CH_3 \\ \\ CH_3 \end{array} \qquad (2.48)$$

The use of dimethyl(chloromethyl)chlorosilane instead of dimethyldichlorosilane leads to the formation of 2,2-dimethyl-2-sila-1,4-dithiacyclohexane [712]:

$$\begin{array}{cc} CH_2SH & ClCH_2 \\ | & | \\ CH_2SH & ClSi(CH_3)_2 \end{array} \xrightarrow[-2R_3N \cdot HCl]{2R_3N} \qquad (2.49)$$

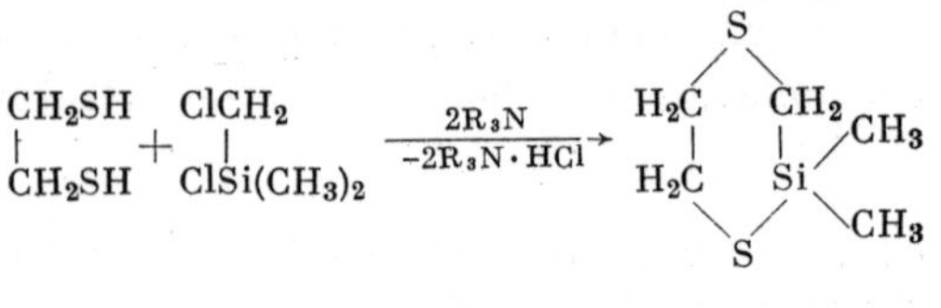

The Si$-$N bond in amino- and alkylaminosilanes is cleaved by the action of alkanethiols and thiophenols with the formation of the corresponding alkyl- and arylthio derivatives [129, 137, 186, 449, 589]:

$$R_{4-n}Si(NR'R'')_n + nHSR''' \longrightarrow R_{4-n}Si(SR''')_n + R'R''NH \qquad (2.50)$$

In this way it is also possible to obtain compounds containing unlike alkylthio groups [449], for example:

$$(CH_3)_2Si(NHCH_3)_2 + C_5H_{11}SH + C_6H_5SH \xrightarrow{-2CH_3NH_2} (CH_3)_2Si(SC_5H_{11})SC_6H_5 \quad (2.51)$$

The reaction of alkanethiols with alkylaminosilanes (but not with arylamino derivatives) is reversible, since the free amine formed is capable of cleaving the Si$-$S bond [130]. Therefore the synthesis of alkylthiosilanes from alkylamino derivatives is possible only with the removal of the amine formed from the reaction mixture. This may be achieved if the latter has a boiling point lower than that of the original thiol [129]. Hexaalkyldisilazans are not cleaved by alkanethiols even in the presence of trialkyl-chlorosilanes [442]. They also behave analogously toward hydrogen sulfide [442]. However, according to data in [669], allyl mercaptan reacts smoothly with an equimolecular mixture of hexamethyldisilazan and trimethylchlorosilane to form trimethyl-(allylthio)silane:

$$[(CH_3)_3Si]_2NH + (CH_3)_3SiCl + 3CH_2=CHCH_2SH \longrightarrow$$
$$\longrightarrow 3(CH_3)_3SiSCH_2CH=CH_2 + NH_4Cl \qquad (2.52)$$

When cysteine hydrochloride is boiled with hexamethyldisilazan there is silylation of the amino and carboxyl groups and also the mercapto group [195]. Cleavage of the silazan grouping occurs when 1,3-bis(chloromethyl)tetramethyldisilazan is treated with hydrogen sulfide, since the product of their interaction in the presence of triethylamine is 2,2,5,5-tetramethyl-1,4-dithia-2,5-disilacyclohexane [628]:

$$[(CH_3)_2SiCH_2Cl]_2NH + 2H_2S \xrightarrow[\substack{-(C_2H_5)_3N\cdot HCl; \\ -NH_4Cl}]{(C_2H_5)_3N} \begin{array}{c} S \\ (CH_3)_2Si \quad CH_2 \\ | \qquad | \\ H_2C \quad Si(CH_3)_2 \\ S \end{array} \qquad (2.53)$$

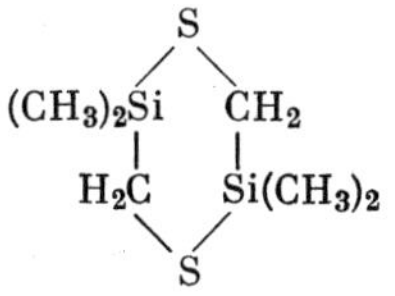

A convenient method for the synthesis of triorgano(alkylthio)-silanes is the reaction of alkyl halides with alkali metal triorgano-silianethiolates [196, 354, 750]:

$$R_3SiSM + R'X \xrightarrow[-MX]{} R_3SiSR' \quad (2.54)$$

Acylthio derivatives of silanes may be synthesized analogously [354]:

$$(C_6H_5)_3SiSLi + C_6H_5COCl \xrightarrow[-LiCl]{} (C_6H_5)_3SiSCOC_6H_5 \quad (2.55)$$

The reaction of triphenylsilyllithium with diphenyl sulfide forms triphenyl(phenylthio)silane [721], which, however, was not isolated in a free form:

$$(C_6H_5)_3SiLi + C_6H_5SC_6H_5 \longrightarrow (C_6H_5)_3SiSC_6H_5 + C_6H_5Li \quad (2.56)$$

The Si–P bond is cleaved by alkanethiols in the same way as the Si–N bond. Thus, for example, when 1-butanethiol is boiled for 12 h with the dibutyl ester of trimethylsilylphosphinic acid, trimethyl(butylthio)silane is formed (16% yield) [402]:*

$$(CH_3)_3SiP(O)(OC_4H_9)_2 + C_4H_9SH \longrightarrow (CH_3)_3SiSC_4H_9 + HP(O)(OC_4H_9)_2 \quad (2.57)$$

The product from the reaction of trimethyl(diethylamino)-silane with carbon disulfide is the trimethylsilyl ester of diethyl-dithiocarbamic acid [212, 213]. The addition of diethylamine accelerates this reaction, while the presence of dimethyldichloro-silane has no effect at all. Diethylaminotrichlorosilane likewise does not react with carbon disulfide. On the basis of these observations a hypothesis was put forward [213] that the reaction begins with the interaction of CS_2 with diethylamine (added beforehand or traces of it present in the starting trimethyldiethylamino-silane):

$$CS_2 + HN(C_2H_5)_2 \longrightarrow HS\underset{\underset{S}{\|}}{C}N(C_2H_5)_2 \quad (2.58)$$

Then there follows the reaction of the diethyldithiocarbamic acid formed with the trimethyl(diethylamino)silane:

$$(CH_3)_3SiN(C_2H_5)_2 + HS\underset{\underset{S}{\|}}{C}N(C_2H_5)_2 \longrightarrow (CH_3)_3SiS\underset{\underset{S}{\|}}{C}N(C_2H_5)_2 + (C_2H_5)_2NH \quad (2.59)$$

*For more details on this see Part I, section 1.1.1.

The structure of the compound obtained was demonstrated by confirmatory synthesis:

$$(CH_3)_3SiCl + (C_2H_5)_2NH \cdot HSC(S)N(C_2H_5)_2 \longrightarrow$$
$$\longrightarrow (CH_3)_3SiSC(S)N(C_2H_5)_2 + (C_2H_5)_2NH \cdot HCl \qquad (2.60)$$

The reaction of sodium or potassium silicofluoride with aluminum alkanethiolates, which proceeds 130–300°C in accordance with the following scheme, has been described [617]:

$$3Na_2SiF_6 + 4Al(SR)_3 \longrightarrow 3Si(SR)_4 + 4AlF_3 + 6NaF \qquad (2.61)$$

The reaction of a trialkylsulfonium iodide with silver hexafluorosilicate forms a trialkylsulfonium hexafluorosilicate [295].

An attempt to prepare trialkyl(organothio)silanes by the reaction of trialkylsilanes with alkanethiols [69, 962] was unsuccessful, however, triphenylsilane and thiocresol react in accordance with the scheme (2.62) on prolonged boiling [362, 720]:

$$R_3SiH + R'SH \longrightarrow R_3SiSR' + H_2O \qquad (2.62)$$

On heating to 190–200°C in the presence of zinc chloride, triethylsilane cleaves the S–S bond in diphenyl disulfide to form thiophenol and triethyl(phenylthio)silane [795]. In this reaction triphenylsilane forms triphenyl(phenylthio)silane [477].* The latter could not be obtained by the reaction of triphenylsilane with C_6H_5SNa or by the action of hydrogen sulfide on triphenyl(phenoxy)-silane at 320°C [359].

In the presence of zinc chloride, triethylsilane reduces the C–S bond in 1,2-dithia-4-cyclopentene-3-thiones to CH_2 with simultaneous cleavage of the S–S bond and the formation of Si–S–C groupings [200, 840].

2.2. Physical Properties

The physical constants of organothiosilanes are given in Table 16; available data [166, 169, 170, 415, 416] on their crystalline form are given in the same table.

*In contrast to this, according to data in [359] triphenylsilane does not react with $(C_6H_5)_2S_2$ at 260°C.

TABLE 15. NMR Spectra of Organothiosilanes
(Negative values indicate a shift (in ppm) to low fields relative to $(CH_3)_4Si$ as an internal standard)

Compound	δ_{H-C-Si}	δ_{H-C-S}	Literature
$(CH_3)_3SiSCH_3$	−0.275	−1.93	385
$(CH_3)_2Si(SCH_3)_2$	−0.50	−2.00	385
$CH_3Si(SCH_3)_3$	−0.70	−2.08	385
$(CH_3S)_4Si$	—	−2.16	385
$(CH_3)_3SiSC_6H_5$	−0.24	—	385
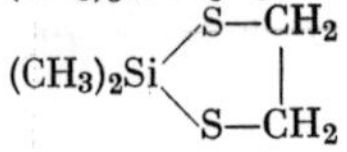	−0.61	−3.1	715
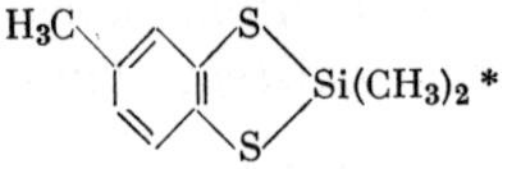 *	−0.73	—	717
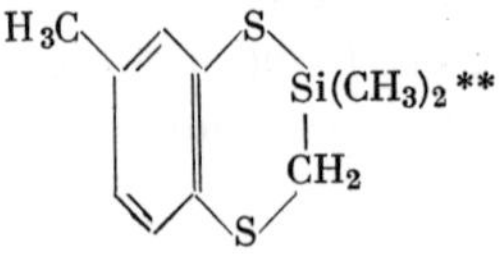 **	−0.38	−2.05	717
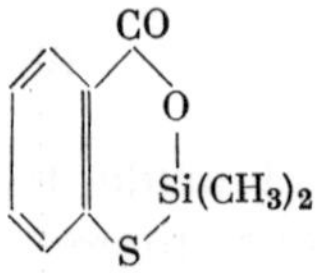	−0.66	—	713
CF_3SSiH_3 ***	—	—	276

* $\delta_{H_{ar}} = -7.0$; $\delta_{CH_3\text{-}C} = -2.22$.
** $\delta_{H_{ar}} = -7.1$; $\delta_{CH_3\text{-}C} = -2.28$.
*** $\delta_{Si\text{-}H} = -5.58$.

In the infrared spectra of trialkyl(organothio)silanes an absorption maximum is observed in the region of 460–490 cm^{-1}, which corresponds to the bond vibrations of the Si – S bond [129, 385]. The frequencies of 507 and 514 cm^{-1} in the spectrum of (trifluoromethylthio)silane correspond to this [276]. The infrared spectrum of dimethyldi(methylthio)silane contains two maxima at 431 and 512 cm^{-1}, while the spectrum of tri(methylthio)silane has two maxima at 522 and 540 cm^{-1} [385]. They also may be assigned in some measure to ν Si – S. The question of the frequency ν S – C in the grouping Si – S – C is less clear, since the frequency of 639–641 cm^{-1} in the infrared spectra of trialkyl(organothio)silanes is assigned to it [129], the frequency of 695 cm^{-1} in the spectrum of methylthiosilane [673] and the frequency of 440-450 cm^{-1} or 751 cm^{-1} in the spectrum of (trifluoromethylthio)silane [276]. A spectroscopic investigation of the reaction of phenol with trimethyl-(tert-butylthio)silane shows the latter is an extremely weak base [706]. The same result was obtained in a spectroscopic investigation of the basicity of trimethyl(methylthio)silane and trimethyl-(ethylthio)silane [133a].

In the ultraviolet spectra of methyl(methylthio)silanes $(CH_3)_{4-n}Si(SCH_3)_n$ two absorption bands are observed with maxima at 203-204.5 and 224.5-226.5 mμ, and are increased in intensity with an increase in n. The first of these maxima is characteristic of all compounds of divalent sulfur, and its value remains constant with a change to the analogous derivatives of carbon and germanium. The intensity of the second maximum increases in the series C < Si < Ge. In the spectrum of $(CH_3)_3CSCH_3$ this band is not observed at all, while in the spectrum of $(CH_3)_3SiSCH_3$ there is an absorption band with λ_{max} = 224.5 mμ (log ε = 2.15). Moreover, in the case of methylthiosilanes there is a hypsochromic shift of this maximum relative to the spectra of analogous derivatives of carbon (234.5-244.5 nm) and germanium (~247.5 nm). All this may indicate an interaction of the free electron pair of the sulfur atom with vacant 3d-orbitals of silicon [262].

A comparison of the experimental values of the dipole moments of methyl(methylthio)silanes $(CH_3)_{4-n}Si(SCH_3)_n$ (1.73 D with n = 1; 1.25 D with n = 2; 1.84 D with n = 3; 1.41 D with n = 4) with the values calculated for fixed conformations and those calculated taking into account free rotation about the Si−S bond shows that free rotation is strongly hindered in methyl(methylthio)silane [264]. The moment of the group $(CH_3)_3SiS$ was calculated to equal 1.44 D from the experimental value of the dipole moment of trimethyl-(methylthio)silane [263].

In the proton magnetic resonance spectra of compounds of the type $(CH_3S)_nSi(CH_3)_{4-n}$, a characteristic relation is observed between the chemical shifts of the methyl proton and the total inductive effect of the substituents. With an increase in n all signals are shifted to lower fields [385, 824, 858]. The results of investigating the NMR spectra of organic thiosilanes are given in Table 15.

TABLE 16. Compounds Containing the Grouping Si–S–C (Organosilanes)

Empirical formula	Compound	B.p., °C (mm)	M.p., °C (crystal form)	n_D^{20}	d_4^{20}	Literature
CH_3F_3SSi	CF_3SSiH_3	13.6±0.2	−127±0.5	—	—	276
CH_6SSi	CH_3SSiH_3	—	−116.7±0.4	—	—	673
$C_2H_5Cl_3SSi$	$C_2H_5SSiCl_3$	31 (0.1)	—	—	—	610
		151	—	—	—	608
$C_2H_7BrS_2Si$	$(CH_3S)_2SiHBr$	70—72 (8)	—	1.5660^{25}	1.5041_{25}^{25}	729
C_2H_8SSi	$C_2H_5SSiH_3$*	15—20	—	—	—	610
$C_3H_8Cl_2SSi$	$CH_3(C_2H_5S)SiCl_2$	50—51 (18)	—	—	—	608
$C_3H_9BrS_3Si$	$(CH_3S)_3SiBr$	80—81 (1)	—	1.5978^{25}	1.4988_{25}^{25}	729
		119—121 (6)	—	—	—	726
$C_3H_{10}S_3Si$	$(CH_3S)_3SiH$	76—78 (2.5)	—	1.5761	1.1423^{25}	729
		90—91 (7)				
$C_4H_8S_4Si$	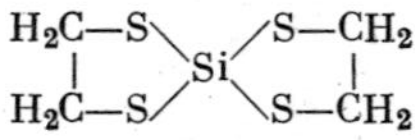	—	144	—	—	165
$C_4H_9Cl_3SSi$	tert-$C_4H_9SSiCl_3$	174—177	—	—	—	166
$C_4H_{10}S_2Si$	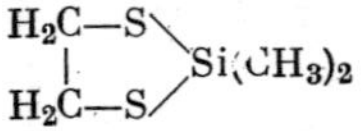	54 (2)	—	1.5571	1.1077	575
		75—77 (12)	8—10	—	—	715
		188	10	1.5534	—	136
$C_4H_{10}Br_2S_2Si$	$(C_2H_5S)_2SiBr_2$	115—115.5 (2.5)	—	1.5658^{25}	1.6541_{25}^{25}	729
$C_4H_{10}Cl_2S_2Si$	$(C_2H_5S)_2SiCl_2$	76 (0.1)	—	—	—	610
		236	—	—	—	608
$C_4H_{10}F_2S_2Si$	$(C_2H_5S)_2SiF_2$	163—164	—	—	—	454
$C_4H_{10}OSSi$	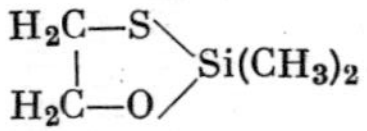	95 (17)	—	1.49274	1.0742	575
$C_4H_{11}BrS_2Si$	$(C_2H_5S)_2SiHBr$	81—84 (3)	—	1.5408^{25}	1.3717_{25}^{25}	729
$C_4H_{11}ClS_2Si$	$(C_2H_5S)_2SiHCl$	63—64 (2.5)	—	1.5160^{25}	1.1250_{25}^{25}	729
$C_4H_{12}SSi$	$(CH_3)_3SiSCH_3$	110—111	—	—	—	262, 385

$C_4H_{12}S_2Si$	$(CH_3)_2Si(SCH_3)_2$	52—54 (7)	—	—	—	385
		58 (9)	—	—	—	262
	$(C_2H_5S)_2SiH_2$	30—40 (0.1)	—	—	—	610
$C_4H_{12}S_3Si$	$CH_3Si(SCH_3)_3$	45 (0.2)	—	—	—	385
		100 (5)	—	—	—	262
$C_4H_{12}S_4Si$	$(CH_3S)_4Si$	118 (5)	—	—	—	726
		98—100 (2.5)	—	—	—	726
		144—146 (12)	31	1.5989^{25}	1.1888^{35}	167
		121 (3.8)	—	—	—	262
		66 (0.3)	30	—	—	385
$C_5H_{12}S_2Si$	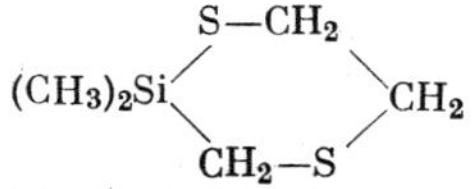	88—91 (12)	—17 to —14	1.5471	1.098	712
$C_5H_{13}ClS_2Si$	$CH_3(C_2H_5S)_2SiCl$	111—113 (28)	—	—	—	608
$C_5H_{14}SSi$	$(CH_3)_3SiSC_2H_5$	130	—	1.4512	0.832	129
		110—112	—	—	—	608
$C_6H_5Cl_3SSi$	$C_6H_5SSiCl_3$	110 (12)	—	—	—	398
		113 (17)	—	—	—	608
$C_6H_{14}O_2SSi$	$(CH_3)_3SiSCH_2COOCH_2$	87 (10)	—	1.4678	1.0097	577
$C_6H_{14}O_2S_2Si$	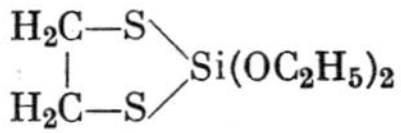	129 (19)	—	1.49564	1.344	575
$C_6H_{14}SSi$	$(CH_3)_3SiSCH_2CH{=}CH_3$	150 (740)	—	—	—	669
$C_6H_{14}S_2Si$	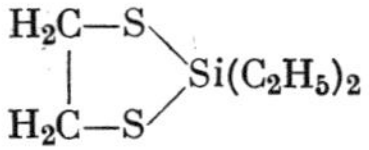	78—80 (5)	—	1.53503	1.0524	575
	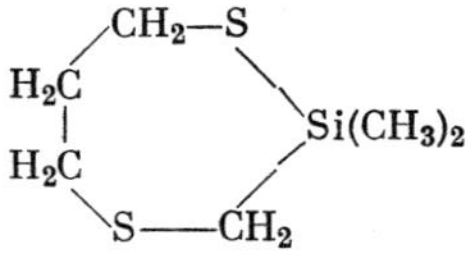	61—63 (1)	—	—	—	712

* The formation of this compound by the reduction of $C_2H_5SSiCl_3$ with lithium aluminum hydride [610] seems doubtful. Judging by the boiling point the substance obtained is silanethiol H_3SiSH.

TABLE 16 (Continued)

Empirical formula	Compound	B.p., °C (mm)	M.p., °C (crystal form)	n_D^{20}	d_4^{20}	Literature
$C_6H_{15}BrS_2Si$	$(C_3H_7S)_2SiHBr$	102—103 (2)	—	1.5265^{25}	1.2774^{25}_{25}	726
	(*изо*-$C_3H_7S)_2SiHBr$	83—85 (2)	—	1.5195^{25}	1.2720^{25}_{25}	729
$C_6H_{15}BrS_3Si$	$(C_2H_5S)_3SiBr$	155—158 (2.5)	—	1.5650^{25}	1.3508^{25}_{25}	729
$C_6H_{15}ClS_3Si$	$(C_2H_5S)_3SiCl$	97 (0.1)	—	—	—	608, 610
$C_6H_{15}FS_3Si$	$(C_2H_5S)_3SiF$	114—115 (10)	—	—	—	454
$C_6H_{16}SSi$	$(CH_3)_3SiSC_3H_7$	151	—	1.4524	0.844	129
	$(CH_3)_3SiSC_3H_7$-iso	142	—	1.4497	0.824	129
$C_6H_{16}S_2Si$	$(CH_3)_2Si(SC_2H_5)_2$	81 (10)	—	—	—	608
$C_6H_{16}S_2Si_2$	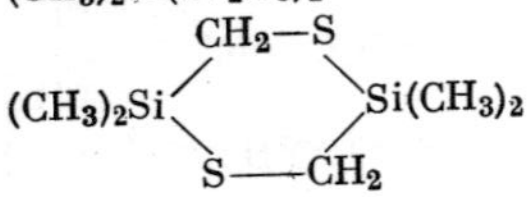	—	81—83	—	—	628
$C_6H_{16}S_3Si$	$(C_2H_5S)_3SiH$	87—88 (1) 104—105 (3) 108 (6)	—	1.5440^{25}	1.0484	727
$C_7H_8Cl_2SSi$	$CH_3(C_6H_5S)SiCl_2$	84—85 (3—4)	—	—	—	608
$C_7H_{16}S_2Si$	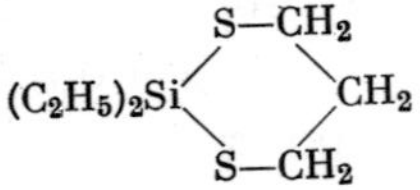	110—120 (6)	—	—	—	575
$C_7H_{18}SSi$	$(CH_3)_3SiSC_4H_9$	171—172.5 (740)	—	1.4540	0.842^{20} 0.837^{25}	442
		168	—	1.4550	0.854	129
	$(CH_3)_3SiSC_4H_9$-tert	157	—	1.4570	0.834	129
$C_8H_{11}NSSi$	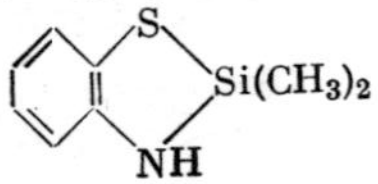	—	95	—	—	716

$C_8H_{14}OSSi$	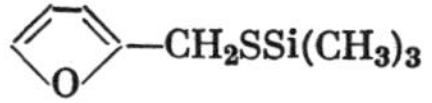	94—95 (11)	—	1.5020	0.9937	69
$C_8H_{18}Br_2S_2Si$	(iso-$C_4H_9S)_2SiBr_2$	76—79 (3.5)	—	1.497^{25}	1.3566^{25}_{25}	729
$C_8H_{18}Cl_2S_2Si$	(tert-$C_4H_9S)_2SiCl_2$	133.5—135 (13)	—	1.522^{16}	—	166
$C_8H_{19}BrS_2Si$	(iso-$C_4H_9S)_2SiHBr$	121—125 (3.5)	—	1.5159^{25}	1.2481^{25}_{25}	729
$C_8H_{19}ClS_2Si$	$(C_4H_9S)_2SiHCl$	121—123 (7)	—	1.5030^{25}	1.0358^{25}_{25}	727
	(tert-$C_4H_9S)_2SiHCl$	78—80 (4)	—	1.5040^{25}	1.0222^{25}_{4}	727
$C_8H_{19}NS_2Si$	$(CH_3)_3SiSC(S)N(C_2H_5)_2$	92—97 (0.7)	—	1.5481	—	179
		98 (1)				
$C_8H_{20}O_2SSi_2$	$(CH_3)_3SiSCH_2COOSi(CH_3)_3$	105 (2.5)	—	1.5640	—	213
$C_8H_{20}SSi$	$(C_2H_5)_3SiSC_2H_5$	108 (11.5)	—	1.4552	0.9689	577
$C_8H_{20}S_2Si$	$(C_2H_5)_2Si(SC_2H_5)_2$	56—58 (0.1)	—	—	—	610
$C_8H_{20}S_3Si$	$C_2H_5Si(SC_2H_5)_3$	81—82 (0.1)	—	—	—	610
$C_8H_{20}S_4Si$	$(C_2H_5S)_4Si$	115 (0.1)	—	—	—	610
		145—147 (3)	—	—	—	726
		169—171 (12)	—5.8	1.5638^{25}	1.0860^{25}_{4}	167
				1.5591^{35}	1.0785^{35}_{4}	
$C_9H_{10}O_2SSi$	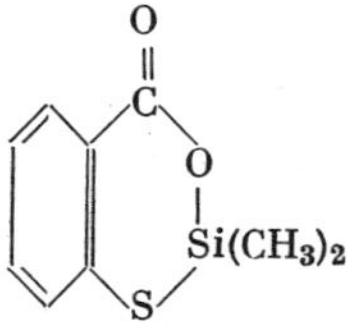	—	100	—	—	713
$C_9H_{12}S_2Si$	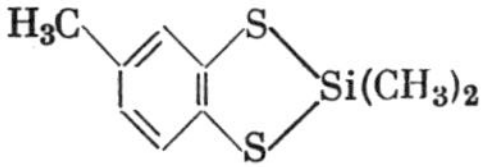	108 (2)	—	1.6142^{25}	—	717
	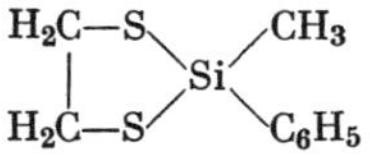	98 (0.05)	—	1.6180	—	137

TABLE 16 (Continued)

Empirical formula	Compound	B.p., °C (mm)	M.p., °C (crystal form)	n_D^{20}	d_4^{20}	Literature
$C_9H_{14}SSi$	$(CH_3)_3SiSC_6H_5$	40 (0.3)	—	—	—	385
		71(3)	—	—	—	608
		72(8)	—	—	—	187
$C_9H_{21}BrS_3Si$	$(C_3H_7S)_3SiBr$	136—138 (1.5)	—	1.5418^{25}	1.2444_{25}^{25}	729
	(iso-$C_3H_7S)_3SiBr$	132—134 (2.5)	—	1.5410^{25}	1.2244_{25}^{25}	729
$C_9H_{22}O_2SSi_2$	$(CH_3)_3SiSCH(CH_3)COOSi(CH_3)_3$	100 (10)	—	1.4515	0.9382_4^{25}	577
$C_9H_{22}S_3Si$	$(C_3H_7S)_3SiH$	120—121 (3)	—	1.5278^{25}	0.991	727
	(iso-$C_3H_7S)_3SiH$	100 (2.5)	—	1.5221^{25}	0.9864_4^{25}	727
$C_{10}H_{14}S_2Si$	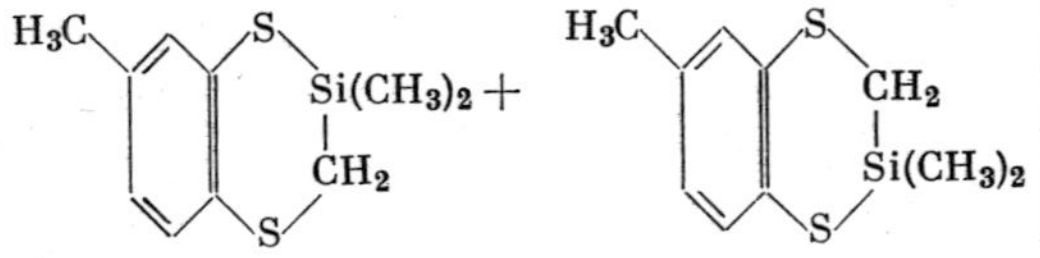	138—140 (2)	—	1.6136^{25}	—	717
$C_{12}H_{10}Cl_2S_2Si$	$(C_6H_5S)_2SiCl_2$	211 (15)	—	—	—	608
$C_{12}H_{16}O_2S_2Si$	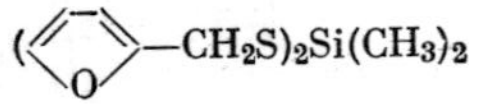	130—135 (2)	—	1.5050	1.0321	69
$C_{12}H_{27}BrS_3Si$	(iso-$C_4H_9S)_3SiBr$	143—144 (1)	—	1.5282^{25}	1.1823_{25}^{25}	729
$C_{12}H_{27}ClS_3Si$	(tert-$C_4H_9S)_3SiCl$	161—164 (3—4)	71—71.5 (hexagon.)	—	—	168, 169
$C_{12}H_{28}S_3Si$	$(C_4H_9S)_3SiH$	180—182 (9)	—	1.5160^{25}	0.9819_4^{25}	727
		132 (2)	—	—		726
	(iso-$C_4H_9S)_3SiH$	135—138 (3)	—	1.5160^{25}	0.9694_4^{25}	727
	(tert-$C_4H_9S)_3SiH$	116—120 (4)	47—48	—	—	727
$C_{12}H_{28}S_4Si$	$(C_3H_7S)_4Si$	204—206 (17)	—	1.5431^{25}	1.0328_4^{25}	167
				1.5379^{35}	1.0252_4^{35}	
	(iso-$C_3H_7S)_4Si$	176—178 (13)	33.5 (tricl.)	1.5350^{35}	1.0099_4^{35}	167

$C_{12}H_{28}OS_3Si$	(tert-$C_4H_9S)_3SiOH$	—	90—91 (tricl.)	—	—	169
$C_{12}H_{31}NO_2SSi_3$	$(CH_3)_3SiSCH_2CH[NHSi(CH_3)_3]COOSi(CH_3)_3$	84 (0.2)	—	1.4559	—	195
		78 (0.1)		1.4559	0.9334	
$C_{13}H_{13}ClS_2Si$	$CH_3(C_6H_5S)_2SiCl$	192—194 (5)	—	—	—	608
$C_{13}H_{30}S_4Si$	(tert-$C_4H_9S)_3SiSCH_3$	159—160 (4)	43—44	—	—	169
	(iso-$C_3H_7S)_3SiSC_4H_9$ tert	160—162 (3)	23—23.5	—	—	166
$C_{14}H_{32}S_4Si$	(tert-$C_4H_9S)_3SiSC_2H_5$	163—164 (4)	26—27	—	—	169
		189—194 (12)				
	(tert-$C_4H_9S)_2Si(SC_3H_7$-iso$)_2$	147—148 (2)	61.5—62.5 (rhomb. pseudo-tetragon.)	—	—	166
$C_{15}H_{26}SSi$	$(C_3H_7)_3SiSC_6H_5$	149—150 (10)				449
$C_{15}H_{34}SSi$	$(CH_3)_3SiSC_{12}H_{25}$	210—215 (0.1)	~20	1.4750	—	199
$C_{15}H_{34}S_4Si$	(tert-$C_4H_9S)_3SiSC_3H_7$	—	62—62.5 (rhomb.)	—	—	166
	(tert-$C_4H_9S)_3SiSC_3H_7$-iso	161—163 (3—4)	105 (tetragon.)	—	—	169
$C_{15}H_{34}OS_3Si$	(tert-$C_4H_9S)_3SiOC_3H_7$-iso	183—186 (15)	55—56 (monocl.)	—	—	166
$C_{16}H_{36}S_4Si$	$(C_4H_9S)_4Si$	200 (2.5)	—	1.5292^{25}	0.9958_4^{25}	168
		210 (4)	—	—	—	
	(tert-$C_4H_9S)_3SiSC_4H_9$	153—153.5 (1)	—	—	—	166
	(iso-$C_4H_9S)_4Si$	183 (4)	—	1.5255^{25}	0.9886_4^{25}	168
	(tert-$C_4H_9S)_3SiSC_4H_9$-iso	146—148 (1)	77—77.5 (rhomb.)	—	—	166
	(sec-$C_4H_9S)_4Si$	182 (4)	—	1.5354^{25}	1.0022_4^{25}	168
	(tert-$C_4H_9S)_3SiSC_4H_9$-sec	145—147 (1)	79—80	—	—	166
	(tert-$C_4H_9S)_4Si$	—	160—161 (tetragon.)	—	—	166, 169
$C_{17}H_{36}S_4Si$	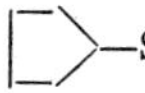—$SSi(SC_4H_9$-tert$)_3$	—	104—105 (tetragon.)	—	—	166

TABLE 16 (Continued)

Empirical formula	Compound	B.p., °C (mm)	M.p., °C (crystal form)	n_D^{20}	d_4^{20}	Literature
$C_{17}H_{38}S_4Si$	$(tert\text{-}C_4H_9S)_3SiSCH(C_2H_5)_2$	169—170 (2)	27—29	—	—	166
	$(tert\text{-}C_4H_9S)_3SiSC_5H_{11}\text{-}tert$	—	111—114 (tetragon.)	—	—	166
$C_{18}H_{15}ClS_3Si$	$(C_6H_5S)_3SiCl$	255 (2)	—	—	—	608
$C_{18}H_{18}SSi$	$\alpha\text{-}C_{10}H_7C_6H_5CH_3SiSCH_3$	—	57—59	—	—	664
$C_{18}H_{38}S_4Si$	$(tert\text{-}C_4H_9S)_3SiS$— 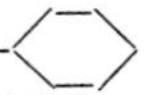	—	64—65	—	—	166
$C_{18}H_{40}S_3Si$	$(C_2H_5S)_3SiC_{12}H_{25}$	188 (2.5)	—	1.5133	—	199
$C_{19}H_{18}SSi$	$(C_6H_5)_3SiSCH_3$	145—150 (0.06)	83—84	—	—	354
		148—156 (0.04)	83—84	—	—	196
$C_{20}H_{20}SSi$	$(C_6H_5)_3SiSC_2H_5$	155—162 (0.04)	87—88	—	—	196
$C_{20}H_{44}S_4Si$	$(C_5H_{11}S)_4Si$	230—232 (3—4)	—	1.5250^{15}	0.9810_4^{15}	170
				1.5212^{25}	0.9739_4^{25}	
				1.5170^{35}	0.9666_4^{35}	
$C_{21}H_{20}SSi$	$(C_6H_5)_3SiSCH_2CH{=}CH_2$	166—173 (0.03)	73—74	—	—	196
$C_{21}H_{22}SSi$	$(C_6H_5)_3SiSC_3H_7$	162—172 (0.04)	70—71	—	—	196
	$(C_6H_5)_3SiSC_3H_7\text{-}iso$	157—162 (0.03)	84	—	—	196
$C_{24}H_{16}Br_4S_4Si$	$(p\text{-}BrC_6H_4S)_4Si$	—	216—218	—	—	170
$C_{24}H_{20}S_4Si$	$(C_6H_5S)_4Si$	—	114.5—115 (rhomb.)	—	—	168

$C_{24}H_{44}S_4Si$	$(C_6H_{11}S)_4Si$	—	101.5—102.5 (tetragon.)	—	—	170
$C_{24}H_{54}OS_6Si_2$	$[(tert\text{-}C_4H_9S)_3Si]_2O$	—	248—249 (tricl.)	—	—	166, 169
$C_{25}H_{20}OSSi$	$(C_6H_5)_3SiSCOC_6H_5$	183—189 (0.01)	128—129	—	—	354, 453
$C_{25}H_{22}SSi$	$(C_6H_5)_3SiSCH_2C_6H_5$	172—177 (0.015)	92—94	—	—	354
	$(C_6H_5)_3SiSC_6H_4CH_3\text{-}p$	250—265 (3)	79—80	—	—	720
$C_{26}H_{56}S_2Si$	$(CH_3)_2Si(SC_{12}H_{25})_2$	210 (0.01)	29.5—30	—	—	199
$C_{28}H_{28}SSi$	$(o\text{-}CH_3C_6H_4)_3SiSC_6H_4CH_3\text{-}p$	—	112—113	—	—	359
$C_{28}H_{28}S_4Si$	$(p\text{-}CH_3C_6H_4S)_4Si$	—	128.5—129 (rhomb.)	—	—	168
$C_{28}H_{44}O_2S_2Si$	$(C_8H_{17}S)_2Si(OC_6H_5)_2$	—	—54 *	—	—	539
$C_{30}H_{62}OS_3Si$	$(C_8H_{17}S)_3SiO$— 	—	—54 *	—	—	539
$C_{32}H_{68}OS_3Si$	$(C_8H_{17}S)_3SiOC_8H_{17}$	—	—55 *	—	—	539
$C_{37}H_{30}SSi$	$(C_6H_5)_3SiSC(C_6H_5)_3$	—	169—170	—	—	196
$C_{37}H_{78}S_3Si$	$CH_3Si(SC_{12}H_{25})_3$	220—240 (10^{-4})	33—34	—	—	199
$C_{40}H_{52}S_4Si$	$(p\text{-}tert\text{-}C_4H_9C_6H_4S)_4Si$	—	185—186 (tetragon.)	—	—	170
$C_{40}H_{84}O_2S_2Si$	$(C_8H_{17}S)_2Si(OC_{12}H_{25})_2$	—	—29 *	—	—	539
$C_{44}H_{92}O_3SSi$	$(C_{12}H_{25}O)_3SiSC_8H_{17}$	—	—10 *	—	—	539
$C_{48}H_{100}S_3Si$	$C_{12}H_{25}Si(SC_{12}H_{25})_3$	130—140 (0.001)	12—14	1.4825	—	199
$C_{48}H_{100}S_4Si$	$(C_{12}H_{25}S)_4Si$	—	43—44	—	—	199
$C_{64}H_{132}S_4Si$	$(C_{16}H_{33}S)_4Si$	—	50—51	—	—	170

* Solidification point.

For mass spectrometric analysis of trimethyl(butylthio)-silane see [639]; for details of the chromatographic analysis of organothiosilanes see [563].

It is reported that tri(alkylthio)silanes have a soporific action and produce headaches and irritation of the eyes [727].

2.3. Chemical Properties

Alkylthiosilanes are more reactive than their oxygen analogs, namely alkoxysilanes. Most of them are readily hydrolyzed by water with the formation of the corresponding alkanethiols and silanols [129, 359, 385, 577, 682, 712, 715, 717]:

$$\gt Si-SR + H_2O \longrightarrow \gt Si-OH + R-SH \quad (2.63)$$

Even tetra(tert-butylthio)silane is sensitive to atmospheric moisture [416], while tetra(tert-butoxy)silane is hydrolytically stable. However, triphenyl(ethylthio)silane is not changed in aqueous dioxane in 24 h [196].

When the hydrolysis is carried out in an alkaline medium it is the corresponding alkanethiolates which are formed and not the free alkanethiols [720].

$$\gt Si-SR + KOH \longrightarrow \gt Si-OH + R-SK \quad (2.64)$$

In contrast to their organic analogs, namely, the dialkyl sulfides, alkylthiosilanes do not have basic properties. Thus, for example, methylthiosilane does not form an addition product with diborane even at −78°C [673]. On the other hand, dimethyl sulfide under these conditions forms a stable complex with the composition $(CH_3)_2S \cdot BH_3$. (Trifluoromethylthio)silane CF_3SSiH_3 does not react even with BF_3 [276].

The Si–S bond in alkylthiosilanes is readily cleaved by many reagents. Thus, under the action of alcohols they give exceptional yields of the corresponding alkoxysilanes [129, 195, 628]:

$$\gt Si-SR + ROH \longrightarrow \gt Si-OR + R-SH \quad (2.65)$$

tert-Alkylthio groups attached to a silicon atom are resistant to the action of alcohol and even to alcoholic alkali. As a result of this tetra(tert-alkylthio)silane may be recrystallized smoothly from alcohols [169, 416]. Moreover, by the action of alcholic alkali on tri(tert-butylthio)chlorosilane it is possible to obtain the corresponding silanol and disiloxane [169]:

$$(RS)_3SiCl + KOH \longrightarrow (RS)_3SiOH + KCl \quad (2.66)$$

$$2(RS)_3SiCl + 2NaOH \longrightarrow (RS)_3SiOSi(SR)_3 + 2NaCl + H_2O \quad (2.67)$$

$$(R = \text{tert-}C_4H_9)$$

Alkoxyl derivatives are obtained by the action of sodium alcoholates [166]:

$$(RS)_3SiCl + R'ONa \longrightarrow (RS)_3SiOR' + NaCl \quad (2.68)$$

A reaction similar to transetherficiation can be achieved in the case of thiols [129]:

$$\gt Si{-}SR + HSR' \rightleftarrows \gt Si{-}SR' + HSR \quad (2.69)$$

Bromine cleaves the Si−S bond in alkylthiosilanes even at −70°C [726, 729], forming the corresponding bromosilanes and alkanesulfenyl bromides:

$$\gt Si{-}SR + Br_2 \longrightarrow \gt Si{-}Br + BrSR \quad (2.70)$$

The Si−S bond is also cleaved readily by inorganic halides such as hydrogen chloride [628, 712], hydrogen iodide [276], mercury halides [276, 825], boron trichloride [136, 137], phenyldichloroborane [136], halides of phosphorus [135, 137], arsenic, and antimony [132, 137], silicon tetrachloride [609], and tin tetrachloride [137]:

$$\gt Si{-}SR + HX \longrightarrow \gt Si{-}X + HSR \quad (2.71)$$

$$n \gt Si{-}SR + MX_n \longrightarrow n \gt Si{-}X + M(SR)_n \quad (2.72)$$

($X = Cl$, Br or I; n = valence of the element M)

At a higher temperature there is cleavage of the Si – S bond in dialkyldi(alkylthio)silanes R_2SiX_2 by bifunctional organosilicon compounds R_2SiY_2 (dialkyldihalosilanes [701], dialkyldialkoxysilanes [701], dialkyldiisocyanosilanes [502], and dialkyldiisothiocyanatosilanes [502]), leading to an equilibrium mixture of R_2SiXY, R_2SiX_2, and R_2SiY_2 [501].

Reactions with acid halides of carboxylic acids proceed analogously [282, 578, 728]:

$$\rangle Si{-}SR + RCOX \longrightarrow \rangle Si{-}X + RCOSR \qquad (2.73)$$

The remarkable reactivity of alkylthiosilanes is clearly illustrated by their conversion to the corresponding halosilanes by the action of alkyl halides [129, 134, 385]:

$$\rangle Si{-}SR + R'X \longrightarrow \rangle Si{-}X + RSR' \qquad (2.74)$$

This reaction proceeds most rapidly in the case of alkyl and cycloalkyl iodides. In the reaction of trimethyl(ethylthio)silane with 3-chloro-3-ethylpentane or with iodocyclohexane they are dehydrogenated with the formation of olefins. The hydrogen halide eliminated then cleaves the Si – S bond to form trimethylhalosilane and ethanethiol. In contrast to iodocyclohexane, bromocyclohexane reacts with trimethyl(butylthio)silane normally in accordance with scheme (2.74). Iodobenzene does not react with trimethyl(butylthio)silane even on prolonged heating [134]. Methylene iodide reacts with trimethyl(butylthio)silane with the participation of both C – I bonds [134]:

$$2(CH_3)_3SiSC_4H_9 + CH_2I_2 \longrightarrow 2(CH_3)_3SiI + (C_4H_9S)_2CH_2 \qquad (2.75)$$

To explain the mechanism of the cleavage of the C – S bond in organothiosilanes by alkyl halides, the hypothesis was put forward that the reaction proceeds with the formation of unstable silylalkylsulfonium salts [129]; some of these have even been isolated in a free state [134]. However, the very low basicity of organothiosilanes, which was established spectroscopically, and the unsatisfactory analytical data for the compounds isolated raised doubts on their structure. An experimental check did not confirm the formation of silylalkylsulfonium salts either in reaction (2.74) or when dimethyl sulfide was heated in a sealed ampoule with trimethylchlorosilane at 120°C [385]. It was shown that the interaction of trimethyl(methylthio)silane with excess methyl iodide proceeds according to the scheme

$$(CH_3)_3SiSCH_3 + 2CH_3I \longrightarrow (CH_3)_3SiI + [(CH_3)_3S]^+I^- \qquad (2.76)$$

Acetic anhydride converts alkylthiosilanes into acetoxy derivatives [728]:

$$\geq Si{-}SR + (CH_3CO)_2O \longrightarrow \geq SiOCOCH_3 + CH_3COSR \qquad (2.77)$$

The acetoxysilanes formed in this way may then be condensed with the starting alkylthiosilanes to form the corresponding siloxanes and S-alkylthioacetates:

$$\geq Si{-}SR + CH_3COOSi\leq \longrightarrow \geq Si{-}O{-}Si\leq + CH_3COSR \qquad (2.78)$$

The reaction with acetic anhydride proceeds quantitatively so it may be used for the quantitative determination of organothiosilanes [187].

Carboxylic acids also react readily with alkylthiosilanes with rupture of the Si–S bond [130, 577]:

$$\geq Si{-}SR + R'COOH \longrightarrow \geq SiOCOR' + RSH \qquad (2.79)$$

Phenyllithium and triphenylsilyllithium cleave triphenylorganothiosilanes. In the first case tetraphenylsilane is formed [354] and in the second, hexaphenyldisilane [720]. The other reaction product in both cases is a lithium organothiolate:

$$(C_6H_5)_3SiSR + R'Li \longrightarrow (C_6H_5)_3SiR' + RSLi \qquad (2.80)$$
$$(R = \text{alkyl, aryl;} \quad R' = \text{phenyl, triphenylsilyl})$$

A very curious reaction, which includes cleavage of the Si–S bond, is observed when (trifluoromethylthio)silane is treated with trimethylamine [276]:

$$CF_3SSiH_3 + 2(CH_3)_3N \longrightarrow H_3SiF \cdot N(CH_3)_3 + CSF_2 \cdot N(CH_3)_3 \qquad (2.81)$$

Organothiosilanes are oxidized by potassium permanganate. However, the oxidation products are not the corresponding sulfoxides and sulfones, but silanols and siloxanes [359, 607]. Thus, the product from the oxidation of triphenyl(cresylthio)silane with $KMnO_4$ in acetic acid is triphenylsilanol [607]. On the other hand, when oxygen is passed into a boiling benzene solution of trimethyl(methylthio)silane no changes in the latter are observed. In contrast to this, sulfur reacts readily with alkylthiosilanes on heating above 150°C to form alkylpolythiosilanes [359]:

$$\gt Si-SR + nS \longrightarrow \gt Si-(S)_n-R \qquad (2.82)$$

The thermal stability of alkylthiosilanes is considerably lower than that of their oxygen analogs, alkoxysilanes. While pyrolysis of tetraethoxysilane proceeds at an appreciable rate only at temperatures of the order of 600-700°C, tetra(ethylthio)silane decomposes smoothly even at 250-300°C in accordance with the scheme [606, 610]*:

$$Si(SR)_4 \longrightarrow SiS_2 + 2RSR \qquad (2.83)$$

The reaction proceeds in stages, as is confirmed by the isolation of intermediate pyrolysis products, for example:

$$2Si(SR)_4 \longrightarrow (RS)_2Si\langle{}^{S}_{S}\rangle Si(SR)_2 + 2RSR \qquad (2.84)$$

$$(R = C_2H_5)$$

Further thermal decomposition leads to the formation of oligomers and polymers $Si_nS_{2n-2}(SR)_4$, from which, in particular, a substance with the composition $Si_{22}S_{42}(SR)_4$ was isolated [609].

When circulated through a tube heated to 300°C triethyl(ethylthio)silane and diethyldi(ethylthio)silane form hexaethyldisilthiane

* The addition of sulfur makes it possible to lower the temperature of reaction (2.83) to 200°C [182].

and tetraethylcyclodisilthiane, respectively:

$$2R_3SiSR \longrightarrow R_3SiSSiR_3 + RSR \qquad (2.85)$$

$$2R_2Si(SR)_2 \longrightarrow R_2Si\underset{S}{\overset{S}{\diagdown}}SiR_2 + 2RSR \qquad (2.86)$$

Di(ethylthio)dichlorosilane begins to decompose analogously at 250°C:

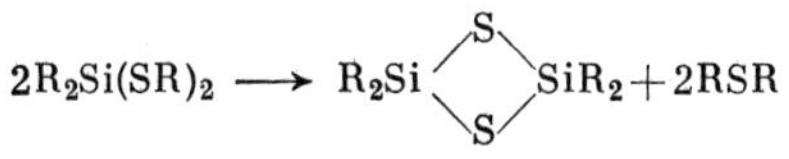

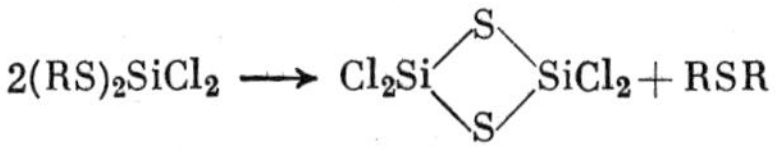

$$2(RS)_2SiCl_2 \longrightarrow Cl_2Si\underset{S}{\overset{S}{\diagdown}}SiCl_2 + RSR \qquad (2.87)$$

The tetrachlorocyclodisilthiane formed then decomposes readily to $SiCl_4$ and SiS_2 [606, 609].

The thermal decomposition of (trifluoromethylthio)silane, which is stable at room temperature only in the absence of impurities, proceeds by a different scheme:

$$CF_3SSiH_3 \longrightarrow FSiH_3 + CSF_2 \qquad (2.88)$$

When $Hg(SCF_3)_2$ is used as a catalyst, this reaction is observed even at −45°C [276].

Organosilicon esters of diethyldithiocarbamic acid have a low thermal stability and begin to decompose even at 100°C [212]:

$$R_3SiSC(S)N(C_2H_5)_2 \longrightarrow R_3SiN(C_2H_5)_3 + CS_2 \qquad (2.89)$$

When treated with thionyl chloride, these compounds lose a trialkylsilyl group and are converted into N,N-diethylthiocarbamoyl chloride [837].

Trimethyl(methylthio)silane adds to the carbonyl group of bis(trifluoromethyl) ketone to form 1,1,1,3,3,3-hexafluoro-2-methylthio-2-trimethylsiloxypropane [828].

The reduction of optically active α-naphthylphenylmethyl-(methylthio)silane with lithium aluminum hydride in boiling ether to α-naphthylphenylmethylsilane proceeds stereospecifically and, in contrast to the corresponding oxygen analog, with inversion of the configuration [664, 964].

3. COMPOUNDS CONTAINING THE GROUPING Si – S – Si (SILTHIANES)

3.1. Silicon Disulfide and Its Use in the Synthesis of Organosilicon Compounds

The simplest compound containing the grouping Si–S–Si is silicon disulfide SiS_2* [15, 65, 104, 211, 377, 403, 489, 552, 703, 704]. It is formed by high temperature reactions of silicon with sulfur [189, 326, 332, 378, 470, 571, 582, 595, 596, 644, 693]; silicon with hydrogen sulfide [595, 596, 597, 685]; silicates and SiO_2 with CS_2 [326, 327, 334-337, 450, 685], sulfur [708], and a mixture of S_2Cl_2 with chlorine [191]; SiO_2 with hydrogen sulfide in the presence of charcoal [685]; lithium, calcium, copper, magnesium, and iron silicides with sulfur [188, 244, 489, 506, 507]; SiO with SO_2 [736]; and sulfides of iron [141, 293, 576, 738] and aluminum [192, 193, 420, 421, 423, 636, 642, 685-687, 737] with silicon dioxide; the latter reaction is usually used for the preparation of considerable amounts of silicon disulfide. It is also formed by the thermal decomposition of $(SiSCl_2)_n$ [545] and tetra-(ethylthio)silane [182, 606, 610]. It was not possible to obtain SiS_2 by a wet method [190].

Investigation of the structure of SiS_2 showed that this compound has a rhombic cell with lattice parameters of a = 5.60, b = 5.53, c = 9.55 Å. Its structure consists of a chain of tetrahedra connected to each other through opposite edges (see Fig. 1). The Si–S distances equal 2.16(2.14) Å [232, 737], and the S–S distances 3.24, 3.5, and 3.62 Å [232].

The reaction of silicon with sulfur at high temperature (~1300°C) and pressure forms another modification of SiS_2, in which the $(SiS_4)^{4-}$ tetrahedra, connected through the apices, form a three-dimensional lattice [566, 644].

Silicon disulfide reacts with a series of inorganic compounds: ammonia [182, 202], carbon dioxide [256, 257], water [326, 332,

*Silicon monosulfide SiS has also been synthesized and its properties studied [1, 38, 173-177, 211, 238, 247, 248, 274, 313, 521, 579, 635, 643, 692].

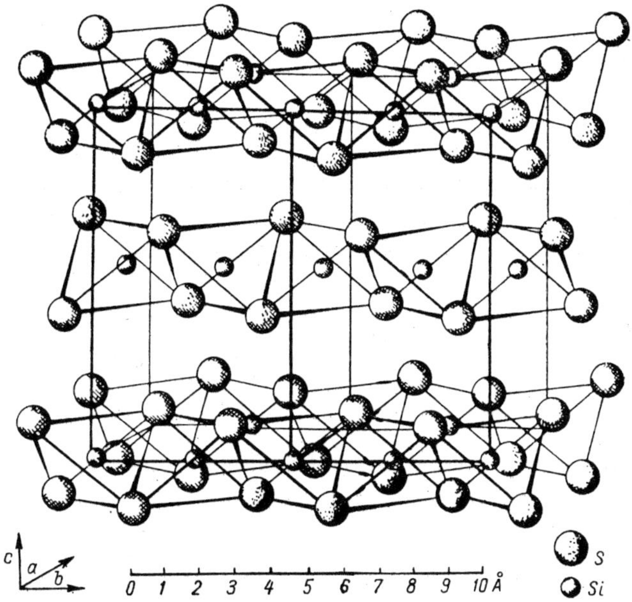

Fig. 1. SiS_2 lattice (two cells with double the b axis).

636], and metal sulfides [240, 501, 702]. However, the reactions of SiS_2 with organic compounds containing active hydrogen are most important for organosilicon chemistry. These reactions have been used as a basis for the development of preparative methods for obtaining organoxy, acyloxy, and aminosilanes, and also organoxysilanethiols.

The reaction of SiS_2 with stoichiometric amounts of primary alcohols or with excess secondary or tertiary alcohols proceeds in accordance with scheme (2.90) with the formation of tetraalkoxysilanes [258, 260, 470-472, 581, 653, 725]:

$$SiS_2 + 4ROH \longrightarrow Si(OR)_4 + 2H_2S \qquad (2.90)$$

With excess silicon disulfide, sulfur-containing substances are obtained [326, 471, 724], and the following structure has been assigned to these: $(RO)_3SiS[(RO)_2SiS]_nSi(OR)_3$ [471]. However, these compounds were not isolated, and the only sulfur-containing organosilicon product of the reaction under these conditions which was characterized was trimethoxysilanethiol [724].

The products from the reaction of SiS_2 with secondary and tertiary alcohols (even in the presence of excess alcohol) are trialkoxysilanethiols [553, 722-725]:

$$SiS_2 + 3ROH \longrightarrow (RO)_3SiSH + H_2S \qquad (2.91)$$

It may be assumed that in these reactions the alcohol molecule attacks the Si–S bond with cleavage of the four-membered SiS_2 ring, forming a silanethiol grouping:

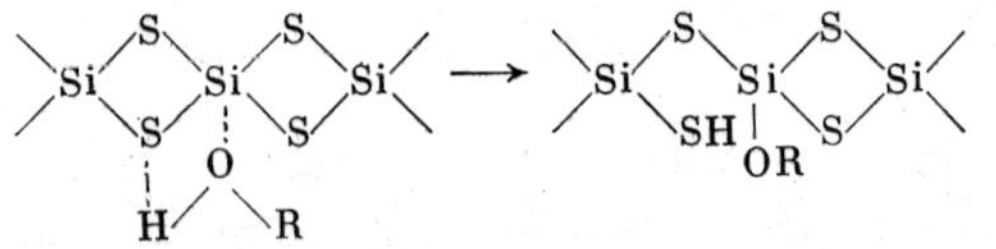

(2.92)

The attack of a second alcohol molecule may lead to both monothiols and to dithiols:

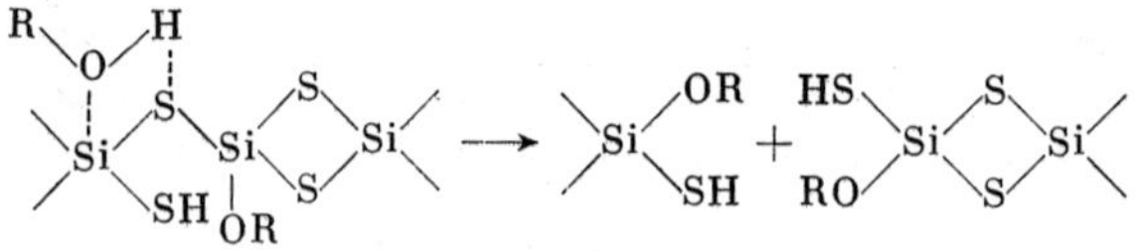

(2.93)

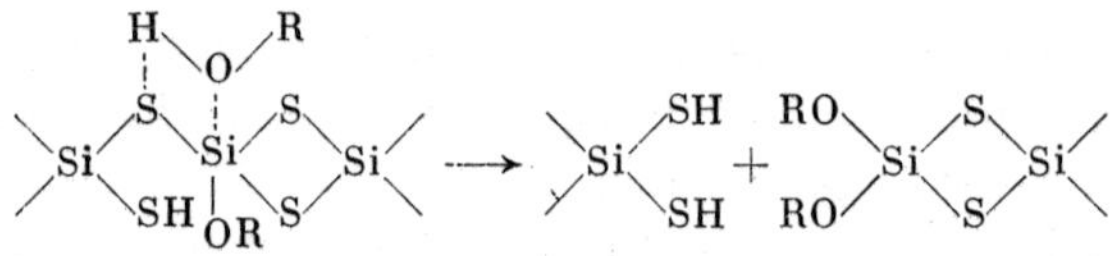

(2.94)

There is then the possibility of a large number of different reactions including cleavage of the rings, condensation of the silanethiols formed, and their reaction with the alcohol. Of all the variety of intermediate silanethiols formed, the most stable are trialkoxysilanethiols, which do not condense to disilthianes and do not react with excess alcohol under the reaction conditions when they contain secondary and tertiary alkyl radicals. Therefore it was possible to isolate them from the reaction mixture. In this way it was possible to obtain tri(isopropoxy)silanethiol, tri(sec-butoxy)silanethiol, tri(sec-amoxy)silanethiol, tri(tert-butoxy)silanethiol, and tri(tert-amoxy)silanethiol.

When the reaction of tert-butyl alcohol is carried out with excess SiS_2, it is possible to isolate a small amount of tetra(tert-butoxy)cyclodisilthiane in addition to tri(tert-butoxy)silanethiol [553]. It is possible that this compound is formed as a result of the condensation of di(tert-butoxy)silanedithiol formed initially:

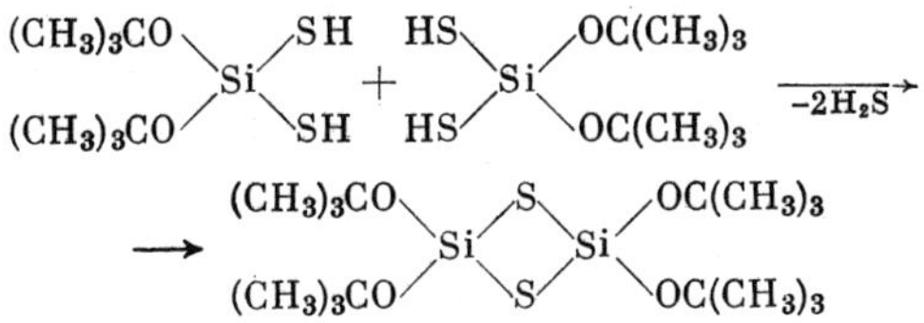

$$
\begin{array}{c}
\underset{(CH_3)_3CO}{\overset{(CH_3)_3CO}{>}}Si\underset{SH}{\overset{SH}{<}} + \underset{HS}{\overset{HS}{>}}Si\underset{OC(CH_3)_3}{\overset{OC(CH_3)_3}{<}} \xrightarrow{\overline{-2H_2S}} \\
\longrightarrow \underset{(CH_3)_3CO}{\overset{(CH_3)_3CO}{>}}Si\underset{S}{\overset{S}{<}}Si\underset{OC(CH_3)_3}{\overset{OC(CH_3)_3}{<}}
\end{array} \qquad (2.95)
$$

However it is equally probable that this is the product of partial alcoholysis of SiS_2, in which one cyclodisilthiane ring is retained.

Silicon disulfide reacts with phenol with the formation of tetraphenoxysilane [259, 471]:

$$SiS_2 + 4C_6H_5OH \longrightarrow (C_6H_5O)_4Si + 2H_2S \qquad (2.96)$$

When a mixture of phenol and ethyl alcohol is added to SiS_2 the main reaction product is diethoxydiphenoxysilane. With the reverse order of mixing of the reagents triethoxyphenoxysilane and tetraethoxysilane are formed [581].

The reaction of carboxylic acids with SiS_2 leads to tetra-acyloxysilanes [471]:

$$SiS_2 + 4RCOOH \longrightarrow (RCOO)_4Si + 2H_2S \qquad (2.97)$$

The reaction of primary amines with SiS_2 forms tetra(organo-amino)silanes [471]:

$$SiS_2 + 4RNH_2 \longrightarrow (RNH)_4Si + H_2S \qquad (2.98)$$

The reaction is carried out in the cold or with moderate heating as polymers are formed at higher temperatures.

The reaction of SiS_2 with 2-nonanone, which proceeds at 100-110°C with the liberation of H_2S, has also been described. However, no organosilicon products were identified in this case [733].

3.2. Methods of Preparing Organosilicon Compounds Containing the Grouping Si − S − Si*

Trialkylsilanethiols condense readily to the corresponding hexaalkyldisilthianes. Therefore, in many reactions for the preparation of trialkylsilanethiols mixtures of them with the corresponding disilthianes are formed, and sometimes it is not possible to isolate trialkylsilanethiols at all and silthianes are the only reaction product. Thus, for example, the reaction of dimethyl-(trimethylsilylmethyl)bromosilane with hydrogen sulfide in the presence of pyridine forms a mixture of dimethyl(trimethylsilylmethyl)silanethiol and 1,1,3,3-tetramethyl-1,3-bis(trimethylsilylmethyl)disilthiane in a molar ratio of 1:1.15 [89]. The reaction of iodosilane with sodium bisulfide leads to the formation of disilthiane, which does not contain even traces of silanethiol [294]. In the reaction of triethylaminosilane or tripropylaminosilane with hydrogen sulfide, together with triethyl- and tripropylsilanethiols there is also a considerable amount of the corresponding hexaalkyldisilthiane formed [448], while the reaction of N-trimethylsilylimidazole with H_2S is a convenient preparative method of obtaining hexamethyldisilthiane [901]. In the reactions of lower trialkylhalosilanes with hydrogen sulfide in the presence of tertiary amines, only hexaalkyldisilthianes are formed. Therefore, the reaction of organohalosilanes with hydrogen sulfide in the presence of tertiary amines is of greater value for preparing various silthianes than for the corresponding silanethiols.

The reaction of triorganohalosilanes with hydrogen sulfide in the presence of tertiary amines forms the corresponding hexaalkyldisilthianes [90, 246, 262, 664]:

$$2R_3SiCl + H_2S + 2R_3N \longrightarrow R_3SiSSiR_3 + 2R_3N \cdot HCl \qquad (2.99)$$

In cases where this reaction leads to the formation of triorganosilanethiols [89, 301], the latter may be converted into hexaorganodisilthianes readily by the action of a triorganohalosilane

*See [114] for the nomenclature of cyclosilthianes.

in the presence of tertiary amines [89]:

$$R_3SiCl + HSSiR_3 \xrightarrow[-R_3N \cdot HCl]{R_3N} R_3SiSSiR_3 \qquad (2.100)$$

The same may be achieved by the reaction of triorganohalo-silanes with alkali metal triorganosilanethiolates [196, 868]:

$$R_3SiSNa + ClSiR_3 \longrightarrow R_3SiSSiR_3 + NaCl \qquad (2.101)$$

Diorganodihalosilanes react with hydrogen sulfide in the presence of tertiary amines with the formation of four-membered cyclodisilthianes or six-membered cyclotrisilthianes. Thus, for example, the reaction of dimethyldichlorosilane with hydrogen sulfide in the presence of pyridine at room temperature leads to the formation of hexamethylcyclotrisilthiane [2, 532]:

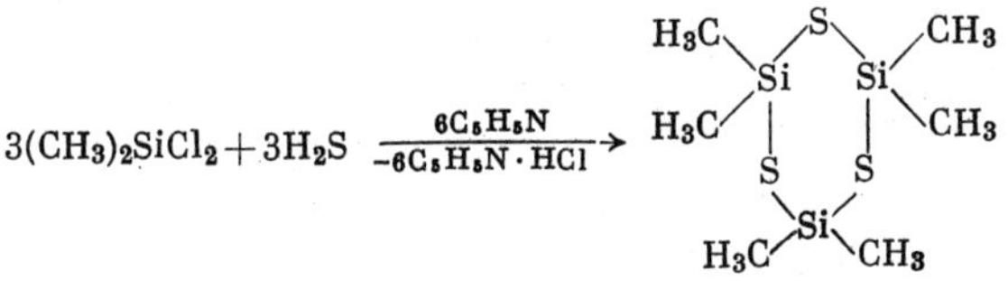

$$3(CH_3)_2SiCl_2 + 3H_2S \xrightarrow[-6C_5H_5N \cdot HCl]{6C_5H_5N} \qquad (2.102)$$

When heated in a stream of H_2S or when distilled at atmospheric pressure, this trimer is converted into a cyclic dimer, namely, tetramethylcyclodisilthiane [136, 429, 504, 532]:

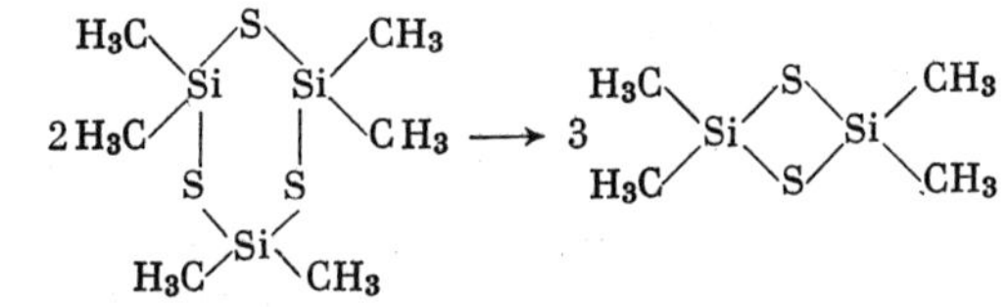

$$\qquad (2.103)$$

Hexaphenylcyclotrisilthiane behaves analogously [481].

The reaction of alkyldichlorosilanes with hydrogen sulfide in the presence of pyridine forms six-membered cyclic trimers $(RSiHS)_3$, which are thermally stable up to 250°C. In the case of methyldichlorosilane it is not only the $Si-Cl$ bonds which react with H_2S, but also the $Si-H$ bonds. As a result of this the reaction products include both the cyclic trimer and crystalline $(CH_3Si)_4S_6$, which has the structure of the product from the reaction of hydrogen sulfide with methyltrichlorosilane [102].

In some cases tetraorganocyclodisilthianes may also be obtained by the direct reaction of diorganodichlorosilanes with hydrogen sulfide in the presence of tertiary amines* [246, 300]:

$$2R_2SiCl_2 + 2H_2S \xrightarrow[-4R_3N \cdot HCl]{4R_3N}$$

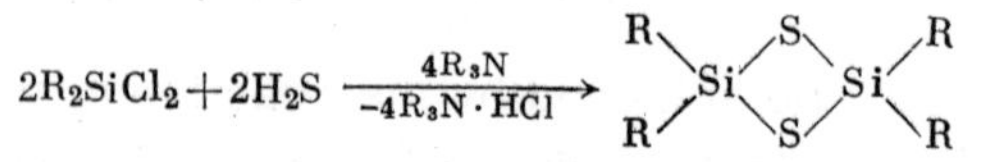

(2.104)

H$_2$S also reacts analogously with the complex of pyridine and silicon tetrachloride [301]:

$$SiCl_4 \cdot 2C_5H_5N + 2H_2S \xrightarrow[-4C_5H_5N \cdot HCl]{2C_5H_5N}$$

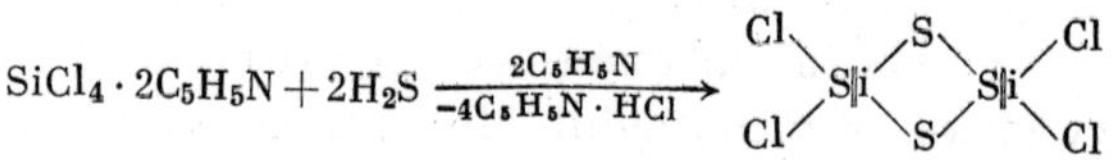

(2.105)

The complex of pyridine with SiF$_4$ [301] does not react with hydrogen sulfide and neither does diethyldifluorosilane in the presence of pyridine [300]; tetrafluorocyclodisilthiane is formed from SiF$_4$ and SiS$_2$ at 1000°C [371].

Tetrachlorocyclodisilthiane is obtained from thiohydrolysis of SiCl$_4$, which occurs at 650°C in the absence of hydrogen chloride acceptors [369, 545]. The tetramer (SiCl$_2$S)$_4$ was isolated from the reaction products [634], and also a compound to which was assigned the structure

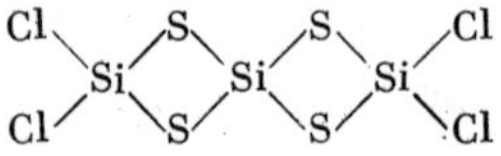

Tetrachlorocyclodisilthiane is also formed by thermolysis of hexachlorodisilthiane (at 950°C) [545] and di(ethylthio)dichlorosilane (at 250°C) [608, 609]:

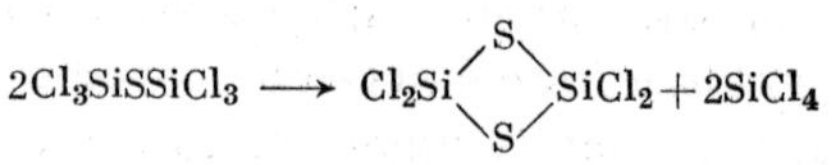

$$2Cl_3SiSSiCl_3 \longrightarrow Cl_2Si\underset{S}{\overset{S}{<>}}SiCl_2 + 2SiCl_4$$

(2.106)

$$2(C_2H_5S)_2SiCl_2 \longrightarrow Cl_2Si\underset{S}{\overset{S}{<>}}SiCl_2 + 2C_2H_5SC_2H_5$$

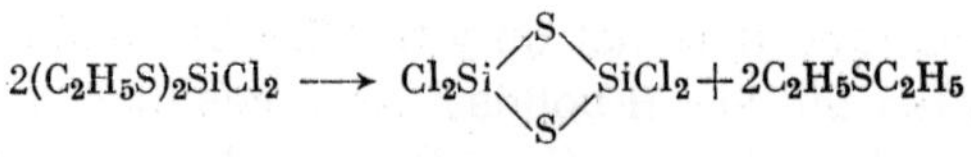

(2.107)

The reaction of organotrichlorosilanes with hydrogen sulfide in the presence of pyridine gives compounds of the type [(RSi)$_2$S$_3$]$_2$

---

*The compound [(CH$_3$)$_2$SiS]$_n$, which was described as a trimer in [515, 516], is also a dimer judging by its melting point.

[300, 301, 321], to which may be assigned two structural formulas:

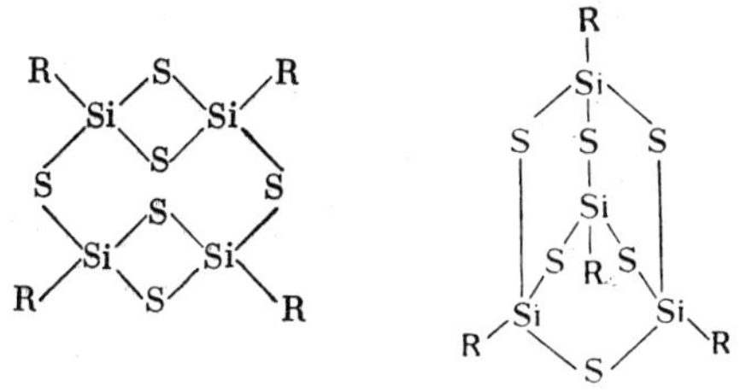

I II

The first of these consists of four- and eight-membered cyclosilthiane rings [102, 463], while the second consists of only six-membered rings (like adamantane) [321].

The use of organosilicon compounds containing two silicon atoms attached to chlorine atoms and separated from each other by methylene groups or an oxygen atom in reaction (2.99) leads to the formation of various heterocyclic systems containing the disilthiane grouping. Thus, for example, the reaction of 1,5-dichlorohexamethyltrisiloxane with H_2S in the presence of pyridine forms hexamethyl-2-thiacyclotrisiloxane [12]:

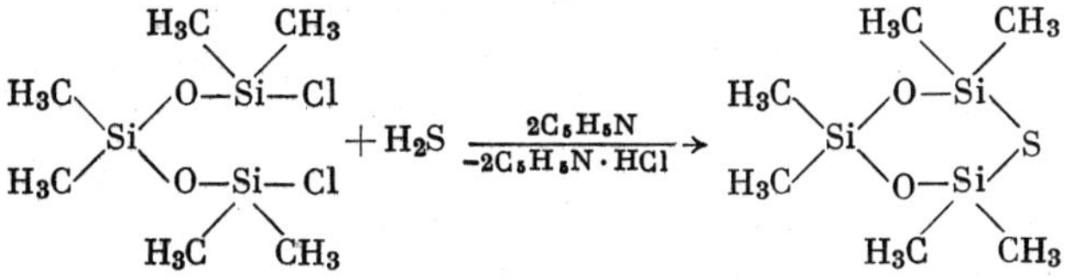

(2.108)

In an analogous reaction 1,3-dichlorotetramethyldisiloxane is converted into octamethyl-2,6-dithiacyclotetrasiloxane [12]:

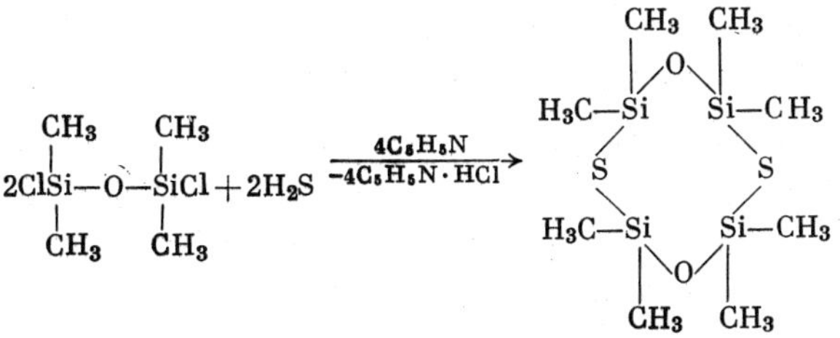

(2.109)

Cyclosilthianes are also formed by the reaction of 1,3-bis-(diorganochlorosilyl)propane [695] and 1,1,2,2-tetramethyl-1,2-

dichlorodisilane with hydrogen sulfide in the presence of pyridine [697]:

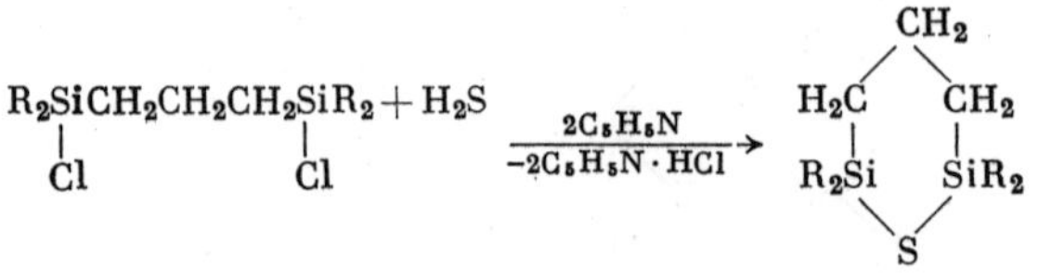

(2.110)

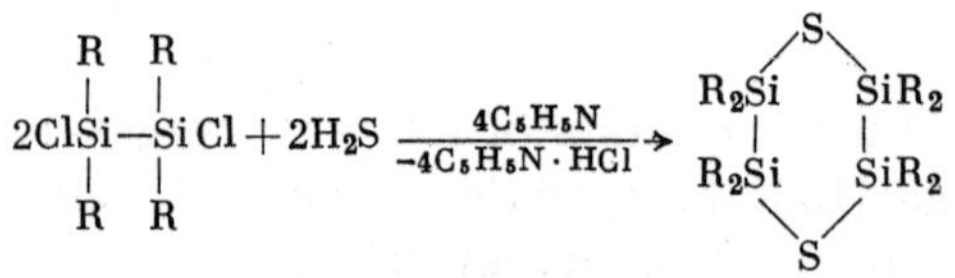

(2.111)

A general method for the synthesis of hexaorganodisilthianes is the reaction of triorganoiodosilanes with silver sulfide [11, 152, 156, 160, 285, 300, 634] and mercury sulfide [292, 294, 296-299, 457, 634, 700, 975, 979]:

$$2R_3SiI + Ag_2S \longrightarrow R_3SiSSiR_3 + 2AgI \qquad (2.112)$$

$$2R_3SiI + HgS \longrightarrow R_3SiSSiR_3 + HgI_2 \qquad (2.113)$$

It is not possible to carry out this synthesis with chlorosilanes and bromosilanes, since the silver or mercury chloride and bromide formed cleaves the Si—S bond in hexaorganodisilthianes [285] and the equilibrium of the reaction is displaced completely to the left. On this basis it was possible to synthesize hexachlorodisilthiane by heating trichloroiodosilane with silver sulfide [634]:

$$2Cl_3SiI + Ag_2S \longrightarrow Cl_3SiSSiCl_3 + 2AgI \qquad (2.114)$$

Heating trimethylchlorosilane with potassium sulfide at 400°C for 4 h in a melt of a eutectic mixture of LiCl and KCl gave hexamethyldisilthiane in 81% yield [314, 675].

The formation of hexabutyldisilthiane was observed when tetrabutylsilane was heated with sulfur at 240°C. It is possible that the first stage of the reaction leads to tributyl(butylthio)-silane, which then decomposes to hexabutyldisilthiane and dibutyl sulfide [626]:

$$(C_4H_9)_4Si + S \longrightarrow (C_4H_9)_3SiSC_4H_9 \qquad (2.115)$$

$$2(C_4H_9)_3SiSC_4H_9 \longrightarrow (C_4H_9)_3SiSSi(C_4H_9)_3 + (C_4H_9)_2S \qquad (2.116)$$

Tetraphenylsilane reacts with sulfur only on heating to 380°C. There is then cleavage of the Si$-C_6H_5$ bond. Only diphenyl sulfide was identified among the products of this reaction [626]. On heating with sulfur, allyltrichlorosilane forms an inseparable mixture of polysulfides [83].

Disildithianes are formed by the action of iodine on sodium triorganosilanethiolates [868, 902], while α,ω-disilylpolysulfanes $R_3SiS_nSiR_3$ are formed by the reaction of sodium triorganosilanethiolates with sulfur chlorides S_nCl_2 (n = 3-9) [868].

3.3. Physical Properties

Organosilthianes are colorless liquid or crystalline substances. Their physical constants are given in Table 17.

In the Raman spectrum of disilthiane there is an intense and polarized line at 480 cm^{-1}, which belongs to the symmetrical bond vibration of the Si$-$S$-$Si skeleton [292, 457]. The corresponding frequency in the infrared spectrum is ~480 cm^{-1} [292], indicating the nonlinearity of the skeleton since in the case of a linear structure this frequency would have been forbidden in the infrared spectra. On the other hand, an intense absorption band in the infrared spectrum at 517 cm^{-1}, which belongs to the antisymmetrical bond vibration of the Si$-$S$-$Si skeleton, corresponds to a weaker line (501 cm^{-1}) in the Raman spectrum, which also indicates the nonlinearity of the Si$-$S$-$Si group of disilthiane [292, 456, 457]. Calculations based on spectroscopic data show that the Si$-$S$-$Si group in the disilthiane molecule is actually bent and the Si$-$S$-$Si angle equals 101° [457]. According to electron diffraction data [140], this angle equals 97.4 ± 0.7°. The Raman and infrared spectra of other silthianes have also been investigated [291, 292, 296, 298, 369, 425, 429, 457, 481, 626, 634, 700, 743, 750, 758-760, 849-851]. From spectroscopic data it follows that the symmetrical bond vibrations ν_s Si$-$S$-$Si in the spectra of silthianes correspond to frequencies in the range of 420-440 cm^{-1} with a fall to 407 cm^{-1} in the case of $(Cl_2SiS)_2$ and 395 cm^{-1} for $(Cl_3Si)_2S$. The antisymmetrical bond vibrations ν_{as} Si$-$S$-$Si in the spectra of organosilthianes correspond to frequencies in the range of 490-460 cm^{-1}. The assignment of frequencies corresponding to deformational vibrations of Si$-$S$-$Si is less cer-

TABLE 17. Compounds Containing the Grouping Si—S—Si (Silthianes)

Empirical formula	Compound	B.p., °C (mm)	M.p., °C	n_D^{20}	d_4^{20}	Literature
Br_4S_2Si	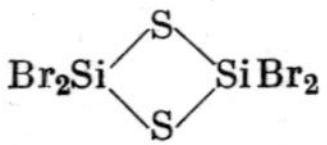	150 (18)	93	—	—	201
$Br_4S_4Si_3$	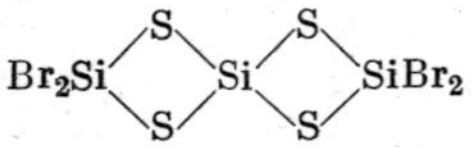	164 (12)	108	—	—	201
$Br_4S_6Si_4$	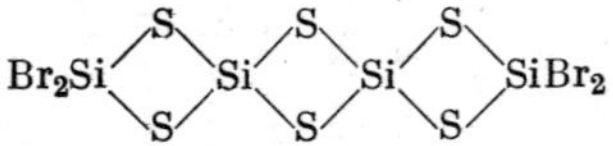	171 (12)	148 (decomp.)	—	—	201
$Cl_4S_2Si_2$	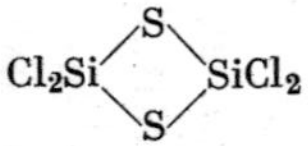	—	75	—	—	301
		—	80	—	—	545
Cl_6SSi_2	$Cl_3SiSSiCl_3$	187—191 (decomp.)	—	—	—	545
		100—100.5 (50)	—45	—	—	634
		72—73 (14)	—	—	—	369
		70 (10)	—	—	—	545
$Cl_6S_2Si_3$	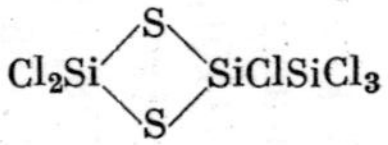	91.5—93 (2.5)	—	—	—	231
$Cl_8S_3Si_4$	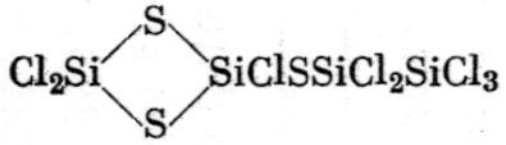	136—138 (0.01)	—	—	—	231
$F_4S_2Si_2$	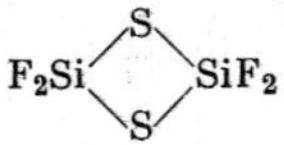	—18	—80	—	—	371
H_6SSi_2	$H_3SiSSiH_3$	58.8±0.7	—70.0±0.2	1.52±0.03	0.9296	296
$H_{10}SSi_4$	$(H_3SiSiH_2)_2S$	—	—70.4±0.1	—	0.950^{0}	700

$C_2H_{10}SSi_2$	$(CH_3SiH_2)_2S$	105±0.5	−120±0.5	—	0.885	298
		0 (9)	—	—	—	297
$C_3H_{12}S_3Si_3$	$CH_3SiH(SSiHCH_3)_2S$ (ring closed between first Si and last S)	89—92 (3)	—	1.5894	1.1632	102
$C_4H_{12}S_2Si_2$	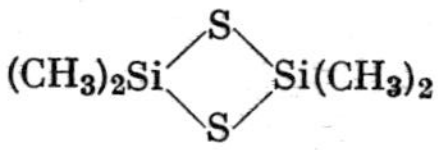	180—182	—	—	—	429
		133 (17)	108—110	—	—	429
		172—173	105.5	—	—	532
		—	113	—	—	300
		20 (0.01)	—	—	—	136
$C_4H_{12}S_6Si_4$	$[(CH_3Si)_2S_3]_2$ (see I and II on p. 197)	—	275	—	—	301
		—	272—275	—	—	321
		—	268—296	—	—	102
$C_4H_{14}SSi_2$	$[(CH_3)_2SiH]_2S$	144—146	−146±0.1	—	—	299
$C_6H_{14}Cl_2S_2Si_2$	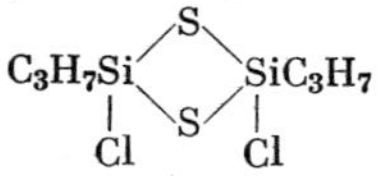	261—263	—	—	—	300
$C_6H_{18}SSi_2$	$(CH_3)_3SiSSi(CH_3)_3$	161—162	—	—	—	262
		162—163	—	—	—	314, 369
		162.5	—	—	—	284
		162.5—163.5	—	—	—	285
		163—164	—	—	—	246
		164	—	1.4598	—	425
$C_6H_{18}S_3Si_3$	$(CH_3)_2Si[SSi(CH_3)_2]_2S$ (ring closed between first Si and last S)	238—240	—	—	—	429
		151—153 (38)	17—18	—	—	429, 532
	$C_2H_5SiH(SSiHC_2H_5)_2S$ (ring closed between first Si and last S)	103—105 (2)	—	1.5672	1.1071	102
$C_8H_{18}SSi_2$	$[(CH_3)_2(CH_2{=}CH)Si]_2S$	—	170	—	—	382
$C_8H_{20}S_2Si_2$	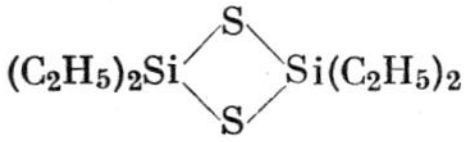	160—162 (46)	20.1	—	—	246
		—	21.1	—	—	606
$C_8H_{20}S_6Si_2$	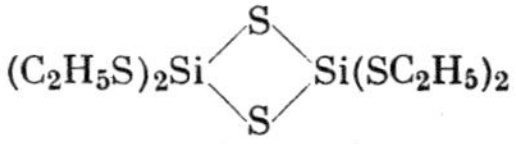	296 (0.1)	—	—	—	609
$C_8H_{20}S_6Si_4$	$[(C_2H_5Si)_2S_3]_2$ (see I and II on p. 197)	—	140	—	—	301
$C_8H_{24}O_2S_2Si_4$	$[(CH_3)_2SiOSi(CH_3)_2S]_2$	116—122 (2)	38—42	—	—	12

TABLE 17 (Continued)

Empirical formula	Compound	B.p., °C (mm)	M.p., °C	n_D^{20}	d_4^{20}	Literature
$C_8H_{24}S_2Si_4$	$[(CH_3)_2SiSi(CH_3)_2S]_2$	—	111—112	—	—	697
$C_{12}H_{20}SSi_2$	$(CH_3)_2\underbrace{SiCH_2CH_2CH_2Si}_{S}(CH_3)C_6H_5$	89 (1·5)	—	1·5340	0·9770	695
$C_{12}H_{28}S_2Si_2$	$(C_3H_7)_2Si{<}^{S}_{S}{>}Si(C_3H_7)_2$	176 (21)	23·5	—	—	300
$C_{12}H_{28}S_6Si_4$	$[(C_3H_7Si)_2S_3]_2$ (see I and II on p. 197)	—	86	—	—	300
$C_{12}H_{30}SSi_2$	$(C_2H_5)_3SiSSi(C_2H_5)_3$	274	—	—	—	449
		277—279	—	1·492[12]	∼0·9	246, 285
		278—279 (750)	—	—	—	284
		128 (7)	—	—	—	448
$C_{12}H_{34}SSi_4$	$[(CH_3)_3SiCH_2Si(CH_3)_2]_2S$	117—118 (3·5)	—98	1·4777	0·8774	89
$C_{16}H_{36}O_4S_2Si_2$	$(tert\text{-}C_4H_9O)_2Si{<}^{S}_{S}{>}Si(OC_4H_9\text{-}tert)_2$	—	121·5—123	—	—	553
$C_{18}H_{42}SSi_2$	$[(C_3H_7)_3Si]_2S$	168 (7)	—	—	—	448
$C_{22}H_{54}SSi_4$	$[(C_2H_5)_3SiCH_2Si(C_2H_5)_2]_2S$	202—204 (3)	—85	1·4980	0·9108	89
$C_{24}H_{20}S_2Si_2$	$(C_6H_5)_2Si{<}^{S}_{S}{>}Si(C_6H_5)_2$	—	145—147	—	—	515, 516
		—	163—165	—	—	481
$C_{24}H_{20}S_6Si_4$	$[(C_6H_5Si)_2S_3]_2$ (see I and II on p. 197)	—	216	—	—	301
$C_{24}H_{54}SSi_2$	$(C_4H_9)_3SiSSi(C_4H_9)_3$	160—163 (1)	—	1·—65	—	626
$C_{26}H_{46}SSi_4$	$[(CH_3)_3Si(CH_2)_3Si(CH_3)C_6H_5]_2S$	275—280 (2)	—	51	0·9426	90
$C_{34}H_{30}S_2Si$	$(\alpha\text{-}C_{10}H_7C_6H_5CH_3Si)_2S$	—	115—116	—	—	664
$C_{34}H_{62}SSi_4$	$[(C_2H_5)_3Si(CH_2)_3Si(C_2H_5)C_6H_5]_2S$	290—300 (2)	—	—	—	90
$C_{36}H_{30}SSi_2$	$(C_6H_5)_3SiSSi(C_6H_5)_3$	—	142	—	—	196
$C_{36}H_{30}S_2Si_2$	$(C_6H_5)_3SiSSSi(C_6H_5)_3$	—	181—182	—	—	196
$C_{36}H_{30}S_3Si_3$	$[(C_6H_5)_2SiS]_3$	—	188—189	—	—	481
$C_{48}H_{40}SSi_2$	$(C_6H_5)\underbrace{Si[(C_6H_5)_2Si]_3S}$	—	151—153	—	—	379

tain. Even in the case of the same compound different authors assign very different frequencies to δ Si−S−Si [289, 425]. The bond vibrations of the Si−H bond are represented by the frequency of 2180 cm^{-1} in the spectrum of disilthiane [292] and 2160 cm^{-1} in the spectrum of 1,3-dimethyldisilthiane [291, 298] and bis-(disilanyl) sulfide [700]. The surface tension of disilthiane equals 22.31 dyn/cm [465].

From vibration spectral data on hexamethyldisilthiane, the force constant k (Si−S) was calculated as equal to 2.20 mdyn/Å and the Si−S−Si angle determined as 104° [428].

In a determination of the dipole moments of methylcyclosilthianes it was found that for tetramethylcyclodisilthiane μ = 0.0D, while for hexamethylcyclotrisilthiane μ = 1.03D [439]. This indicates that the four-membered disilthiane ring is planar, while the six-membered trisilthiane ring is nonplanar. The exact configuration of the ring of cyclotrisilthiane has not been established [429, 732]. The dipole moment of hexamethyldisilthiane equals 1.85D [263]. The interatomic distances and bond angles in tetramethylcyclodisilthiane were determined by electron diffraction: Si−S length 2.18 ± 0.03 Å; S−Si−S angle 105° and Si−S−Si angle 75°. The following values were obtained for hexamethylcyclotrisilthiane: Si−S length 2.15 ± 0.03 Å; S−Si−S angle 115° and Si−S−Si angle 110° [730-732]. The cyclic silthiane $(CH_3Si)_4S_6$ has an adamantane-like structure: Si−S length 2.128 Å, Si−C 1.854 Å; Si−Si−S angle 104.5°, S−Si−S angle 111.8°, S−Si−C angle 107.0° [836].

Crystallographic investigations of phenylcyclosilthianes showed that crystals of hexaphenylcyclotrisilthiane are monoclinic [481]. Investigations of the NMR spectra of compounds of the type $[(CH_3)_3Si]_2X$, where X = O, S, or Se, show that with a change from X = O to X = S and Se the signals from the methyl protons are found at lower fields and not the reverse, as would have been expected from the series of electronegativities of these elements. This may indicate a decrease in the $d_\pi - p_\pi$ interaction of the p-electrons of the atom X with the 3d orbitals of the silicon atom with an increase in the atomic number of the element X. The same conclusion may be drawn on examining the nature of the changes in the spin-spin coupling constants $J(H-C^{13})$ and $J(H-C-Si^{29})$

[611, 614].* The NMR spectra of $(H_3SiSiH_2)_2S$ and $(CH_3SiH_2)_2S$ are given in [283].

Examination of the physical properties of silthianes shows that in contrast to siloxanes they evidently do not form long linear chains and large rings, or form them with difficulty. Thus, for example, $[(CH_3)_2SiS]_n$ has been obtained only in the form of the dimer or trimer.

Cyclic dimers form four-membered disilthiane rings with the Si−S−Si angle equal to 75°, which corresponds to 80° in four-membered rings of silicon disulfide [732]. In the siloxane series, where there is delocalization of the free p-electron pairs of the oxygen atom due to the vacant 3d-orbitals of the silicon atom, the formation of a compound with such valence angles is impossible [349]. The absence or the decrease of $d_\pi-p_\pi$ interaction between the sulfur and silicon atoms undoubtedly affects the chemical properties of silthianes, which show a greater tendency for cleavage of the Si−S bond by nucleophilic agents than siloxanes. On the other hand, x-ray structural investigations indicate a substantial difference in the interaction of silicon with bonding and unshared electrons of sulfur in comparison with carbon [16]. Therefore, silthianes differ in chemical properties from both siloxanes and also their carbon analogs, organic sulfides.

3.4. Chemical Properties

Like disilthiane itself [294, 296], organosilthianes [12, 285, 298, 481, 515, 516, 664] are extremely sensitive to moisture and are hydrolyzed rapidly by water even at room temperature, with the formation of the corresponding silanols or siloxanes and hydrogen sulfide:

$$R_3SiSSiR_3 + H_2O \longrightarrow R_3SiOSiR_3 + H_2S \qquad (2.117)$$

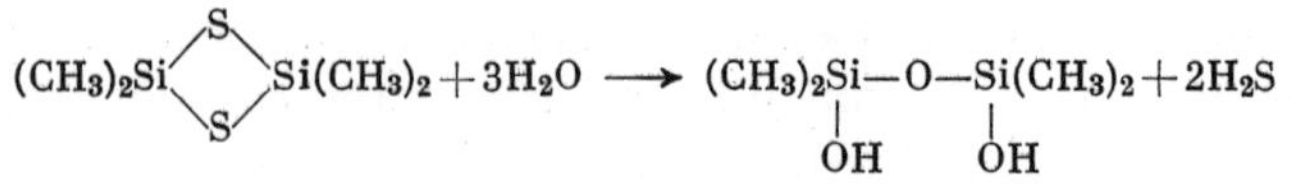

$$(CH_3)_2Si\underset{S}{\overset{S}{<\,>}}Si(CH_3)_2 + 3H_2O \longrightarrow (CH_3)_2Si\underset{OH}{\overset{|}{-}}O-Si(CH_3)_2\underset{OH}{\overset{|}{}} + 2H_2S \quad (2.118)$$

_____

*The hypsochromic shift in the absorption maximum in the ultraviolet spectrum of $[(CH_3)_3Si]_2S$ (202.5 mμ) in comparison with the analogous derivatives of carbon (215.5 mμ) and germanium (213.0 mμ) may still be ascribed to the interaction of the free electron pair of sulfur with the 3d orbitals of silicon [262].

Alkanols [799, 974, 975] and alkanethiols [137] also cleave the Si −S−Si bond in cyclosilthianes.

Disilthiane reacts with iodine with the formation of iodosilane and sulfur [294, 296]:

$$(H_3Si)_2S + I_2 \longrightarrow 2H_3SiI + S \tag{2.119}$$

Bromine cleaves tetrachlorocyclodisilthiane analogously, with the formation of dichlorodibromosilane [545]. In contrast to organic sulfides, silthianes do not add HI with the formation of sulfonium salts, but react with it with cleavage of the Si −S bond and are converted into the corresponding iodosilanes [296, 297]:

$$R_3SiSSiR_3 + 2HI \longrightarrow 2R_3SiI + H_2S \tag{2.120}$$

Disilthiane and 1,3-dimethyldisilthiane do not form adducts with BF_3 and BCl_3 at −78°C, as occurs in the case of dimethyl sulfide. With a rise in temperature to 100°C, there is a vigorous reaction leading to a complex mixture of unidentified organosilicon compounds. Hexamethyldisilthiane also reacts vigorously with BCl_3. There is then rupture of both Si −S bonds, and trimethylchlorosilane and boron sulfide are formed [136]:

$$3(CH_3)_3SiSSi(CH_3)_3 + 2BCl_3 \longrightarrow 6(CH_3)_3SiCl + B_2S_3 \tag{2.121}$$

When the molar ratio of hexamethyldisilthiane and BCl_3 is 3:1, a compound which contains the grouping Si −S−B is formed as an intermediate product and then decomposes to hexamethyldisilthiane and B_2S_3:

$$3(CH_3)_3SiSSi(CH_3)_3 + BCl_3 \longrightarrow 3(CH_3)_3SiCl + [(CH_3)_3SiS]_3B$$

$$\downarrow$$

$$B_2S_3 + [(CH_3)_3Si]_2S \tag{2.122}$$

The Si −S bond in silthianes is cleaved analogously by phenyldichloroborane. In the case of cyclic tetramethylcyclodisilthiane and hexamethylcyclotrisilthiane, dimethyldichlorosilane and six-membered triphenylcyclotriborothiane is formed [136]:

$$+ 3C_6H_5BCl_2 \longrightarrow 3(CH_3)_2SiCl_2 + \tag{1.123}$$

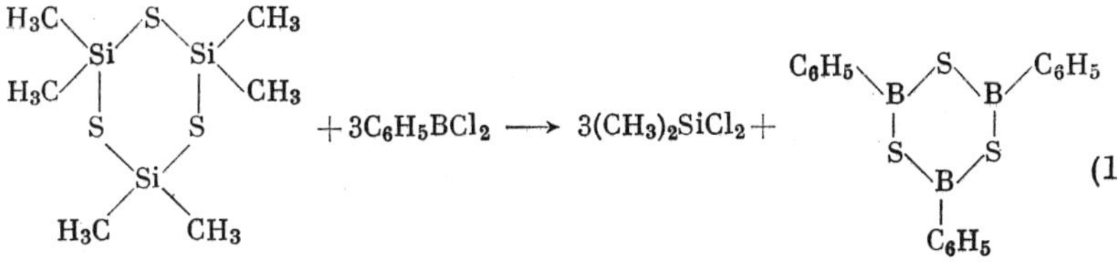

Organosilthianes are cleaved with the formation of the corresponding organochloro- and organobromosilanes under the action of many inorganic halides: PBr_3 [135], $TiCl_4$ [299], $PdCl_2$ [299], $HgCl_2$ [294, 296, 825], $PbCl_2$ [285], $CdCl_2$ [285, 825], AgCl [152, 156, 284, 285], AgBr [285], $AsCl_3$ [132], $SbCl_3$ [132], NH_4F + HF [517]. The Si–S bond in hexaorganodisilthiane is cleaved analogously by silver salts (AgCN, AgNCS, AgNCO, $AgOCOCH_3$) [160, 285]. The reaction is carried out by heating the hexaorganodisilthiane with the silver salt (AgX) with simultaneous distillation of the triorganosilyl derivative formed R_3SiX. Hexaalkyldisilthianes do not react with silver iodide [285].

On the basis of the reactions of silthianes and other silicon-functional compounds with silver salts, a series of conversions of trimethyl- and triethylsilyl derivatives was carried out in which each compound could be converted into the compound to the right of it in the series given by the action of the corresponding silver salt:

$$R_3SiI \longrightarrow (R_3Si)_2S \longrightarrow R_3SiBr \longrightarrow R_3SiNC \longrightarrow R_3SiCl \longrightarrow R_3SiNCS \longrightarrow$$
$$\longrightarrow R_3SiNCO$$

Hexaethyldisilthiane may be converted smoothly into hexaethyldisiloxane (90% yield) by simple heating with HgO for 10 min:

$$(C_2H_5)_3SiSSi(C_2H_5)_3 + HgO \longrightarrow (C_2H_5)_3SiOSi(C_2H_5)_3 + HgS \qquad (2.124)$$

Disilthiane [294, 296] and 1,3-dimethyldisilthiane [297] do not react with methyl iodide and, in contrast to organic sulfides, they do not form sulfonium salts. When heated with hexaalkyldisilthianes and hexaalkylcyclotrisilthianes, higher alkyl halides cleave the Si–S bond, converting them into the corresponding alkylhalosilanes and organic sulfides [134]:

$$R_3SiSSiR_3 + 2R'X \longrightarrow 2R_3SiX + R'_2S \qquad (2.125)$$

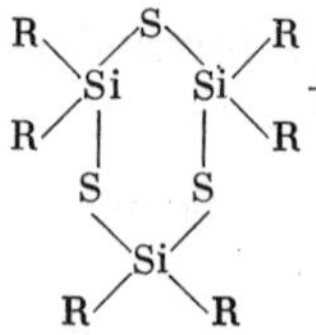

$$+ 6R'X \longrightarrow 3R_2SiX_2 + 3R'_2S \qquad (2.126)$$

(X = Cl, Br or I)

Thus, for example, when hexamethyldisilthiane is heated with heptyl iodide at 160°C for 100 h, trimethyliodosilane (88% yield) and diheptyl sulfide (96.5%) are formed. When α,ω-dihaloalkanes are used, in addition to organohalosilanes the products of this reaction include cyclic sulfides, e.g., tetramethylene sulfide (in the case of 1,4-dibromobutane) or ethylene sulfide (in the case of 1,2-dibromoethane):

$$R_3SiSSiR_3 + Br(CH_2)_nBr \longrightarrow 2R_3SiBr + (CH_2)_nS \qquad (2.127)$$

However, in the case of 1,3-dibromopropane instead of trimethylene sulfide a polymer of it is formed [134].

Hexamethylcyclotrisilthiane is readily cleaved by benzoyl chloride in accordance with the scheme [131]

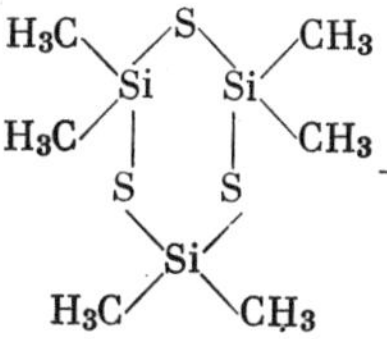

$$+ 6C_6H_5COCl \longrightarrow 3(CH_3)_2SiCl_2 + 3(C_6H_5CO)_2S \qquad (2.128)$$

1,3-Dimethyldisilthiane does not react with trimethylborane either on cooling to −78°C or on heating to 150°C, although dimethyl sulfide readily forms the addition product $(CH_3)_2S \cdot B(CH_3)_3$ with this reagent [297].

The only known case so far of complex formation of silthianes is observed when disilthiane reacts with ammonia or trimethylamine. However, these complexes are most likely formed as a result of the coordination of the ligand with the silicon atom rather than through addition to the sulfur atom [467].

When hexamethyl- and hexaethyldisilthianes are heated to 200°C with sulfur, organosilicon polysulfanes containing from 2 to 9 atoms [607] are formed:

$$R_3SiSSiR_3 + (n-1)\,S \longrightarrow R_3SiS_nSiR_3 \qquad (2.129)$$
$$(n=2-9)$$

Passing a mixture of nitrogen oxides through an ether solution of hexamethyl- or hexaethyldisilthiane at −20°C leads to the formation of yellowish white crystalline bis(trialkylsilyl) sulfones $R_3SiSO_2SiR_3$ ($R = CH_3$ or C_2H_5) [607].

Of the other reactions of silthianes we should mention the conversion of disilthiane into silanethiol [296]:

$$(H_3Si)_2S + H_2S \longrightarrow 2H_3SiSH \qquad (2.130)$$

cleavage of hexaphenyldisilthiane by α-chloro-N-phenylbenzalimine on heating in xylene [475]:

$$2C_6H_5C(Cl){=}NC_6H_5 + (C_6H_5)_3SiSSi(C_6H_5)_3 \longrightarrow 2(C_6H_5)_3SiCl + C_6H_5C({=}NC_6H_5){-}S{-}C({=}NC_6H_5)C_6H_5 \qquad (2.131)$$

and the thermal conversions of cyclosilthianes [136, 369, 429, 481, 532, 545, 634], which were examined in Section 3.2.

The reaction of hexaalkyldisilthianes with a diarylmercury leads to the formation of trialkylarylsilanes and HgS [789].

The reduction of optically active 1,3-di(α-naphthyl)-1,3-diphenyl-1,3-dimethyldisilthiane with lithium aluminum hydride in ether at room temperature forms optically active α-naphthylphenylmethylsilane. In this case the reduction of the first sulfur atom proceeds with inversion of the configuration, which is in sharp contrast to the reaction of the analogous disiloxane, where the reduction of both Si*–O bonds proceeds with the retention of the configuration [664, 964].

4. COMPOUNDS CONTAINING THE GROUPING Si–O–S

4.1. Preparation Methods

4.1.1. Reactions of Organosilicon Compounds with Sulfuric Acid

Compounds containing the grouping Si–O–S are formed readily by the action of sulfuric acid on very varied organosilicon compounds (halosilanes, silanols, organoxysilanes, siloxanes, tetraorganosilanes, and acyloxysilanes). The sulfuric acid cleaves the bond of the silicon to the other atom, forming a compound containing the grouping Si–O–S, namely, organosilyl sulfates or organosilyl hydrosulfates. Therefore, one of the general methods of preparing derivatives of this type is the reaction of concentrated sulfuric acid with organosilicon compounds. Depending on their struc-

ture, this reaction may proceed by different routes and lead to
different products. The reactions of sulfuric acid with different
classes of organosilicon compounds are examined below separately.

The reaction of concentrated H_2SO_4 with trimethylchlorosilane
in benzene for 12 h forms bis(trimethylsilyl) sulfate in 76% yield.
The reaction of H_2SO_4 with triphenylchlorosilane in boiling ether
proceeds analogously but much more slowly [125, 280]:

$$2R_3SiCl + H_2SO_4 \longrightarrow (R_3Si)_2SO_4 + 2HCl \qquad (2.132)$$

Dimethyldichlorosilane does not dissolve in concentrated sulfuric acid. When these substances are mixed two layers are
formed, and at the interface the liberation of hydrogen chloride begins gradually. Heating increases the rate of reaction and after
some time the solution becomes homogeneous. From the reaction
products (if 100% H_2SO_4 was used), it was possible to isolate di-
meric (dimethylsilylene) sulfate, to which a cyclic structure was
assigned [620, 623]:

$$2(CH_3)_2SiCl_2 + 2H_2SO_4 \longrightarrow \underset{\underset{O-SO_2-O}{\overset{O-SO_2-O}{\overset{H_3C}{\diagdown}}}}{\overset{}{}} Si \underset{CH_3}{\overset{CH_3}{\diagup}} Si \overset{CH_3}{\underset{CH_3}{\diagdown}} + 4HCl \qquad (2.133)$$

If reaction (2.133) is carried out with 95-96% sulfuric acid,
then with a molar ratio $(CH_3)_2SiCl_2 : H_2SO_4 = 2:1$ a product of the reaction is a crystalline compound, to which the following structure
has been assigned [72, 75]:

$$
\begin{array}{c}
(CH_3)_2Si-O-SO_2-O-Si(CH_3)_2-O \\
\ \ \ | \qquad\qquad\qquad\qquad\qquad\qquad | \\
\ \ \ O \qquad\qquad\qquad\qquad\qquad\qquad SO_2 \\
\ \ \ | \qquad\qquad\qquad\qquad\qquad\qquad | \\
(CH_3)_2Si-O-SO_2-O-Si(CH_3)_2-O
\end{array}
$$

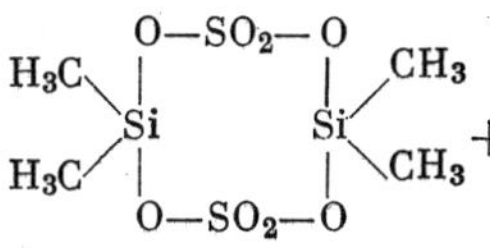

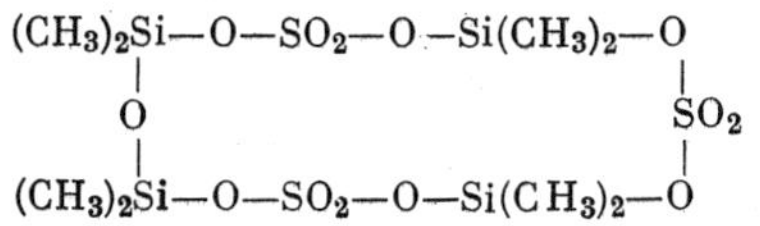

The reaction of a mixture of trimethylchlorosilane, dimethyldichlorosilane, and 100% sulfuric acid in a molar ratio of 2:1:2
forms a mixture of trimethylsilyl sulfate, dimethylsilylene sulfate,
and the compound $(CH_3)_3SiOSO_2O[(CH_3)_2SiOSO_2O]_n Si(CH_3)_3$ [620, 623].
Methyltrichlorosilane reacts with sulfuric acid more slowly than
dimethyldichlorosilane. When a mixture of them was heated at
100°C, the theoretical amount of hydrogen chloride was liberated

only after 7 days [623]:

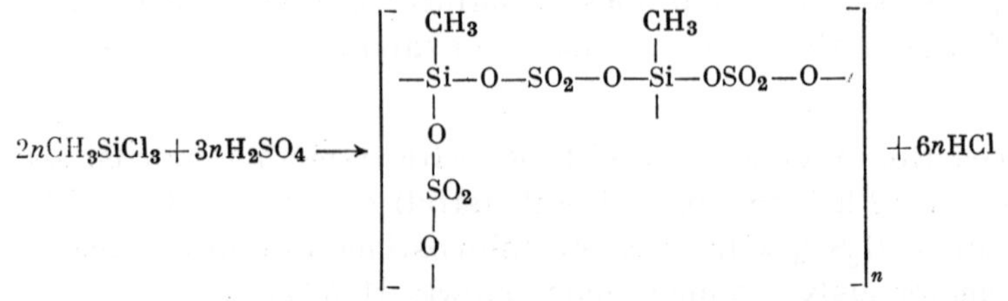

$$2n\text{CH}_3\text{SiCl}_3 + 3n\text{H}_2\text{SO}_4 \longrightarrow \begin{bmatrix} \begin{array}{cc} \text{CH}_3 & \text{CH}_3 \\ | & | \\ -\text{Si}-\text{O}-\text{SO}_2-\text{O}-\text{Si}-\text{OSO}_2-\text{O}-\prime \\ | & | \\ \text{O} \\ | \\ \text{SO}_2 \\ | \\ \text{O} \\ | \end{array} \end{bmatrix}_n + 6n\text{HCl} \quad (2.134)$$

The reaction of dimethyldichlorosilane with sulfuric acid with the formation of the fourteen-membered heterocyclic compound given above [74] has been proposed [73] for freeing dimethyldichlorosilane from traces of methyltrichlorosilane and also for direct conversion of dimethyldichlorosilane containing up to 10% methyltrichlorosilane into dimethylpolysiloxane elastomers. However, a critical evaluation of this proposal showed that in simplicity of technology and efficiency this method is clearly inferior to the synthesis of dimethylpolysiloxane elastomers using efficient fractionation of dimethyldichlorosilane with subsequent conversion to purified cyclosiloxanes, which are then used for polymerization [18].

The reaction of triethylsilanol with concentrated sulfuric acid for two hours in the presence of ammonium sulfate gives bis-(triethylsilyl) sulfate in 50% yield [666]. However, from the results of cryoscopic measurements it follows that in solutions of triethylsilanol in sulfuric acid the factor i, which shows the number of particles formed in solution per particle of solute, equals 3 [316, 523]. The properties of a solution of triethylsilanol in sulfuric acid differ markedly from the properties of such solutions of tertiary alcohols (i $\approx$ 2, but it changes with time) and approach the behavior of primary alcohols in this solvent. Therefore, to explain the conversions which occur when triethylsilanol is dissolved in sulfuric acid a scheme was proposed for their interaction which is analogous to the reaction of H_2SO_4 with primary alcohols and includes protonization of the triethylsilanol with further attack of the bisulfate ion and liberation of a water molecule [523]:

$$\text{R}_3\text{SiOH} + \text{H}_2\text{SO}_4 \rightleftarrows \text{R}_3\text{Si}\overset{+}{\text{O}}\text{H}_2 + \text{HSO}_4^- \quad (2.135)$$

$$\text{HSO}_4^- + \text{R}_3\text{Si}\overset{+}{\text{O}}\text{H}_2 \rightleftarrows \text{R}_3\text{SiHSO}_4 + \text{H}_2\text{O} \quad (2.136)$$

$$\text{H}_2\text{O} + \text{H}_2\text{SO}_4 \rightleftarrows \text{H}_3\text{O}^+ + \text{HSO}_4^- \quad (2.137)$$

Overall this leads to the factor i = 3:

$$R_3SiOH + 2H_2SO_4 \longrightarrow R_3SiHSO_4 + H_3O^+ + HSO_4^- \qquad (2.138)$$

Thus, triethylsilyl hydrosulfate is present in a solution of triethylsilanol in sulfuric acid and not bis(triethylsilyl) sulfate, though the former has never been isolated in a free form.

Trialkylsilyl hydrosulfates are also formed when trimethylethoxy- [316] and triethylethoxysilanes [523] are dissolved in sulfuric acid:

$$R_3SiOC_2H_5 + 3H_2SO_4 \longrightarrow R_3SiHSO_4 + C_2H_5HSO_4 + H_3O^+ + HSO_4^- \qquad (2.139)$$

The reaction of dimethyldiethoxysilane with sulfuric acid leads to the formation of dimethylsilylene bishydrosulfate. Cryoscopic investigations showed that seven molecules (experimentally 7.3) are formed from one molecule of dimethyldiethoxysilane in H_2SO_4 solution. This corresponds to the reaction equation [316]

$$(CH_3)_2Si(OC_2H_5)_2 + H_2SO_4 \longrightarrow (CH_3)_2Si(HSO_4)_2 + 2C_2H_5HSO_4 + 2H_3O^+ + 2HSO_4^- \qquad (2.140)$$

Analogous investigations of the solution of methyltriethoxysilane in sulfuric acid show that no compound with three hydrosulfate groups at one silicon atom is present in it, but polymers are formed instead of this. It is possible that the polymers contain structural units with six-membered rings [316]:

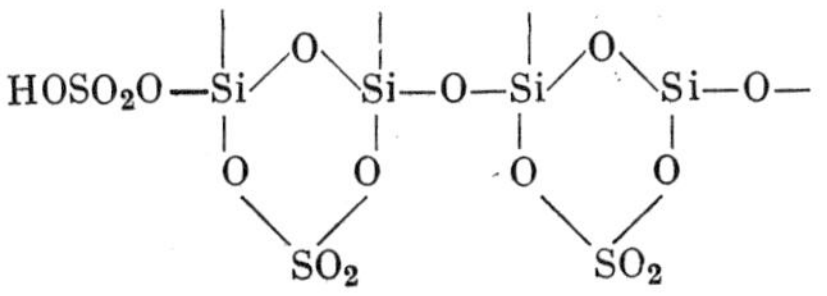

The reaction of 100% H_2SO_4 with trialkylmethoxysilanes with distillation of the methanol formed leads to the formation of the corresponding bis(trialkylsilyl) sulfates [30, 34].

Heating triethyl(acetoxy)silane with sulfuric acid with simultaneous distillation of the acetic acid formed leads analogously to bis(triethylsilyl) sulfate [163]:

$$2(C_2H_5)_3SiOCOCH_3 + H_2SO_4 \longrightarrow [(C_2H_5)_3SiO]_2SO_2 + 2CH_3COOH \qquad (2.141)$$

The use of triethyl(trichloroacetoxy)silane in reaction (2.141) gives much lower yields of bis(triethylsilyl) sulfate.

Concentrated sulfuric acid readily cleaved the siloxane bond in hexaalkyldisiloxanes. Cryoscopic and conductometric investigations showed that triorganosilyl hydrosulfate is formed [316, 523, 567], and this is a nonelectrolyte [316]:

$$R_3SiOSiR_3 + 3H_2SO_4 \rightleftarrows 2R_3SiHSO_4 + H_3O^+ + HSO_4^- \qquad (2.142)$$

However, the reaction products extracted in a free form (extraction with pentane) are bis(trialkylsilyl) sulfates [548, 662, 666]. The yields of bis(trimethylsilyl) and bis(triethylsilyl) sulfates obtained by heating a benzene solution of the corresponding hexaalkyldisiloxane with sulfuric acid with azeotropic distillation of the water are 78 and 80%, respectively [280].

Triorganosilyl hydrosulfates or triorganosilyl sulfates are evidently intermediates in syntheses of triorganohalosilanes from hexaorganodisiloxanes, sulfuric acid, and ammonium halides [315, 363, 436, 448, 565, 657, 662, 666, 667].

Sulfuric acid cleaves the siloxane bond in octamethylcyclotetrasiloxane with the formation of a compound which contains two hydrosulfate groups at one silicon atom [567, 680]:

$$[(CH_3)_2SiO]_4 + 12H_2SO_4 \longrightarrow 4(CH_3)_2Si(HSO_4)_2 + 4H_3O^+ + 4HSO_4^- \qquad (2.143)$$

In contradiction to this, the literature [149] contains a report that it is impossible to obtain $[(CH_3)_2SiOSO_2O]_n$ by heating dimethylpolysiloxane with sulfuric acid to 140°C.

It should be noted that compounds containing the grouping Si–O–S are formed in many rearrangements and polymerizations of siloxanes which proceed under the influence of sulfuric acid. The first stage of all these reactions is cleavage of the siloxane group with the formation of a silyl hydrosulfate group [3, 10, 19, 20, 40, 50, 52, 66, 67, 103, 105-109, 265, 266, 333, 381, 383, 389, 391, 422, 431-433, 451, 452, 484, 518, 530, 549, 570, 587, 637, 652, 660, 815, 830, 956; 957].

The action of sulfuric acid on polyorganosiloxanes may also be accompanied by the elimination of organic radicals from silicon atoms. An investigation of the destruction of polyorganosiloxane films chemisorbed on glass under the effect of H_2SO_4 of different concentrations showed that the stability of organic radicals in them falls inthe series [35]:

$$CH_3 > CH_2{=}CH > CH_2{=}CHCH_2 > C_2H_5 > C_6H_5$$

Trialkylsilanes dissolve readily and completely in concentrated sulfuric acid and are converted to bis(trialkylsilyl) sulfates [33, 440]:

$$2R_3SiH + H_2SO_4 \longrightarrow (R_3SiO)_2SO_2 + H_2 \qquad (2.144)$$

This reaction, which is catalyzed by mercury salts, may be used for the synthesis of bis(trialkylsilyl) sulfates and hexaalkyldisiloxanes which are formed by hydrolysis of the latter [29]. It can also be used for analysis of mixtures of tri- and tetraalkylsilanes [33].

Concentrated sulfuric acid is capable of cleaving the Si–C bond in tetraorganosilanes. The rate of this reaction depends on the structure of the organic radicals attached to the silicon atom. Elimination of an allyl group proceeds extremely readily. Thus, for example, trimethylallylsilane gives off propylene and is converted into bis(trimethylsilyl) sulfate when treated with sulfuric acid even at −20°C [668].

A phenyl group is also removed very readily from a silicon atom by the action of H_2SO_4 [3, 280, 307, 408, 678]. In the case of tetraphenylsilane, benzenesulfonic acid is formed and partly sulfated polysilicic acid [307, 678]. Alkyl groups are eliminated with much more difficulty than aryl groups. Thus, for example, after trimethylphenylsilane had been treated with sulfuric acid for 15 h, benzene (97% yield), and bis(trimethylsilyl) sulfate (85%) were formed [280]:

$$2(CH_3)_3SiC_6H_5 + H_2SO_4 \longrightarrow [(CH_3)_3SiO]_2SO_2 + 2C_6H_6 \qquad (2.145)$$

The Si–C bond is cleaved with the greatest difficulty in tetraalkylsilanes. It was assumed for a long time that tetramethylsilane does not react at all with sulfuric acid. However, investigation of the NMR spectra of a mixture of tetramethylsilane with sulfuric acid showed that there is cleavage of the Si–C bond [572, 832]. This observation is extremely important for spectroscopists, since the use of tetramethylsilane as the normal internal standard in the study of NMR spectra in sulfuric acid solutions may lead to serious errors.

When the reaction of tetramethylsilane with concentrated sulfuric acid is carried out for 70 h, cleavage of the Si–C bond is quantitative [572]. When tetraethylsilane is shaken with H_2SO_4 at room temperature for 9.5 h only 4% of the starting $(C_2H_5)_4Si$ reacts [97].* Of the four different radicals in methylethylpropylbenzylsilane, the methyl radical is eliminated first by the action of sulfuric acid [409]. The same rule is observed in the case of the elimination of an alkyl radical from different β-(methyldialkylsilyl)propionic acids [46, 47]. Elimination of the methyl radical by the action of H_2SO_4 also occurs in the case of trimethylsilyl derivatives of alkyl halides [659], alcohols [386], ethers [800], amines [663, 921, 960], and ketones [663, 667, 821, 959, 961].

When four- and five-membered 1,1-dimethylsilacyclanes are treated with sulfuric acid, opening of the heterocycle occurs more readily than elimination of methyl groups attached to the silicon atom. Thus, for example, the four-membered ring of 1,1-dimethylsilacyclobutane is cleaved completely by concentrated sulfuric acid even at 0°C [659]. The reaction proceeds with the formation of the corresponding silyl hydrosulfate, which, after hydrolysis, gives 1,3-dipropyl-1,1,3,3-tetramethyldisiloxane:

$$\begin{array}{l} CH_2{-}CH_2 \\ | \qquad\ \ | \\ CH_2{-}Si(CH_3)_2 \end{array} + H_2SO_4 \longrightarrow C_3H_7(CH_3)_2SiOSO_3H \xrightarrow[-H_2SO_4]{H_2O}$$

$$\longrightarrow C_3H_7(CH_3)_2SiOSi(CH_3)_2C_3H_7 \qquad (2.146)$$

The reaction of 1,1,3,3-tetramethyl-1,3-disilacyclobutane with sulfuric acid proceeds with cleavage of one Si–C bond of the ring [878].

1,1-Dimethyl- and 1,1-diethylsilacyclopentanes are cleaved by sulfuric acid at room temperature and converted after hydrolysis of the intermediate hydrosulfate into 1,3-dibutyltetramethyl- and 1,3-dibutyltetraethyldisiloxanes, respectively [92, 97]. When 1-methyl-1-chloromethylsilacyclopentane is treated with sulfuric acid there is only opening of the ring, and even this is several times slower than 1,1-dimethylsilacyclopentane [25]. Silaspirans containing five-membered rings are cleaved by the action of H_2SO_4 like other five-membered silacyclanes, but with opening of only one

*Treatment with sulfuric acid has been used repeatedly for purification of tetraethylsilane [33, 393].

ring [98]:

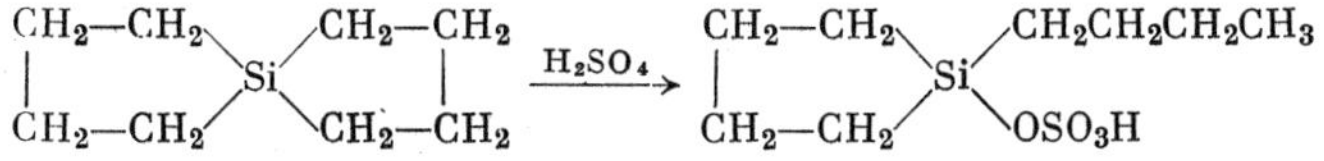

$$\begin{array}{c} \text{CH}_2\text{—CH}_2 \\ | \quad\quad\quad \text{Si} \\ \text{CH}_2\text{—CH}_2 \end{array} \begin{array}{c} \text{CH}_2\text{—CH}_2 \\ | \\ \text{CH}_2\text{—CH}_2 \end{array} \xrightarrow{\text{H}_2\text{SO}_4} \begin{array}{c} \text{CH}_2\text{—CH}_2 \\ | \quad\quad\quad \text{Si} \\ \text{CH}_2\text{—CH}_2 \end{array} \begin{array}{c} \text{CH}_2\text{CH}_2\text{CH}_2\text{CH}_3 \\ \\ \text{OSO}_3\text{H} \end{array} \quad (2.147)$$

Opening of the second ring is possible under more drastic conditions.

Six-membered silacyclanes react with sulfuric acid with approximately 20 times as much difficulty as the corresponding five-membered rings. In the case of 1,1-dimethylsilacyclohexane, in addition to opening of the six-membered ring there is also elimination of a methyl group [98]:

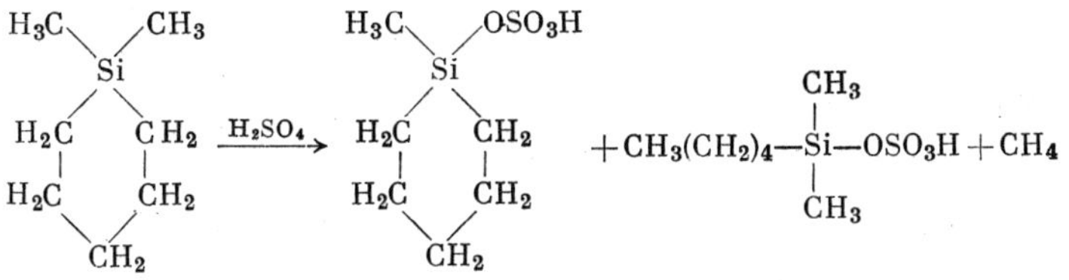

$$\begin{array}{c} \text{H}_3\text{C} \quad \text{CH}_3 \\ \text{Si} \\ \text{H}_2\text{C} \quad \text{CH}_2 \\ | \quad\quad | \\ \text{H}_2\text{C} \quad \text{CH}_2 \\ \text{CH}_2 \end{array} \xrightarrow{\text{H}_2\text{SO}_4} \begin{array}{c} \text{H}_3\text{C} \quad \text{OSO}_3\text{H} \\ \text{Si} \\ \text{H}_2\text{C} \quad \text{CH}_2 \\ | \quad\quad | \\ \text{H}_2\text{C} \quad \text{CH}_2 \\ \text{CH}_2 \end{array} + \begin{array}{c} \text{CH}_3 \\ | \\ \text{CH}_3(\text{CH}_2)_4\text{—Si—OSO}_3\text{H} \\ | \\ \text{CH}_3 \end{array} + \text{CH}_4 \quad (2.148)$$

Concentrated sulfuric acid cleaves a methyl group from a molecule of α,ω-bis(trimethylsilyl)alkanes $(\text{CH}_3)_3\text{Si}(\text{CH}_2)_n\text{Si}(\text{CH}_3)_3$ [437, 656]. The rate of the reaction falls with an increase in the length of the hydrocarbon chain. Thus, when n = 1-3 rupture of the Si—CH$_3$ bond occurs even at 30°C, while when n = 4 the reaction temperature must be raised to 40-45°C, and if n = 5-6, it must be raised to 60-65°C. In the case of 1,4-bis(trimethylsilyl)-butane there is also some cleavage of the Si—CH$_2$ bond [437].

In the reaction of sulfuric acid with hexamethyldisilane there may be elimination of one or two methyl groups at neighboring silicon atoms [438, 897]:

$$(\text{CH}_3)_3\text{SiSi}(\text{CH}_3)_3 \xrightarrow[-\text{CH}_4]{\text{H}_2\text{SO}_4} (\text{CH}_3)_3\text{SiSi}(\text{CH}_3)_2\text{OSO}_3\text{H} +$$
$$+ \text{HO}_3\text{SO}(\text{CH}_3)_2\text{SiSi}(\text{CH}_3)_2\text{OSO}_3\text{H} \quad (2.149)$$

A more detailed investigation has been made of the cleavage of the Si—C bond in ω-(trialkylsilyl)alkanoic acids by sulfuric acid, which proceeds by the scheme* [46, 47, 64, 640, 641, 658,

---

*Acids with n = 2 and 3 were investigated since when n = 1 there is β-decomposition of the acid under the reaction conditions with rupture of the Si—CH$_2$ bond and not elimination of the radical R.

663, 667, 739, 801, 958, 963]

$$R_3Si(CH_2)_nCOOH + H_2SO_4 \longrightarrow R_2Si(OSO_3H)(CH_2)_nCOOH + RH \qquad (2.150)$$

The rate of reaction (2.150) falls with dilution of the sulfuric acid; with an H_2SO_4 concentration below 76% there is no cleavage of the Si–C bond at all [641]. The radicals R in β-(methyldialkylsilyl)propionic acids, $CH_3R_2SiCH_2CH_2COOH$, may be arranged in the following series with respect to resistance to the action of sulfuric acid [46, 47]:

$$C_2H_5 > C_3H_7 > CH_3 > C_5H_{11} > C_4H_9 \text{ and iso-}C_5H_{11}$$

When ω-(trimethylsilyl)alkanoic acids are treated with sulfuric acid only one methyl group is eliminated in each case. Therefore, the hypothesis was put forward that the slow, rate-determining stage of the reaction is the electrophilic attack on the carbon atom of the methyl group, leading to its elimination as methane. Then the corresponding silyl hydrosulfate is formed rapidly, and the strongly electronegative substituent in this prevents elimination of the next CH_3 group* [658]. However, eliminations of methyl groups from β-(trimethylsilyl)propionic and from γ-(trimethylsilyl)butyric acids have different mechanisms [640, 641, 658]. In the first case methane is liberated quantitatively, while the addition of sodium hydrosulfate does not affect the reaction rate and the reaction proceeds as a first-order reaction (first-order relative to the carboxylic acid and first order relative to the sulfuric acid). In the case of γ-(trimethylsilyl)butyric acid the formation of CH_4 is not quantitative, the addition of $NaHSO_4$ reduces the reaction rate, and the reaction is third-order (first with respect to the carboxylic acid and second with respect to sulfuric acid). It is possible that in this case one molecule of sulfuric acid makes an electrophilic attack on the methyl group, while the second makes a nucleophilic attack on the silicon atom.

The product of the reaction of trimethylisopropenylsilane with sulfuric acid (after hydrolysis) is tert-butyldimethylsilanol. The reaction begins with attack by a proton on the β-unsaturated carbon atom of the isopropenyl group and then there follows a re-

*In the reaction of trimethyl(chloromethyl)silane with H_2SO_4, also a methyl group is eliminated, not a chloromethyl group [658].

arrangement of the carbonium cation formed with addition of a hydrosulfate ion to the silicon atom [434, 661]:

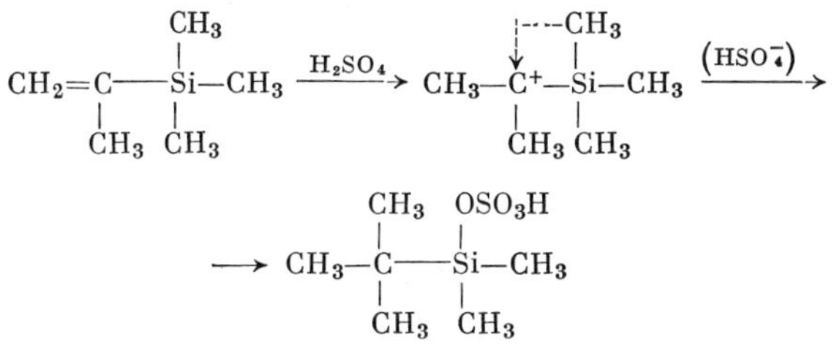

$$
\begin{array}{c}
\underset{\substack{| \\ CH_3}}{CH_2=C}\!\!-\!\!\underset{\substack{| \\ CH_3}}{Si}\!\!-\!\!CH_3 \xrightarrow{H_2SO_4}
CH_3-\underset{\substack{| \\ CH_3}}{\overset{\substack{----CH_3 \\ \downarrow}}{C^+}}\!\!-\!\!\underset{\substack{| \\ CH_3}}{Si}\!\!-\!\!CH_3 \xrightarrow{(HSO_4^-)}
\end{array}
$$

$$
\longrightarrow CH_3-\underset{\substack{| \\ CH_3}}{\overset{\substack{CH_3 \\ |}}{C}}\!\!-\!\!\underset{\substack{| \\ CH_3}}{\overset{\substack{OSO_3H \\ |}}{Si}}\!\!-\!\!CH_3 \qquad (2.151)
$$

If the starting unsaturated compound contains a substituent which increases the degree of $d_\pi-p_\pi$ interaction of the π-electrons of the double bonds with the 3d-orbitals of the silicon, then the proton attacks the α-carbon atom with subsequent elimination of an alkenyl group. Thus, for example, in the reaction of dimethyl(chloromethyl)isopropenylsilane with sulfuric acid the main process is elimination of propylene, while rearrangement to the tert-butyl derivative proceeds to a much lesser extent [434]. In the case of trimethyl(vinyl)silane [400] and dimethyl(benzyl)(vinyl)-silane [26] there is also only elimination of a vinyl group. A rearrangement analogous to (2.151) occurs in the reaction of sulfuric acid with pentamethylisopropenyldisilane. However, in this case it is a trimethylsilyl group which migrates and not a methyl group [436].

4.1.2. Reactions of Organosilicon Compounds

with Sulfur Trioxide

A second general method of synthesizing substances containing the grouping Si $-$O$-$S is the reaction of organosilicon compounds with sulfur trioxide. Depending on their structure, the reaction of organosilicon compounds with SO_3 may lead to very different types of sulfur-containing compounds.* Thus, the reaction of organochlorosilanes with SO_3 at $-30°C$ forms organosilyl esters of chlorosulfonic acids containing one or two chlorosulfonyloxy groups [621,

---

*Compounds with the grouping Si$-$O$-$S are not formed in the reaction of $SiCl_4$ with SO_3. The products of this reaction are hexachlorodisiloxane and the acid chloride of pyrosulfuric acid [41-43].

622, 625]:

$$(CH_3)_3SiCl + SO_3 \longrightarrow (CH_3)_3SiOSO_2Cl \quad (2.152)$$

$$(CH_3)_2SiCl_2 + SO_3 \longrightarrow (CH_3)_2SiClOSO_2Cl \quad (2.153)$$

$$(CH_3)_2SiCl_2 + 2SO_3 \longrightarrow (CH_3)_2Si(OSO_2Cl)_2 \quad (2.154)$$

$$CH_3SiCl_3 + SO_3 \longrightarrow CH_3SiCl_2OSO_2Cl \quad (2.155)$$

The same esters are obtained by the reaction of organohalosilanes with chlorosulfonic acid [278, 280, 621, 622, 625] (see 4.1.4).

Organosilyl esters of fluorosulfonic acid are obtained by the reaction of alkylfluorosilanes with sulfur trioxide [613]. Thus, for example, trimethylfluorosilane reacts with SO_3 at room temperature to form the trimethylsilyl ester of fluorosulfonic acid in 89% yield:

$$(CH_3)_3SiF + SO_3 \longrightarrow (CH_3)_3SiOSO_2F \quad (2.156)$$

Initially, the reaction of dimethylfluorochlorosilane with SO_3 proceeds with rupture of the Si−Cl bond:

$$(CH_3)_2SiFCl + SO_3 \longrightarrow (CH_3)_2SiFOSO_2Cl \quad (2.157)$$

Only with further treatment of the product obtained with sulfur trioxide is the Si−F bond broken [613]:

$$(CH_3)_2SiFOSO_2Cl + SO_3 \longrightarrow (CH_3)_2Si(OSO_2F)OSO_2Cl \quad (2.158)$$

The reaction of sulfur trioxide with dimethyldifluorosilane proceeds analogously to its reactions with dimethyldichlorosilane (2.153 and 2.154), but much more slowly [613].

The reaction of sulfur trioxide with triorganoalkoxysilanes at −78°C forms triorganosilyl esters of alkylsulfuric acids [619]:

$$R_3SiOCH_3 + SO_3 \longrightarrow R_3SiOSO_2OCH_3 \quad (2.159)$$

Their structure has been demonstrated by confirmatory synthesis from triorganochlorosilanes and methylsulfuric acid [619]:

$$R_3SiCl + HOSO_2OCH_3 \longrightarrow R_3SiOSO_2OCH_3 + HCl \quad (2.160)$$

Trialkylsilyl esters of methylsulfuric acid are also formed by the reaction of trialkylmethoxysilanes with the complex of sulfur trioxide and N,N-diethylbenzamide [277].

Sulfur trioxide cleaves organosiloxanes with the formation of silyl sulfates [30, 623, 627, 665]. Thus, for example, the reaction of dimethylpolysiloxane with SO_3 forms polymeric dimethylsilylene sulfate in 97.5% yield [623]:

$$[-(CH_3)_2SiO-]_n + nSO_3 \longrightarrow [-(CH_3)_2SiOSO_2O-]_n \tag{2.161}$$

Sulfur trioxide reacts analogously with trimethylsiloxytrimethylgermane at −35°C to form the corresponding sulfate in almost quantitative yield [615]:

$$(CH_3)_3SiOGe(CH_3)_3 + SO_3 \longrightarrow (CH_3)_3SiOSO_2OGe(CH_3)_3 \tag{2.162}$$

The exothermic reaction between SO_3 and trimethyl(diethylamino)silane also proceeds quantitatively [621]:

$$(CH_3)_3SiN(C_2H_5)_2 + SO_3 \longrightarrow (CH_3)_3SiOSO_2N(C_2H_5)_2 \tag{2.163}$$

Trialkylsilyl esters of aromatic sulfonic acids are formed by the reaction of SO_3 with trialkyl(aryl)silanes. Thus, for example, sulfur trioxide reacts with trimethyl(phenyl)silane in carbon tetrachloride with the formation of the trimethylsilyl ester of benzenesulfonic acid [205, 288]:

$$(CH_3)_3SiC_6H_5 + SO_3 \longrightarrow (CH_3)_3SiOSO_2C_6H_5 \tag{2.164}$$

Instead of SO_3 it is possible to use its complex with N,N-diethylacetamide in this reaction [277]. Tetraalkylsilanes will also undergo reaction (2.164) [945].

In carrying out the reaction of sulfur trioxide with bis(trialkylsilyl)benzenes it is possible to select conditions under which derivatives of monosulfonic acids are formed. This then makes it possible to obtain trialkylsilyl derivatives of aromatic sulfonic acids by hydrolyzing these products [205, 288]:

$$(CH_3)_3SiC_6H_4Si(CH_3)_3 \xrightarrow{SO_3} (CH_3)_3SiC_6H_4SO_2OSi(CH_3)_3 \xrightarrow{H_2O} (CH_3)_3SiC_6H_4SO_2OH + (CH_3)_3SiOH \tag{2.165}$$

4.1.3. Synthesis from Arenesulfonic Acids

Another method of preparing trialkylsilyl esters of aromatic sulfonic acids is the reaction of these acids with trialkylarylsilanes and trialkylhydrosilanes, which occurs when the reagents are boiled in toluene for 10 h. The yield is about 70% [281].

$$C_6H_5SO_2OH + C_6H_5Si(CH_3)_3 \longrightarrow (CH_3)_3SiOSO_2C_6H_5 + C_6H_6 \quad (2.166)$$

$$C_6H_5SO_2OH + (C_2H_5)_3SiH \longrightarrow (C_2H_5)_3SiOSO_2C_6H_5 + H_2 \quad (2.167)$$

Trialkylsilyl esters of benzenesulfonic acid are among the products from the reduction of benzenesulfonyl chloride by trialkylsilanes, and they are also formed by the reaction of benzenesulfonyl chloride with hexamethyldisilane [209]:

$$5(CH_3)_3SiSi(CH_3)_3 + 3C_6H_5SO_2Cl \longrightarrow 3(CH_3)_3SiCl + 3[(CH_3)_3Si]_2O + $$
$$+ C_6H_5SO_2OSi(CH_3)_3 + C_6H_5SSC_6H_5 \quad (2.168)$$

4.1.4. Synthesis from Halosulfonic Acids

Halosulfonic acids may be used instead of sulfur trioxide and sulfuric acid for the synthesis of compounds containing the grouping Si–O–S. The reaction of chlorosulfonic acid with organochlorosilanes forms trialkylsilyl esters [278, 280, 621, 622, 625]:

$$(CH_3)_3SiCl + HOSO_2Cl \longrightarrow (CH_3)_3SiOSO_2Cl + HCl \quad (2.169)$$

In an analogous reaction fluorosulfonic acid forms organosilicon esters of chlorosulfonic acid and the corresponding organofluorosilanes [625]:

$$2(CH_3)_3SiCl + HOSO_2F \longrightarrow (CH_3)_3SiF + (CH_3)_3SiOSO_2Cl + HCl \quad (2.170)$$

$$2(CH_3)_2SiCl_2 + 2HOSO_2F \longrightarrow (CH_3)_2SiF_2 + (CH_3)_2Si(OSO_2Cl)_2 + 2HCl \quad (2.171)$$

The reaction of organosiloxanes with chlorosulfonic acid leads to the formation of silyl sulfates [583, 624]. Thus, for example, chlorosulfonic acid reacts with hexamethyldisiloxane even at room temperature to give an 80-85% yield of bis(trimethylsilyl) sulfate:

$$(CH_3)_3SiOSi(CH_3)_3 + ClSO_2OH \longrightarrow [(CH_3)_3SiO]_2SO_2 + HCl \quad (2.172)$$

The analogous reaction with the methyl ester of chlorosulfonic acid proceeds when the mixture is heated to 100°C [624]:

$$(CH_3)_3SiOSi(CH_3)_3 + ClSO_2OCH_3 \longrightarrow [(CH_3)_3SiO]_2SO_2 + CH_3Cl \quad (2.173)$$

In the reaction of fluorosulfonic acid with hexaalkyldisiloxanes, the latter are converted to trialkylfluorosilanes [624]:

$$R_3SiOSiR_3 + 2FSO_2OH \longrightarrow 2R_3SiF + H_2S_2O_7 \quad (2.174)$$

4.1.5. Synthesis from Salts of Oxygen Acids of Sulfur

A general method of preparing compounds containing the grouping Si$-$O$-$S is the reaction of organochlorosilanes with sodium and silver salts of sulfuric [154, 894], methylsulfuric [619], fluorosulfonic [376], and sulfamic [524] acids. Thus, when a mixture of triethylchlorosilane and silver sulfate in nitromethane is boiled for 5 h, bis(triethylsilyl) sulfate is formed in 60% yield. The yield of bis(trimethylsilyl) sulfate under analogous conditions is considerably lower (30%) [154]:

$$2R_3SiCl + Ag_2SO_4 \longrightarrow (R_3SiO)_2SO_2 + 2AgCl \quad (2.175)$$

Trialkylsilyl esters of sulfamic acid are formed similarly from trialkylchlorosilanes and silver sulfamate, and these then react with a second molecule of trialkylchlorosilane [524]:

$$R_3SiCl + AgOSO_2NH_2 \longrightarrow R_3SiOSO_2NH_2 + AgCl \quad (2.176)$$

$$R_3SiOSO_2NH_2 + ClSiR_3 \xrightarrow[\substack{-HOSO_2NH_2 \\ -AgCl}]{+AgOSO_2NH_2} R_3SiOSO_2NHSiR_3 \quad (2.177)$$

Sodium salts are found to be less active in this reaction than silver salts. However, sodium methylsulfate may be used successfully to prepare trialkylsilyl esters of methylsulfuric acid [619]:

$$(CH_3)_3SiCl + NaOSO_2OCH_3 \longrightarrow (CH_3)_3SiOSO_2OCH_3 + NaCl \quad (2.178)$$

4.1.6. Other Synthesis Methods

Of the other reactions for the formation of the grouping Si$-$O$-$S we should mention the reaction of the ethyl ester of chlorosulfonic acid with hexamethyldisilazan, which proceeds according to the scheme [180]

$$[(CH_3)_3Si]_2NH + C_2H_5OSO_2Cl \longrightarrow (CH_3)_3SiNHSO_2OSi(CH_3)_3 + C_2H_5Cl \quad (2.179)$$

and the reaction of triethylgermyl ethanesulfonate with triphenylsilyl isocyanate [159]:

$$(C_2H_5)_3GeOSO_2C_2H_5 + (C_6H_5)_3SiNCO \longrightarrow (C_6H_5)_3SiOSO_2C_2H_5 + (C_2H_5)_3GeNCO \quad (2.180)$$

Siloxysulfanes $R_3SiOS_nOSiR_3$ are formed by the reaction of sodium triorganosilanolates with sulfur chloride S_nCl_2 [869].

It is considered that organosilicon compounds containing the grouping Si–O–S are formed as intermediate products in the ammonium sulfate catalyzed transamination of polyalkylcyclosilazans with aliphatic amines, [87], and also in the rearrangement of cyclosilazans in the presence of sulfuric acid [49].

4.2. Physical Properties

The physical constants of compounds containing the grouping Si–O–S are given in Table 19. Most of these substances are readily soluble in ethyl ether, benzene, and tetrahydrofuran. They are all hydrolyzed by atmospheric moisture.

In the NMR spectrum of bis(trimethylsilyl) sulfate dissolved in CCl_4, there is a signal from the proton of the methyl groups, which is shifted by 23 Hz to lower fields relative to the signal of the internal standard $(CH_3)_4Si$ (H_0 = 60 MHz). The spin-spin constant $J(C^{13}-H)$ = 121.0 Hz [612]. The results of investigating the NMR spectra of methylsilyl halosulfates are given in Table 18 [613].

4.3. Chemical Properties

Compounds containing the Si–O–S bond are readily hydrolyzed by water with cleavage of this bond and the formation of siloxanes and sulfuric acid [71, 524, 615, 619, 621, 623, 665, 684]:

$$2\,{>}Si{-}O{-}SO_2OH + H_2O \longrightarrow {>}Si{-}O{-}Si{<} + 2H_2SO_4 \qquad (2.181a)$$

$$>Si{-}O{-}SO_2{-}O{-}Si{<} + H_2O \longrightarrow {>}Si{-}O{-}Si{<} + H_2SO_4 \qquad (2.181b)$$

TABLE 18. NMR Spectra of Methylsilyl Halosulfates
Negative values of δ indicate a shift (in Hz) to lower fields relative to $(CH_3)_4Si$ as an internal standard.

Compound	δ	$J(H^1-C^{13})$	$J(H^1-C-Si^{29})$	$J(H^1-C-Si-F^{19})$
$(CH_3)_3SiOSO_2F$	–15.0	121.0	7.40	—
$(CH_3)_3SiOSO_2Cl$	–20.5	121.5	7.30	—
$(CH_3)_2FSiOSO_2Cl$	–29.5	123.5	7.85	6.6
$(CH_3)_2FSiOSO_2F$	–21.5	123.0	8.00	6.76
$(CH_3)_2Si(OSO_2F)_2$	–37.0	122.5	7.80	—

TABLE 19. Compounds Containing the Grouping Si—O—S

Empirical formula	Compound	B.p., °C (mm)	M.p., °C	n_D^{20}	d_4^{20}	Literature
$CH_3Cl_3O_3SSi$	$CH_3SiCl_2(OSO_2Cl)$	66 (12)	—	—	—	621
$C_2H_6ClFO_3SSi$	$(CH_3)_2SiF(OSO_2Cl)$	52—54 (13)	—50	—	—	613
$C_2H_6Cl_2O_3SSi$	$(CH_3)_2SiCl(OSO_2Cl)$	68 (12)	—	—	—	621
$C_2H_6Cl_2O_6S_2Si$	$(CH_3)_2Si(OSO_2Cl)_2$	173	—	—	—	621
		101 (12)				
$C_2H_6F_2O_3SSi$	$(CH_3)_2SiF(OSO_2F)$	33—36 (14)	—	—	—	613
$C_2H_6F_2O_6S_2Si$	$(CH_3)_2Si(OSO_2F)_2$	decomp.	—	—	—	613
$C_3H_9ClO_3SSi$	$(CH_3)_3SiOSO_2Cl$	168—169	—	—	—	621
		74 (20)	—	1.4225	1.220	280
		70 (16)				
		68 (16)	—	1.4230	—	278
		66 (12)	—26	—	—	621
$C_3H_9ClO_4SSi$	$(CH_3)_2SiCl(OSO_2OCH_3)$	65 (1)	—	—	—	619
$C_3H_9FO_3SSi$	$(CH_3)_3SiOSO_2F$	36.5 (12)	—45	—	—	613
$C_4H_{12}O_3SSi$	$(CH_3)_3SiOSO_2CH_3$	103—104 (25)	—	1.4235	1.102	235
$C_4H_{12}O_4SSi$	$(CH_3)_3SiOSO_2OCH_3$	201	—	—	—	619
		93—94 (15)	—	1.4065	1.171	279, 280
$C_4H_{12}O_8S_2Si$	$(CH_3)_2Si(OSO_2OCH_3)_2$	118 (1)	—	—	—	619
$C_4H_{12}O_8S_2Si_2$	$[(CH_3)_2SiOSO_2O]_2$	—	101—118	—	—	623
		—	100—116	—	—	624
$C_4H_{12}O_{12}S_3Si$	$CH_3Si(OSO_2OCH_3)_3$	138 (1)	—	—	—	619
$C_5H_{12}O_3SSi$	$(CH_3)_3SiOSO_2CH{=}CH_2$	112—113 (28)	—	1.4355	1.090	235
$C_5H_{14}O_3SSi$	$(CH_3)_3SiOSO_2C_2H_5$	112—114 (22)	—	1.4261	1.077	235
$C_5H_{15}NO_3SSi$	$(CH_3)_3SiOSO_2N(CH_3)_2$	71 (2)	7.3—8.1	—	—	524
$C_6H_{16}O_5SSi$	$(CH_3)_2(C_2H_5O)SiOSO_2OC_2H_5$	66—67 (0.6)	—	1.4085^{25}	—	205

TABLE 19 (Continued)

Empirical formula	Compound	B.p., °C (mm)	M.p., °C	n_D^{20}	d_4^{20}	Literature
$C_6H_{17}NO_3SSi$	$(CH_3)_3SiOSO_2NHC_3H_7$	115 (2)	−15.5 to −14.5	—	—	524
$C_6H_{18}GeO_4SSi$	$(CH_3)_3SiOSO_2OGe(CH_3)_3$	—	93—96	—	—	615
$C_6H_{18}O_3SSi_2$	$(CH_3)_3SiSi(CH_3)_2OSO_2CH_3$	125—127 (20)	—	1.4485	1.037	207
$C_6H_{18}O_4SSi_2$	$[(CH_3)_3SiO]_2SO_2$	118 (18)	56—58	—	—	280
		115 (15)				
		95—97 (9)	45	—	—	583
		87—90 (4)	45—46	—	—	548
		—	49—52	—	—	668
		—	55—57	—	—	624
		—	56—58	—	—	615, 665
$C_6H_{18}O_6S_2Si_2$	$[CH_3SO_2OSi(CH_3)_2]_2$	170—172 (3)	120—122	—	—	207
$C_6H_{18}O_6S_3Si_2$	$[(CH_3)_3SiOSO_2]_2S$	—	−70 to −69	—	—	627
$C_7H_{18}O_3SSi$	$(CH_3)_2(C_4H_9)SiOSO_2CH_3$	92—94 (7)	—	1.4334	1.039	235
$C_7H_{19}NO_3SSi$	$(CH_3)_3SiOSO_2N(C_2H_5)_2$	76 (1.8)	−10.5 to −9.8	—	—	524
		108 (13)	—	—	—	621
$C_8H_{18}O_3SSi$	$(C_2H_5)_3SiOSO_2CH{=}CH_2$	101—102 (1)	—	1.4510	1.062	235
$C_8H_{20}O_3SSi$	$(C_2H_5)_3SiOSO_2C_2H_5$	90—91 (6)	—	1.4423	1.042	235
$C_8H_{21}NO_3SSi$	$(C_2H_5)_3SiOSO_2N(CH_3)_2$	101.5—103 (1.5)	−51 to −50	—	—	524
$C_9H_{14}O_3SSi$	$(CH_3)_3SiOSO_2C_6H_5$	155 (18)	—	1.4938	1.418	280, 281
$C_9H_{27}NO_3SSi_3$	$(CH_3)_3SiOSO_2N[Si(CH_3)_3]_2$	78 (0.005)	—	—	—	180
$C_{10}H_{16}O_3SSi$	$(CH_3)_3SiOSO_2C_6H_4CH_3$-*m*	124—125 (1)	—	1.4916	—	179, 205
		179—180 (30)	—	1.4962	1.114	208
	$(CH_3)_3SiOSO_2C_6H_4CH_3$-*p*	164 (16)	—	1.4960	1.120	280
		163 (17)	—	—	1.126	279
$C_{10}H_{16}O_4SSi$	$(CH_3)_3SiOSO_2C_6H_4OCH_3$-*o*	157—159 (2)	—	1.5088	1.186	210
	$(CH_3)SiOSO_2C_6H_4OCH_3$-*p*	150—152 (2)	—	1.5065	1.165	210
$C_{10}H_{25}NO_3SSi$	$(C_2H_5)_3SiOSO_2N(C_2H_5)_2$	108—109 (1.8)	−78	—	—	524
$C_{10}H_{26}O_3SSi_2$	$CH_3(C_2H_5)_2SiSi(C_2H_5)_2OSO_2CH_3$	121—127 (3)	—	1.4725	1.0098	207
$C_{12}H_{20}O_3SSi$	$(C_2H_5)_3SiOSO_2C_6H_5$	189 (18)	—	1.4938	1.0963	281

$C_{12}H_{22}O_3SSi_2$	$(CH_3)_3SiOSO_2C_6H_4Si(CH_3)_3$-*o*	161—163 (2)	—	1·4872	1·118	210
	$(CH_3)_3SiOSO_2C_6H_4Si(CH_3)_3$-*m*	135—136 (1·5)	—	$1·4966^{25}$	—	179, 205
	$(CH_3)_3SiOSO_2C_6H_4Si(CH_3)_3$-*p*	130—131 (0·6)	83—86	—	—	288
		—	83—85	—	—	208
		150—151 (2)	—	—	—	205
		163—164 (2)	—	1·4860	1·101	210
$C_{12}H_{22}O_6S_2Si_2$	$(CH_3)_3SiOSO_2C_6H_4OSO_2Si(CH_3)_3$-*p*	—	121—123	—	—	208, 235
$C_{12}H_{30}O_4SSi_2$	$[(C_2H_5)_3SiO]_2SO_2$	278	—	1·4440	—	163
		279	—	1·443	—	154
		175 (15)	—	—	—	280
		170 (12)	—	1·4442	—	665, 666
		128 (2)	—	—	—	666
$C_{12}H_{31}NO_3SSi_2$	$(C_2H_5)_3SiOSO_2NHSi(C_2H_5)_3$	141—143·5 (1·5)	—1 to +0·5	—	—	524
$C_{13}H_{24}O_3SSi_2$	$(CH_3)_3SiCH_2C_6H_4SO_2OSi(CH_3)_3$-*o*	111—112 (0·5)	—	$1·4958^{25}$	—	205
	$(CH_3)_3SiCH_2C_6H_4SO_2OSi(CH_3)_3$-*m*	133—134 (0·6)	—	$1·4922^{25}$	—	205
	$(CH_3)_3SiCH_2C_6H_4SO_2OSi(CH_3)_3$-*p*	151—152 (1)	114—115	—	—	205
$C_{14}H_{16}O_3SSi$	$(CH_3)_2(C_6H_5)SiOSO_2C_6H_5$	169—171 (2)	—	1·5430	1·191	235
$C_{16}H_{36}O_3SSi$	$(C_4H_9)_3SiOSO_2C_4H_9$	154—156 (1)	—	1·4517	0·963	235
$C_{19}H_{18}O_3SSi$	$CH_3(C_6H_5)_2SiOSO_2C_6H_5$	220—222 (2)	—	$1·5795^{25}$	$1·234^{25}$	235
$C_{27}H_{28}O_3SSi_2$	$(C_6H_5)_3SiC_6H_4SO_2OSi(CH_3)_3$-*m*	—	127(decomp.)	—	—	205
	$(C_6H_5)_3SiC_6H_4SO_2OSi(CH_3)_3$-*p*	—	132(decomp.)	—	—	205
$C_{36}H_{30}O_4SSi_2$	$[(C_6H_5)_3SiO]_2SO_2$	—	134—136	—	—	279, 280
$C_{45}H_{34}O_{10}SSi$	$((C_6H_5CO)_2CH]_3SiOSO_2OH$	—	242	—	—	272, 273

If the hydrolysis is carried out in an alkaline medium it is possible to isolate the corresponding silanols which are formed as intermediates [662, 666].

The reaction of the trimethylsilyl ester of chlorosulfonic acid with hydrogen sulfide in ether at −78°C for 2 h forms the O-trimethylsilyl ester of thiosulfuric acid in 45% yield [627]:

$$(CH_3)_3SiOSO_2Cl + H_2S \longrightarrow (CH_3)_3SiOSO_2SH + HCl \qquad (2.182)$$

The ester formed will react with a second molecule of $(CH_3)_3SiOSO_2Cl$ to form the bis(trimethylsilyl) ester of trithionic acid (41.5% yield):

$$(CH_3)_3SiOSO_2SH + ClSO_2OSi(CH_3)_3 \xrightarrow[-HCl]{} [(CH_3)_3SiOSO_2]_2S \qquad (2.183)$$

Together with this there is oxidation of the O-trimethylsilyl ester of thiosulfuric acid with the formation (11.4%) of a disulfide compound $[(CH_3)_3SiOSO_2S]_2$ [627].

The compounds obtained in this way are extremely unstable thermally and decompose even at room temperature:

$$2(CH_3)_3SiOSO_2SH \longrightarrow 2(CH_3)_3SiOH + 2SO_2 + 2S \qquad (2.184)$$

$$[(CH_3)_3SiOSO_2)]_2S \longrightarrow (CH_3)_3SiOSi(CH_3)_3 + SO_3 + SO_2 + S \qquad (2.185)$$

In the latter case the sulfur trioxide formed initially then reacts with hexamethyldisiloxane to form bis(trimethylsilyl) sulfate.

Organosilyl sulfates react with hydrogen chloride with the formation of chlorosilanes. This reaction is carried out in the presence of ammonium sulfate, which prevents the reverse reaction of the organochlorosilane with sulfuric acid:

$$(R_3SiO)_2SO_2 + 2HCl \longrightarrow 2R_3SiCl + H_2SO_4 \qquad (2.186)$$

Reaction (2.186) is used to prepare trialkylhalosilanes, when the bis(triorganosilyl) sulfate is treated with an ammonium halide instead of a hydrogen halide [9, 315, 363, 436, 448, 565, 657, 662, 666, 667, 679]. This eliminates the need to prepare the dry hydrogen halide beforehand, and the need to add ammonium sulfate, since the latter is formed directly in the reaction mixture during the process itself. This reaction is usually used for the synthesis of trialkylchlorosilanes and trialkylfluorosilanes. However, the latter are obtained more conveniently and more efficiently by using ammonium bifluoride (or KHF_2) instead of ammonium fluoride [28].

Dimethylsilylene sulfate (cyclic dimer) is only partly cleaved by hydrogen chloride in the presence of ammonium sulfate with the formation of dimethyldichlorosilane and sulfuric acid [623]. Cleavage of trimethylsilyl methyl sulfate by hydrogen chloride to trimethylchlorosilane and methylsulfuric acid has been proposed for the separation of a mixture of trimethylchlorosilane and silicon tetrachloride. For this purpose the two compounds are mixed with methylsulfuric acid, the trimethylsilyl methyl sulfate distilled in vacuum from the involatile product from the reaction of methylsulfuric acid and $SiCl_4$, and, finally, the purified trimethylsilyl methyl sulfate decomposed with hydrogen chloride [619].

Passing dry ammonia into an ether solution of bis(trimethylsilyl) sulfate for 3 h gives a 71% yield of hexamethyldisilazan [662]:

$$[(CH_3)_3SiO]_2SO_2 + 3NH_3 \longrightarrow (CH_3)_3SiNHSi(CH_3)_3 + (NH_4)_2SO_4 \qquad (2.187)$$

The products from the analogous reaction of bis(trimethylsilyl) sulfate and triethyl(amino)silane are 1,1,1-trimethyl-3,3,3-triethyldisilazan, hexaethyldisilazan, and ammonium sulfate [5, 6]:

$$[(CH_3)_3SiO]_2SO_2 + 6(C_2H_5)_3SiNH_2 \longrightarrow 2(CH_3)_3SiNHSi(C_2H_5)_3 + \\ + 2(C_2H_5)_3SiNHSi(C_2H_5)_3 + (NH_4)_2SO_4 \qquad (2.188)$$

Bis(trimethylsilyl)sulfate reacts with ethylmagnesium bromide and propylmagnesium bromide to form trimethylethylsilane and trimethylpropylsilane [662]:

$$[(CH_3)_3SiO]_2SO_2 + 2RMgBr \longrightarrow 2(CH_3)_3SiR + (BrMg)_2SO_4 \qquad (2.189)$$

In the case of isopropylmagnesium bromide the yield of trimethylisopropylsilane does not exceed 34% even after the reaction mixture has been heated for eight days, while the main reaction product is trimethylbromosilane. In accordance with a scheme analogous to (2.189), bis(trimethylsilyl) sulfate reacts with sodium acetylide to form trimethylethynylsilane [125, 822].

The Si–O–S bond in silyl sulfates is cleaved by the action of alcohols [280, 619, 623]. Thus, for example, the methanolysis of trimethylsilyl methanesulfonate forms trimethyl(methoxy)silane and methanesulfonic acid [619]:

$$(CH_3)_3SiOSO_2CH_3 + CH_3OH \longrightarrow (CH_3)_3SiOCH_3 + CH_3SO_2OH \qquad (2.190)$$

The methanolysis of bis(trimethylsilyl) sulfate and the trimethylsilyl ester of chlorosulfonic acid proceeds analogously with the formation of dimethyl sulfate and trimethylsilanol (which condenses to hexamethyldisiloxane) [280]. However, when the reactions are carried out with equimolar amounts of the reagents it is possible to obtain a compound in which the Si–O–S grouping is preserved, namely trimethylsilyl methyl sulfate [280]:

$$[(CH_3)_3SiO]_2SO_2 + CH_3OH \longrightarrow (CH_3)_3SiOSO_2OCH_3 + (CH_3)_3SiOH \quad (2.191)$$

$$(CH_3)_3SiOSO_2Cl + CH_3OH \longrightarrow (CH_3)_3SiOSO_2OCH_3 + HCl \quad (2.192)$$

A reaction analogous to (2.191) occurs when bis(trimethylsilyl)sulfate is treated with triphenylsilanol. In this case the trimethylsilyl group in the silyl sulfate molecule is replaced by a triphenylsilyl group, and the reaction products are bis(triphenylsilyl) sulfate and trimethylsilanol [280]:

$$[(CH_3)_3SiO]_2SO_2 + (C_6H_5)_3SiOH \longrightarrow [(C_6H_5)_3SiO]_2SO_2 + 2(CH_3)_3SiOH \quad (2.193)$$

Triethylsilane and triphenylsilane reduce bis(trimethylsilyl) sulfate with the formation of SO_2, trimethylsilanol, and 1,1,1-trimethyl-3,3,3-triethyl (or -3,3,3-triphenyl)disiloxane [280]:

$$[(CH_3)_3SiO]_2SO_2 + R_3SiH \longrightarrow R_3SiOSi(CH_3)_3 + SO_2 + (CH_3)_3SiOH \quad (2.194)$$

In this reaction the trimethylsilyl ester of chlorosulfonic acid forms bis(triphenylsilyl) sulfate, trimethylchlorosilane, trimethylsilanol, HCl, and SO_2 [279, 280]:

$$3(CH_3)_3SiOSO_2Cl + 2(C_6H_5)_3SiH \longrightarrow [(C_6H_5)_3SiO]_2SO_2 + \\ + (CH_3)_3SiOH + 2(CH_3)_3SiCl + 2SO_2 + HCl \quad (2.195)$$

The trimethylsilyl ester of chlorosulfonic acid is completely cleaved by pyridine with the formation of trimethylchlorosilane and the complex of pyridine with sulfur trioxide (pyridine–sulfur trioxide) [277]:

$$(CH_3)_3SiOSO_2Cl + C_5H_5N \longrightarrow (CH_3)_3SiCl + C_5H_5N \cdot SO_3 \quad (2.196)$$

When treated with phosphorus pentachloride, bis(trimethylsilyl) sulfate is converted into the trimethylsilyl ester of chlorosulfonic acid. This reaction is carried out in cyclohexane and the yield of the ester after heating for half an hour is 73% [278, 280]:

$$[(CH_3)_3SiO]_2SO_2 + PCl_5 \longrightarrow (CH_3)_3SiOSO_2Cl + (CH_3)_3SiCl + POCl_3 \quad (2.197)$$

When a benzene solution of $(CH_3)_3SiOSO_2Cl$ is heated to boiling the trimethylsilyl ester of benzenesulfonic acid is obtained in good yield [279, 280]:

$$(CH_3)_3SiOSO_2Cl + C_6H_6 \longrightarrow (CH_3)_3SiOSO_2C_6H_5 + HCl \qquad (2.198)$$

The trimethylsilyl ester of chlorosulfonic acid also participates in a series of other reactions which proceed with retention of the grouping Si–O–S [236, 280, 841, 852, 864]:

$$(CH_3)_3SiOSO_2Cl + CH_3COOC_4H_9 \longrightarrow (CH_3)_3SiOSO_2OC_4H_9 + CH_3COCl \qquad (2.199)$$

$$(CH_3)_3SiOSO_2Cl + CH_3COOSi(CH_3)_3 \longrightarrow [(CH_3)_3SiO]_2SO_2 + CH_3COCl \qquad (2.200)$$

$$(CH_3)_3SiOSO_2Cl + [(CH_3)_3Si]_2NCH_3 \longrightarrow (CH_3)_3SiOSO_2N(CH_3)Si(CH_3)_3 + + (CH_3)_3SiCl \qquad (2.201)$$

When the trimethylsilyl ester of chlorosulfonic acid is heated with hexamethyldisilane for 2 h, trimethylchlorosilane and the pentamethyldisilanyl ester of methanesulfonic acid are formed [207]:

$$(CH_3)_3SiOSO_2Cl + (CH_3)_3SiSi(CH_3)_3 \longrightarrow (CH_3)_3SiCl + CH_3SO_2OSi(CH_3)_2Si(CH_3)_3 \qquad (2.202)$$

The reactions of the trimethylsilyl ester of chlorosulfonic acid with trimethylbutylsilane [207], triethylphenylsilane [210], and other tetraorganosilanes [235] proceed analogously:

$$(CH_3)_3SiOSO_2Cl + (CH_3)_3SiC_4H_9 \longrightarrow (CH_3)_3SiCl + CH_3SO_2OSi(CH_3)_2C_4H_9 \qquad (2.203)$$

$$(CH_3)_3SiOSO_2Cl + (C_2H_5)_3SiC_6H_5 \longrightarrow (CH_3)_3SiCl + C_6H_5SO_2OSi(C_2H_5)_3 \qquad (2.204)$$

The reaction of the trimethylsilyl ester of fluorosulfonic acid with trimethylchlorosilane forms trimethylfluorosilane and the trimethylsilyl ester of chlorosulfonic acid [613]:

$$(CH_3)_3SiOSO_2F + (CH_3)_3SiCl \longrightarrow (CH_3)_3SiF + (CH_3)_3SiOSO_2Cl \qquad (2.205)$$

Analogous transsulfonations occur in the reaction of the dimethylfluorosilyl ester of chlorosulfonic acid with trimethylchlorosilane and the reaction of the dimethylfluorosilyl ester of fluorosulfonic acid with trimethylfluorosilane [613].

The reaction of the trimethylsilyl ester of fluorosulfonic acid with hexamethyldisiloxane leads to the formation of trimethylfluorosilane and bis(trimethylsilyl) sulfate [613]:

$$(CH_3)_3SiOSO_2F + (CH_3)_3SiOSi(CH_3)_3 \longrightarrow (CH_3)_3SiF + [(CH_3)_3SiO]_2SO_2 \quad (2.206)$$

5. COMPOUNDS CONTAINING THE GROUPING Si − N − S

Organosilicon compounds with the grouping Si−N−S have been investigated little up to now. In most cases they are formed by the reaction of hexaorganodisilazans or their N-sodium derivatives with various sulfur-containing substances. Thus, when sodium bis(trimethylsilyl)amide is heated with sulfur a mixture of polysulfane bis(trialkylsilyl)diamides is formed and from this it is possible to isolate derivatives of di- and trisulfane and also derivatives of tetrasulfane which decompose on distillation [602, 604]. With a large excess of hexaalkyldisilazanylsodium the reaction proceeds according to the scheme

$$8[(CH_3)_3Si]_2NNa + S_8 \longrightarrow 8[(CH_3)_3Si]_2NSNa \quad (2.207)$$

Sulfur dioxide reacts with sodium bis(trimethylsilyl)amide analogously to CO_2 to form bis(trimethylsilyl)thiodiimide (but in a lower yield than the corresponding carbodiimide), hexamethyldisiloxane, and sodium sulfite [596, 698]:

$$2[(CH_3)_3Si]_2NNa + 2SO_2 \longrightarrow (CH_3)_3SiN{=}S{=}NSi(CH_3)_3 + [(CH_3)_3Si]_2O + Na_2SO_3 \quad (2.208)$$

Bis(trimethylsilyl)thiodiimide is formed together with sulfur tetrakis(trimethylsilyl)diamide in the reaction of sodium bis(trimethylsilyl)amide with sulfur dichloride [698]:

$$[(CH_3)_3Si]_2NNa \xrightarrow{SCl_2} (CH_3)_3SiN{=}S{=}NSi(CH_3)_3 + [(CH_3)_3Si]_2NSN[Si(CH_3)_3]_2 + [(CH_3)_3Si]_3N \quad (2.209)$$

A disulfane derivative, formed in accordance with the following scheme, was isolated from the products of the reaction of sodium bis(trimethylsilyl)amide with sulfur chloride:

$$2[(CH_3)_3Si]_2NNa + S_2Cl_2 \xrightarrow[-2NaCl]{} [(CH_3)_3Si]_2NSSN[Si(CH_3)_3]_2 \quad (2.210)$$

Sodium bis(trimethylsilyl)amide reacts with thiocyanogen in ether even at 0°C. However, instead of the expected bis(trimethylsilyl)aminothiocyanogen, disulfane bis(trimethylsilyl)diamide, bis-(trimethylsilyl)carbodiimide, and sodium thiocyanate are formed

[605]:

$$4[(CH_3)_3Si]_2NNa + 3(SCN)_2 \longrightarrow [(CH_3)_3Si]_2NSSN[Si(CH_3)_3]_2 +$$
$$+ 2(CH_3)_3SiN=C=NSi(CH_3)_3 + 4NaSCN \qquad (2.211)$$

The reaction of sodium bis(trimethylsilyl)amide with organic disulfides in benzene proceeds quantitatively in accordance with the scheme [476, 602, 603, 943]

$$(R'_3Si)_2NNa + RSSR \longrightarrow RSN(SiR'_3)_2 + RSNa \qquad (2.212)$$

Arylhalosulfanes react analogously [602]. The rate of the reaction depends on the structure of the radicals R and R' and increases with a change in R in the following series: $n-C_4H_9 < C_2H_5 < C_6H_5$; it falls when R' = CH_3 is replaced by an isopropoxyl group [603].

Compounds containing the grouping $Si-N-S$ are formed when hexamethyldisilazan reacts with sulfuryl chloride, the ethyl ester of chlorosulfonic acid, or benzenesulfamide [180]:

$$[(CH_3)_3Si]_2NH + SO_2Cl_2 \longrightarrow [(CH_3)_3SiNH]_2SO_2 + 2(CH_3)_3SiCl \qquad (2.213)$$

$$[(CH_3)_3Si]_2NH + C_2H_5OSO_2Cl \longrightarrow (CH_3)_3SiNHSO_2OSi(CH_3)_3 + C_2H_5Cl \qquad (2.214)$$

$$[(CH_3)_3Si]_2NH + 2C_6H_5SO_2NH_2 \longrightarrow 2(CH_3)_3SiNHSO_2C_6H_5 + NH_3 \qquad (2.215)$$

In ligroin, N-lithio-N-methylaminotrimethylsilane adds to phenylsulfinylimine to form the N-lithio-N-phenyl-N'-methyl-N'-trimethylsilyldiamide of sulfurous acid, which on reaction with trimethylchlorosilane gives the N-methyl-N'-phenyl-N,N'-bis-(trimethylsilyl)diamide of sulfurous acid, while with aluminum chloride it is cleaved to N-methyl-N'-phenylthiodiimide and tris-(trimethylsiloxy)alumane [601]:

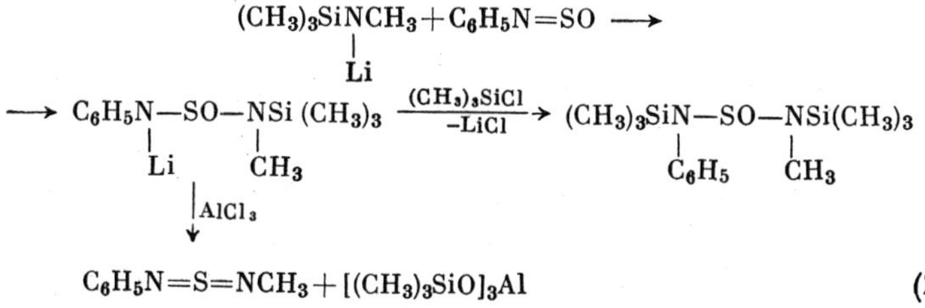

$$(CH_3)_3SiNCH_3 + C_6H_5N=SO \longrightarrow$$
$$\underset{\displaystyle Li}{|}$$

$$\longrightarrow C_6H_5N\underset{\underset{\displaystyle Li}{|}}{-}SO\underset{\underset{\displaystyle CH_3}{|}}{-}NSi(CH_3)_3 \xrightarrow[-LiCl]{(CH_3)_3SiCl} (CH_3)_3SiN\underset{\underset{\displaystyle C_6H_5}{|}}{-}SO\underset{\underset{\displaystyle CH_3}{|}}{-}NSi(CH_3)_3$$

$$\Big\downarrow AlCl_3$$

$$C_6H_5N=S=NCH_3 + [(CH_3)_3SiO]_3Al \qquad (2.216)$$

The reaction of triorganochlorosilanes with sulfamides in benzene in the presence of triethylamine leads to the formation

TABLE 20. Compounds Containing the Grouping Si−N−S

Empirical formula	Compound	B.p., °C (mm)	M.p., °C	n_D^{20}	d_4^{20}	Literature
C_3H_9NOSSi	$(CH_3)_3SiN{=}SO$	108—110	< −78	—	—	601
$C_4H_{13}NO_2SSi$	$(CH_3)_3SiNHSO_2CH_3$	89—90 (0.012)	68—70	—	—	45
$C_5H_{15}NO_2SSi$	$(CH_3)_3SiNHSO_2C_2H_5$	83—84 (0.013)	—	1.4530	1.0901	45
$C_6H_{17}NO_2SSi$	$(CH_3)_3SiNHSO_2C_3H_7$	93—94 (0.013)	—	1.4520	1.0046	45
$C_6H_{17}NO_3SSi$	$(CH_3)_3SiN(CH_3)SO_2OC_2H_5$	72 (0.005)	—	—	—	180
$C_6H_{18}N_2SSi_2$	$(CH_3)_3SiN{=}S{=}NSi(CH_3)_3$	73—74 (24)	—	—	—	698
$C_6H_{19}NO_3SSi_2$	$(CH_3)_3SiNHSO_2OSi(CH_3)_3$	84 (0.001)	44	—	—	180
		110—112 (2.8)	50—51	—	—	524
$C_6H_{20}N_2O_2SSi_2$	$[(CH_3)_3SiNH]_2SO_2$	—	104—105	—	—	180
$C_7H_{19}NO_2SSi$	$(CH_3)_3SiNHSO_2C_4H_9$	105—106 (0.013)	—	1.4525	1.0440	45
$C_7H_{21}NO_3SSi_2$	$(CH_3)_3SiN(CH_3)SO_2OSi(CH_3)_3$	72 (0.05)	—	—	—	180
$C_8H_{21}NO_2SSi$	$(CH_3)_3SiNHSO_2C_5H_{11}$	110—112 (0.012)	—	1.4532	1.0218	45
$C_8H_{23}NSSi_2$	$[(CH_3)_3Si]NSC_2H_5$	36—37 (1)	—	—	—	602
$C_9H_{14}ClNO_2SSi$	$(CH_3)_3SiNHSO_2C_6H_4Cl$-p	128—129 (0.013)	95—97	—	—	45
$C_9H_{14}N_2O_4SSi$	$(CH_3)_3SiNHSO_2C_6H_4NO_2$-m	150—151 (0.014)	100—102	—	—	45
	$(CH_3)_3SiNHSO_2C_6H_4NO_2$-p	156—158 (0.013)	144—146	—	—	45
$C_9H_{15}NO_2SSi_2$	$(CH_3)_3SiNHSO_2C_6H_5$	—	63	—	—	180
		128—130 (0.013)	63—65	—	—	45
$C_9H_{21}NO_2SSi$	$(CH_3)_3SiNHSO_2C_6H_{13}$	113—114 (0.012)	—	1.4543	1.0109	45
$C_{10}H_{17}NO_2SSi$	$(CH_3)_3SiNHSO_2C_6H_4CH_3$-p	127—129 (0.013)	75—77	—	—	45
$C_{10}H_{27}NSSi_2$	$[(CH_3)_3Si]_2NSC_4H_9$	71—72 (1)	—	—	—	602
$C_{12}H_{21}NO_2SSi$	$(C_2H_5)_3SiNHSO_2C_6H_5$	135—137 (0.025)	53—55	—	—	45
$C_{12}H_{23}NO_2SSi$	$[(CH_3)_3Si]_2NSO_2C_6H_5$	105—107 (0.05)	58—60	—	—	45
$C_{12}H_{23}NSSi_2$	$[(CH_3)_3Si]_2NSC_6H_5$	87—88 (1)	—	—	—	602
$C_{12}H_{31}NO_3SSi_2$	$(C_2H_5)_3SiNHSO_2OSi(C_2H_5)_3$	141—143.5 (1.5)	−1 to +0.5	—	—	524
$C_{12}H_{36}N_2SSi$	$[(CH_3)_3Si]_2NSN[Si(CH_3)_3]_2$	142—143 (15)	65	—	—	698
$C_{12}H_{36}N_2S_2Si$	$[(CH_3)_3Si]_2NSSN[Si(CH_3)_3]_2$	115—116 (0.5)	—	—	—	698
$C_{13}H_{17}NO_2SSi$	$(CH_3)_3SiNHSO_2C_{10}H_7$-β	160 (0.014)	150—152	—	—	45
$C_{13}H_{26}N_2SSi$	$(CH_3)_3SiN(C_6H_5)SON(CH_3)Si(CH_3)_3$	—	27—29	—	—	601
$C_{14}H_{28}N_2SSi$	$(CH_3)_3SiN(C_6H_5)SON(C_2H_5)Si(CH_3)_3$	—	25—27	—	—	601

of N-triorganosilylsulfamides even at room temperature [45]:

$$RSO_2NH_2 + ClSiR_3 \xrightarrow[-R_3'N \cdot HCl]{+R_3'N} RSO_2NHSiR_3 \qquad (2.217)$$

In the case of benzenesulfamide it is possible to introduce two trimethylsilyl radicals into the amide group, but p-acetamidobenzenesulfamide does not react at all with trimethylchlorosilane under these conditions [45].

5,5-Dimethylsulfodiimine reacts with trialkylchlorosilanes by a scheme analogous to (2.217) [833].

The physical constants of compounds containing the grouping Si−N−S are given in Table 20. All these compounds are hydrolytically unstable.

6. COMPOUNDS CONTAINING THE GROUPING Si − NCS (ISOTHIOCYANATOSILANES)

6.1. Preparation Methods

The general method which is used most frequently for the synthesis of isothiocyanatosilanes is the reaction of halosilanes with metal or ammonium thiocyanates:

$$\equiv Si{-}X + MSCN \longrightarrow \equiv Si{-}NCS + MX \qquad (2.218)$$

$$(X = Cl, Br, I)$$

Depending on the number of halogen atoms attached to the silicon, by this method it is possible to obtain compounds containing from one to four NCS groups. Donors of these groups used include thiocyanates of sodium [32, 203, 215, 366, 399, 490, 682], potassium [32, 80, 675], ammonium [32, 353, 366, 522, 681, 917], silver [32, 143, 145-148, 150, 157, 158, 160, 162, 285, 286, 290, 307, 318, 353, 370, 430, 464, 468, 482, 522, 672], copper [32, 285], and lead [32, 241, 353, 370, 498-500, 573, 574, 580, 631]. Reaction (2.218) is carried out in benzene, carbon tetrachloride, or in the absence of a solvent. A slight excess of the thiocyanate (20-30%) is usually used. In many cases heating the reaction mixture for

15-20 min is sufficient for complete replacement of halogen atoms by isothiocyanate groups. The yields of isothiocyanatosilanes, which reach 80-95%, depend little on the thiocyanate used. Only in individual cases has it been reported that reaction (2.218) proceeds quite slowly and the yields are not so high (50-60%) [353, 366]. Thus, for example, the reaction of triphenylchlorosilane with lead thiocyanate forms triphenylisothiocyanatosilane in 97% yield. At the same time, with the use of ammonium thiocyanate the reaction of the reagents is much slower and the yield of the reaction products does not exceed 56% [353]. By gradual addition of silver thiocyanate to silicon tetrachloride it is possible to obtain trichloroisothiocyanatosilane as well as tetraisothiocyanatosilane [143, 370].

Isothiocyanatosilane [290, 464] and methylisothiocyanatosilane [430] were obtained from iodosilane and methyliodosilane and silver thiocyanate by means of reaction (2.218). In both cases the Si–H bond remained untouched. In contrast to this, triethylsilane reduces silver thiocyanate with the formation of silver, triethylisothiocyanatosilane, and hydrogen [285]:

$$2(C_2H_5)_3SiH + 2AgSCN \longrightarrow 2(C_2H_5)_3SiNCS + 2Ag + H_2 \qquad (2.219)$$

Analogous reductions occur in the reaction of cyclohexylsilane with silver thiocyanate [157] and the reaction of benzylsilane [161] and triethylsilane [30] with mercury thiocyanate. In the reaction of tributoxysilane with dithiocyanogen there is replacement of one butoxyl group by an isothiocyanate group, with the Si–H bond retained [690]. Replacement of an alkoxyl group by an isothiocyanate group also occurs in the reaction of ethoxytrichlorosilane and diethoxydichlorosilane with silver thiocyanate [162]. It is possible that in this case the exchange is connected with disproportionation of ethoxythiocyanatosilanes formed initially.

A second method used for the synthesis of triorganoisothiocyanatosilanes is the reaction of triorganochlorosilanes with thiourea [353, 370]:

$$R_3SiCl + H_2N{-}\underset{\underset{\displaystyle S}{\|}}{C}{-}NH_2 \longrightarrow R_3SiNCS + NH_4Cl \qquad (2.220)$$

The reaction is carried out in sealed ampoules at 250°C. The triorganoisothiocyanatosilane, which is formed in ~70% yield, is separated from the ammonium chloride by extraction with ligroin.

Triorganoisothiocyanatosilanes are formed by heating triorganoisocyanosilanes with sulfur [198, 370, 482, 483, 970]:

$$R_3SiNC + S \longrightarrow R_3SiNCS \qquad (2.221)$$

Thus, for example, heating dimethylphenylisocyanosilane with sulfur at 250°C for half an hour leads to the formation of dimethylphenylisothiocyanatosilane in 57% yield [482].

This reaction provides a demonstration of the isothiocyanate structure of the organosilicon compounds examined in this section.

We should also examine the methods of the synthesis of some compounds containing the Si–NCS grouping formed as a result of the reaction of organoisothiocyanates with compounds containing the Si–N bond.

The reaction of trimethyl(diethylamino)silane with organic isothiocyanates forms organosilicon derivatives of thiourea [536, 888, 889], which are converted by hydrolysis to trimethylsilanol and the corresponding N,N,N'-trisubstituted derivatives of thiourea [536]:

$$(CH_3)_3SiN(C_2H_5)_2 + RNCS \longrightarrow (CH_3)_3SiN(R)\text{—}C(=S)\text{—}N(C_2H_5)_2 \xrightarrow{H_2O} (CH_3)_3SiOH + RNHCSN(C_2H_5)_2 \qquad (2.222)$$

The reaction of hexamethyldisilazan with phenylisothiocyanate leads to the formation of N,N'-bis(trimethylsilyl)-N-phenylthiourea [48, 88]:

$$[(CH_3)_3Si]_2NH + C_6H_5NCS \longrightarrow (CH_3)_3SiNH\text{—}C(=S)\text{—}N(C_6H_5)Si(CH_3)_3 \qquad (2.223)$$

This reaction proceeds much more slowly than the analogous reaction with phenyl isocyanate. While in the latter case the yield of the reaction products is quantitative, under the same conditions the yield of N,N'-bis(trimethylsilyl)-N-phenylthiourea reaches only 5-7%. However, the use of twice the amount of hexamethyldisilazan raises the yield to 70%.

The reactions of phenyl isothiocyanate with N-methylhexamethyldisilazan [884, 889] and of methyl isothiocyanate with N-

methyl-N,N'-bis(trimethylsilyl)hydrazine and N-methyl-N-tri-
methylsilylhydrazine [396] proceed analogously with rupture of the
Si −N bond in the starting nitrogen-containing organosilicon compound. In contrast to this, it is considered that trifluoromethane-
sulfenyl isocyanate reacts with hexamethyldisilazan with the formation of a compound in which the Si −N−Si grouping is preserved
[373]:

$$[(CH_3)_3Si]_2NH + CF_3SNCO \longrightarrow [(CH_3)_3Si]_2NCONHSCF_3 \qquad (2.224)$$

We cannot exclude the possibility that in reactions (2.223) and
(2.224) an equilibrium system is formed initially:

$$
\begin{array}{ccc}
(CH_3)_3Si-N-C-N-R & \rightleftharpoons & (CH_3)_3Si-N=C-N-R \\
\;\; | \;\; \| \;\; | & & \qquad | \quad | \\
(CH_3)_3Si \;\; X \;\; H & & (CH_3)_3Si-X \;\; H
\end{array}
$$

$$(X = O \;\; or \;\; S)$$

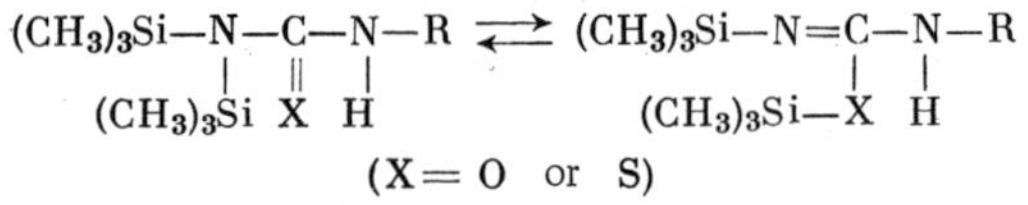

When the temperature is raised above 120°C there is decomposition with the formation of trimethylisocyanato- or trimethylisothiocyanatosilane:

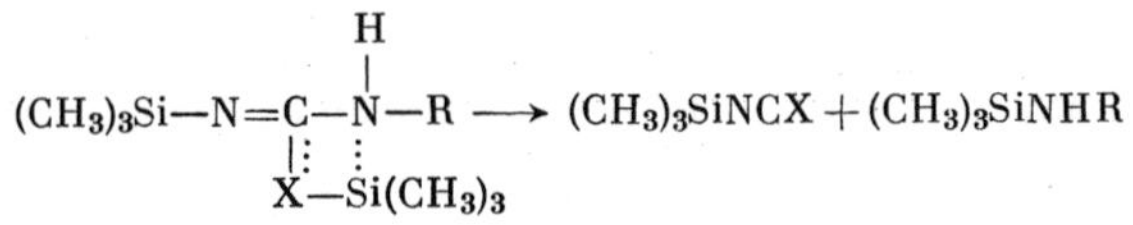

$$
\begin{array}{c}
\qquad\qquad H \\
\qquad\qquad | \\
(CH_3)_3Si-N=C-N-R \longrightarrow (CH_3)_3SiNCX + (CH_3)_3SiNHR \qquad (2.225) \\
\quad\; \vdots\; \vdots \\
X-Si(CH_3)_3
\end{array}
$$

In actual fact, the formation of trimethylisothiocyanatosilane
was observed both on heating hexamethyldisilazan with phenyl isothiocyanate at 120°C and in the thermal decomposition of their reaction product obtained with a reaction temperature of 30°C. From
this it follows that in all probability reaction (2.220) proceeds
with the intermediate formation of an organosilicon derivative of
thiourea, which decomposes under the reaction conditions (200-
250°C) to triorganoisothiocyanatosilane and ammonium chloride.

6.2. Physical Properties

In the first work on the reaction of chlorosilanes with metal
thiocyanates the reaction products were assigned the structure of
thiocyanatosilanes containing the grouping Si −S−C ≡ N [498-500,
573]. However, even the first refractometric investigations [145,
307] and then spectroscopic investigations [241, 290, 307, 320, 347,
370, 394, 395, 455, 464, 468, 495, 600, 682] showed that these com-

TABLE 21. Compounds Containing the Grouping Si – NCS

Empirical formula	Compound	B.p., °C (mm)	M.p., °C	n_D^{20}	d_4^{20}	Literature
CCl_3NSSi	Cl_3SiNCS	129.5	—75 ± 2	1.5091	1.461[24]	143
CH_3NSSi	H_3SiNCS	84 ± 4	—51.8 ± ± 0.2	—	1.05	464
$C_4N_4S_4Si$	$Si(NCS)_4$	314.2	143.8	—	—	573
		313.8	143.8	—	1.409	145
		182 (15)	—	—	—	32
		—	143	—	—	241
		—	144	—	—	522
		—	144—145	—	—	370
$C_4H_3N_3S_3Si$	$CH_3Si(NCS)_3$	268—270	70—71	—	—	215
		267.5 146 (15)	72.4	—	1.304	32, 145
		266.8 ± 1	72.5	—	—	522
$C_4H_6N_2S_2Si$	$(CH_3)_2Si(NCS)_2$	217.5	—	—	—	32
		217.3	—	1.5677	1.142	145
		101 (15)	—	1.5675	1.1416	32, 370
		217—218	—	1.5661	1.1330	203
		64 (1)	—	—	—	522
C_4H_9NSSi	$(CH_3)_3SiNCS$	144.2	—	—	—	370
		143.8	—33	1.4809	0.9308	32
		143—143.5	—	—	—	522
		143.1 ± 0.3	—32.8	1.4820	0.931	145, 682
		141.8—142	—	1.4809	0.9257	203
$C_4H_9NO_3SSi$	$(CH_3O)_3SiNCS$	171	—	1.4426	1.134	144
		110 (70)			1.119[30]	
$C_5H_3N_3S_3Si$	$CH_2{=}CHSi(NCS)_3$	276	—	1.6409	1.292	158
		117—118 (1)				
$C_5H_5N_3S_3Si$	$C_2H_5Si(NCS)_3$	134—138 (3)	—	1.6350[26]	—	366
		276.5	—	—	—	32, 147
		151 (15)	—	—	1.2658	32
		173.2—174.2 (38)	—	1.6195	1.264	147
$C_5H_{11}F_3N_2OSSi$	$CF_3SNHCONHSi(CH_3)_3$	—	89	—	—	373

TABLE 21 (Continued)

Empirical formula	Compound	B.p., °C (mm)	M.p., °C	n_D^{20}	d_4^{20}	Literature
$C_5H_{11}NSSi$	$(CH_3)_2(C_2H_5)SiNCS$	167.5	—	1.484^{25}	—	286
$C_6N_6S_6Si_2$	$O[Si(NCS)_3]_2$	347 200 (3)	120—121	—	—	318
$C_6H_5N_3S_3Si$	$CH_2{=}CHCH_2Si(NCS)_3$	126—128 (2.5)	—	1.6140^{26}	—	366
$C_6H_7N_3S_3Si$	$C_3H_7Si(NCS)_3$	289.5 160 (14)	—	1.6014	1.2248	32
		152—153 (13)	—	—	—	150
	iso-$C_3H_7Si(NCS)_3$	279	—	1.6066	1.2177	150
		157 (14)	16.5	1.6050	1.2158	32
$C_6H_{10}N_2S_2Si$	$(C_2H_5)_2Si(NCS)_2$	245.5 ± 2 73.5—74.5 (3)	—	1.5540	1.089	147
$C_7H_9N_3S_3Si$	$C_4H_9Si(NCS)_3$	300.6 ± 1	—	—	—	148
		298.8	—	—	—	32
		166 (15)	—	1.5927	1.1888	32
		135—136 (3)	—0.5	1.5928	1.189	148
$C_7H_{13}NSSi$	$C_6H_{11}SiH_2NCS$	231 91—91.8 (1)	—	1.5336	1.018	157
$C_7H_{15}NSSi$	$(C_2H_5)_3SiNCS$	128.3—128.9 (72)	—	1.4944	0.934	147
		210.5 ± 0.5	—	1.4948	0.9385	32
$C_7H_{15}NO_3SSi$	$(C_2H_5O)_3SiNCS$	122.2—122.8 (53) 205.5 ± 0.5	—	1.4431	1.036	148
$C_7H_{20}N_2SSi_2$	$(CH_3)_3SiNHCSNHSi(CH_3)_3$	—	156	—	—	197
C_8H_9NSSi	$C_6H_5CH_2SiH_2NCS$	253 87—88 (1)	—	1.5929	1.092	161
$C_8H_{17}NSSi$	$C_7H_{15}SiH_2NCS$	235	—	1.4939	0.921	162
$C_9H_5N_3S_3Si$	$C_6H_5Si(NCS)_3$	153.5—154.5 (3) 339.6 ± 1	52 ± 1	—	—	146
$C_9H_5NOS_3Si$	$C_6H_5OSi(NCS)_3$	165—166 (3)	85	—	1.270	580

$C_9H_{11}NSSi$	$(CH_3)_2(C_6H_5)SiNCS$	257—259	—	—	—	286
		252—254	—	$1{\cdot}5556^{30}$	$1{\cdot}0384^{30}_{4}$	368
$C_9H_{11}N_3S_3Si$	$C_6H_{11}Si(NCS)_3$	348	—	1·6179	1·231	157
		172—173 (1)				
$C_9H_{19}NSSi$	$(C_4H_9)_2SiHNCS$	98—99 (6)	—	1·4110	—	691
$C_9H_{19}NO_2SSi$	$(C_4H_9O)_2SiHNCS$	98—99 (6)	—	—	—	690
$C_9H_{22}N_2SSi$	$(CH_3)_3SiN(CH_3)CSN(C_2H_5)_2$	88 (0·15)	—	1·5177	—	536
$C_{10}H_7N_3S_3Si$	$C_6H_5CH_2Si(NCS)_3$	348·9 ± 1	36	—	1·275 (supercool.)	148
		171—172 (3)				
		169 (2)	—	—	—	32
	$p\text{-}CH_3C_6H_4Si(NCS)_3$	182—185 (3·5)	—	$1{\cdot}6490^{26}$	—	366
$C_{10}H_{15}N_3S_3Si$	$C_7H_{15}Si(NCS)_3$	347	—	1·5739	1·129	162
$C_{10}H_{18}N_2S_2Si$	$(C_4H_9)_2Si(NCS)_2$	154·5 (6)	—	1·5106	0·9950	32
$C_{11}H_{27}N_3SSi$	$(CH_3)_2SiN(C_6H_5)CSN(CH_3)N(CH)_3$	—	98	—	—	590
$C_{13}H_{27}NSSi$	$(C_4H_9)_3SiNCS$	137·5 (2)	—	1·4893	0·9063	32
$C_{14}H_{10}N_2S_2Si$	$(C_6H_5)_2Si(NCS)_2$	371 ± 1·5	46 ± 1	—	—	146
		173—174 (3)				
$C_{14}H_{10}N_2O_2S_2Si$	$(C_6H_5O)_2Si(NCS)_2$	205—206 (3)	29·5	—	1·245	580
$C_{15}H_{29}N_3SSi_2$	$(CH_3)_3SiN(C_6H_5)CSN(CH_3)N(CH_3)Si(CH_3)_3$	—	63—65	—	—	590
$C_{16}H_{14}N_2S_2Si$	$(C_6H_5CH_2)_2Si(NCS)_2$	348·9	36 ± 1	—	1·275	148
		171—172 (3)				
$C_{18}H_{22}ClN_3OSSi$	$C_6H_5CON(CH_3)N(CH_3)CSN(C_6H_5)SiCl(CH_3)_2$	—	125—127	—	—	590
$C_{19}H_{15}NSSi$	$(C_6H_5)_3SiNCS$	188·5—189·5 (3)	76 ± 1	—	—	146
		396·6 ± 1·5	98—99	—	—	487
		—	100	—	—	198
		—	98—101	—	—	681
		—	100—101	—	—	353
$C_{19}H_{15}NO_3SSi$	$(C_6H_5O)_3SiNCS$	230—231 (3)	—	—	$1{\cdot}201^{20}$	508
$C_{21}H_{21}N_3SSi$	$(CH_3)_2SiN(C_6H_5)CSN(C_6H_5)N(C_6H_5)$	—	138	—	—	590
	$(C_6H_5)_2SiN(C_6H_5)CSN(CH_3)N(CH_3)$	—	82—85	—	—	590

pounds have the structure of isothiocyanatosilanes containing the grouping Si – N = C = S.

The $Si(NCS)_4$ molecule is a regular tetrahedron with a linear skeleton Si–N=C=S [236, 495]. Taking the length of the C–S bond to be 1.560 Å and determining the length of the Si–H bond as 1.489 Å and the H–Si–H and H–Si–N angles as 111°22' and 107°30', respectively, from infrared absorption spectra, it was possible to calculate the lengths of the bonds in H_3SiNCS from data from microwave spectra: Si–N 1.714 ± 0.010 Å and N–C 1.211 ± 0.010 Å [395]. The Si–N=C=S group is linear in this molecule also and in it there is an interaction of the p-electrons of the nitrogen atom with the 3d-orbitals of the silicon atom [395, 455]. According to data in [404], the Si–N–C angle in $(CH_3)_3SiNCS$ equals 154 ± 2° and the length of the Si–N bond is 1.78 ± 0.02 Å.

In the series of electronegativities (χ) of substituents calculated from vibration-spectra data on compounds of the type R_3SiY (Y = NCS, Cl, OH, etc.), the isothiocyanate group (χ_{NCS} = 2.93) attached to a silicon atom occupies a position between chlorine (χ_{Cl} = 3.00) and a hydroxyl group (χ_{OH} = 2.70) [319, 320].

The physical constants of compounds containing the grouping Si–NCS are given in Table 21.

6.3. Chemical Properties

Isothiocyanatosilanes are hydrolyzed by water to form silanols and isothiocyanic acid [32, 146, 147, 150, 241, 285, 318, 353, 370, 430, 464, 485, 498, 580]:

$$\geq Si{-}NCS + H_2O \longrightarrow \geq Si{-}OH + HNCS \qquad (2.226)$$

This reaction may be used for analytical purposes, titrating the acid liberated with alcoholic alkali [147]. The hydrolysis rate depends on the number of isothiocyanate groups in the molecule and the nature of the substituents attached to the silicon atom. Thus, for example, tetraisothiocyanatosilane is hydrolyzed rapidly even by atmospheric moisture [241, 318, 498], while triphenylisothiocyanatosilane is practically inert toward water [146]. Trialkylisothiocyanatosilanes are hydrolyzed slowly at room temperature and somewhat more rapidly on heating to 80°C [32, 147]. The hydrolysis rate increases if the hydrolysis is carried out in acetone or dioxane [285, 353]. In aqueous acetone triphenylisothiocyanatosilane is hydrolyzed at an appreciable rate [353] to form triphenylsilanol. An increase in the number of isothiocyanate groups in the molecule of isothiocyanatosilanes also increases their tendency for hydrolysis.

Analogous rules are observed in the alcoholysis of isothiocyanatosilanes [62, 144, 499, 573, 580, 795, 907], which proceeds according to the scheme

$$\geq Si{-}NCS + ROH \longrightarrow \geq Si{-}OR + HNCS \qquad (2.227)$$

Reaction (2.227) is also used for quantitative determination of isothiocyantosilanes. For this purpose a solution of isothiocyanatosilane in acetonitrile is titrated with a solution of sodium methylate in methyl alcohol with the equivalence point determined potentiometrically [62] or by means of color indicators [62, 63].

The reaction of tetraisothiocyanatosilanes with alcohols usually forms tetraalkoxysilanes. However, the careful addition of methyl alcohol to tetraisothiocyanatosilane makes it possible to obtain trimethoxyisothiocyanatosilane, which contains methoxyl and isothiocyanate groups at the same time [144].

Isothiocyanatosilanes react vigorously with primary and secondary amines in anhydrous benzene to form after hydrolysis unsymmetrically substituted thioureas [153, 522, 573]:

$$\geq Si{-}NCS + RR'NH \longrightarrow \left[\geq SiNHCSNRR'\right] \xrightarrow{H_2O} \geq SiOH + RR'NHCSNH_2 \qquad (2.228)$$

The activity of amines in this reaction falls in the following series:

$$RNH_2 > R_2NH > ArNH_2 \gg ArNHR \gg Ar_2NH$$
$$(R = \text{alkyl};\ Ar = \text{aryl})$$

In the reaction with nitrogen-containing aromatic heterocycles, tetraisothiocyanatosilane gives the complexes $Si(NCS)_4 \cdot 2X$ (X = pyridine, quinoline, and isoquinoline). These complexes, in contrast to $Si(NCS)_4$ itself, give a red color with a solution of ferric chloride benzene [631].

By hydrolysis of the product from the reaction of triphenylisothiocyanatosilane with phenyllithium, which is formed immediately after they are mixed, it is possible to obtain triphenylsilanol (56%), tetraphenylsilane (17%), and thiobenzamide (57%). Stirring the reaction mixture for 16 h leads to an increase in the tetraphenylsilane content of the hydrolysis product to 74.2%. In this case thiobenzamide is not obtained, but 44% of sulfur (relative to the amount of it in the starting material) is found in the gases liberated during the reaction [353]. The reaction of triphenylisothiocyanatosilane with phenylmagnesium bromide forms hydrogen sulfide (4.5%) and 70% of triphenylsilanol (after hydrolysis) [353].

The Si–NCS bond is cleaved by the action of a whole series of organic and inorganic halides: COF_2 [973], benzoyl chloride [142, 145], phenyldichlorophosphine [145], aluminum chloride [145], sulfur chloride [145], mercuric chloride [145], silicon tetrachloride [317] and tetrabromide [145], dimethyldichlorosilane [502], and titanium tetrachloride [145]:

$$\rangle Si{-}NCS + M{-}X \longrightarrow \rangle Si{-}X + M{-}NCS \qquad (2.229)$$

At the same time, tetraisothiocyanatosilane does not react with antimony trichloride even on heating to 220°C [148], while methyltriisothiocyanatosilane is not cleaved by antimony trichloride or germanium tetrabromide [155].

Triorganoisothiocyanatosilanes react vigorously with silver and mercuric oxides to form the corresponding hexaorganodisiloxanes [285, 370]:

$$2R_3SiNCS + HgO \longrightarrow R_3SiOSiR_3 + Hg(SCN)_2 \qquad (2.230)$$

Heating trimethylisothiocyanatosilane with hexaethyldistannoxane forms hexamethyldisiloxane and triethylisothiocyanatostannane (86% yield) [155a].

Heating trialkylisothiocyanatosilanes or dialkyldiisothiocyanatosilanes with silver isocyanate, acetate, and trifluoroacetate gives good yields of the corresponding isocyanato-, acetoxy-, and trifluoroacetoxysilanes [160, 285]:

$$(CH_3)_3SiNCS + AgNCO \longrightarrow (CH_3)_3SiNCO + AgNCS \quad (2.231)$$

$$(C_2H_5)_2Si(NCS)_2 + 2AgOCOCH_3 \longrightarrow (C_2H_5)_2Si(OCOCH_3)_2 + 2AgNCS \quad (2.232)$$

Tetraisothiocyanatosilane does not react with silver isocyanate when heated for 1 h in boiling benzene [148].

In contrast to its oxygen analog, trimethylisothiocyanatosilane reacts with sodium bis(trimethylsilyl)amide only on heating to boiling. The reaction products are bis(trimethylsilyl)carbodiimide, hexamethyldisilthiane, tris(trimethylsilyl)amine, and sodium thiocyanate. The last two compounds may be formed in accordance with scheme (2.233) if the isothiocyanate group behaves like a halogen [699]:

$$(CH_3)_3SiNCS + NaN[Si(CH_3)_3]_2 \longrightarrow [(CH_3)_3Si]_3N + NaSCN \quad (2.233)$$

An interesting reaction, which proceeds with cleavage of the Si – N bond and the formation of β-thiocyanatoethoxysilanes, occurs when isothiocyanatosilanes are treated with ethylene oxide [564]:

$$\rangle Si{-}NCS + \underset{\diagdown O \diagup}{CH_2{-}CH_2} \longrightarrow \rangle Si{-}OCH_2CH_2SCN \quad (2.234)$$

6.4. Biological Action

Many isothiocyanatosilanes have an appreciable antifungal action. The addition of 0.3-1% of butyltriisothiocyanatosilane, dimethyldiisothiocyanatosilane, diethyldiisothiocyanatosilane, and triethylisothiocyanatosilane to a nutrient medium inoculated with a culture with the mold Macrosporium or Cladosporium strongly suppresses or even completely stops its development for a layer or more, and also prevents molds and bacteria from the surrounding air from settling in this medium. Dimethyldiisothiocyanatosilane has a considerable fumigant action. In particular, as a result of exposure for 6-15 h in an atmosphere with a comparatively low concentration of dimethyldiisothiocyanatosilane there is 100% mortality of the agricultural pest shield bugs [71].

7. COMPOUNDS WITH THE GROUPING Si − $(C)_n$ − S

7.1. Compounds Containing the Grouping Si − $(C)_n$ − SH (Organosilicon Thiols)

7.1.1. Preparation Methods

Organosilicon thiols (mercaptans) are formed by the reaction of haloalkylsilanes with alkali metal hydrosulfides:

$$\geq Si-(\overset{|}{\underset{|}{C}})_n-X + NaSH \longrightarrow \geq Si-(\overset{|}{\underset{|}{C}})_n-SH + NaX \qquad (2.235)$$

Thus, for example, the reaction of trimethyl(chloromethyl)-silane with potassium hydrosulfide gives a 42% yield of trimethyl-silylmethanethiol. However, this method is used little for the preparation of organosilicon thiols. This is explained by the fact that their yields in reaction (2.235) are very low and considerable amounts of sulfides are often formed together with the thiols [55]. Sulfides are sometimes the only reaction product [618].

In the synthesis of organosilicon thiols by hydrolysis of organosilicon isothiuronium salts the yields of the thiols formed vary greatly, depending on the structure of the isothiuronium salt:

$$\left[\geq Si-(\overset{|}{\underset{|}{C}})_n-SC\begin{matrix}\nearrow NH_2\\ \searrow NH_2\end{matrix}\right]^+ X^- \xrightarrow[-NH_2CONH_2]{NaOH} \geq Si-(\overset{|}{\underset{|}{C}})_n-SH \qquad (2.236)$$

Thus, the hydrolysis of (trimethylsilylmethyl)isothiuronium bromide in an NaOH solution gives trimethylsilylmethanethiol in 26.7% yield [531]. Secondary triorganosilylalkynethiols cannot be obtained at all by this method. In contrast to this, alkyldiethoxy-silylalkanethiols are obtained in 66–80% yield by treatment of the corresponding isothuronium salt with ammonia and sodium ethylate [419].

High yields of organosilicon thiols are obtained by alkaline hydrolysis of acetates of organosilicon thiols, which are obtained readily by addition of thioacetic S-acid to alkenylsilanes [83, 84, 204, 338, 480, 876]:

$$\underset{\diagup}{\overset{\diagdown}{-Si}}-\left(\overset{|}{\underset{|}{C}}\right)_n-SCOCH_3 \xrightarrow{H_2O} \underset{\diagup}{\overset{\diagdown}{-Si}}-\left(\overset{|}{\underset{|}{C}}\right)_n-SH+CH_3COOH \qquad (2.237)$$

Reaction (2.237) may be carried out without preliminary isolation of the thiol acetate, and this makes it possible to obtain organosilicon thiols with a yield of 70-80%.

Hydrogen sulfide adds to triorgano(allyl)silanes at −70°C under the effect of ultraviolet radiation to form 3-(triorganosilyl)propanethiols in about 75% yield [671]:

$$\underset{\diagup}{\overset{\diagdown}{-Si}}CH_2CH{=}CH_2+H_2S \longrightarrow \underset{\diagup}{\overset{\diagdown}{-Si}}CH_2CH_2CH_2SH \qquad (2.238)$$

The addition of H_2S to triorgano(vinyl)silanes proceeds with more difficulty and the yield of 2-(trimethylsilyl)ethanethiol is only 21% [671]. A side process in this case is further addition of the organosilicon thiol formed to the starting alkenylsilane. The addition of hydrogen sulfide to vinyltrichloro- and vinyltriethoxysilane at 220°C in an autoclave leads to the formation of the corresponding sulfides and not thiols [302]. However, there are reports that 2-mercaptoethyl(triethoxy)silane may be obtained by this method at 100° [905].

Organosilicon thiols may be obtained by the reaction of Grignard reagents with organosilicon thiocyanates [55, 57, 81]:

$$\underset{\diagup}{\overset{\diagdown}{-Si}}-\left(\overset{|}{\underset{|}{C}}\right)_n-SCN \xrightarrow{RMgX} \underset{\diagup}{\overset{\diagdown}{-Si}}-\left(\overset{|}{\underset{|}{C}}\right)_n-SH \qquad (2.239)$$

By-products of this reaction are organosilicon sulfides $\underset{\diagup}{\overset{\diagdown}{-Si}}-(\overset{|}{\underset{|}{C}})_n-S-R$. In the case of α-silicoorganic thiocyanates only a small amount of sulfide is formed and the yields of organosilicon thiols reached 70%. In the reaction with γ-thiocyanates the yield of sulfides reaches 40% while that of thiols falls to 37% [81]. An attempt to synthesize an organosilicon dithiol starting with $(CH_3)_2Si(CH_2SCN)_2$ and C_3H_7MgBr led to the formation of $(CH_3)_2Si(CH_2SH)CH_2SC_3H_7$, which contains both thiol and sulfide groups [81].

TABLE 22. Compounds Containing the Grouping Si$-(C)_n-$SH (Organosilicon thiols)

Empirical formula	Compound	B.p., °C (mm)	M.p., °C	n_D^{20}	d_4^{20}	Literature
$C_4H_{12}SSi$	$(CH_3)_3SiCH_2SH$	115—115.5 (749)	—	1.4468^{25}	0.8320^{25}	531
		55 (93)	—	1.4502	0.8430	250
	derivative with $HgCl_2$	—	142	—	—	271
$C_5H_{13}ClS_2Si$	$(CH_3)_2SiClCH_2SCH_2CH_2SH$	124—126 (12)	—	—	—	712
$C_5H_{14}SSi$	$(CH_3)_3SiCH_2CH_2SH$	143	—	—	—	671
		62—63 (43)	—	1.4540	0.8488	83
		40 (18)	—	1.4515^{25}	—	671
$C_6H_{12}SSi$	$(CH_3)_3SiC{\equiv}CCH_2SH$	58.5—60 (7)	—	1.4782	0.8869	57
$C_6H_{16}SSi$	$(CH_3)_3SiCH_2CH_2CH_2SH$	164.2	—	1.4539	—	81, 671
		97 (30)	—	1.4559	0.8496	83
		53 (16)	—	1.4538	—	671
	$CH_3(C_2H_5)_2SiCH_2SH$	171.1 (735)	—	1.4658	0.8751	81
$C_6H_{18}OSSi$	$(CH_3)_3SiOSi(CH_3)_2CH_2SH$	99 (88)	—	1 4308	0.8909	250
$C_7H_{14}SSi$	$(CH_3)_2(C_2H_5)SiC{\equiv}CCH_2SH$	75—78 (8)	—	1.4828	0.8990	57
	$(CH_3)_3SiC{\equiv}CCH(CH_3)SH$	54—55 (8)	—	1.4664	0.8650	55
	derivative with $HgCl_2$	—	132	—	—	55
$C_7H_{18}SSi$	$(CH_3)_3SiCH_2CH(CH_3)CH_2SH$	69.5—70.5 (16)	—	1.4576^{25}	0.8597_4^{25}	338
	$(C_2H_5)_3SiCH_2SH$	110 (50)	—	1.4678^{25}	0.8750^{25}	271
	derivative with $HgCl_2$	—	105—106	—	—	270
	$CH_3(C_2H_5)_2SiCH_2CH_2SH$	82 (15)	—	1.4690	0.8751	84
$C_7H_{18}S_2Si$	$(CH_3)_2Si(CH_2SH)CH_2SC_3H_7$	74.5 (2)	—	—	0.8674	81
$C_8H_{16}SSi$	$(CH_3)_2(C_2H_5)SiC{\equiv}CCH(CH_3)SH$	74 (10)	—	1.4745	0.8700	55
	derivative with $HgCl_2$	—	145	—	—	55
$C_8H_{20}SSi$	$CH_3(C_3H_7)_2SiCH_2SH$	43 (2)	—	1.4642	0.8689	81
	$(CH_3)_2(C_2H_5)Si(CH_2)_4SH$	73.5 (6)	—	1.4636	0.8625	83
	$(CH_3)_2(C_3H_7)Si(CH_2)_3SH$	60 (2)	—	1.4630	0.8586	83

	$CH_3(C_2H_5)_2SiCH_2CH_2CH_2SH$	100 (18)	—	1.4684	0.8718	84
	$(C_2H_5)_3SiCH_2CH_2SH$	113 (23)	—	1.4730^{25}	0.8789_4^{25}	338
$C_8H_{20}S_2Si$	$(CH_3)_2Si(CH_2CH_2CH_2SH)_2$	119—119.5 (6)	—	1.5034^{25}	0.9631_4^{25}	338
$C_8H_{20}O_2SSi$	$CH_3(C_2H_5O)_2SiCH_2CH_2CH_2SH$	95—100 (12)	—	—	—	419
$C_8H_{22}OS_2Si$	$[(CH_3)_2SiCH_2CH_2SH]_2O$	94 (1)	—	1.4781^{25}	—	204
$C_8H_{24}O_4SSi_4$	$O[(CH_3)_2SiO]_3Si(CH_3)CH_2SH$	98.5 (10)	—	1.4280	1.0310	250
$C_9H_{14}SSi$	m-$(CH_3)_3SiC_6H_4SH$	118—120 (25)	—	—	—	598
	p-$(CH_3)_3SiC_6H_4SH$	115—117 (20)	—	—	—	598
$C_9H_{22}SSi$	$CH_3(C_2H_5)_2Si(CH_2)_4SH$	89.5 (6)	—	1.4690	0.8716	83
	$(C_2H_5)_3SiCH_2CH_2CH_2SH$	82 (3)	—	1.4738	0.8794	83
$C_9H_{22}O_2SSi$	$CH_3(C_2H_5O)_2SiCH_2CH(CH_3)CH_2SH$	105—108 (10)	—	—	—	419
$C_9H_{22}O_3SSi$	$(C_2H_5O)_3SiCH_2CH_2CH_2SH$	112—115 (14)	—	—	—	419
$C_{10}H_{24}SSi$	$CH_3(C_4H_9)_2SiCH_2SH$	68 (1)	—	1.4640	0.8658	81
	$(C_3H_7)_3SiCH_2SH$	89 (3)	—	1.4676^{25}	0.8614_4^{25}	271
	derivative with $HgCl_2$	—	134—136	—	—	270
$C_{10}H_{24}S_2Si$	$(CH_3)_2Si[CH_2CH(CH_3)CH_2SH]_2$	147—148 (8)	—	1.5031^{25}	0.9534_4^{25}	338
$C_{10}H_{24}O_3SSi$	$(C_2H_5O)_3Si(CH_2)_4SH$	127—130 (10)	—	—	—	419
	$(C_2H_5O)_3SiCH_2CH(CH_3)CH_2SH$	120—125 (10)	—	—	—	419
$C_{10}H_{26}OS_2Si_2$	$[(CH_3)_2SiCH_2CH_2CH_2SH]_2O$	87—89 (0.25)	—	1.4739	—	480
$C_{11}H_{18}SSi$	$(CH_3)_2(C_6H_5)SiCH_2CH_2CH_2SH$	110 (2)	—	1.5335	0.9755	83
$C_{11}H_{27}NO_2SSi$	$CH_3(C_2H_5O)_2Si(CH_2)_4NHCH_2CH_2SH$	112—114 (0.06)	—	1.4600^{26}	1.00^{26}	222
$C_{11}H_{27}NO_3SSi$	$(C_2H_5O)_3Si(CH_2)_3NHCH_2CH_2SH$	97—98 (0.2)	—	1.4500^{26}	0.96^{26}	222
$C_{13}H_{22}S_2Si$	$CH_3(C_6H_5)Si(CH_2CH_2CH_2SH)_2$	161—162 (2.5)	—	1.5580^{25}	1.0396_4^{25}	338
$C_{13}H_{22}O_2SSi$	$C_6H_5(C_2H_5O)_2SiCH_2CH_2CH_2SH$	156—160 (10)	—	—	—	419

A compound which contains at the same time thiol and sulfide groups was also obtained by the cleavage of 2,2-dimethyl-2-sila-1,4-dithiacyclohexane with dry hydrogen chloride [712]:

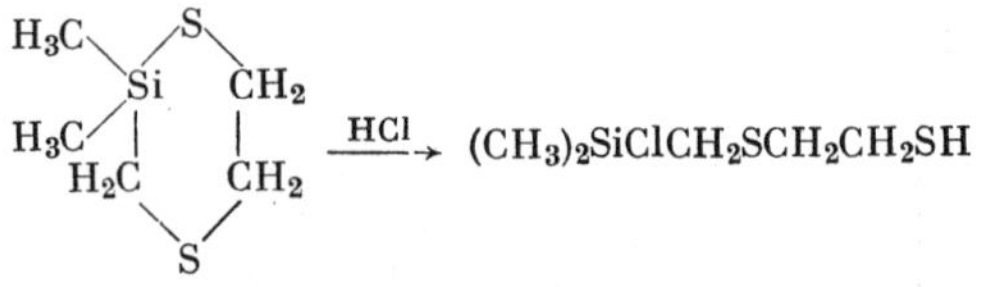

$$\text{(2,2-dimethyl-2-sila-1,4-dithiacyclohexane)} \xrightarrow{\text{HCl}} (CH_3)_2SiClCH_2SCH_2CH_2SH \qquad (2.240)$$

Triorganosilyl-substituted thiophenols are formed by the reaction of triorganosilylphenylmagnesium bromide with sulfur [598]:

$$R_3SiC_6H_4MgBr + S \xrightarrow[-Mg(OH)Br]{H_2O} R_3SiC_6H_4SH \qquad (2.241)$$

7.1.2. Physical Properties

The physical constants of organosilicon thiols are given in Table 22.

The molar refraction of the mercaptomethyl group at a silicon atom equals 15.22 ml/mole [81].

The bond vibrations νSH in the infrared spectra of organosilicon thiols correspond to an absorption band at 2568 ± 4 cm^{-1} [57, 271]. In Raman spectra, the frequencies of the fully symmetrical bond vibrations $\nu Si-C$ of α-silicon substituted thiols (556 cm^{-1}) are low in comparison with the frequencies of β- and γ-silicon substituted thiols (579 cm^{-1}) [84]. The NMR spectra of organosilicon thiols are examined in [757].

7.1.3. Chemical Properties

Organosilicon thiols undergo all the reactions which are characteristic of their purely carbon analogs and the presence of a silicon atom in them affects only the rates of these reactions.

Organosilicon alkanethiols [497] and thiophenols [598] react with metallic sodium to form the corresponding sodium thiolates, which may be converted to organosilicon sulfides by the action of organic halides [270, 497, 598]:

$$\ce{>Si-\left(\overset{|}{\underset{|}{C}}\right)_n-SNa + X-\overset{|}{\underset{|}{C}}- \xrightarrow{NaX} >Si-\left(\overset{|}{\underset{|}{C}}\right)_n-S-\overset{|}{\underset{|}{C}}-} \qquad (2.242)$$

Organosilicon thiols are alkylated at the mercapto group by 2-chloroethyl vinyl ether in the presence of KOH [765, 769].

Organosilicon thiols are oxidized readily by oxygen in the presence of various catalysts [204], hydrogen peroxide [497], sodium hypoiodite [271], and sulfuryl chloride to the corresponding disulfides [709]:

$$2\,{\geq}Si-\left(\overset{|}{\underset{|}{C}}\right)_n-SH \xrightarrow[-H_2O]{O} {\geq}Si-\left(\overset{|}{\underset{|}{C}}\right)_n-S-S-\left(\overset{|}{\underset{|}{C}}\right)_n-Si{\leq} \qquad (2.243)$$

Thus, for example, the oxidation of trimethylsilylmethanethiol by sodium hypoiodite leads to bis(trimethylsilylmethyl)disulfide (57.8% yield) [271]. Dimethylchlorosilylmethanethiol is converted to bis(dimethylchlorosilylmethyl) disulfide by the action of sulfuryl chloride, while chlorosulfanes give derivatives of higher sulfanes [709]:

$$(CH_3)_2SiClCH_2SH + ClS_xCl \xrightarrow[-2HCl]{} (CH_3)_2SiClCH_2S-S_x-SCH_2SiCl(CH_3)_2 \qquad (2.244)$$

In the presence of triethylamine, organosilicon thiols react with phenyl isocyanate in accordance with the scheme

$${\geq}Si-\left(\overset{|}{\underset{|}{C}}\right)_n-SH + C_6H_5NCO \longrightarrow {\geq}Si-\left(\overset{|}{\underset{|}{C}}\right)_n-SCONHC_6H_5 \qquad (2.245)$$

In this reaction, (methyldiethylsilyl)methanethiol ($n = 1$) is less reactive by a factor of 10 than its closest homologs with the thiol group in the β- and γ-positions and less active by a factor of 2 than 1-butanethiol [84].

Organosilicon thiols add readily to various unsaturated compounds: acetylene [123, 765, 818, 820], diacetylene [764], vinylethynylcarbinol [767], alkenylsilanes [275, 480, 671, 780, 803-807, 814], vinyl ethers [55, 57], vinyl sulfides [123, 819, 820], acrolein [418], acrylonitrile [57, 81, 83, 428], styrene [428], polybutadiene [861], vinyl acetate [428], and ketene and diketene [53, 786, 787].

The vinylation of α-silicon organic thiols by acetylene proceeds more readily than the analogous reactions of the β- and γ-derivatives, and this is evidently connected with the increase in the nucleophilicity of the organosilicon thiolate ion due to the very great inductive effect of the triorganosilyl group in the α-position relative to the sulfur atom [56, 58, 123]:

$$\geqslant Si-\left(\overset{|}{\underset{|}{C}}\right)_n-SH+CH\equiv CH \longrightarrow \geqslant Si-\left(\overset{|}{\underset{|}{C}}\right)_n-SCH=CH_2 \qquad (2.246)$$

The cyanoethylation of organosilicon thiols (n = 1-4) proceeds rapidly and vigorously in accordance with the scheme [81, 83, 428]

$$\geqslant Si-\left(\overset{|}{\underset{|}{C}}\right)_n-SH+CH_2=CH-CN \longrightarrow \geqslant Si-\left(\overset{|}{\underset{|}{C}}\right)_n-SCH_2CH_2CN \qquad (2.247)$$

Organosilicon thiols with the R_3Si-group in the α- and β-positions add to acrylonitrile in the presence of sodium methylate more vigorously than the corresponding organosilicon amines and alcohols [81].

The addition of organosilicon thiols to vinyl butyl ether in the presence of SO_2 proceeds in accordance with Markownikov's rule and in the presence of atmospheric oxygen it proceeds contrary to this rule, while both isomeric adducts are formed under the action of azobisisobutyronitrile [55, 57]:

$$\geqslant Si-\left(\overset{|}{\underset{|}{C}}\right)_n-SH+CH_2=CHOC_4H_9 - \begin{cases} \xrightarrow{(SO_2)} \geqslant Si-\left(\overset{|}{\underset{|}{C}}\right)_n-SCH(CH_3)OC_4H_9 \\ \xrightarrow{(O_2)} \geqslant Si-\left(\overset{|}{\underset{|}{C}}\right)_n-SCH_2CH_2OC_4H_9 \end{cases} \qquad (2.248)$$

Thus, the addition of organosilicon thiols to unsaturated compounds may proceed both by ionic and radical mechanisms. The radical addition of organosilicon dithiols to dialkenylsilanes leads to the formation of sulfur-containing organosilicon polymers [480].

7.2. Compounds Containing the Grouping Si − (C)$_n$ − S − C (Organosilicon Sulfides and Their Derivatives)

7.2.1. Preparation Methods

Organosilicon sulfides may be obtained from sulfur-containing organosilicon compounds (mainly from the corresponding thiols), from carbofunctional organic compounds and inorganic and organic derivatives of sulfur which do not contain the grouping C−S−C, and, finally, from organosilicon compounds and organic substances which contain the sulfide grouping C−S−C already formed.

The first group of synthesis methods for organosilicon sulfides includes reactions of organosilicon thiols (2.242-2.248), which lead to the formation of sulfides, disulfides, and polysulfides. These reactions have already been examined in sufficient detail in Section 7.1.3. Here we should only note that the most valuable of these reactions for the preparation of organosilicon sulfides is the addition of organosilicon thiols to unsaturated compounds (2.246 and 2.247), which make it possible to synthesize the corresponding sulfides containing various functional groups in the organic radical.

The reactions used most widely for the synthesis of organosilicon sulfides are those of the second group, which includes the reaction of haloalkylsilanes with alkali metal thiolates and the addition of mercapto compounds to alkenylsilanes.

The reaction of triorgano(haloalkyl)silanes with sodium sulfide is carried out in an aqueous alcohol solution. Bis(triorganosilylalkyl) sulfides are formed in 60-80% yield as a result [271, 309, 310, 629]:

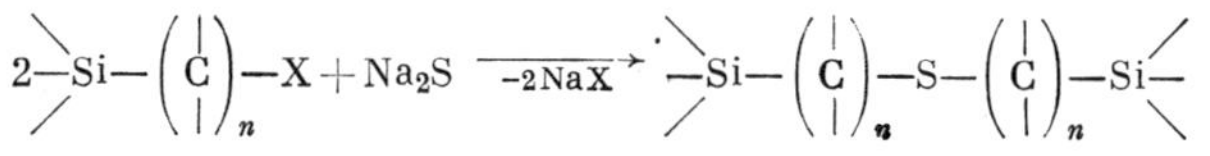

$$2 \underset{/}{\overset{\backslash}{\text{Si}}} - \left(\underset{|}{\overset{|}{\text{C}}} \right)_n - X + Na_2S \xrightarrow{-2NaX} \underset{/}{\overset{\backslash}{\text{Si}}} - \left(\underset{|}{\overset{|}{\text{C}}} \right)_n - S - \left(\underset{|}{\overset{|}{\text{C}}} \right)_n - \underset{\backslash}{\overset{/}{\text{Si}}} \quad (2.249)$$

When diorganobis(haloalkyl)silanes are used, the products of reaction (2.249) are cyclic organosilicon sulfides. Thus, for example, heating 3-bromopropyl(bromomethyl)dimethylsilane with an aqueous alcohol solution of sodium sulfide for 16 h forms 3,3-dimethyl-3-sila-1-thiacyclohexane (69% yield) [309]:

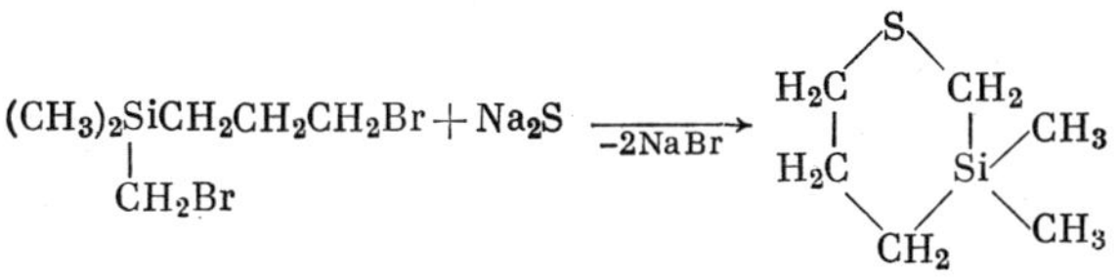

$$(CH_3)_2 \underset{|}{\text{SiCH}_2CH_2CH_2Br} + Na_2S \xrightarrow{-2NaBr} \begin{array}{c} S \\ H_2C \quad CH_2 \\ | \quad | \quad CH_3 \\ H_2C \quad Si \\ CH_2 \quad CH_3 \end{array} \quad (2.250)$$
$$CH_2Br$$

When 3-chloropropyl(chloromethyl)dimethylsilane is used instead of the bromo derivative, the yield of the reaction product is reduced to 47%. Under analogous conditions, 4-chlorobutyl(chloromethyl)dimethylsilane gives 3,3-dimethyl-3-sila-1-thiacycloheptane. The yield of this seven-membered heterocyclic compound is considerably lower than that of the six-membered compound, and does not exceed 11.6% even with an increase in the reaction time to 20 h [310].

In some cases a sulfide group is formed instead of a thiol group in the reaction of chloroalkylsilanes with alkali metal hydrosulfides. Thus, for example, when 1,3-bis(chloromethyl)tetramethyldisiloxane is heated with an alcohol solution of potassium hydrosulfide for 4 h, hydrogen sulfide is liberated and 2,2,6,6-tetramethyl-1-oxa-4-thia-2,6-disilacyclohexane is formed in 60% yield [618]:

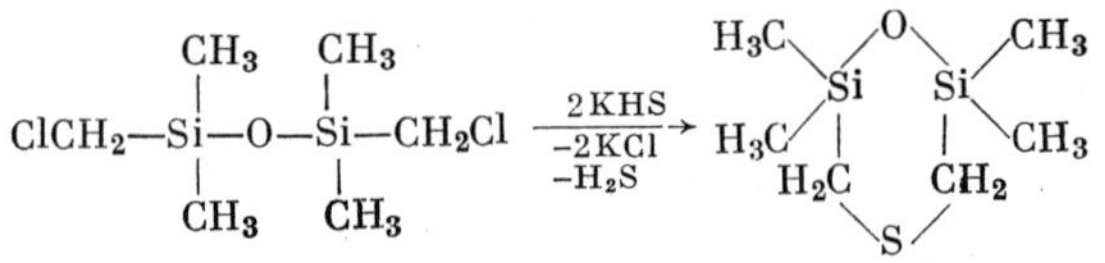

$$\begin{array}{c} \text{CH}_3 \quad \text{CH}_3 \\ | \qquad | \\ \text{ClCH}_2-\text{Si}-\text{O}-\text{Si}-\text{CH}_2\text{Cl} \\ | \qquad | \\ \text{CH}_3 \quad \text{CH}_3 \end{array} \xrightarrow[\substack{-2\,\text{KCl} \\ -\text{H}_2\text{S}}]{2\,\text{KHS}} \begin{array}{c} \text{H}_3\text{C} \diagdown \quad \diagup \text{O} \diagdown \quad \diagup \text{CH}_3 \\ \text{Si} \qquad \text{Si} \\ \text{H}_3\text{C} \diagup | \qquad | \diagdown \text{CH}_3 \\ \text{H}_2\text{C} \qquad \text{CH}_2 \\ \diagdown \quad \text{S} \diagup \end{array} \qquad (2.251)$$

Whether there is the replacement of both chlorine atoms by mercapto groups with subsequent liberation of a molecule of hydrogen sulfide and cyclization or whether there is replacement of one chlorine atom and cyclization occurs as the result of a subsequent intramolecular elimination of a molecule of hydrogen chloride has not been established as yet. An analogous reaction may be used for vulcanization of poly(chloromethyl)siloxane, heating of which with KSH leads to the liberation of hydrogen sulfide and crosslinking of the polyorganosiloxane chains by sulfide bridges [618]:

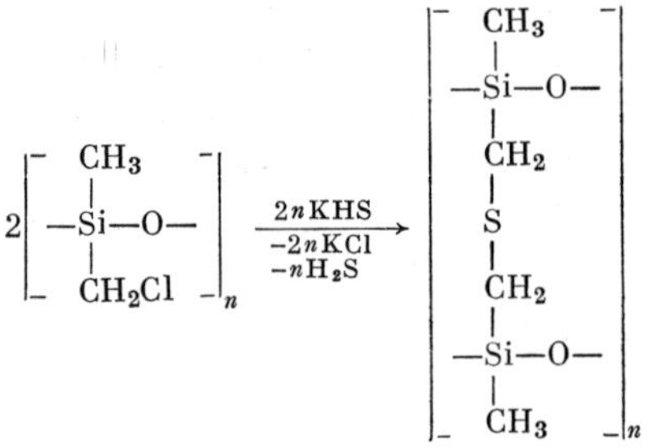

$$2\begin{bmatrix} \text{CH}_3 \\ | \\ -\text{Si}-\text{O}- \\ | \\ \text{CH}_2\text{Cl} \end{bmatrix}_n \xrightarrow[\substack{-2n\,\text{KCl} \\ -n\,\text{H}_2\text{S}}]{2n\,\text{KHS}} \begin{bmatrix} \text{CH}_3 \\ | \\ -\text{Si}-\text{O}- \\ | \\ \text{CH}_2 \\ | \\ \text{S} \\ | \\ \text{CH}_2 \\ | \\ -\text{Si}-\text{O}- \\ | \\ \text{CH}_3 \end{bmatrix}_n \qquad (2.252)$$

The reaction of bis(chloromethyl)diorganosilanes [520] and bis(chloromethyl)tetraorganodisiloxanes [618] with sodium poly-

sulfides leads to the formation of polymeric organosilicon polysulfides.

The reaction of haloalkylsilanes and haloalkylsiloxanes with
alkali metal alkanethiolates is used widely for the preparation of
organosilicon sulfides [83, 251, 348, 435, 443-445, 488, 497, 531,
550, 551, 707]:

$$\diagup^{\diagdown}Si-\left(\overset{|}{\underset{|}{C}}\right)_n-X+NaSR \longrightarrow \diagup^{\diagdown}Si-\left(\overset{|}{\underset{|}{C}}\right)_n-S-R+NaX \qquad (2.253)$$

It is usually carried out in absolute ethyl alcohol and sometimes in dimethylformamide [488]. The use of a slight excess of
alkanethiolate increases the yield of the organosilicon sulfide [531].
Better yields can also be achieved by using the analogous bromo
derivatives instead of chloromethylsilanes [497, 531, 550, 551]. It
is possible to carry out reaction (2.253) with salts of mercapto
acids [443-445, 550, 551] and mercapto amino acids (cysteine and
homocysteine) [325, 348]. In this case the reaction products contain a carboxyl or amino and carboxyl groups as well as the grouping C−S−C.

Sometimes instead of thiolates it is possible to use the thiols
themselves, using tertiary amines as hydrogen halide acceptors
2,2-dimethyl-1-oxa-4-thia-2-silacyclohexane [645]:

$$\diagup^{\diagdown}Si-\left(\overset{|}{\underset{|}{C}}\right)_n-X+HSR+R_3'N \longrightarrow \diagup^{\diagdown}Si-\left(\overset{|}{\underset{|}{C}}\right)_n-S-R+R_3'N \cdot HX \qquad (2.254)$$

If the starting haloalkylsilane contains a functional group at the
silicon atom (halogen or alkoxyl) while the starting thiol contains a
second functional group (OH, SH, NH_2, COOH), then as a result of
their interaction cyclic organosilicon sulfides are formed in which
the silicon is attached to another heteroatom (O, N, or S). Thus,
for example, the reaction of dimethyl(bromomethyl)ethoxysilane
with 2-mercaptoethanol in the presence of triethylamine forms
2,2-dimethyl-1-oxa-4-thia-2-silacyclohexane [645]:

$$(CH_3)_2\underset{\underset{CH_2Br}{|}}{Si}OC_2H_5+HOCH_2CHSH \xrightarrow[\substack{-R_3N \cdot HBr \\ -C_2H_5OH}]{R_3N} \begin{array}{c} H_3C \diagdown \diagup^O \diagdown \\ Si \qquad CH_2 \\ H_3C \diagup | \qquad | \\ H_2C \qquad CH_2 \\ \diagdown_S\diagup \end{array} \qquad (2.255)$$

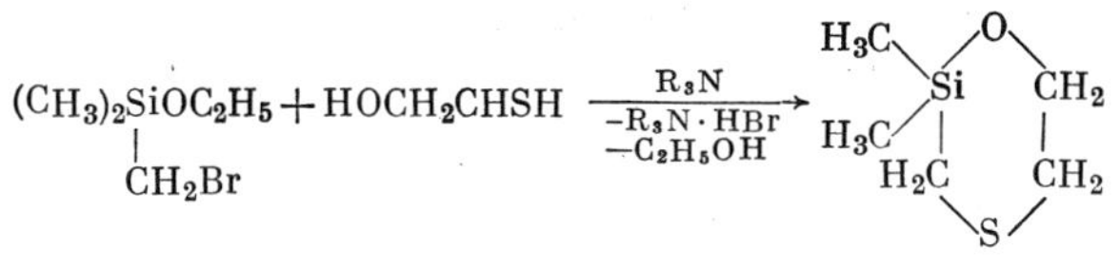

This compound may be converted by the action of acids into a siloxane containing hydroxyethylthiomethyl groups [303, 305, 648].

Dimethyl(chloromethyl)chlorosilane reacts analogously with mercaptoacetic acid in the presence of triethylamine to form 2,2-dimethyl-1-oxa-4-thia-2-silacyclohexan-6-one [578]:

$(CH_3)_2SiClCH_2Cl + HSCH_2COOH \xrightarrow[-2R_3N\cdot HCl]{2R_3N}$ 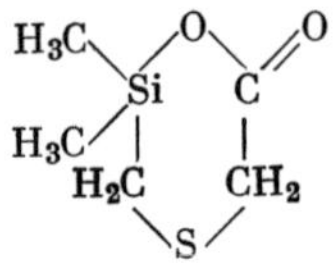 (2.256)

Heating vinyltrichlorosilane with hydrogen sulfide in an autoclave at 220°C forms bis(β-trichlorosilylethyl) sulfide [302, 734]:

$$2CH_2{=}CHSiCl_3 + H_2S \longrightarrow [Cl_3SiCH_2CH_2]_2S \quad (2.257)$$

Alkanethiols add readily to vinylsilanes under the action of ultraviolet radiation [339, 340, 397, 401, 569, 671] or peroxides [707]. Thus, for example, UV irradiation of a mixture of methyl(vinyl)diethoxysilane and methanethiol at 75°C gives 95% of the addition product [339, 569]. The addition of alkanethiols and thiophenol to vinylsilanes proceeds with much more difficulty in the presence of sodium thiolate [339, 568]. The radical reaction in the presence of peroxides [339, 707] or with UV radiation [340, 401] gives a higher yield of adducts (65-90%) than heterolytic addition, which is catalyzed by trimethylbenzylammonium hydroxide (14%) [707]. Thiophenol and phenylmethanethiol add to alkyldialkoxy- and vinyltrialkoxysilanes with simple heating of the reaction mixture to 80-130°C in the absence of catalysts [542]. However, while the yield of the product from the addition of thiophenol to vinyltriethoxysilane is 75%, in other cases it does not exceed 40%. The yield of adducts is not increased by increasing the heating time (from 8 to 48 h) or by using a large excess of alkanethiol [542]. The case of the addition of 1-tetradecanethiol to methyl(vinyl)diethoxysilane in the presence of catalytic amounts of sulfur has been described [514].

It is usually considered that in all the above cases of the addition of thiols to vinylsilanes the reaction proceeds contrary to Markownikov's rule with the formation of β-isomers. However, more accurate investigations show that in the radical addition of phenylmethanethiol and thiophenol to trimethyl(vinyl)silane the reaction products contain 5 and 3%, respectively, of the corresponding α-isomer:

$$R_3SiCH{=}CH_2 + HSR' \longrightarrow R_3SiCH_2CH_2SR' + R_3SiCH(CH_3)SR' \quad (2.258)$$

The addition of thiophenol to trimethylvinylsilane in the presence of trimethylbenzylammonium hydroxide forms only the β-isomer [707].

Triorgano(allyl)silanes are more reactive in the addition of thiols than triorgano(vinyl)silanes. Thus, for example, ethanethiol adds to trimethyl(allyl)silane exothermally without any catalysts or initiators to give a good yield of γ-trimethylsilylpropyl ethyl sulfide [83]. p-Thiocresol adds equally readily to trimethyl- and triethoxy(allyl)silane [216, 225, 226]. The addition of alkanethiols to allylsilanes may also be carried out in the presence of peroxides [225, 230, 473] or with UV radiation [671]. The addition of dithiols to diorganodiallylsilanes leads to the formation of sulfur-containing organosilicon polymers [230, 480]:

$$CH_2{=}CHCH_2{-}\overset{|}{\underset{|}{Si}}{-}CH_2CH{=}CH_2 + HSYSH \longrightarrow$$

$$\longrightarrow \left[-YSCH_2CH_2CH_2{-}\overset{|}{\underset{|}{Si}}{-}CH_2CH_2CH_2S-\right]_n$$

$$(Y = \text{alkylene}) \quad (2.259)$$

Mercaptoacetic (thioglycolic) acid [214, 217, 225, 228, 229, 446, 550, 707], its ethyl ester [223, 225, 227], and β-mercaptopropionic acid [707] add smoothly to alkenylsilanes. The reaction with allylsilane sometimes proceeds exothermally [225], but sometimes it requires initiation by peroxides [225, 707]. The products of the addition of thioglycolic acid and β-mercaptopropionic acid to trimethyl(vinyl)silane, which are formed by heating mixtures of the reactants with di-tert-butyl peroxide in sealed ampoules at 100°C, are mainly the β-isomers. However, a certain amount of the α-isomers is obtained together with the β-isomers [707]:

$$R_3SiCH{=}CH_2 + HS(CH_2)_nCOOH \longrightarrow$$

$$\longrightarrow \underset{94\%}{R_3SiCH_2CH_2S(CH_2)_nCOOH} + \underset{6\%}{R_3SiCH(CH_3)S(CH_2)_nCOOH}$$

$$(n = 1 \text{ or } 2) \quad (2.260)$$

Thioacetic S-acid adds to alkenylsilanes even more readily [876, 952]. It is sufficient to heat a mixture of it with the alkenylsilane in cyclohexane for 2 h to obtain the corresponding acetate of the organosilicon thiol in about 80% yield. The rate of the addition reaction depends on the structure of the alkenylsilane. Thus, in the reaction with thioacetic S-acid, 3-butenyltrimethylsilane is ~1.5 times as reactive as trimethylvinylsilane [83]. Allylsilanes add most readily to thioacetic S-acid [83, 84, 95, 338, 480]. The reaction rate and yield of addition products of thioacetic S-acid and vinylsilanes may be increased by adding benzoyl peroxide [204]. Thus, for example, the yield of 1,3-bis(2-acetylthioethyl)tetramethyldisiloxane, formed by boiling a cyclohexane solution of 1,3-divinyltetramethyldisiloxane and thioacetic S-acid for 8 h in the presence of benzoyl peroxide, reaches 93.3% [204]. The addition of alkanethiols [121, 772, 793, 813], thiophenol [59, 116, 121], thioacetic S-acid [770, 772], and mercaptoacetic acid [60] to ethynylsilane proceeds equally readily.

An organosilyl group may be introduced into the molecule of an organic sulfide by hydrosilylation of unsaturated sulfides [526, 562, 773, 774, 802], the reaction of organolithium and magnesium derivatives of sulfides and sulfones with halosilanes [351, 537], and the reaction of alkanethiols with organosilicon epoxides [562].

Hydrosilanes add to alkenyl sulfides in the presence of platinum on charcoal, H_2PtCl_6, and peroxides, and under the action of UV radiation:

$$\equiv Si{-}H + CH_2{=}CH(CH_2)_nSR \longrightarrow \equiv SiCH_2CH_2(CH_2)_nSR \qquad (2.261)$$

The addition of diorganosilanes to dialkenyl sulfides leads to the formation of polymers [562]. Diphenylsilane does not add to thianthrene, but on heating to 250-260°C it displaces one sulfur atom in the form of hydrogen sulfide to form 10,10-diphenylphenothiasiline (4.7% yield) [362, 720]:

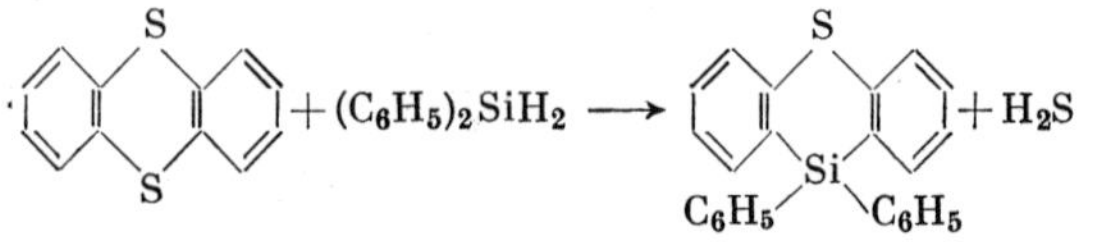

(2.262)

It is not possible to replace the second sulfur atom by silicon even by the use of a two-fold excess of diphenylsilane. The structure of the product obtained was demonstrated by the identity of its sulfone and a sulfone obtained by the action of 2,2'-dilithiodiphenyl sulfone with diphenylchlorosilane [537]:

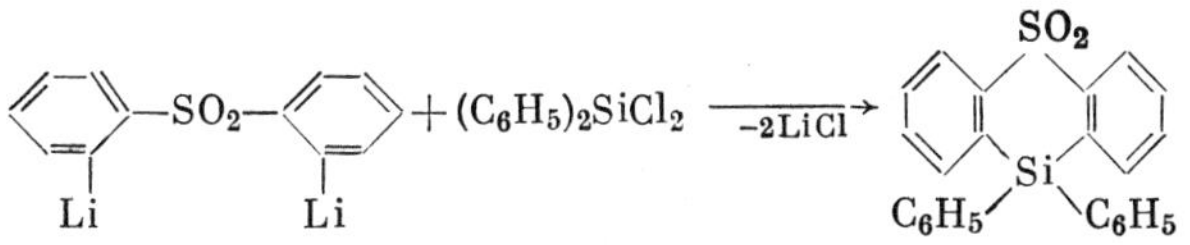

(2.263)

An analogous reaction may be used for the synthesis of some triorganosilylmethyl aryl sulfones [351]. However, it cannot be used as a general method for preparing organosilicon sulfones, as in a number of cases there is no reaction between an arylsulfonylmethyllithium and a triorganochlorosilane. Thus, for example, when p-tolylsulfonylmethyllithium $CH_3C_6H_4SO_2CH_2Li$ is treated with trimethylchlorosilane, no organosilicon sulfone is formed [351]. In contrast to this, the lithium derivative of p-tolyl methyl sulfide reacts smoothly with trimethylchlorosilane and is converted into (trimethylsilylmethyl) p-tolyl sulfide in 78% yield [351]:

$$p\text{-}CH_3C_6H_4SCH_2Li + ClSi(CH_3)_3 \xrightarrow[-LiCl]{} p\text{-}CH_3C_6H_4SCH_2Si(CH_3)_3 \quad (2.264)$$

Chlorosilanes react analogously with lithium derivatives of dimethyl sulfide [925], 1,3-dithiacyclopentanes [844, 856], and 1,3-dithiacyclohexane [845, 846, 855]. Magnesium derivatives are also used instead of lithium derivatives [843, 936]. In the case of sodium derivatives of organomethyl sulfoxides the silicon-containing sulfoxides $R_3SiCH_2S(O)R$ formed initially rearrange to siloxy derivatives R_3SiOCH_2SR [843].

The reaction of haloalkylsilanes with thiourea in ethanol or butanol forms organosilicon isothiuronium salts [250, 419, 531, 533, 594]:

$$\geq Si-\left(\overset{|}{\underset{|}{C}}\right)_n-X + H_2NCSNH_2 \longrightarrow \left[\geq Si-\left(\overset{|}{\underset{|}{C}}\right)_n-SC\begin{matrix}\nearrow NH_2 \\ \searrow NH_2\end{matrix}\right]^+ X^- \quad (2.265)$$

Bromoalkylsilanes react more readily than chloroalkylsilanes. In the latter case the reaction is catalyzed by sodium iodide [419].

Heating a solution of equimolecular amounts of trimethyl(chloromethyl)silane, benzene, and $AlCl_3$ in excess carbon disulfide forms trimethylsilylmethyl dithiobenzoate, whose yield is 43% after heating for 4 h [344, 345]:

$$C_6H_6 + CS_2 + ClCH_2Si(CH_3)_3 \xrightarrow[-HCl]{AlCl_3} C_6H_5C(S)SCH_2Si(CH_3)_3 \qquad (2.266)$$

An analogous result is obtained when an equimolecular amount of CS_2 but an excess of benzene are used. If only catalytic amounts of $AlCl_3$ are used, this reaction product is not formed.

7.2.2. Physical Properties

The physical properties of organosilicon sulfides are given in Table 23. The proton magnetic spectra of some of the organosilicon sulfides synthesized were plotted to confirm the structures of these compounds [646, 650, 707, 714].

7.2.3. Chemical Properties

Organosilicon sulfides add alkyl halides and are converted into the corresponding sulfonium salts [251, 271, 618]:

$$\geq Si-\left(\overset{|}{\underset{|}{C}}\right)_n-S-\overset{|}{\underset{|}{C}}- + RX \longrightarrow \left[\geq Si-\left(\overset{|}{\underset{|}{C}}\right)_n-\underset{R}{\underset{|}{S}}-\overset{|}{\underset{|}{C}}-\right]^+ X^- \qquad (2.267)$$

When dimethyl(pentamethyldisilanylmethyl)sulfonium iodide is heated with excess methyl iodide, dimethyl sulfide is eliminated and the compound converted to pentamethyl(iodomethyl)disilane [937].

An interesting case of the formation of a sulfonium salt was observed in the reaction of dimethyl(bromomethyl)chlorosilane with β-mercaptoethanol, whose products are 1,4-dithiane and 2,2,6,6-tetramethyl-4-(β-hydroxyethyl)-1-oxa-4-thionia-2,6-disilacyclohexane bromide. It is considered that the 1,4-dithiane is obtained as a result of the condensation of two molecules of β-mercaptoethanol, while the water liberated in this case converts the dimethyl(bromomethyl)chlorosilane into 1,3-bis(bromomethyl)tetramethyldisiloxane, which then reacts with β-mercaptoethanol

TABLE 23. Compounds Containing the Grouping $Si-(C)_n-S-C$ (Organosilicon sulfides and their derivatives)

Empirical formula *	Compound	B.p., °C (mm)	M.p., °C	n_D^{20}	d_4^{20}	Literature
$C_4H_8Cl_2SSi_2$	$(Cl_3SiCH_2CH_2)_2S$	114—116 (2)	—	—	—	302
$C_5H_{10}O_2SSi$	$OCOSi(CH_3)_2CH_2SCH_2$ (ring)	95—97 (2)	—	—	—	714
$C_5H_{12}OSSi$	$(CH_3)_2SiOCH_2CH_2SCH_2$ (ring)	41 (2.2)	—	1.4800	—	645, 647
$C_5H_{14}SSi$	$(CH_3)_3SiCH_2SCH_3$	70 (93)	—	1.4505	0.8399	251
$C_5H_{14}O_2SSi$	$(CH_3)_3SiCH_2SO_2CH_3$	—	78—79	—	—	251
$C_5H_{15}BrN_2SSi$	$[(CH_3)_3SiCH_2SC(=NH_2)NH_2]Br$	—	172—173	—	—	531
$C_5H_{15}ClN_2SSi$	$[(CH_3)_3SiCH_2SC(=NH_2)NH_2]Cl$	—	141.5—143	—	—	250
		—	144—144.5	—	—	533
	$[(CH_3)_3SiCH_2SC(=NH_2)NH_2]OC_6H_2(NO_2)_3$-2,4,6	—	200	—	—	533
$C_6H_{14}SSi$	$(CH_3)_2SiCH_2CH_2CH_2SCH_2$ (ring)	181	—	1.4978^{26}	—	309
$C_6H_{14}O_2SSi$	$(CH_3)_3SiCH_2SCH_2COOH$	109—110 (2)	—	1.4778	1.0262	27
$C_6H_{16}Cl_2S_2Si_2$	$[(CH_3)_2SiClCH_2S]_2$	140 (1)	—	—	—	709
$C_6H_{16}SSi$	$(CH_3)_3SiCH_2SC_2H_5$	43.5 (9)	—	1.4512^{25}	0.8402_4^{25}	531
	$(CH_3)_3SiCH=S(CH_3)_2$	51—53 (12)	—51 to —47	—	—	496
$C_6H_{16}OSSi_2$	$(CH_3)_2SiCH_2SCH_2Si(CH_3)_2$ (ring closed by —O—)	186 (720)	10—12	1.4722	0.985	496, 618
$C_6H_{16}O_3SSi_2$	$(CH_3)_2SiCH_2SO_2CH_2Si(CH_3)_2$ (ring closed by —O—)	—	115—116	—	—	618
$C_6H_{17}ISSi$	$[(CH_3)_3SiCH_2S(CH_3)_2]I$	—	106	—	—	251
$C_7H_{15}ClOSSi$	$(CH_3)_3Si(CH_2)_3SCOCl$	44—45 (2)	—	1.4323	0.9942	54, 79
$C_7H_{15}NSSi$	$(CH_3)_3SiCH_2SCH_2CH_2CN$	250 (748)	—	1.4768	0.9443	83
		—	—	—	0.9420	81
$C_7H_{16}SSi$	$(CH_3)_2SiCH_2CH_2CH_2CH_2SCH_2$ (ring)	82—83 (13)	—	$1.5054^{24.5}$	—	310
$C_7H_{16}OSSi$	$(CH_3)_3SiCH_2CH_2SCOCH_3$	82—84 (6)	—	1.4660	0.9308	83
$C_7H_{16}O_2SSi$	$(CH_3)_3SiCH_2CH_2SCH_2COOH$	143—144 (7)	—	1.4811	1.0139	217, 225
		94—95 (0.1)	—	—	—	707

*As an exception, thiuronium nitrates and picrates are not included in the column "Empirical formula" but for convenience in comparison they are placed together with the corresponding halides.

TABLE 23 (Continued)

Empirical formula *	Compound	B.p., °C (mm)	M.p., °C	n_D^{20}	d_4^{20}	Literature
	$(CH_3)_3SiCH(CH_3)SCH_2COOH$	88—89 (0.1)	—	1.4783^{25}	1.0144_4^{25}	707
$C_7H_{17}NSSi$	$(CH_3)_3Si(CH_2)_3SCONH_2$	—	73—74	—	—	79
$C_7H_{18}SSi$	$(CH_3)_3SiCH_2SC_3H_7$	60 (9)	—	1.4518^{25}	0.8377_4^{25}	531
	$(CH_3)_3SiCH_2SC_3H_7$-iso	53 (9)	—	1.4494^{25}	0.8321_4^{25}	531
$C_7H_{20}OSSi_2$	$(CH_3)_3SiOSi(CH_3)_2CH_2SCH_3$	114 (96)	—	1.4342	—	251
$C_7H_{20}O_3SSi_2$	$(CH_3)_3SiOSi(CH_3)_2CH_2SO_2CH_3$	91 (1)	—	—	—	251
$C_8H_7Cl_3SSi$	$Cl_3SiCH{=}CHSC_6H_5$	107.5—110 (2)	—	1.5838	—	116
$C_8H_{11}O_2SSi$	$(CH_3)_3SiCH_2SCH_2COOC_2H_5$	90—91 (6)	—	—	—	27
$C_8H_{17}NSSi$	$(CH_3)_3SiCH_2CH_2SCH_2CH_2CN$	147—148 (6)	—	1.4774	0.9371	83
	$(CH_3)_3SiCH(CH_3)SCH_2CH_2CN$	157 (7)	—	1.4879	0.9386	83
$C_8H_{18}SSi$	$(CH_3)_3Si(CH_2)_3SCH{=}CH_2$	74.5 (7)	—	1.4785	0.8736	123
	$CH_3(C_2H_5)_2SiCH_2SCH{=}CH_2$	76 (8)	—	1.4812	0.8770	58, 123
$C_8H_{18}OSSi$	$(CH_3)_3Si(CH_2)_3SCOCH_3$	92 (12)	—	1.4655	0.9198	83
		126—127 (60)	—	1.4652	0.9198	95
	$(CH_3)_3Si(CH_2)_3S(O)CH{=}CH_2$	119 (4.5)	—	1.4805	0.9520	123
		111—112 (1.5)	—	—	—	124
$C_8H_{18}O_2SSi$	$(CH_3)_3Si(CH_2)_3SCH_2COOH$	164—166 (9)	—	1.4790	1.0099	217, 225
	$(CH_3)_3SiCH(CH_3)SCH_2CH_2COOH$	125—126 (1.5)	—	1.4787^{25}	1.0028_4^{25}	707
	$(CH_3)_3Si(CH_2)_3SO_2CH{=}CH_2$	116 (1)	—	1.4682	1.0055	123, 124
$C_8H_{20}SSi$	$(CH_3)_3SiCH_2SC_4H_9$	75—75.5 (9)	—	1.4530^{25}	0.8352_4^{25}	531
	$(CH_3)_3SiCH_2SC_4H_9$-iso	65.5—66 (9)	—	1.4502^{25}	0.8314_4^{25}	531
	$(CH_3)_3SiCH_2SC_4H_9$-tert	55.5—56.5 (9)	—	1.4496^{25}	0.8270_4^{25}	531
	$(CH_3)_3Si(CH_2)_3SC_2H_5$	71 (7)	—	1.4581	0.8479	83
$C_8H_{20}O_2SSi$	$CH_3(C_2H_5O)_2SiCH_2CH_2SCH_3$	127.5—129 (59)	—	1.4454	0.954	334, 401, 569
$C_8H_{21}ClN_2SSi$	$[(CH_3)_2(C_4H_9)SiCH_2SC({=}NH_2)NH_2]Cl$	—	62—64	—	—	533
	$[(CH_3)_2(C_4H_9)SiCH_2SC({=}NH_2)NH_2]OC_6H_2(NO_2)_3$-2,4,6	—	90	—	—	533
$C_8H_{22}SSi_2$	$[(CH_3)_3SiCH_2]_2S$	129 (95)	—	1.4570	0.800	251
		71—73 (5)	—	1.4543^{25}	0.8401_4^{25}	497

$C_8H_{22}HgI_2SSi_2$	$[(CH_3)_3SiCH_2]_2S \cdot HgI_2$	—	63	—	—	270, 271
	$(CH_3)_3SiSi(CH_3)_2CH_2SC_2H_5$	94 (18)	—	1.4778	0.8523	435
$C_8H_{22}S_2Si_2$	$[(CH_3)_3SiCH_2S]_2$	82 (1)	—	1.4874^{25}	0.9147^{25}	270, 271
		71—73 (0.5)	—	1.4906^{25}	0.9203_4^{25}	497
$C_8H_{22}OS_2Si_2$	$[CH_3SCH_2Si(CH_3)_2]_2O$	87 (2)	—	1.4742	—	251
$C_8H_{22}O_2SSi$	$[(CH_3)_3SiCH_2]_2SO_2$	—	47—48	—	—	251
$C_8H_{24}Cl_2N_4OS_2Si_2$	$\{[(CH_3)_2SiCH_2SC(=NH_2)NH_2]O\}Cl_2$	—	164—165	—	—	533
	$\{[(CH_3)_2SiCH_2SC(=NH_2)NH_2]_2O\}(NO_3)_2$	—	84—85	—	—	533
	$\{[CH_3)_2SiCH_2SC(=NH_2)NH_2]_2O\}[OC_6H_2(NO_2)_3\text{-}2,4,6]_2$	—	231—232	—	—	533
$C_9H_{12}Cl_2SSi$	$CH_3SiCl_2CH_2CH_2SC_6H_5$	140.5—143 (4)	—	—	—	339, 401
$C_9H_{15}NSSi$	$(CH_3)_3SiC{\equiv}CCH_2SCH_2CH_2CN$	104 (1.5)	—	1.4952	0.9599	57
$C_9H_{19}NSSi$	$(CH_3)_3Si(CH_2)_3SCH_2CH_2CN$	142 (8)	—	—	—	83
		108—108.5 (1)	—	1.4770	0.9287	81
	$CH_3(C_2H_5)_2SiCH_2SCH_2CH_2CN$	100.5 (2)	—	1.4836	0.9443	81
$C_9H_{20}SSi$	$(CH_3)_2(C_2H_5)Si(CH_2)_3SCH{=}CH_2$	72 (1.5)	—	1.4790	0.8729	123
$C_9H_{20}OSSi$	$(CH_3)_2(C_2H_5)Si(CH_2)_3SCOCH_3$	99 (8)	—	1.4706	0.9226	83
	$(CH_3)_2(C_2H_5)SiCH_2CH(CH_3)SCOCH_3$	80 (3)	—	1.4743	0.9383	83
	$(CH_3)_3SiCH_2CH(CH_3)CH_2SCOCH_3$	99—100 (9)	—	1.4664^{25}	0.9158_4^{25}	338
	$CH_3(C_2H_5)_2SiCH_2CH_2SCOCH_3$	122 (19)	—	1.4752	0.9345	84
	$(CH_3)_2(C_2H_5)Si(CH_2)_3S(O)CH{=}CH_2$	120 (2)	—	1.4832	0.9583	123, 124
$C_9H_{20}O_2SSi$	$(CH_3)_3Si(CH_2)_4SCH_2COOH$	172—175 (14)	—	1.4775	—	550
	$(CH_3)_3SiCH_2CH_2CH(CH_3)SCH_2COOH$	172 (16)	—	1.4778	—	551
	$(CH_3)_2(C_2H_5)Si(CH_2)_3SO_2CH{=}CH_2$	145 (4.5)	—	1.4720	1.0055	123, 124
$C_9H_{22}SSi$	$(CH_3)_3SiCH_2SC_5H_{11}$	89 (9)	—	1.4542^{25}	0.8374_4^{25}	531
	$(CH_3)_3SiCH_2SC_5H_{11}\text{-iso}$	85—86 (9)	—	1.4530^{25}	0.8347_4^{25}	531
	$(CH_3)_3Si(CH_2)_3SC_3H_7$	90 (11)	—	1.4557	0.8518	81
$C_9H_{25}ISSi_2$	$\{[(CH_3)_3SiCH_2]_2SCH_3\}I$	—	147	—	—	270, 271

*See footnote on p. 259.

TABLE 23 (Continued)

Empirical formula *	Compound	B.p., °C (mm)	M.p., °C	n_D^{20}	d_4^{20}	Literature
$C_9H_{26}O_4SSi_4$	$O[(CH_3)_2SiO]_3Si(CH_3)CH_2SCH_3$ (ring)	110 (11)	—	1.4318	1.0203	251
$C_9H_{26}O_6SSi_4$	$O[(CH_3)_2SiO]_3Si(CH_3)CH_2SO_2CH_3$ (ring)	—	61—62.5	—	—	251
$C_{10}H_{16}SSi$	$(CH_3)_3SiCH_2SC_6H_5$	48 (0.04)	—	1.5372^{25}	0.9668_4^{25}	497
		158.5 (52)	—	1.5380	0.9671	251
	p-$(CH_3)_3SiC_6H_4SCH_3$	121 (10)	—	1.5465	—	179
$C_{10}H_{16}O_2SSi$	$(CH_3)_3SiCH_2SO_2C_6H_5$	160 (6)	—	1.5250	—	251
$C_{10}H_{17}ClN_2SSi$	$[(CH_3)_2(C_6H_5)SiCH_2SC(=NH_2)NH_2]Cl$	—	113—114	—	—	533
	$[(CH_3)_2(C_6H_5)SiCH_2SC(=NH_2)NH_2]OC_6H_2(NO_2)_3$-2,4,6	—	93	—	—	533
$C_{10}H_{17}NSSi$	$(CH_3)_2(C_2H_5)SiC{\equiv}CCH_2SCH_2CH_2CN$	127 (4)	—	1.4970	0.9500	57
$C_{10}H_{21}NS_2Si$	$(CH_3)_2Si(CH_2SC_3H_7)CH_2SCH_2CH_2CN$	143 (2)	—	1.5110	1.0059	81
$C_{10}H_{22}SSi$	$(C_2H_5)_3SiCH_2CH_2SCH{=}CH_2$	84 (1)	—	1.4913	0.8927	56, 123
$C_{10}H_{22}OSSi$	$(CH_3)_2(C_2H_5)Si(CH_2)_4SCOCH_3$	86 (1)	—	1.4699	0.9181	83
	$(CH_3)_2(C_3H_7)Si(CH_2)_3SCOCH_3$	78 (2)	—	1.4702	0.9148	83
	$CH_3(C_2H_5)_2Si(CH_2)_3SCOCH_3$	143 (26)	—	1.4749	0.9277	84
	$(C_2H_5)_3SiCH_2CH_2S(O)CH{=}CH_2$	108.5 (0.5)	—	1.4945	0.9583	123, 124
	$(C_2H_5)_3SiCH_2CH_2SCOCH_3$	117—119 (6.5)	—	1.4779^{25}	0.9284_4^{25}	338
			—			
$C_{10}H_{22}O_2SSi$	$(CH_3)_3Si(CH_2)_3SCH_2COOC_2H_5$	179—182 (45)	—	—	0.9943	217
		148—149 (24)	—	1.4630	0.9493	225
	$(C_2H_5)_3Si(CH_2)_2SCH_2COOH$	184—186 (12)	—	—	—	446
	$(C_2H_5)_3SiCH(CH_3)SCH_2COOH$	173—175 (9)	—	—	—	443
		175—177 (10)	—	—	—	445
	$(C_2H_5)_3SiCH_2CH_2SO_2CH{=}CH_2$	134 (1.5)	—	1.4823	1.0089	123, 124
$C_{10}H_{22}O_5S_2Si_2$	$[HOOCCH_2SCH_2Si(CH_3)_2]_2O$	—	—	1.4437	—	649
$C_{10}H_{23}ClSSi$	$(C_2H_5)_3Si(CH_2)_2SCHClCH_3$	112—113 (2)	—	1.4870	0.9462	56
$C_{10}H_{24}O_3SSi_2$	$(CH_3)_3SiOSi(CH_3)_2(CH_2)_3SCH_2COOH$	149—150 (2)	—	1.4588	0.9902	225
$C_{10}H_{24}SSi$	$(CH_3)_3SiCH_2SC_6H_{13}$	82 (4)	—	1.4560^{25}	0.8385_4^{25}	497
$C_{10}H_{26}O_2SSi_2$	$[(CH_3)_3SiCH_2CH_2]_2SO_2$	—	123	—	—	671
$C_{10}H_{26}SSi_2$	$[(CH_3)_3SiCH_2CH_2]_2S$	65 (1)	—	—	—	302
		39 (0.05)	—	1.4620^{25}	—	671
$C_{10}H_{28}N_2O_5S_2Si_2$	$[NH_4OOCCH_2SCH_2Si(CH_3)_2]_2O$	—	115—120 (decomp.)	—	—	714

$C_{10}H_{29}IO_4SSi_4$	$\{O[(CH_3)_2SiO]_3Si(CH_3)CH_2S(CH_3)_2\}I$	—	129—130	—	—	251
$C_{11}H_{15}ClS_2Si$	$ClC_6H_4C(S)SCH_2Si(CH_3)_3$ †	115—141 (0.1)	—	1.5567	—	344
$C_{11}H_{16}SSi$	$(CH_3)_3SiCH{=}CHSC_6H_5$	118—120 (9)	—	1.5556	0.9607	59
$C_{11}H_{16}S_2Si$	$(CH_3)_3SiCH_2SC(S)C_6H_5$	95 (0.1)	—	1.5987	—	344
$C_{11}H_{18}SSi$	$(CH_3)_3SiCH_2SCH_2C_6H_5$	124 (8)	—	1.5242^{25}	0.9507^{25}_{4}	531
	$(CH_3)_3SiCH_2SCH_2C_6H_4CH_3$-o	64 (0.3)	—	1.5389^{25}	0.9683^{25}	497
	$(CH_3)_3SiCH_2SCH_2C_6H_4CH_3$-m	74 (0.5)	—	1.5343^{25}	0.9560^{25}_{4}	497
	$(CH_3)_3SiCH_2SC_6H_4CH_3$-p	79 (0.06)	—	1.5343^{25}	0.9557^{25}_{4}	497
		83—85 (1)	—	1.5356	0.9545^{20}_{20}	351
	$(CH_3)_3SiCH(CH_3)SC_6H_5$	88—89 (1.7)	—	1.5318^{25}	0.9569^{25}_{4}	707
$C_{11}H_{22}F_2SSi$	$(C_2H_5)_3SiCF{=}CFSC_3H_7$	65 (0.35)	—	1.4640^{25}	—	179
$C_{11}H_{23}NO_2SSi$	$CH_3(C_2H_5O)_2Si(CH_2)_3SCH_2CH_2CN$	130—137 (0.3)	—	—	—	418
$C_{11}H_{23}NSSi$	$CH_3(C_3H_7)_2SiCH_3SCH_2CH_2CN$	124—125 (2)	—	1.4800	0.9301	81
	$(CH_3)_2(C_2H_5)Si(CH_2)_4SCH_2CH_2CN$	153.5 (6)	—	1.4802	0.9251	83
	$(CH_3)_2(C_3H_7)Si(CH_2)_3SCH_2CH_2CN$	139 (3)	—	1.4790	0.9241	83
$C_{11}H_{24}SSi$	$(C_2H_5)_3Si(CH_2)_3SCH{=}CH_2$	76 (1)	—	1.4850	0.8880	123
$C_{11}H_{24}OSSi$	$CH_3(C_2H_5)_2Si(CH_2)_4SCOCH_3$	93—94 (1)	—	1.4740	0.9211	83
	$(C_2H_5)_3Si(CH_2)_3SCOCH_3$	122 (5)	—	1.4783	0.9274	83
	$(C_2H_5)_3SiCH_2CH_2CH_2S(O)CH{=}CH_2$	128 (1)	—	1.4852	0.9574	123
		128—130 (1)	—	—	—	124
$C_{11}H_{24}O_2SSi$	$(C_2H_5)_3SiCH(CH_3)SCH_2CH_2COOH$	187—188 (8)	—	—	—	443
$C_{11}H_{24}O_3SSi$	$CH_3(C_2H_5O)_2Si(CH_2)_3SCH_2CH_2CHO$	120—130 (0.1)	—	—	—	418
$C_{11}H_{26}SSi$	$(CH_3)_3SiCH_2SC_7H_{15}$	84 (1)	—	1.4571^{25}	0.8400	497
$C_{11}H_{26}O_3SSi$	$(C_2H_5O)_3SiCH_2CH_2SC_3H_7$-iso	111.5—113 (6)	—	1.4378	—	397
$C_{12}H_{18}SSi$	$(CH_3)_2(C_2H_5)SiCH{=}CHC_6H_5$	115—118 (5.5)	—	1.5505	0.9615	59
$C_{12}H_{20}SSi$	$(CH_3)_3SiCH(CH_3)SCH_2C_6H_5$	97—98 (1.6)	—	1.5214^{25}	0.9455^{25}_{4}	707
$C_{12}H_{22}OSSi_2$	$(CH_3)_3SiOSi(CH_3)_2CH_2SC_6H_5$	170.5 (37)	—	1.5014	0.9705	251
$C_{12}H_{22}SSi_2$	$(CH_3)_3SiSi(CH_3)_2CH_2SC_6H_5$	109 (2)	—	1.5419	0.9485	435
$C_{12}H_{24}OSSi$	$(CH_3)_3SiC{\equiv}CCH_2SCH_2CH_2OC_4H_9$	97 (1)	—	1.4719	0.9032	57

*See footnote on p. 259.

†Whether it is the o-, m-, or p-isomer is not reported in the original.

TABLE 23 (Continued)

Empirical formula *	Compound	B.p., °C (mm)	M.p., °C	n_D^{20}	d_4^{20}	Literature
$C_{12}H_{24}O_2S_2Si$	$(CH_3)_2Si[(CH_2)_3SCOCH_3]_2$	127.5—129.5 (0.35)	—	1.5068	—	480
$C_{12}H_{25}NSSi$	$CH_3(C_2H_5)_2Si(CH_2)_4SCH_2CH_2CN$	162.5 (5)	—	1.4823	0.9278	83
	$(C_2H_5)_3Si(CH_2)_3SCH_2CH_2CN$	152 (3)	—	1.4869	0.9365	83
$C_{12}H_{26}O_3S_2Si_2$	$[CH_3COSCH_2CH_2Si(CH_3)_2]_2O$	149—150 (1.6)	—	1.4839^{25}	—	204
	$[HOCH_2CH_2SCH_2Si(CH_3)_2]_2O$	—	—	1.4581	—	649
$C_{12}H_{26}O_4SSi$	$CH_3(C_2H_5O)_2Si(CH_2)_3SCH_2CH_2OCOCH_3$	110—114 (0.2)	—	—	—	418
$C_{12}H_{28}O_3SSi$	$(C_2H_5O)_3SiCH_2CH_2SC_4H_9$	138—140 (8)	—	1.4422	0.952	302, 339, 397, 568
	$(C_2H_5O)_3SiCH_2CH_2SC_4H_9$-iso	73—74 (5)	—	1.4388	—	597
$C_{12}H_{29}ClN_2SSi$	$[(CH_3)_2(C_8H_{17})SiCH_2SC({=}NH_2)NH_2]Cl$	—	62—63	—	—	533
	$[(CH_3)_2(C_8H_{17})SiCH_2SC({=}NH_2)NH_2]OC_6H_2(NO_2)_3$-2,4,6	—	87	—	—	533
$C_{12}H_{30}SSi_2$	$[(CH_3)_3SiCH_2CH_2CH_2]_2S$	117 (4)	—	—	—	302
		66 (0.5)	—	—	—	671
		46 (0.05)	—	1.4621^{25}	—	671
$C_{13}H_{20}OSSi$	$(CH_3)_2(C_6H_5)Si(CH_2)_3SCOCH_3$	121 (2)	—	1.5311	1.0129	83
$C_{13}H_{21}NO_2SSi$	o-$(CH_3)_3SiC_6H_4CH_2SCH_2CH(NH_2)COOH$	—	183	—	—	348
	p-$(CH_3)_3SiC_6H_4CH_2SCH_2CH(NH_2)COOH$	—	191	—	—	325, 348
$C_{13}H_{22}SSi$	$(CH_3)_3Si(CH_2)_3SC_6H_4CH_3$-p	206—209 (46)	—	1.5267	0.9408	225
$C_{13}H_{22}OSSi$	$(CH_3)_2(C_2H_5O)SiCH_2CH_2SCH_2C_6H_5$	107 (0.5)	—	1.5490	1.1029	519, 542
$C_{13}H_{22}O_2SSi$	$CH_3(C_2H_5O)_2SiCH_2CH_2SC_6H_5$	155—156 (4)	—	—	—	339, 569
$C_{13}H_{26}OSSi$	$(CH_3)_3SiC{\equiv}CCH(CH_3)SCH(CH_3)OC_4H_9$	62 (5)	—	1.4418	0.8431	55
	$(CH_3)_3SiC{\equiv}CCH(CH_3)SCH_2CH_2OC_4H_9$	107—108 (2)	—	1.4665	0.8798	55
	$(CH_3)_2(C_2H_5)SiC{\equiv}CCH_2SCH_2CH_2OC_4H_9$	80 (1)	—	1.4777	0.9022	57
	$(CH_3)_2(C_2H_5)SiC{\equiv}CCH_2SCH(CH_3)OC_4H_9$	52—53 (1)	—	1.4580	0.8775	57
$C_{13}H_{27}NSSi$	$CH_3(C_4H_9)_2SiCH_2SCH_2CH_2CN$	147 (2)	—	1.4795	0.9172	81
$C_{13}H_{27}NO_3SSi$	$(C_2H_5O)_3SiCH_2CH(CH_3)CH_2SCH_2CH_2CN$	148—155 (0.2)	—	—	—	418
$C_{13}H_{28}SSi$	$(C_3H_7)_3Si(CH_2)_2SCH{=}CH_2$	117 (2.5)	—	1.4868	0.8781	123
		107—108.5 (0.5)	—	1.4858	—	56

$C_{13}H_{28}OSSi$	$(C_2H_5O)_3Si(CH_2)_3SCH_2COOC_2H_5$	200—201.5 (50)	—	1.4479	1.0301	225, 227
$C_{14}H_{17}NSSi$	m-$(CH_3)_3SiC_6H_4S$— 	—	88—89	—	—	598
$C_{14}H_{18}N_2O_2SSi$	p-$(CH_3)_3SiC_6H_4CH_2$— 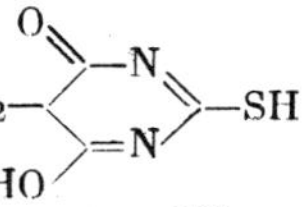	—	264	—	—	323
$C_{14}H_{18}N_2SSi$	p-$(CH_3)_3SiC_6H_4S$— 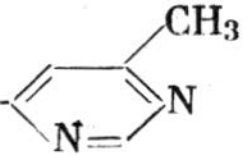	—	89—90	—	—	598
$C_{14}H_{19}NO_3SSi$	p-$(CH_3)_3SiC_6H_4CH_2SCH_2CHCOOCONH$ (ring)	—	89	—	—	348
	o-$(CH_3)_3SiC_6H_4CH_2SCH_2CHCOOCONH$ (ring)	—	88	—	—	348
$C_{14}H_{20}F_2SSi$	$(C_2H_5)_3SiCF{=}CFSC_6H_5$	101.5 (0.3)	—	1.5217^{25}	—	179
$C_{14}H_{21}NO_3SSi$	p-$(CH_3)_3SiC_6H_4CH_2SCH_2CH(NHCHO)COOH$	—	120	—	—	348
	o-$(CH_3)_3SiC_6H_4CH_2SCH_2CH(NHCHO)COOH$	—	117	—	—	348
$C_{14}H_{21}NSSi$	$(CH_3)_2(C_6H_5)Si(CH_2)_3SCH_2CH_2CN$	208 (9)	—	1.5360	1.0164	83
$C_{14}H_{23}O_2SSi$	p-$(CH_3)_3SiC_6H_4CH_2SCH_2CH_2CH(NH_2)COOH$	—	188	—	—	348
	o-$(CH_3)_3SiC_6H_4CH_2SCH_2CH_2CH(NH_2)COOH$	—	184	—	—	348
$C_{14}H_{24}O_2SSi$	$CH_3(C_2H_5O)_2SiCH_2CH_2SCH_2C_6H_5$	114 (0.2)	—	—	—	542
$C_{14}H_{24}O_3SSi$	$(C_2H_5O)_3SiCH_2CH_2SC_6H_5$	130 (0.8)	—	—	—	542
		137—140 (1.5)	—	1.4946^{25}	—	397, 542
$C_{14}H_{26}SSi_2$	$[(CH_3)_3SiC{\equiv}CCH(CH_3)]_2S$	115—116 (8)	—	1.4945	0.8924	55
$C_{14}H_{28}OSSi$	$(CH_3)_2(C_2H_5)SiC{\equiv}CCH(CH_3)SCH(CH_3)OC_4H_9$	60 (1)	—	1.4529	0.8650	55
	$(CH_3)_2(C_2H_5)SiC{\equiv}CCH(CH_3)SCH_2CH_2OC_4H_9$	121 (2.5)	—	1.4700	0.8839	55
$C_{14}H_{28}O_2S_2Si$	$(CH_3)_2Si[CH_2CH(CH_3)CH_2SCOCH_3]_2$	99—100 (9)	—	1.5480^{25}	1.0819^{25}_{4}	179
$C_{14}H_{28}O_4SSi_4$	$O[(CH_3)_2SiO]_3Si(CH_3)CH_2SC_6H_5$ (ring)	178 (12)	—	1.4807	1.0664	251

*See footnote on p. 259.

TABLE 23 (Continued)

Empirical formula	Compound	B.p., °C (mm)	M.p., °C	n_D^{20}	d_4^{20}	Literature
$C_{14}H_{30}O_3S_2Si_2$	$[CH_3COS(CH_2)_3Si(CH_3)_2]_2O$	125—127 (0·15)	—	1·4816	—	480
$C_{14}H_{34}SSi$	$[(C_2H_5)_3SiCH_2]_2S$	111 (0·8)	—	$1{\cdot}4771^{25}$	$0{\cdot}8802_4^{25}$	271
$C_{14}H_{34}S_2Si_2$	$[(C_2H_5)_3SiCH_2S]_2$	122 (0·02)	—	$1{\cdot}5020^{25}$	$0{\cdot}9334_4^{25}$	271
$C_{14}H_{34}OS_2Si_2$	$[C_4H_9SCH_2Si(CH_3)_2]_2O$	130—150 (3)	—	1·4718	—	647
		140 (1)	—	—	—	649
$C_{15}H_{16}N_2O_4SSi$	m-$(CH_3)_3SiC_6H_4S$— 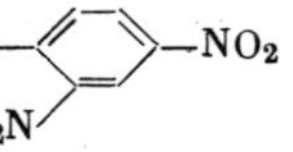	—	92—93	—	—	598
	p-$(CH_3)_3SiC_6H_4S$— 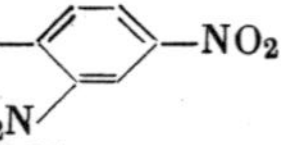	—	160·5—161·5	—	—	598
$C_{15}H_{18}SSi$	o-$(CH_3)_3SiC_6H_4SC_6H_5$	132—136 (2)	—	1·5908	—	179
	p-$(CH_3)_3SiC_6H_4SC_6H_5$	167—168 (3)	—	1·5903	—	179
$C_{15}H_{20}N_2OSSi$	p-$(CH_3)_3SiC_6H_4CH_2$— 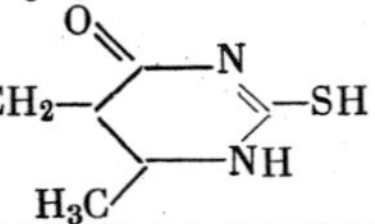	—	224	—	—	323
$C_{15}H_{21}NO_3SSi$	o-$(CH_3)_3SiC_6H_4CH_2SCH_2CH_2CHCOOCONH$ (ring closed between CH and NH)	—	83	—	—	348
$C_{15}H_{23}NO_3SSi$	p-$(CH_3)_3SiC_6H_4CH_2SCH_2CH_2CH(NHCHO)COOH$	—	113	—	—	348
	o-$(CH_3)_3SiC_6H_4CH_2SCH_2CH_2CH(NHCHO)COOH$	—	111—112	—	—	348
$C_{15}H_{25}NOSSi$	$(CH_3)_3Si(CH_2)_3SCH_2CONHC_6H_4CH_3$-p	—	72·5—73	—	—	225
$C_{15}H_{26}O_3SSi$	$(C_2H_5O)_3SiCH_2CH_2SCH_2C_6H_5$	125—126 (0·3—0·4)	—	—	—	542
$C_{15}H_{34}O_2SSi$	$CH_3(C_2H_5O)_2SiCH_2CH_2SCH_2CH(C_2H_5)C_4H_9$	133—135 (2)	—	—	—	339
$C_{15}H_{37}ISSi_2$	$\{[(C_2H_5)_3SiCH_2]_2SCH_3\}I$	—	145	—	—	271
$C_{16}H_{21}N_3O_2SSi$	p-$(CH_3)_3SiC_6H_4CH_2$— 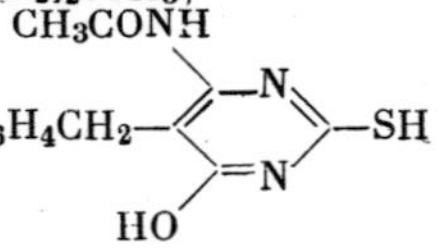	—	—	272	—	323

$C_{16}H_{28}O_2SSi$	$CH_3(C_2H_5O)_2Si(CH_2)_3SCH_2CH_2C_6H_5$	140—145 (0.2)	—	—	—	418
$C_{16}H_{30}SSi_2$	$[(CH_3)_2(C_2H_5)SiC{\equiv}CCH(CH_3)]_2S$	113 (2)	—	1.5085	0.9187	55
$C_{16}H_{30}O_2SSi_2$	$[(CH_3)_2(C_2H_5)SiC{\equiv}CCH(CH_3)]_2SO_2$	—	79	—	—	55
$C_{16}H_{38}O_6SSi_2$	$[(C_2H_5O)_3SiCH_2CH_2]_2S$	156—157 (1)	—	—	—	302
$C_{17}H_{18}N_2SSi$	m-$(CH_3)_3SiC_6H_4S$— 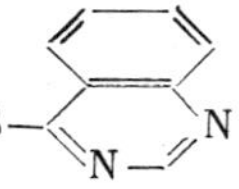 (picrate)	—	181—182	—	—	598
$C_{17}H_{23}S_2Si$	$(CH_3)_3SiCH_2CH(SC_6H_5)_2$	165—166 (1)	—	1.5948	1.0657	59
$C_{17}H_{25}BrO_3SSi$	$(CH_3)_3Si(CH_2)_4SCH_2COOCH_2COC_6H_4Br$ †	—	94—94.5	—	—	550
	$(CH_3)_3Si(CH_2)_2CH(CH_3)SCH_2COOCH_2COC_6H_4Br$†	—	71—71.5	—	—	551
$C_{17}H_{26}O_2S_2Si$	$CH_3(C_6H_5)Si(CH_2CH_2CH_2SCOCH_3)_2$	—	—	1.5480^{25}	1.0819^{25}_{4}	338
$C_{18}H_{16}Cl_{10}OS_2Si_2$	$[C_6Cl_5SCH_2Si(CH_3)_2]_2O$	—	111.5	—	—	488, 493
$C_{18}H_{19}NSSi$	m-$(CH_3)_3SiC_6H_4$—S— 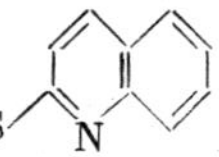 (picrate)	—	152—153	—	—	598
	p-$(CH_3)_3SiC_6H_4$—S— 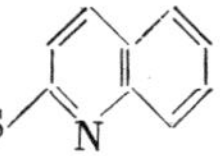 (picrate)	—	162—163	—	—	598
$C_{18}H_{19}NOSSi$	m-$(CH_3)_3SiC_6H_4$—S— 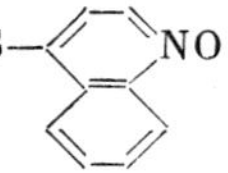	—	101—103	—	—	598
	p-$(CH_3)_3SiC_6H_4$—S— 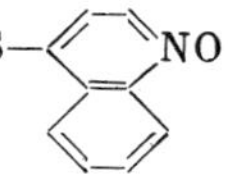	—	126—127	—	—	598
$C_{18}H_{22}O_2SSi$	$CH_3(C_6H_5)_2SiCH_2SCH_2COOC_2H_5$	131 (1)	—	1.5660	1.1172	27
$C_{18}H_{24}N_4O_4SSi$	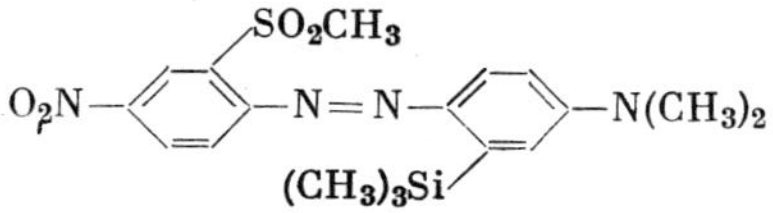	—	243—244	—	—	676

† See second footnote, p. 263

TABLE 23 (Continued)

Empirical formula	Compound	B.p., °C (mm)	M.p., °C	n_D^{20}	d_4^{20}	Literature
$C_{18}H_{24}S_2Si$	$(CH_3)_2(C_2H_5)SiCH_2CH(SC_6H_5)_2$	172—173 (1)	—	1.5965	1.0643	59
$C_{18}H_{26}S_2Si_2$	$[p\text{-}(CH_3)_3SiC_6H_4S]_2$	—	62—63	—	—	598
$C_{18}H_{26}OS_2Si_2$	$[C_6H_5SCH_2Si(CH_3)_2]_2O$	—	—	1.5615	—	647
		—	—	1.5640	—	649
$C_{18}H_{27}N_3O_4S_2Si$	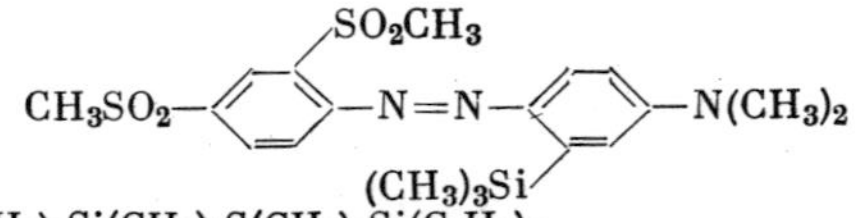	—	217—218	—	—	676
$C_{18}H_{42}S_2Si_2$	$(C_2H_5)_3Si(CH_2)_2S(CH_2)_4Si(C_2H_5)_3$	176 (1)	—	1.5030	0.9254	56
$C_{19}H_{42}O_2SSi$	$CH_3(C_2H_5O)_2SiCH_2CH_2SC_{12}H_{25}$	169—179 (0.5)	—	—	—	339
$C_{20}H_{20}O_2SSi$	$(C_6H_5)_3SiCH_2SO_2CH_3$	—	177—178.5	—	—	494
$C_{20}H_{22}N_2O_2SSi$	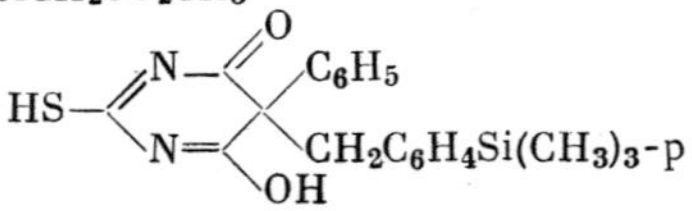	—	232	—	—	323
$C_{20}H_{26}O_5S_2Si_2$	$[o\text{-}C_6H_4(COOH)SCH_2Si(CH_3)_2]_2O$	—	140	—	—	714
$C_{20}H_{28}N_4O_6SSi$	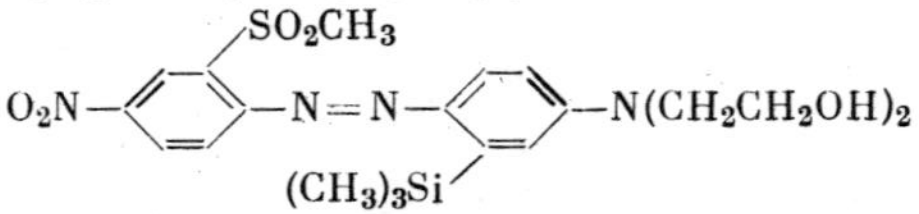	—	195—197	—	—	677
$C_{21}H_{27}NO_4SSi$	$p\text{-}(CH_3)_3SiC_6H_4CH_2SCH_2CH(COOH)NHCOOCH_2C_6H_5$	—	77	—	—	348
	$o\text{-}(CH_3)_3SiC_6H_4CH_2SCH_2CH(COOH)NHCOOCH_2C_6H_5$	—	73	—	—	348
$C_{21}H_{30}N_4O_8S_2Si$	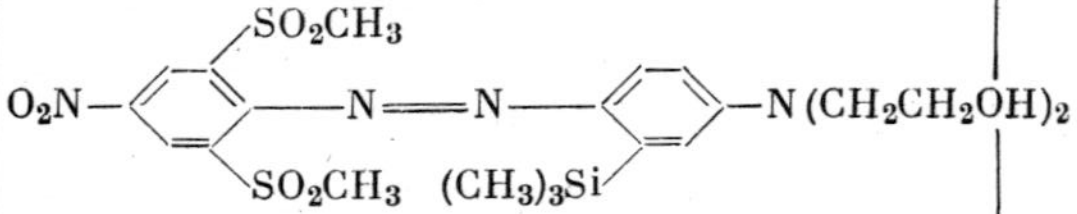	—	164—165	—	—	677

$C_{22}H_{29}NO_4SSi$	o-$(CH_3)_3SiC_6H_4CH_2S(CH_2)_2CH(COOH)NHCOOCH_2C_6H_5$	—	82	—	—	348
$C_{24}H_{54}S_2Si_2$	$(C_3H_7)_3Si(CH_2)_2S(CH_2)_4Si(C_3H_7)_3$	204 (1)	—	1.4952	0.9067	56
$C_{26}H_{24}SSi$	$(C_6H_5)_3SiCH_2SC_6H_4CH_3$-p	—	148—149	—	—	351
	$(C_6H_5)_3SiCH_2CH_2SC_6H_5$	—	99—100	—	—	351
$C_{26}H_{24}O_2SSi$	$(C_6H_5)_3SiCH_2SO_2C_6H_4CH_3$-p	—	175	—	—	351
	$(C_6H_5)_3SiCH_2CH_2SO_2C_6H_5$	—	155	—	—	351
$C_{27}H_{26}SSi$	$(C_6H_5)_3SiCH_2CH_2SCH_2C_6H_5$	—	72—73	—	—	351
	$(C_6H_5)_3SiCH_2CH_2SC_6H_4CH_3$-p	—	97—99	—	—	351
$C_{27}H_{26}O_2SSi$	$(C_6H_5)_3SiCH_2CH_2SO_2CH_2C_6H_5$	—	153—154	—	—	351
	$(C_6H_5)_3SiCH_2CH_2SO_2C_6H_4CH_3$-p	—	152	—	—	351
$C_{28}H_{28}S_4Si$	$(C_6H_5SCH_2)_4Si$	—	67—69	—	—	351
$C_{28}H_{62}O_8SSi_2$	$[(sec\text{-}C_4H_9O)_3SiOCH_2CH_2]_2S$	194 (1)	—	1.4318	0.9636	138 139
$C_{30}H_{42}O_3S_3Si_3$	$[OSi(CH_3)CH_2CH_2SCH_2C_6H_5]_3$	180—200 (10^{-5})	—	—	—	519
$C_{32}H_{36}S_4Si$	(p-$CH_3C_6H_4SCH_2)_4Si$	—	70	—	—	351
$C_{33}H_{30}N_4O_4SSi$	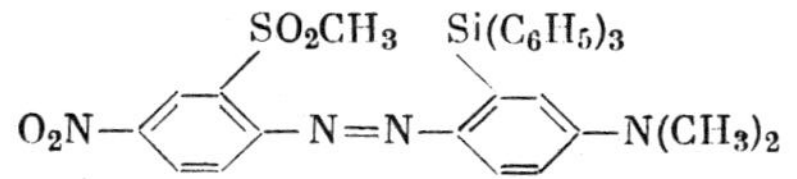	—	263	—	—	676
$C_{34}H_{33}N_3O_4S_2Si$	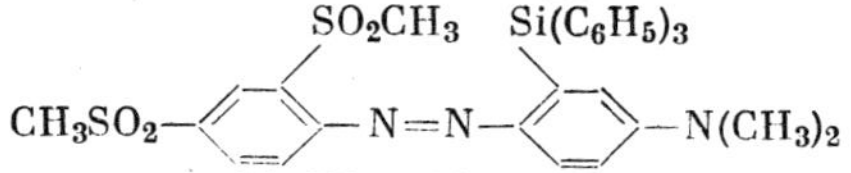	—	235—236	—	—	676
$C_{35}H_{34}N_4O_6SSi$	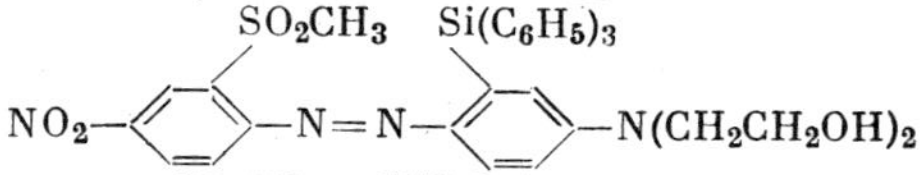	—	209—210	—	—	677
$C_{36}H_{37}N_3O_6S_2Si$	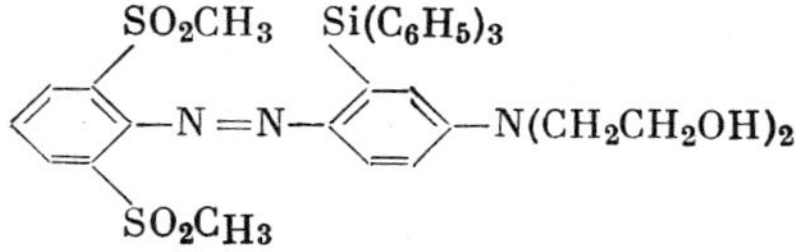	—	211—212	—	—	677

in accordance with the scheme [304, 645]

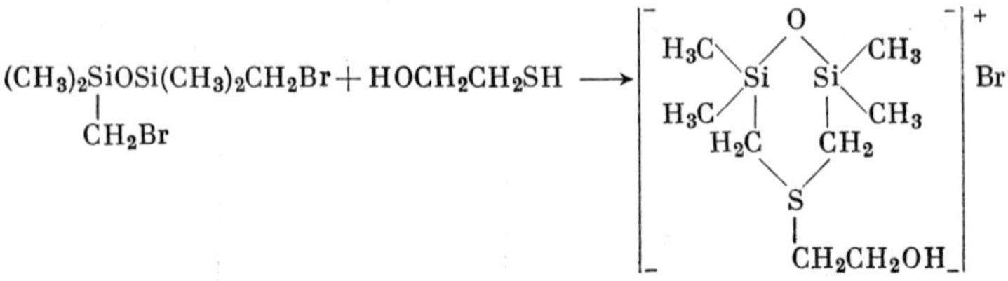

$$(CH_3)_2SiOSi(CH_3)_2CH_2Br + HOCH_2CH_2SH \longrightarrow \begin{bmatrix} H_3C & O & CH_3 \\ & \diagdown Si \diagup \diagdown Si \diagup & \\ H_3C & | & | & \diagdown CH_3 \\ & H_2C & CH_2 & \\ & \diagdown S \diagup & \\ & | & \\ & CH_2CH_2OH & \end{bmatrix}^+ Br^- \quad (2.268)$$

Dimethyl(trimethylsilylmethyl)sulfonium halides are hydrolytically unstable. When treated with 5% NaOH solution they are hydrolyzed rapidly with cleavage of the $Si-C$ bond [251]:

$$[(CH_3)_3SiCH_2S(CH_3)_2]^+ \xrightarrow[H_2O]{OH^-} (CH_3)_3SiOSi(CH_3)_3 + [(CH_3)_3S]^+ \quad (2.269)$$

Water also cleaves these sulfonium salts by an analogous scheme. In this case the reaction proceeds much more slowly, but, nonetheless, if it is allowed to continue for a week (at room temperature) the yield of hexamethyldisiloxane is 80% [251].

Organosilicon sulfides are oxidized by hydrogen peroxide in acetic acid [55, 123, 251, 351, 526, 671] or perphthalic acid in ether [251, 618] to sulfoxides and sulfones:

$$\diagup Si - \left(\begin{matrix} | \\ C \\ | \end{matrix}\right)_n -S-\overset{|}{\underset{|}{C}}- \xrightarrow{H_2O_2} \diagup Si - \left(\begin{matrix} | \\ C \\ | \end{matrix}\right)_n -SO_2-\overset{|}{\underset{|}{C}}- \quad (2.270)$$

The latter have also been obtained by the reaction of lithium [351, 537] or magnesium [351] derivatives of sulfones with chlorosilanes and alkoxysilanes. The preparation of organosilicon sulfones by the reaction of sulfur dioxide with vinylsiloxanes in the presence of silver nitrate has been described [382].

Triorgano(alkanesulfonylmethyl)silanes are readily cleaved by alkalis at the $Si-C$ bond [251]:

$$\diagup SiCH_2SO_2R \xrightarrow[H_2O]{OH^-} \diagup SiOSi\diagdown + RSO_2CH_3 \quad (2.271)$$

In the case of a trimethylsilyl derivative the quantitative formation of hexamethyldisiloxane under the action of a 5% solution of NaOH occurs in one minute. However, the $Si-C$ bond in these sulfones is more resistant to cleavage by acids than in the corresponding organosilicon ketones, nitriles, and esters. For example, trimethyl(methanesulfonylmethyl)silane is not cleaved by the action of 5% hydrochloric acid in 20 h. Sulfones containing the heptamethylcyclotetrasiloxane group instead of the trimethylsilyl group are cleaved completely under these conditions. Boiling them with water leads to the same results.

Sulfuric acid (96%) at 90°C cleaves one methyl group from trimethyl(methanesulfonylmethyl)silane in 30 min. However, after treatment of the reaction mixture with water it is not possible to isolate 1,1,3,3-tetramethyl-1,3-bis(methylsulfonylmethyl)disiloxane, but only dimethyl sulfone (52% yield) [251].

Organosilicon vinyl sulfides undergo all reactions which are characteristic of vinyl sulfides. They are hydrolyzed with the formation of organosilicon thiols, add thiols at the double bond, are polymerized by azobisisobutyronitrile, add hydrogen chloride, and are oxidized to sulfoxides and sulfones by hydrogen peroxide [123, 124]. Their reactivity in radical polymerization and the addition of thiols increases in the series

$$R_3SiCH_2SCH{=}CH_2 < R_3Si(CH_2)_2SCH{=}CH_2 < R_3Si(CH_2)_3SCH{=}CH_2$$

In heterolytic addition reactions (hydration, hydrochlorination, and addition of thiols) there is the reverse relation. In oxidation the reactivity of organosilicon vinyl sulfides falls with a change from a γ-silicon sulfide to a β-silicon sulfide, while decomposition is observed in the case of α-sulfides [123].

Investigation of the addition of organosilicon thiols to organosilicon vinyl sulfides led to the discovery of the isomerization of organosilicon thioacetals (mercaptals) to symmetrical ethers of 1,2-ethanedithiol, which proceeds both in the presence of peroxides and without them, even during distillation in vacuum [123]:

$$CH_3CH[S(CH_2)_nSiR_3]_2 \longrightarrow [R_3Si(CH_2)_nSCH_2]_2 \qquad (2.272)$$

7.3. Organosilicon Derivatives of Sulfur-Containing Heterocyclic Compounds

7.3.1. Preparation Methods

A general method of synthesizing organosilicon derivatives of sulfur-containing heterocyclic compounds is the reaction of lithium derivatives of these heterocycles with organochlorosilanes:

$$R_SLi + Cl{-}Si\lessgtr \longrightarrow R_S{-}Si\lessgtr + LiCl \qquad (2.273)$$

(R_S = sulfur-containing heterocyclic radical)

In this way it was possible to obtain good yields (60-80%) of 2-thienylsilanes [61, 183, 268, 350, 356, 879, 968], 2-benzothienylsilanes [289, 486, 670], and 1-, 2-, 3-, and 4-dibenzothienylsilanes [289, 357, 361, 392, 486, 670]. It was not possible to synthesize N-triphenylsilylphenothiazine by the reaction of N-phenothiazinyllithium with triphenylchlorosilane in diethyl ether (18 h at room temperature) [352], while the yield of 1-thianthrenyltriphenylsilane in the reaction of 1-thianthrenyllithium with triphenylchlorosilane is only 9.5% [360]. Triphenyl(2-benzothiazolyl)silane is also formed in low yield (5%) by the action of 2-benzothiazolyllithium on triphenylchlorosilane. When triphenylbromosilane is used instead of triphenylchlorosilane, the yield is increased to 19%, but when triphenylsilane is used, no interaction of the reagents is observed [486]. Organoalkoxysilanes have also been used instead of organohalosilanes in reaction (2.273) [357].

A second general method of synthesizing organosilicon sulfur-containing heterocyclic compounds is the reaction of organomagnesium derivatives of sulfur-containing heterocycles with halosilanes or alkoxysilanes:

$$R_SMgX + Y{-}Si\lessgtr \longrightarrow R_S{-}Si\lessgtr + MgXY \qquad (2.274)$$

This reaction is used to obtain 2- and 3-thienylsilanes [51, 120, 185, 218, 219, 268, 342, 423, 599] and 3-benzothienylsilanes

[486, 670]. By this method it is possible to synthesize from 2,5-dibromothiophene both monosilicon-substituted thiophenes [51, 185] and 2,5-disilicon-substituted thiophenes [91, 120].

The reaction of halogen-substituted sulfur-containing heterocycles with triphenylsilylpotassium or triphenylsilyllithium proceeds according to the scheme

$$R_SX + MSi(C_6H_5)_3 \longrightarrow R_SSi(C_6H_5)_3 + MX \qquad (2.275)$$

This reaction has been used for the synthesis of triphenyl-(2-dibenzothienyl)silane [486], and also 2-, 3-, and 4-triphenylsilyl substituted 10-ethylphenothiazines and the sulfone of 3,7-bis(triphenylsilyl)-10-ethylphenothiazine [352]. The yields of triphenylsilyl derivatives of 10-ethylphenothiazines vary over the range of 12-40% (45-60% of hexaphenyldisilane is formed in addition to them) [352].

A method has been developed for preparing thienylsilanes which is based on the thermal condensation of hydrosilanes with thiophene and its halogen derivatives [110, 117, 118, 505, 810]:

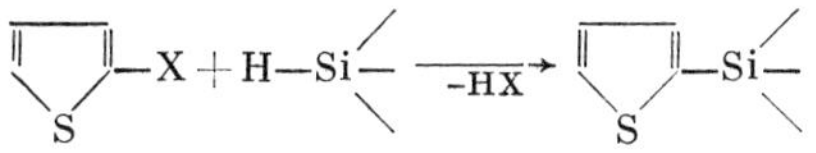

(X = H or halogen) (2.276)

When a mixture of thiophene and trichlorosilane vapors is passed through a tube heated to 600°C, there is obtained 2-thienyltrichlorosilane in 20% yield, as well as a small amount of the 3-isomer. The latter is detected spectroscopically [110, 117]. The reaction of trichlorosilane with 2-methylthiophene proceeds simultaneously in two directions:

a) with replacement of the hydrogen atom in position 5 of the thiophene ring (5-methyl-2-thienyltrichlorosilane is thus formed in 8% yield);

b) with replacement of the methyl group by a trichlorosilyl group (the yield of 2-thienyltrichlorosilane is 13%):

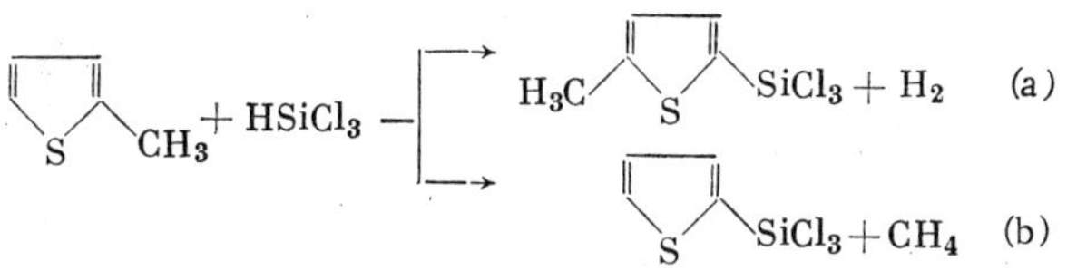

(2.277)

Hexachlorodisilane may also be used instead of trichlorosilane for the thermal condensation reaction with thiophene. In this case the process is carried out in an autoclave at 400°C [694].

When a mixture of 2-chlorothiophene and trichlorosilane is passed through a quartz tube heated to 570°C, three processes occur [110, 117]:

a) condensation with the formation of thienyltrichlorosilane (57% yield) and hydrogen chloride;

b) further condensation of thienyltrichlorosilane with $HSiCl_3$ with the liberation of hydrogen and the formation of 2,5-bis(trichlorosilyl)thiophene (6% yield);

c) reduction of 2-chlorothiophene to thiophene (2%):

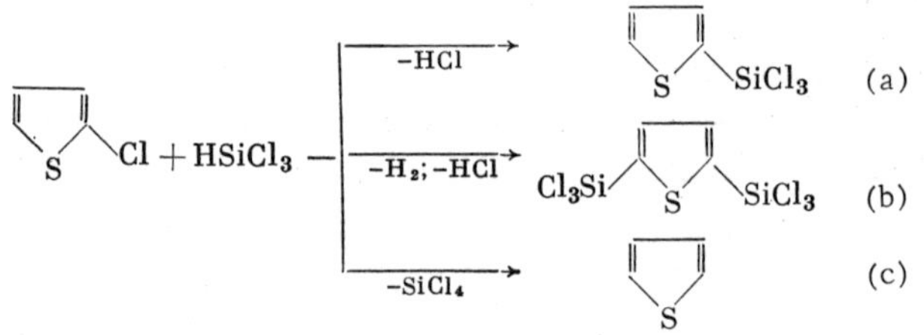

(2.278)

The condensation of trichlorosilane with 2,5-dichlorothiophene proceeds in stages: in addition to the final reaction product, namely, 2,5-bis(trichlorosilyl)thiophene, it is possible to isolate 5-chloro-2-thienylchlorosilane, which is formed as an intermediate [110]. The reaction products of trichlorosilane and tetrachlorothiophene include 2,3,5-tris(trichlorosilyl)thiophene [110]:

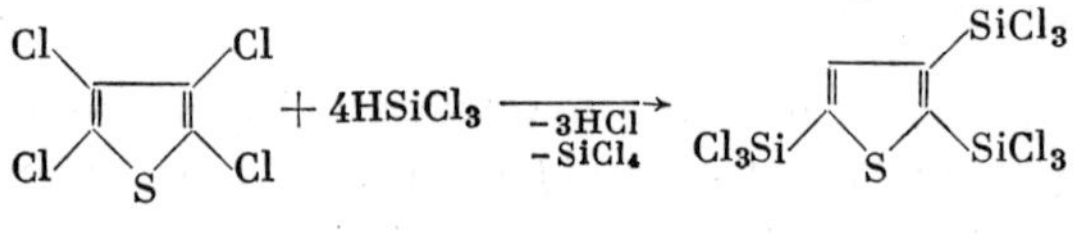

(2.279)

Dibenzothiophene does not condense with diphenylsilane even when a mixture of them is heated to boiling for six days [720].

In addition to organosilicon derivatives of sulfur-containing heterocycles in which the silicon atom is attached directly to the heterocyclic radical, compounds of this type are known with the silicon atom in the side chain which contain yet another heteroatom (for example, organosilicon derivatives of corresponding acids and amines).

The reaction 2-thiophenecarboxylic acid with $SiCl_4$ forms 2-thenoyloxytrichlorosilane [14, 128], which is converted by reaction with acetic acid into 2-thenoyloxytriacetoxysilane [127]:

$$\underset{S}{\overset{\|}{\bigcirc}}\text{—COOH} + SiCl_4 \xrightarrow[-HCl]{} \underset{S}{\overset{\|}{\bigcirc}}\text{—COOSiCl}_3 \xrightarrow[-3HCl]{+3CH_3COOH} \underset{S}{\overset{\|}{\bigcirc}}\text{—COOSi(OCOCH}_3)_3$$

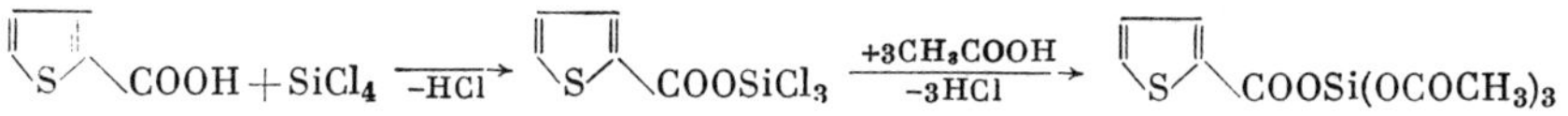

$$(2.280)$$

Silylalkylaminoalkyl derivatives of sulfur-containing hetero-cycles are obtained by condensation of (aminoalkyl)silanes with the corresponding heterocyclic aldehydes and ketones, with subsequent catalytic reduction of the aldimines and ketimines formed to amines:

$$R_S\text{—C}{=}O + H_2N\text{—}\left(\overset{|}{\underset{|}{C}}\right)_n \text{—Si}{\overset{/}{\underset{\backslash}{}}} \xrightarrow[-H_2O]{} R_S\text{—C}{=}N\text{—}\left(\overset{|}{\underset{|}{C}}\right)_n \text{—Si}{\overset{/}{\underset{\backslash}{}}} \xrightarrow{2H}$$

$$\longrightarrow R_S\text{—CH—NH—}\left(\overset{|}{\underset{|}{C}}\right)_n \text{—Si}{\overset{/}{\underset{\backslash}{}}} \qquad (2.281)$$

In this way it was possible to obtain nitrogen-containing organo-silicon derivatives of thiophene [458, 459], benzothiophene [458, 459], thiopyran [458], and thiazole [458].

Organosilicon derivatives of 2-aminothiazole [447] and 6-aminopenicillanic acid [364, 365, 651] containing the Si—N bond may be obtained by the reaction of the latter with triorgano(amino)-silanes or hexaalkyldisilazans:

$$R_S NH_2 + H_2N\text{—Si}{\overset{/}{\underset{\backslash}{}}} \longrightarrow R_S NHSi{\overset{/}{\underset{\backslash}{}}} + NH_3 \qquad (2.282)$$

$$2R_S NH_2 + \overset{\backslash}{\underset{/}{}}Si\text{—NH—Si}{\overset{/}{\underset{\backslash}{}}} \longrightarrow 2R_S NH\text{—Si}{\overset{/}{\underset{\backslash}{}}} + NH_3 \qquad (2.283)$$

Organosilicon esters of 6-aminopenicillanic acid are obtained by three methods: by heating this acid with triorgano(dialkyl-amino)silanes, by the reaction of salts of 6-aminopenicillanic acid with triorganochlorosilanes, and, finally, by silylation with hexa-methyldisilazan [651, 880]. The reaction of 6-aminopenicillanic acid with trimethyl(diethylamino)silane at 80-150°C with simultaneous distillation of the diethylamine formed in the reaction gives a 98.5% yield of the trimethylsilyl ester of 6-trimethylsilylaminopenicillanic acid:

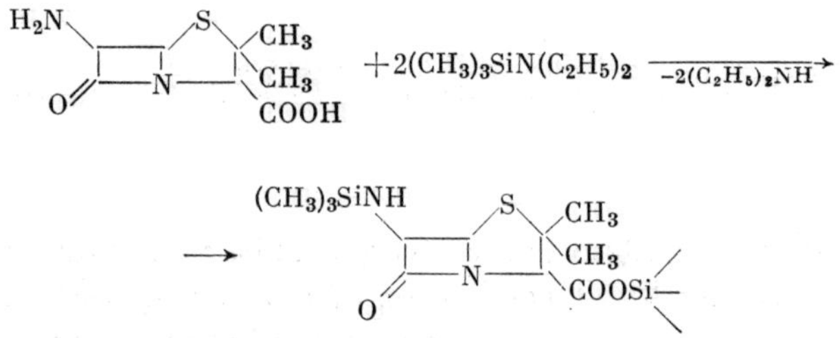

(2.284)

The same compound is formed in 29% yield by heating a suspension of the potassium salt of 6-aminopenicillanic acid in benzene to 60°C for 3 h with trimethylchlorosilane in the presence of triethylamine. When the sodium salt is used instead of the potassium salt the yield of the reaction product is reduced to 25%. Heating of 6-aminopencillanic acid with hexamethyldisilazan in a nitrogen atmosphere at 120°C for 3 h gives the trimethylsilyl ester of 6-trimethylsilylaminopenicillanic acid in 64.5% yield [651]. Treating the trimethylsilyl esters obtained with anhydrides or acid chlorides of carboxylic acids converts them into penicillins which contain no silicon atoms.

Penicillins containing silicon can be obtained by the reaction of 6-aminopenicillanic acid with (trimethylsilylmethylthio)acetic acid in acetone, using N,N'-dicyclohexylcarbodiimide as the condensing agent [27]:

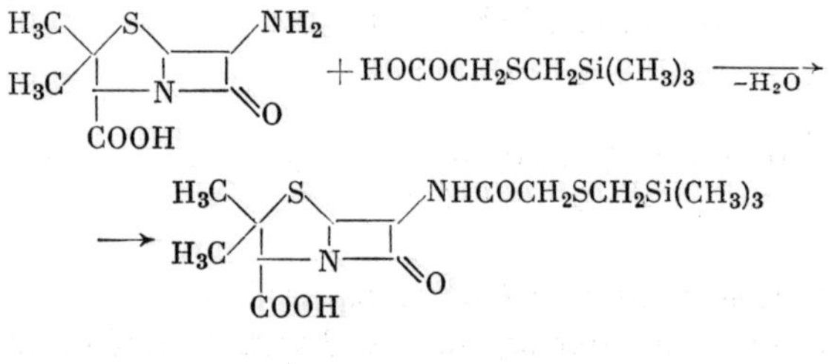

(2.285)

This is as yet the only penicillin containing silicon, and its activity is close to that of benzylpenicillin.*

*The possibility of the silylation of well-known penicillins with hexamethyldisilazan is reported in the patents [194, 669].

7.3.2. Physical Properties

Organosilicon derivatives of sulfur-containing heterocyclic compounds are solids or liquids. Their physical constants are given in Table 24.

The NMR spectra of thienylsilanes are interesting because of the inductive effect and the possibility of $d_\pi - p_\pi$ interaction between the silicon atom and the thiophene ring [755].

7.3.3. Chemical Properties

The reactions of organosilicon derivatives of sulfur-containing heterocyclic compounds may be divided into three groups. The first of these groups includes reactions which involve functional groups at the silicon atom without touching the heterocyclic ring; the second group includes reactions which involve rupture of the bond of the silicon to the heterocycle, and the third group includes reactions which are characteristic of sulfur-containing heterocycles themselves. It should be noted that the most studied reactions are desilylations, while reactions of the first and third group have been investigated mainly only with thienylsilanes.

Thienylsilanes containing functional groups at the silicon atom readily undergo reactions which do not involve the thiophene ring. These include:

a) hydrolysis of thienylchlorosilanes [220, 341] and thienyl-alkoxysilanes [538], leading to the formation of thermally stable polysiloxane fluids and resins:

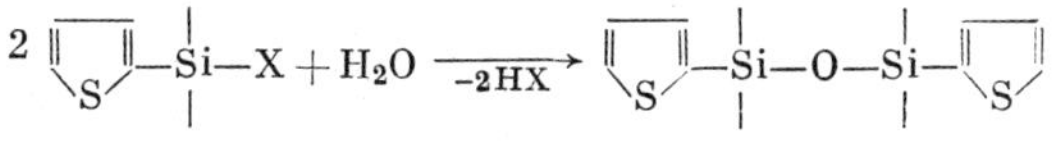

$$2 \; \underset{S}{\boxed{}}\!\!-\!\!\overset{|}{\underset{|}{Si}}\!\!-\!\!X + H_2O \xrightarrow[-2HX]{} \underset{S}{\boxed{}}\!\!-\!\!\overset{|}{\underset{|}{Si}}\!\!-\!\!O\!\!-\!\!\overset{|}{\underset{|}{Si}}\!\!-\!\!\boxed{}_{S} \qquad (2.286)$$

b) the reaction of thienylchlorosilanes with alcohols [342, 343, 538] and thiols [541]:

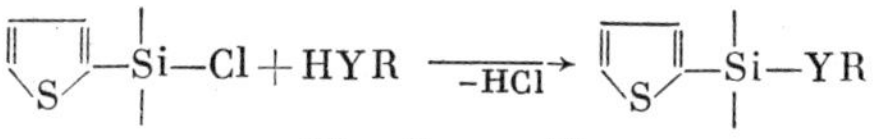

$$\underset{S}{\boxed{}}\!\!-\!\!\overset{|}{\underset{|}{Si}}\!\!-\!\!Cl + HYR \xrightarrow[-HCl]{} \underset{S}{\boxed{}}\!\!-\!\!\overset{|}{\underset{|}{Si}}\!\!-\!\!YR$$

$$(Y = O, \text{ or } S) \qquad (2.287)$$

c) the addition of thienylhydrosilanes to unsaturated compounds in the presence of H_2PtCl_6 [91, 120, 762]:

$$\underset{S}{\boxed{}}\!\!-\!\!\overset{|}{\underset{|}{Si}}\!\!-\!\!H + CH_2\!\!=\!\!CHR \longrightarrow \underset{S}{\boxed{}}\!\!-\!\!\overset{|}{\underset{|}{Si}}\!\!-\!\!CH_2CH_2R \qquad (2.288)$$

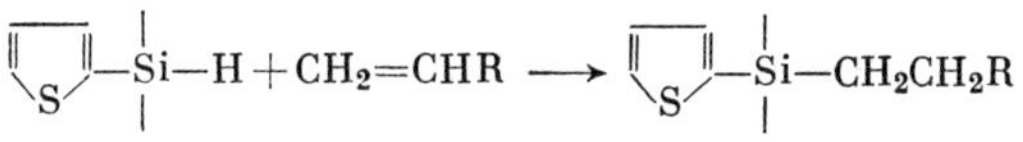

TABLE 24. Organosilicon Derivatives of Sulfur-Containing Heterocycles

R = 2-thienyl 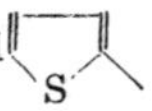

R_b = benzothienyl 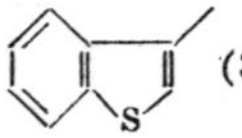 (3- is indicated)

R′ = 3-thienyl 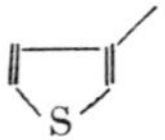

R_d = dibenzothienyl 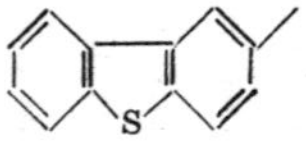 (2- is indicated)

R″ = 2,5-thienylene 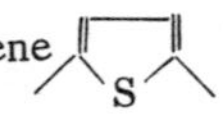

R_f = 10-ethylphenothiazinyl 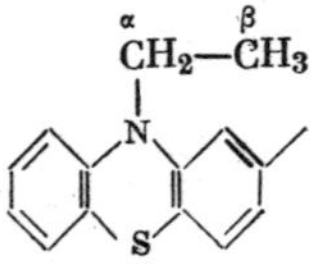 (2- is indicated)

Note: The number in front of the substituent indicates its position in the heterocyclic system, while the symbol O_2 after the radical indicates a sulfone.

Empirical formula	Compound	B.p., °C (mm)	M.p., °C	n_D^{20}	d_4^{20}	Literature
$C_4H_2Cl_4SSi$	$ClR''SiCl_3$	75—76 (4.5)	—	1.5461	1.5071	117
$C_4H_2Cl_6SSi$	$Cl_3SiR''SiCl_3$	112—113 (4)	46	—	—	117
$C_4H_3Cl_3SSi$	$RSiCl_3$	196—197	—45	—	1.45^{25}	343
		73.5—75 (10)	—	—	—	538
		70—71 (11)	—	1.5328	1.4338	117
$C_5H_3Cl_3O_2SSi$	$RCOOSiCl_3$	95—96 (5)	—	—	—	126
$C_5H_5Cl_3SSi$	$ClR''SiCl_2CH_3$	88—90 (6)	—	1.5412	1.3665	117, 118
$C_5H_6Cl_2SSi$	$RSiCl_2CH_3$	199—200	—	—	—	341
		68—69 (11)	—	1.5265	1.2784	117

$C_6H_8Cl_2SSi$	$ClR''SiCl(CH_3)_2$	91—92 (10)	—	1.5331	1.2313	118
$C_6H_8Cl_4SSi$	$CH_3Cl_2SiR''SiCl_2CH_3$	128 (7)	—	1.5322	1.3540	117, 118
C_6H_9ClSSi	$RSiCl(CH_3)_2$	98 (40)	—	—	—	341
$C_7H_{11}BrSSi$	$BrR''Si(CH_3)_3$	218—218.5	—	1.5315	1.298^{20}	185
$C_7H_{12}SSi$	$RSi(CH_3)_3$	165.5	—	1.4960	—	268, 289
		159—160 (748)	—	1.4966	0.945	183
	$R'Si(CH_3)_3$	168	—	1.4993	—	268
$C_7H_{12}O_3SSi$	$RSi(OCH_3)_3$	92 (9)	—	1.4824	—	538
$C_8H_6Cl_2S_2Si$	R_2SiCl_2	305—308	—	—	1.39^{25}	343
$C_8H_{12}OSSi$	$(CH_3)_3SiR''CHO$	—	33—34	—	—	61
$C_8H_{12}OSSi$	$(CH_3)_3SiR''COOH$	—	134—135	—	—	183
$C_8H_{13}NOSSi$	$(CH_3)_3SiR''CH{=}NOH$	—	131—132	—	—	61
$C_8H_{14}Cl_2SSi_2$	$(CH_3)_2ClSiR''SiCl(CH_3)_2$	118—120 (6)	—	1.5312	1.1651	118
$C_8H_{14}SSi$	$CH_3R''Si(CH_3)_3$	101—102 (50)	—	1.4949	0.9274	61
C_9H_9ClSSi	$R_2SiClCH_3$	301—307	—	—	—	342
		138—148 (3)				
$C_9H_{12}O_2SSi$	$(CH_3)_3SiR''COCHO$	89—90 (0.6)	—	—	—	184
$C_9H_{14}OSSi$	$(CH_3)_3SiR''COCH_3$	104—105 (4)	—	1.5289	1.028	183
$C_9H_{16}OSSi$	$(CH_3)_3SiR''CH_2CH_2OH$	144—145 (19)	—	1.5168	1.1977	61
$C_{10}H_8Cl_2SSi$	$RSiCl_2C_6H_5$	117—118 (3)	—	1.5880	1.3011	117
		161—167 (12)	—	—	—	509
$C_{10}H_{12}S_2Si$	$R_2Si(CH_3)_2$	131—133 (8)	—	1.5775	1.1184	61
$C_{10}H_{15}N_3O_2SSi$	$(CH_3)_3SiR''COCH{=}NNHCONH_2$	—	221—224	—	—	184
$C_{10}H_{18}O_3SSi$	$RSi(OC_2H_5)_3$	150 (50)	—	1.4659	1.050	218
$C_{10}H_{20}SSi_2$	$(CH_3)_3SiR''Si(CH_3)_3$	—	31—32	—	—	61
$C_{11}H_{11}BrSSi$	$BrR''SiH(CH_3)C_6H_5$	138—140 (2)	—	1.6050	1.3541	51
$C_{11}H_{12}O_8SSi$	$RCOOSi(OCOCH_3)_3$	—	65—67	—	—	127
$C_{11}H_{14}SSi$	2-$(CH_3)_3SiR_b$	143—144 (17)	—	1.5759	—	670
		123 (7)	—	—	—	289
	3-$(CH_3)_3SiR_b$	120—122 (7)	—	1.5800	—	289
$C_{11}H_{20}OSSi$	$RCH_2OSi(C_2H_5)_3$	90—91 (1.5)	—	1.4886	0.9788	68
$C_{12}H_9ClS_3Si$	R_3SiCl	190—205 (1)	—	—	—	343
$C_{12}H_{10}S_3Si$	R_3SiH	150—155 (0.5)	—	—	—	599
$C_{12}H_{16}S_2Si$	$R_2Si(C_2H_5)_2$	141—142 (6)	—	1.5740	1.1000	61
$C_{12}H_{16}O_2S_2Si$	$R_2Si(OC_2H_5)_2$	209 (50)	—	1.5401	—	218, 343
$C_{14}H_{28}N_2O_3SSi$	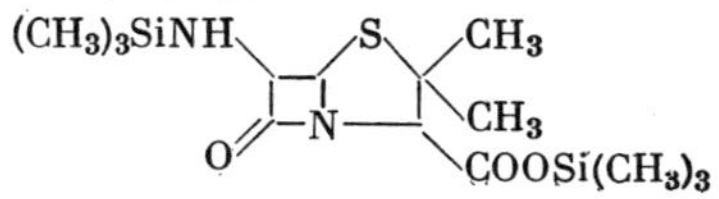	—	67—68	—	—	651

TABLE 24 (Continued)

Empirical formula	Compound	B.p., °C (mm)	M.p., °C	n_D^{20}	d_4^{20}	Literature
$C_{15}H_{15}NO_4SSi$	1-$(CH_3)_3$Si-3-$NO_2R_dO_2$	—	223—224	—	—	357
$C_{15}H_{16}SSi$	4-$(CH_3)_3SiR_d$	215—217 (20)	—	1·6354	1·112	357
		153 (1)	—	1·6342	—	289
	3-$(CH_3)_3SiR_d$	—	103.5—104·5	—	—	392
		—	104—105	—	—	289, 670
	2-$(CH_3)_3SiR_d$	159—160 (1)	48·2—49·2	—	—	392
			49			289
	1-$(CH_3)_3SiR_d$	160—161 (2—3)	—	1·6408	—	289
$C_{15}H_{16}N_2SSi$	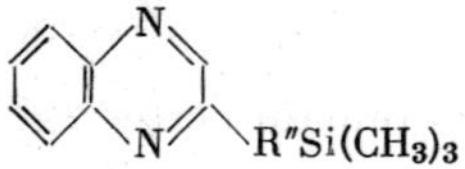	—	99—100	—	—	184
$C_{15}H_{16}O_2SSi$	4-$(CH_3)_3SiR_dO_2$	—	144—147	—	—	357
	3-$(CH_3)_3SiR_dO_2$	—	170·8—171·8	—	—	392
	2-$(CH_3)_3SiR_dO_2$	—	164—165	—	—	392
$C_{15}H_{29}NO_2SSi$	$CH_3R''CH(CH_3)NH(CH_2)_3Si(CH_3)(OC_2H_5)_2$	158—163 (0·5)	—	$1·5060^{25}$	—	344
$C_{16}H_{12}S_4Si$	R_4Si	—	135·5	—	—	423
$C_{16}H_{13}ClSSi$	$RSiCl(C_6H_5)_2$	187—188 (3·5)	54	—	—	117
		162—170 (0·1)	—	—	—	600
$C_{16}H_{21}NO_2SSi$	$(CH_3)_3SiR''CH_2CH_2OCONHC_6H_5$	—	48·5—49·5	—	—	61
$C_{18}H_{20}SSi$	$[CH_3(C_6H_5)SiH]_2R''$	200—205 (2)	—	1·5980	1·085	91
$C_{20}H_{12}O_8S_4Si$	$(RCOO)_4Si$	—	—	—	—	128
$C_{20}H_{22}SSi_2$	$CH_3(C_6H_5)SiHR''Si(CH{=}CH_2)(CH_3)C_6H_5$	225—228 (2)	—	1·5932	1·0670	51
$C_{20}H_{24}SSi_2$	$[C_2H_5(C_6H_5)SiH]_2R''$	228—230 (2)	—	1·5892	1·059	91
$C_{22}H_{18}SSi$	$RSi(C_6H_5)_3$	—	196—198	—	—	350
$C_{23}H_{18}O_2SSi$	$(C_6H_5)_3SiR''COOH$	—	188—190	—	—	350
$C_{25}H_{19}NSSi$	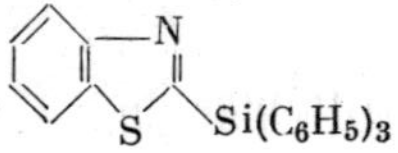	—	141—146	—	—	486

$C_{26}H_{20}SSi$	5-$(C_6H_5)_3SiR_b$	—	148—149	—	—	486
$C_{28}H_{22}O_2SSi$	$R(C_6H_5)_2SiOC_6H_4OC_6H_5$-m	270—272 (1)	—	—	—	509
$C_{28}H_{44}SSi_4$	$[(CH_3)_3SiCH_2CH_2Si(CH_3)C_6H_5]_2R''$	258—260 (2)	—	1·5489	0·9944	91, 120
$C_{30}H_{22}SSi$	4-$(C_6H_5)_3SiR_d$	—	196—198	—	—	486
	2-$(C_6H_5)_3SiR_d$	—	153—154	—	—	486
$C_{30}H_{22}O_2SSi$	4-$(C_6H_5)_3SiR_dO_2$	—	212—213	—	—	357
$C_{30}H_{46}SSi_4$	$[(CH_3)_3Si(CH_2)_3Si(CH_3)C_6H_5]_2R''$	277—280 (2)	—	1·5415	0·9792	91, 120
$C_{31}H_{21}Cl_2NSSi$	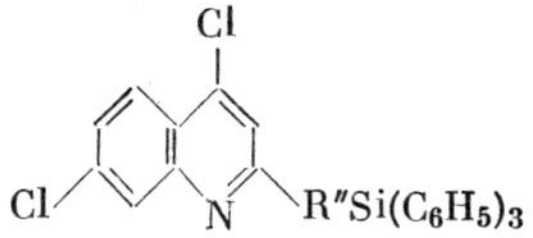	—	200—203	—	—	358
$C_{31}H_{23}NSSi$	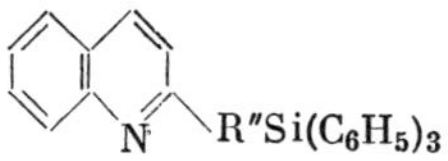	—	168—170	—	—	357
$C_{32}H_{25}NOSi$	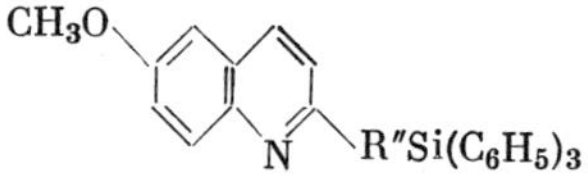	—	227—228	—	—	358
$C_{32}H_{27}NSSi$	2-$(C_6H_5)_3SiR_f$	—	187—189	—	—	352
	3-$(C_6H_5)_3SiR_f$	—	184·5—186	—	—	352
	4-$(C_6H_5)_3SiR_f$	—	166·5—168	—	—	352
	β-$(C_6H_5)_3SiR_f$	—	176—177	—	—	352
$C_{32}H_{52}SSi_4$	$[(CH_3)_3Si(CH_2)_3Si(C_2H_5)C_6H_5]_2R''$	280—285 (2)	—	1·5430	0·9952	91
$C_{34}H_{26}O_4SSi$	$R(C_6H_5)Si(OC_6H_4OC_6H_5\text{-m})_2$	319—323 (1·5)	64—67	—	—	509
$C_{36}H_{60}SSi_4$	$[(C_2H_5)_3SiCH_2CH_2Si(C_2H_5)C_6H_5]_2R''$	300—305 (2)	—	1·5370	1·0080	120
		307—310 (2)	—	—	0·9938	91
$C_{38}H_{64}SSi_4$	$[(C_2H_5)_3Si(CH_2)_3Si(C_2H_5)C_6H_5]_2R''$	325—330 (2)	—	—	—	91
$C_{50}H_{41}NO_2SSi_2$	3,7-$[(C_6H_5)_3Si]_2R_fO_2$	—	271—276·5	—	—	352

d) hydrosilylation of thienylvinylsilanes [120]

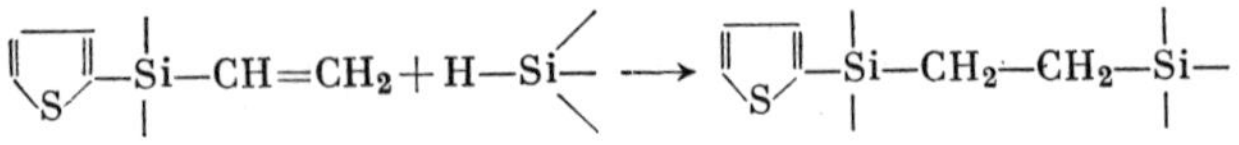

$$\underset{S}{\boxed{}}-Si-CH=CH_2+H-Si\diagup \longrightarrow \underset{S}{\boxed{}}-Si-CH_2-CH_2-Si- \qquad (2.289)$$

e) some conversions of thenoyloxytrichlorosilane [127, 128]

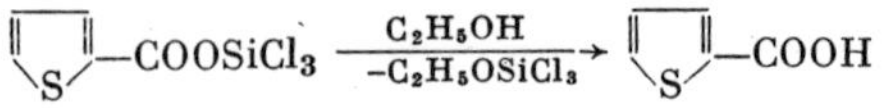

$$\underset{S}{\boxed{}}-COOSiCl_3 \xrightarrow[-C_2H_5OSiCl_3]{C_2H_5OH} \underset{S}{\boxed{}}-COOH \qquad (2.290)$$

$$4\underset{S}{\boxed{}}-COOSiCl_3 \longrightarrow \left(\underset{S}{\boxed{}}-COO\right)_4 Si+3SiCl_4 \qquad (2.291)$$

These reactions include the reaction of 2-benzothienyltri-chlorosilane with the methyl ether of p-chlorothiophenol, which proceeds in the presence of sodium and leads to the formation of tris(p-methylthiophenyl)-2-benzothienylsilane in high yield [632].

The triorganosilyl group is cleaved from the thiophene ring by the action of mineral acids [183, 267, 268, 289, 356], bromine [185], and also aziridine in the presence of a catalytical amount of sodium [969]. All the isomeric trimethyl(thienyl)-, trimethyl-(benzothienyl)-, and trimethyl(dibenzothienyl)silanes may be arranged in the following series with respect to falling reactivity in protódesilylation by perchloric acid in aqueous methanol [268, 289, 670]:

2-thienyl > 3-thienyl > 3-benzothienyl > 2-benzothienyl > 2-dibenzothienyl >
4810   707   40.7   39.6   6.25

1-dibenzothienyl > 3-dibenzothienyl > 4-dibenzothienyl > phenyl
5.53   2.00   1.15   1

Organosilicon derivatives of thiophene undergo many reactions which are characteristic of thiophene compounds, such as metallation, acetylation, bromination, and reduction.

Thienylorganosilanes are metallated by butyllithium with the formation of 5-triorganosilyl-2-thienyllithium [61, 183, 184, 350, 358]:

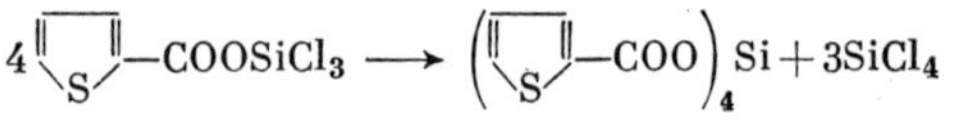

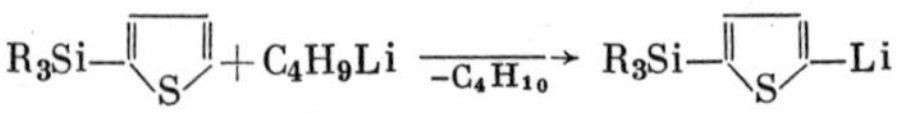

$$R_3Si-\underset{S}{\boxed{}}+C_4H_9Li \xrightarrow[-C_4H_{10}]{} R_3Si-\underset{S}{\boxed{}}-Li \qquad (2.292)$$

Triorgano(thienyl)silanes are acetylated by acetic anhydride like alkylthiophenes [183]. 2-Acetyl-5-trimethylsilylthiophene obtained by this method may be ozidized by SeO_2 to 5-trimethylsilyl-2-thienylglyoxal [184], and by sodium hypoiodite to 5-trimethylsilyl-2-thiophenecarboxylic acid [183]:

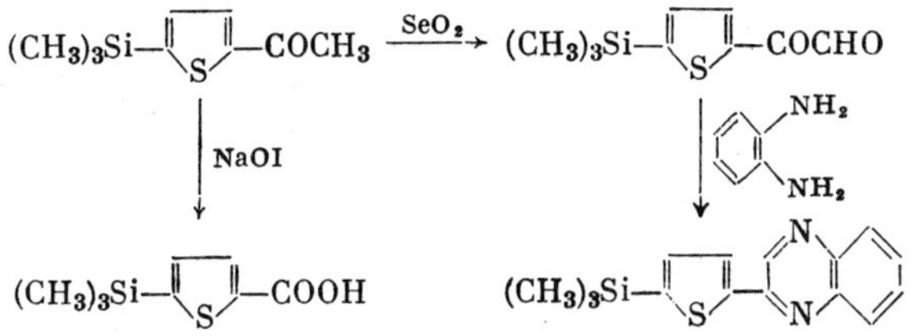

(2.293)

The products of the hydrolysis of 2-thienyltrichlorosilane, octa(2-thienylsilsesquioxane), and dodeca(2-thienylsilsesquioxane) react with bromine in carbon disulfide with the formation of the corresponding tribromothienyl derivatives [538]. When methanol solutions of various thienylsilanes are treated with Raney nickel (W-7) there is reductive elimination of sulfur with the formation of organosilicon compounds of the aliphatic series [61]:

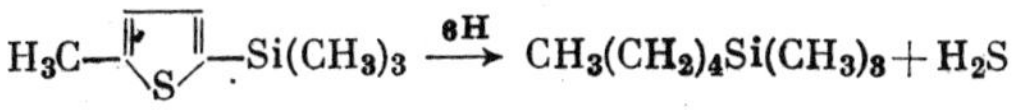

(2.294)

Among the conversions of organosilicon derivatives of other sulfur-containing heterocyclic compounds only the oxidation of dibenzothienyltriorganosilanes to sulfones has been described, and this occurs under the action of hydrogen peroxide or nitric acid in glacial acetic acid [357, 392]:

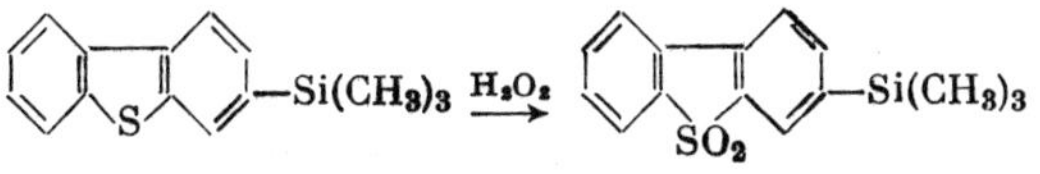

(2.295)

It is interesting to note that these sulfones are resistant to the action of hydrogen chloride in glacial acetic acid, i.e., under conditions under which desilylation of dibenzothienylsilanes themselves occurs [392].

7.4. Compounds Containing the Groupings Si − (C)$_n$ − SCN and Si − (C)$_n$ − NCS (Organosilicon Thiocyanates and Isothiocyanates)

7.4.1. Preparation Methods

The most common method of synthesizing organosilicon thiocyanates is based on the reaction of organo(haloalkyl)silanes with alkali metal or ammonium thiocyanates:

$$\gt Si-\left(\begin{matrix} | \\ C \\ | \end{matrix}\right)_n-X+MSCN \longrightarrow \gt Si-\left(\begin{matrix} | \\ C \\ | \end{matrix}\right)_n-SCN+MX \qquad (2.296)$$

This reaction has been carried out using sodium [55, 57, 78, 81-83, 249], potassium [25, 83, 368, 387, 492, 528], and ammonium [81, 96] thiocyanates. When sodium and potassium thiocyanates are used the yields of organosilicon thiocyanates are higher than with ammonium thiocyanate. Thus, while the yield of trimethylsilylmethyl thiocyanate from the reaction of trimethyl-(chloromethyl)silane with ammonium thiocyanate in acetone over a period of 10 h is 50% [96], when sodium thiocyanate is used the yield reaches 85% [81]. The use of trimethyl(iodomethyl)silane instead of the analogous chloro derivative also increases the yield of trimethylsilylmethyl thiocyanate [81]. Reaction (2.296) is carried out in acetone [55, 96, 368, 387, 492, 528] or in 96% ethanol [25, 55, 57, 81-83]. The use of ethanol as a solvent makes it possible to reduce the reaction volumes considerably, as in the case of acetone these reach 1.6 liter per mole of potassium thiocyanate.

The rate of reaction (2.296) depends on the composition and structure of the chloroalkylsilane $R_3Si(CH_2)_nCl$ used. When methyl radicals at the silicon atom are replaced by ethyl and propyl radicals, the exchange rate falls sharply [96]. The formation of secondary triorganosilylalkyl thiocyanates is even more difficult [55, 83, 96]. Carrying out a competing reaction of equimolecular amounts of trimethyl(chloromethyl)silane and trimethyl(γ-chloropropyl)silane with sodium thiocyanate by heating in ethanol for

84 h shows that it is mainly the α-chloride which reacts, while the γ-chloride is recovered almost completely unchanged [82]. Kinetic investigation established that trimethyl(chloromethyl)silane is almost 20 times as reactive in reaction with potassium thiocyanate in acetone than trimethyl(γ-chloropropyl)silane in butyl chloride [387].

Reaction (2.296) may also be used for introducing two thiocyanate groups into an organosilicon molecule [81, 492, 528]. Thus, for example, when dimethylbis(chloromethyl)silane is heated with sodium thiocyanate in ethanol, dimethylbis(thiocyanatomethyl)silane is formed in 79% yield [81]:

$$(CH_3)_2Si(CH_2Cl)_2 + 2NaSCN \xrightarrow[-2NaCl]{} (CH_3)_2Si(CH_2SCN)_2 \qquad (2.297)$$

A second method of preparing organosilicon thiocyanates is thiocyanation of alkenylsilanes with thiocyanogen [22-24, 93, 100, 119]:

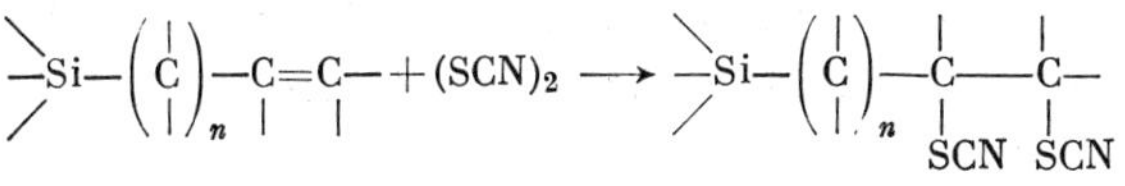

$$\underset{/}{\overset{\backslash}{>}}Si-\left(\overset{|}{\underset{|}{C}}\right)_n-C=C-+(SCN)_2 \longrightarrow \underset{/}{\overset{\backslash}{>}}Si-\left(\overset{|}{\underset{|}{C}}\right)_n-\overset{|}{\underset{SCN}{C}}-\overset{|}{\underset{SCN}{C}}- \qquad (2.298)$$

This reaction makes it possible to determine experimentally the thiocyanate numbers of unsaturated silicohydrocarbons, and these agree well with calculated values. The method of thiocyanate numbers has a series of advantages over the method of determining the bromine numbers of unsaturated compounds which is used widely in organic chemistry. In contrast to bromine, thiocyanogen does not cleave $Si-H$ and $Si-C$ bonds. As a result of this, reaction (2.298) makes it possible to determine the double bond in alkenylhydrosilanes and β-alkenylsilanes. On the other hand, the method of thiocyanate numbers has a drawback which limits its application for analytical purposes. This is the length of the determination of the thiocyanate number of a series of silicohydrocarbons. Thus, for example, triethylvinylsilane adds 100% cyanogen only after 24 h [23]. Therfore, the main value of thiocyanation is that it makes possible the determination of the comparative reactivity of double bonds in different unsaturated silicohydrocarbons. The low rate of addition of thiocyanogen then becomes a positive factor. The investigation of the thiocyanation of alkenylsilanes by reaction (2.298) shows that the lowest reactivity in this reaction is shown by vinylsilanes (n = 0), while the highest is shown by allysilanes (n = 1). 3-Butenylsianes (n = 2) are much more reactive than vinylsilanes, but are still less active than allylsilanes. An analogous picture is observed in ionic additions to alkenylsilanes, so that there is the possibility that the addition of thiocyanogen may be regarded as ionic [23].

3-Vinylheptamethyltrisiloxane adds thiocyanogen more slowly than trimethylvinylsilane, while 3-allylheptamethyltrisiloxane adds it more slowly than trimethylallylsilane [4].

The reactivity of P-trialkylhetero-substituted styrenes in thiocyanation falls in the following series depending on the nature of the hetero atom: Sn > C > Ge > Si [115].

Triorganosilylmethyl isothiocyanates have been obtained from (triorganosilylalkyl)amines. For this purpose the (triorganosilylalkyl)amine is initially treated with carbon disulfide and NaOH to convert it to the sodium N-(triorganosilylmethyl)thiocarbamate [512, 528, 529]. The latter is then treated with the ethyl ester of chlorocarbonic acid and converted to the triorganosilylmethyl isothiocyanate [528, 529]:

$$R_3SiCH_2NH_2 + CS_2 \xrightarrow[-H_2O]{+NaOH} R_3SiCH_2NHCSSNa \qquad (2.299)$$

$$R_3SiCH_2NHCSSNa \xrightarrow[-NaHS]{+ClCOOC_2H_5} R_3SiCH_2NCS \qquad (2.300)$$

When alkoxy(aminomethyl)silanes are used in reaction (2.299) trimethylamine has to be used instead of NaOH [528, 529]. Sodium N-(trimethylsilylmethyl)dithiocarbamate may be converted into the lead salt, which decomposes on heating to form trimethylsilylmethyl isothiocyanate [491, 529]:

$$[(CH_3)_3SiCH_2NHCSS]_2Pb \xrightarrow[-PbS]{} (CH_3)_3SiCH_2NCS \qquad (2.301)$$

Organosilicon isothiocyanates are also formed by the addition of HNCS to alkenylsilanes [101]:

$$R_3SiCH_2CH{=}CH_2 + NaSCN \xrightarrow[-NaHSO_4]{H_2SO_4} R_3SiCH_2CH(NCS)CH_3 \qquad (2.302)$$

7.4.2. Physical Properties

Trialkylsilyl thiocyanates and trialkylsilylalkyl isothiocyanates are liquids which may be distilled in vacuum. The physical con-

stants of these compounds are given in Table 25. The group refraction of Si $-CH_2SCN$, calculated from experimental data, equals
19.93 [96]. The thiocyanate structure of the product from the reaction of chloroalkylsilanes with alkali metal thiocyanates was
confirmed by investigation of their vibration spectra. The latter
contain the frequencies of 285 and 428 cm^{-1}, which are characteristic of the deformation vibrations of $C-Si-X$ and SCN and
also 600 and 737 cm^{-1}, which correspond to the symmetrical and
antisymmetrical bond vibrations of $C-S-C$, and the frequency
2150 cm^{-1} which corresponds to the bond vibrations of $C \equiv N$ [368].

7.4.3. Chemical Properties

The chemical conversions of organosilicon thiocyanates and
isothiocyanates have been studied little. The most detailed investigation was made of the reaction of organosilicon thiocyanates
with Grignard reagents, which proceeds according to scheme
(2.303) and leads to the formation of organosilicon thiols (see Section 7.1) [55, 57, 81, 83]:

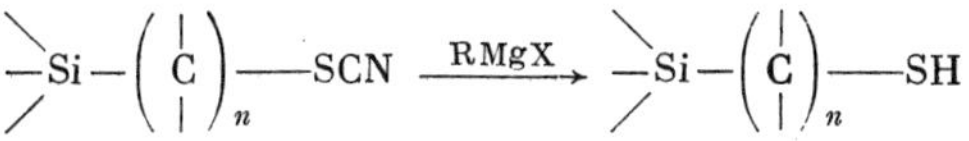

$$\underset{\diagup}{\overset{\diagdown}{\text{Si}}} - \left(\overset{|}{\underset{|}{\text{C}}} \right)_n - \text{SCN} \xrightarrow{\text{RMgX}} \underset{\diagup}{\overset{\diagdown}{\text{Si}}} - \left(\overset{|}{\underset{|}{\text{C}}} \right)_n - \text{SH} \qquad (2.303)$$

While the yields of α-silicon organic thiols may reach 70%,
those of their γ-analogs reach only 40%, since with the latter up
to 40% of organosilicon sulfides are formed together with the thiols
[55].

When chlorine is passed through trimethylsilylmethyl thiocyanate it is absorbed vigorously, but in all probability the Si $-C$
bond is cleaved, since it was not possible to detect the organic
sulfenyl chloride in the reaction products. Chlorine does not react with trimethyl(γ-thiocyanatopropyl)silane [82]. Likewise it is
not possible to obtain S-(trimethylsilylmethyl) thiocarbamate by
the reaction of trimethylsilylmethyl cyanate with concentrated
sulfuric acid [82]. Treatment of a mixture of heptamethylthio-
cyanatomethylcyclotetrasiloxane and hexamethyldisiloxane with
96% sulfuric acid leads to the formation of sulfur-containing polysiloxane oil [249]. The reaction of trimethylsilylmethyl isothiocyanate with aniline leads to the formation of N-phenyl-N'-(tri-

TABLE 25. Compounds Containing the Groupings $Si-(C)_n-SCN$ and $Si-(C)_n-NCS$ (Organosilicon thiocyanates, isothiocyanates, and derivatives of thiourea)

Empirical formula	Compound	B.p., °C (mm)	n_D^{20}	d_4^{20}	Literature
$C_5H_{11}NSSi$	$(CH_3)_3SiCH_2SCN$	199 (741)	1.4650^{25}	0.943^{25}	492
		198.5 (750)	1.4682	0.9476	81, 96
		196—197	1.4676	0.9426	249
		51 (2)	—	—	368
		46 (1)	—	—	82
	$(CH_3)_3SiCH_2NCS$	199 (741)	1.4984^{25}	0.938_4^{25}	528, 529
		90 (25)	—	—	528
$C_6H_{10}N_2S_2Si$	$(CH_3)_2Si(CH_2SCN)_2$	142.5 (2)	1.5361	1.1513	81
		—	1.5278^{25}	1.412^{25}	492
$C_6H_{13}NSSi$	$(CH_3)_3SiCH(CH_3)SCN$	123 (5)	1.4702	0.9380	81
		98 (6)	1.4713	0.9414	83
$C_7H_{11}NSSi$	$(CH_3)_3SiC{\equiv}CCH_2SCN$	72.5 (3)	1.4914	0.9690	57
$C_7H_{12}N_2S_2Si$	$(CH_3)_3SiCH(SCN)CH_2SCN$	m.p. 72—74	—	—	179
$C_7H_{13}NSSi$	$\underline{CH_2(CH_2)_3Si}(CH_3)CH_2SCN$	108—110 (7)	1.5110	1.0295	25
$C_7H_{15}NSSi$	$(CH_3)_3SiCH_2CH_2CH_2SCN$	66.8 (1)	1.4685	0.9278	81
	$(CH_3)_3SiCH_2CH(CH_3)NCS$	92 (15)	1.4875	0.9170	101
	$CH_3(C_2H_5)_2SiCH_2SCN$	72.5 (1.5)	1.4778	0.9489	81, 96
$C_7H_{16}ClNOSSi_2$	$ClCH_2(CH_3)_2SiOSi(CH_3)_2CH_2SCN$	105 (1.5)	1.4691	1.0727	81
$C_7H_{17}NOSSi_2$	$(CH_3)_3SiOSi(CH_3)_2CH_2SCN$	135 (42)	1.4443	0.9518	249
$C_8H_{13}NSSi$	$(CH_3)_3SiC{\equiv}CCH(CH_3)SCN$	80 (2.5)	1.4830	0.9285	55
$C_8H_{14}N_2S_2Si$	$(CH_3)_3SiCH_2CH(SCN)CH_2SCN$	—	$1.4965^{27.5}$	—	179

$C_8H_{16}NOS_2Si_2$	$[(CH_3)_2SiCH_2SCN]_2O$	148 (2)	1.4942	1.0949	81
		—	1.4843^{25}	1.0834^{25}_{4}	492, 528
$C_8H_{17}NSSi$	$(CH_3)_3SiCH_2CH(CH_3)CH_2SCN$	101 (10)	1.4710	0.9239	82
$C_8H_{17}NO_3SSi$	$(C_2H_5O)_3SiCH_2NCS$	120 (3)	1.4558^{25}	1.034^{25}_{4}	528, 529
$C_9H_{15}NSSi$	$(CH_3)_2(C_2H_5)SiC{\equiv}CCH(CH_3)SCN$	90 (2)	1.4889	0.9357	55
	$CH_3(C_2H_5)_2SiC{\equiv}CCH_2SCN$	82 (1)	1.4945	0.9654	57
$C_9H_{19}NSSi$	$CH_3(C_3H_7)_2SiCH_2SCN$	84—86 (2)	1.4760	0.9289	81
	$CH_3(C_2H_5)_2SiCH_2CH_2CH_2SCN$	87.5 (1.5)	1.4780	0.9331	82
	$CH_3(C_2H_5)_2SiCH_2CH(CH_3)NCS$	98 (5)	1.4961	0.9260	101
$C_9H_{23}NO_4SSi_4$	$O[(CH_3)_2SiO]_3Si(CH_3)CH_2SCN$ (cyclic)	168 (47)	1.4370	1.0645	249
$C_{10}H_{16}N_2O_2SSi$	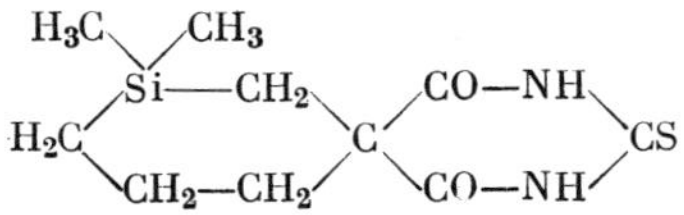	m.p. 196—197	—	—	311
$C_{10}H_{21}NSSi$	$CH_3(C_2H_5)_2SiCH_2CH(CH_3)CH_2SCN$	87 (1.5)	1.4801	0.9320	82
$C_{11}H_{18}N_2SSi$	$(CH_3)_3SiCH_2NHC(S)NHC_6H_5$	m.p. 122—123	—	—	528, 529
$C_{11}H_{18}N_2O_2SSi$	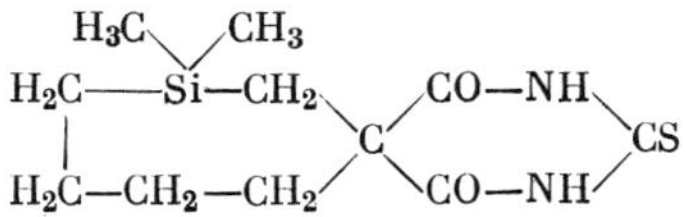	m.p. 191—193	—	—	311
$C_{11}H_{23}NSSi$	$CH_3(C_4H_9)_2SiCH_2SCN$	106.5 (2)	1.4755	0.9174	81
	$(C_3H_7)_3SiCH_2SCN$	120 (2)	1.4781	0.9247	81
	$CH_3(C_3H_7)_2Si(CH_2)_3SCN$	116.5 (1)	1.4760	0.9181	82
$C_{15}H_{15}NSSi$	$CH_3(C_6H_5)_2SiCH_2SCN$	172 (2)	1.5987	1.1204	81

methylsilylmethyl)thiourea [528, 529]:

$$(CH_3)_3SiCH_2NCS + C_6H_5NH_2 \longrightarrow (CH_3)_3SiCH_2NHCSNHC_6H_5 \qquad (2.304)$$

7.5. Organosilicon Sulfonic Acids and Their Derivatives

7.5.1. Preparation Methods

Aryl- and aralkylsilanes or the analogous siloxanes are sulfonated in the para position of the aromatic ring by concentrated sulfuric acid or chlorosulfonic acid [234, 242, 287, 346, 405–409, 411–413, 460, 461, 473]:

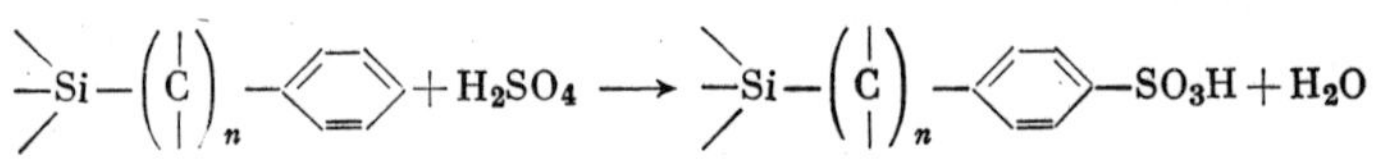

$$\geqslant Si - \left(\begin{matrix} | \\ C \\ | \end{matrix}\right)_n - \langle\!=\!\rangle + H_2SO_4 \longrightarrow \geqslant Si - \left(\begin{matrix} | \\ C \\ | \end{matrix}\right)_n - \langle\!=\!\rangle - SO_3H + H_2O \quad (2.305)$$

A side reaction in this case is elimination of a methyl [406, 409; see also section 4.1.1] or phenyl group [406, 408, 414, 441] attached to the silicon atom.

The reaction with chlorosulfonic acid is a convenient method for preparing organosilicon derivatives of arylsulfonyl chlorides [171, 172]. Thus, for example, the reaction of chlorosulfonic acid with β-phenylethyltrichlorosilane in chloroform at 6–10°C forms p-(β-trichlorosilylethyl)benzenesulfonyl chloride [171, 172]:

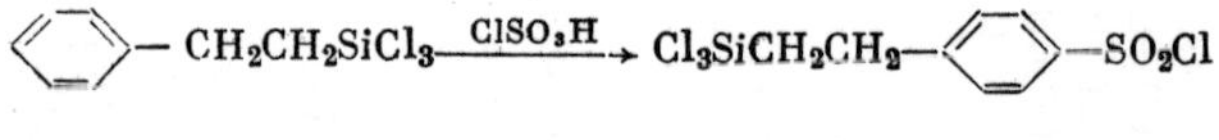

$$\langle\!=\!\rangle - CH_2CH_2SiCl_3 \xrightarrow{ClSO_3H} Cl_3SiCH_2CH_2 - \langle\!=\!\rangle - SO_2Cl \quad (2.306)$$

Sulfur trioxide cleaves the $Si-C_{ar}$ bond in bis(trialkylsilyl)-benzenes with the formation of organosilicon esters of trialkyl-silylarylsulfonic acids [205, 288]:

$$(CH_3)_3SiC_6H_4Si(CH_3)_3 + SO_3 \longrightarrow (CH_3)_3SiC_6H_4SO_2OSi(CH_3)_3 \quad (2.307)$$

Organosilicon derivatives of alkanesulfonyl chlorides are formed by the reaction of tetraalkylsilanes or alkylchlorosilanes with sulfuryl chloride [252–254]:

$$\geqslant Si - \left(\begin{matrix} | \\ C \\ | \end{matrix}\right)_n - H + SO_2Cl_2 \xrightarrow{-HCl} \geqslant Si - \left(\begin{matrix} | \\ C \\ | \end{matrix}\right)_n - SO_2Cl \qquad (2.308)$$

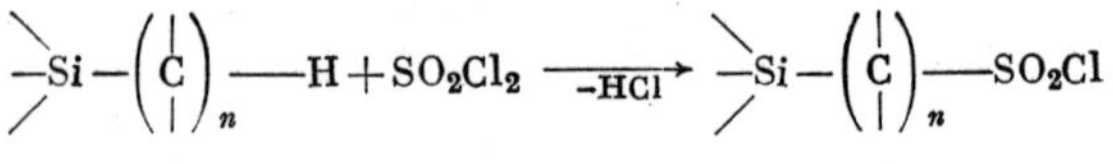

Reaction (2.308) is carried out in UV radiation in the presence of catalytic amounts of pyridine. The yield of trimethyl(chlorosulfonylmethyl)silane is then 53%, while that of the monosulfonyl chloride formed by sulfochlorination of octamethylcyclotetrasiloxane is only 29% [253].*

Salts of organosilicon alkanesulfonic acids are formed as a result of the addition of sodium sulfite to alkenylsilanes [527, 688, 689]:

$$R_3SiCH_2CH{=}CH_2 + NaHSO_3 \longrightarrow R_3SiCH_2CH_2CH_2SO_3Na \qquad (2.309)$$

This reaction is carried out in aqueous methanol solution and is initiated by a mixture of sodium nitrite and nitrate [688, 689, 860]. The salt is converted into the free acid by passing a solution of it through a cationic exchange resin [688].

7.5.2. Physical Properties

The physical constants of organosilicon sulfonic acids, their salts, acid chlorides, and amides are given in Tables 26a and 26b. See [735] for the results of crystallographic investigations of p-(trimethylsilylmethyl)benzenesulfanilide, which forms monoclinic crystals. The monohydrate of sodium 2,2-dimethyl-2-silapentane-5-sulfonate $(CH_3)_3Si(CH_2)_3SO_3Na \cdot H_2O$ is recommended as an internal standard for investigating the nuclear magnetic resonance of aqueous solutions and electrolyte solutions [261, 688, 689].

7.5.3. Chemical Properties and Application

When treated with PCl_5, organosilicon sulfonic acids are converted into sulfonyl chlorides [205, 234]:

$$\rangle Si-\left(\overset{|}{\underset{|}{C}}\right)_n-SO_3H + PCl_5 \longrightarrow \rangle Si-\left(\overset{|}{\underset{|}{C}}\right)_n-SO_2Cl + POCl_3 + HCl \qquad (2.310)$$

Sulfonyl chlorides and esters of organosilicon sulfonic acids are hydrolyzed readily with the formation of free sulfonic acids

*On the other hand, the photochemical chlorination of methyl- and phenylchlorosilanes by sulfuryl chloride leads smoothly to the formation of the corresponding chloro derivative [31].

TABLE 26a. Organosilicon Sulfonic Acids and Their Derivatives

Empirical formula	Compound	B.p., °C (mm)	M.p., °C	n_D^{20}	Literature
$C_2H_4Cl_4O_2SSi$	$Cl_3SiCH_2CH_2SO_2Cl$	71 (4)	29—30	—	254
$C_3H_8Cl_2O_2SSi$	$(CH_3)_2SiClCH_2SO_2Cl$	70—73 (1)	—	1.4780	253
$C_4H_9Cl_3O_2SSi$	$C_2H_5SiCl_2CH_2CH_2SO_2Cl$	85.7 (1)	—	1.4928	253
$C_4H_{11}ClO_2SSi$	$(CH_3)_3SiCH_2SO_2Cl$	57 (1)	—	1.4280	253
		72—74 (2)	—	—	254
$C_4H_{13}NO_2SSi$	$(CH_3)_3SiCH_2SO_2NH_2$	—	122—123	—	253
$C_8H_8Cl_4O_2SSi$	$Cl_3SiCH_2CH_2C_6H_4SO_2Cl$-p	148—154 (0.8—1.5)	40—45	—	171
$C_8H_{21}NO_2SSi$	$(CH_3)_3SiCH_2N(C_2H_5)SO_2C_2H_5$	102.6—103.2 (2)	—	1.4598	77
$C_8H_{23}ClO_6SSi_4$	$O[(CH_3)_2SiO]_3Si(CH_3)CH_2SO_2Cl$ (cyclic)	108—110 (1)	—	1.4350	187, 253
$C_8H_{25}NO_6SSi_4$	$O[(CH_3)_2SiO]_3Si(CH_3)CH_2SO_2NH_2$ (cyclic)	—	61—63	—	253
$C_9H_{11}ClF_2O_2SSi$	$CH_3SiF_2CH_2CH_2C_6H_4SO_2Cl$-p	131—135 (0.7)	—	—	171
$C_9H_{11}Cl_3O_2SSi$	$CH_3SiCl_2CH_2CH_2C_6H_4SO_2Cl$-p	168 (0.8)	—	—	171
	$CH_3SiCl_2CH(CH_3)C_6H_4SO_2Cl$-p	145—155 (0.2—0.85)	50—54	—	171
$C_9H_{13}BrO_2SSi$	$(CH_3)_3SiC_6H_4SO_2Br$ *	—	60—60.5	—	234
$C_9H_{13}ClO_2SSi$	$(CH_3)_3SiC_6H_4SO_2Cl$-o	89—90 (0.3)	—	1.5361	205
	$(CH_3)_3SiC_6H_4SO_2Cl$-m	148—149 (11)	—	1.5304	205, 288
		146—147 (10)			
	$(CH_3)_3SiC_6H_4SO_2Cl$-p	160—162 (11)	57	—	205, 288
$C_9H_{14}O_3SSi$	$(CH_3)_3SiC_6H_4SO_3H$-o · H_2O	—	87—88	—	205
	$(CH_3)_3SiC_6H_4SO_3H$-m · H_2O	—	65—66	—	205
	$(CH_3)_3SiC_6H_4SO_3H$-p · H_2O	—	89—90	—	205, 288
$C_9H_{15}NO_2SSi$	$(CH_3)_3SiC_6H_4SO_2NH_2$-o	—	74—75	—	205
	$(CH_3)_3SiC_6H_4SO_2NH_2$-m	—	112—113	—	205, 288
	$(CH_3)_3SiC_6H_4SO_2NH_2$-p	—	115—116	—	179, 205
$C_{10}H_{15}BrO_2SSi$	$(CH_3)_3SiCH_2C_6H_4SO_2Br$-p	—	60—60.5	—	233
$C_{10}H_{15}ClO_2SSi$	$(CH_3)_3SiCH_2C_6H_4SO_2Cl$-m	111—112 (0.6)	—	1.5280	205
	$(CH_3)_3SiCH_2C_6H_4SO_2Cl$-p	—	45.5—46.5	—	233
$C_{10}H_{16}O_3SSi$	$(CH_3)_3SiCH_2C_6H_4SO_3H$-p · H_2O	—	116	—	233
		—	115—116	—	205
$C_{10}H_{17}NO_2SSi$	$(CH_3)_3SiCH_2C_6H_4SO_2NH_2$-m	—	75—76	—	205
	$(CH_3)_3SiCH_2C_6H_4SO_2NH_2$-p	—	81—81.5	—	206, 233
$C_{10}H_{25}NO_2SSi$	$CH_3(C_2H_5)_2SiCH_2N(C_2H_5)SO_2C_2H_5$	132.4—132.8 (2)	—	1.4689	77
$C_{11}H_{19}NO_2SSi$	$(CH_3)_3SiCH_2C_6H_4SO_2NHCH_3$-p	—	76.4—77.2	—	233

* Whether it is the o-, m-, or p-isomer is not indicated in the original.

TABLE 26b. Organosilicon Sulfonic Acids and Their Derivatives

Empirical formula	Compound	M.p., °C	Literature
$C_{12}H_{20}O_4SSi$	$C_2H_5(C_3H_7)Si(OH)CH_2C_6H_4SO_3H$ *		
	dl -acid + *l* -menthylamine	211—212	408, 409
$C_{12}H_{21}NO_2SSi$	$(CH_3)_3SiCH_2C_6H_4SO_2N(CH_3)_2$-m	108—109	206
	$(CH_3)_3SiCH_2C_6H_4SO_2N(CH_3)_2$-p	112—112.5	206
$C_{12}H_{22}O_3SSi_2$	$(CH_3)_3SiC_6H_4SO_3Si(CH_3)_3$	83—86	205
$C_{12}H_{29}NO_2SSi$	$CH_3(C_3H_7)_2SiCH_2N(C_2H_5)SO_2C_2H_5$	b.p. 141.6—142 (2)	77
$C_{13}H_{22}O_3SSi$	$CH_3(C_2H_5)C_3H_7SiCH_2C_6H_4SO_3H$ *		
	dl -acid + *l* -menthylamine	122—123	409
	dl -acid + cinchonine	110—112	409
	dl -acid + cinchonine + HCl	210—212	409
	dl -acid + narcotine + HCl	—	409
$C_{14}H_{33}NO_2SSi$	$CH_3(C_4H_9)_2SiCH_2N(C_2H_5)SO_2C_2H_5$	b.p. 149—149.5 (2)	77
$C_{15}H_{19}NO_2SSi$	$(CH_3)_3SiC_6H_4SO_2NHC_6H_5$-o	102—103	205
	$(CH_3)_3SiC_6H_4SO_2NHC_6H$-m	83—84	205
	$(CH_3)_3SiC_6H_4SO_2NHC_6H_5$-p	122—123	179, 205
$C_{15}H_{26}O_3SSi$	$C_2H_5(C_3H_7)_2SiCH_2C_6H_4SO_3H$ *		
	dl -acid + *l* -menthylamine	135	473
	dl -acid + *l* -menthylamine · $2H_2O$	62.5—63	473
	dl -acid + cinchonidine	191—192	473
	dl -acid + cinchonidine + HCl	222—224 (decomp.)	473
	dl -acid + quinine	175	473
	dl -acid + quinine + HCl	223—224	473
$C_{16}H_{21}NO_2SSi$	$(CH_3)_3SiCH_2C_6H_4SO_2NHC_6H_5$-m	79—80	205
	$(CH_3)_3SiCH_2C_6H_4SO_2NHC_6H_5$-p	124.2—125	206, 233
$C_{16}H_{28}O_3SSi$	$C_2H_5(C_3H_7)(iso\text{-}C_4H_9)SiCH_2C_6H_4SO_3H$ *		
	dl -acid + *l* -menthylamine·$2H_2O$	127—128	413
	dl -acid + cinchonidine	184	413
	dl -acid + cinchonidine + HCl	218(decomp.)	413
	dl -acid + cinchonine	139—141	413
	dl -acid + quinine	168—171	413
	dl -acid + quinine + HCl	212(decomp.)	413
	dl -acid + brucine	163	413
	dl -acid + strychnine	235—237	413
$C_{17}H_{23}NO_2SSi$	$(CH_3)_3SiCH_2C_6H_4SO_2N(CH_3)C_6H_5$-p	71.5—72.5	233
	$(CH_3)_3SiCH_2C_6H_4SO_2NHCH_2C_6H_5$-p	130.5—131	233
	p-$(CH_3)_3SiCH_2C_6H_4SO_2NHC_6H_4CH_3$-o	153—153.5	233
	p-$(CH_3)_3SiCH_2C_6H_4SO_2NHC_6H_4CH_3$-p	97—98	233
$C_{17}H_{23}N_3O_3SSi$	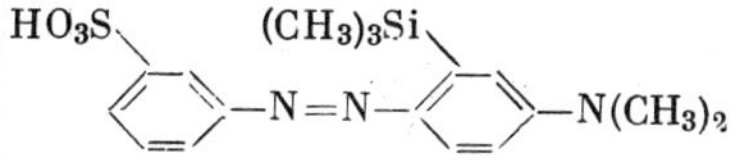	240—241 (decomp.)	676

* Whether it is the o-, m-, or p-isomer is not indicated in the original.

TABLE 26b (Continued)

Empirical formula	Compound	M.p., °C	Literature
	$(CH_3)_3Si$ $HO_3S-\langle\ \rangle-N{=}N-\langle\ \rangle-N(CH_3)_2$	251(decomp.)	676
$C_{17}H_{24}N_2O_3S_2Si$	$[o\text{-}(CH_3)_3SiC_6H_4SO_3]^- \quad \left[C_6H_5CH_2SC{\Large\langle}{}^{NH_2}_{NH_2}\right]^+$	161—162	205
	$[p\text{-}(CH_3)_3SiC_6H_4SO_3]^- \quad \left[C_6H_5CH_2SC{\Large\langle}{}^{NH_2}_{NH_2}\right]^+$	212—213	205
$C_{17}H_{24}N_4O_2SSi$	$(CH_3)_3Si$ $H_2NSO_2-\langle\ \rangle-N{=}N-\langle\ \rangle-N(CH_3)_2$	213	676
$C_{19}H_{26}O_3SSi$	$C_2H_5(C_3H_7)C_6H_5CH_2SiCH_2C_6H_4SO_3H$ *		
	$\quad l$ -acid + strychnine	197—198	243
	$\quad dl$ -acid + strychnine·$3H_2O$	199	242
	$\quad l$ -acid + l -menthylamine	100.5—101.5	243
	$\quad dl$ -acid + l -menthylamine	99—99.5	243
	$\quad l$ -acid + quinine	133—135	243
	$\quad dl$ -acid + quinine	132—133	243
	$\quad l$ -acid + morphine	160—165 (decomp.)	243
	$\quad dl$ -acid + morphine	155—160	243
	$\quad l$ -acid + cinchonine	158—159	243
	$\quad dl$ -acid + cinchonine	157—159	243
$C_{24}H_{19}ClO_2SSi$	$(C_6H_5)_3SiC_6H_4SO_2Cl\text{-}m$	123—124	179, 205
	$(C_6H_5)_3SiC_6H_4SO_2Cl\text{-}p$	180—183	288
		185—186	205
$C_{24}H_{20}O_3SSi$	$(C_6H_5)_3SiC_6H_4SO_3H\text{-}m$	143(decomp.)	205
	$(C_6H_5)_3SiC_6H_4SO_3H\text{-}p$	138(decomp.)	205, 288
$C_{24}H_{38}O_7S_2Si_2$	$[C_2H_5(C_3H_7)SiCH_2C_6H_4SO_3H]_2O$ *		
	$\quad d$ -acid + d -bornylamine	210—211	410
	$\quad l$ -acid + d -bornylamine	211—213	410
	$\quad dl$ -acid + d -bornylamine	207—208	410
	$\quad d$ -acid + cinchonidine	154—159	410
	$\quad l$ -acid + cinchonidine	149—150	410
	$\quad dl$ -acid + cinchonidine	148—150	410
	$\quad d$ -acid + cinchonidine $\quad +HCl$	222—224 (decomp.)	410
	$\quad l$ -acid + cinchonidine $\quad +HCl$	225—227 (decomp.)	410
	$\quad dl$ -acid + cinchonidine $\quad +HCl$	~220	410
	$\quad d$ -acid + l -menthylamine	228—229	410
		235—236 **	410

*Whether it is the o-, m-, or p-isomer is not indicated in the original.

**The compound is dimorphous.

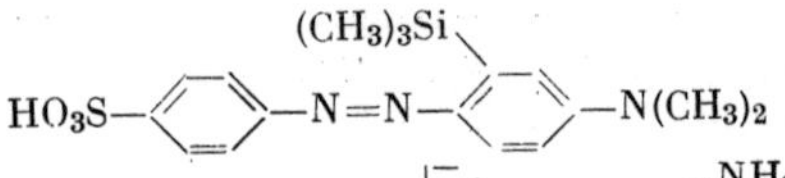

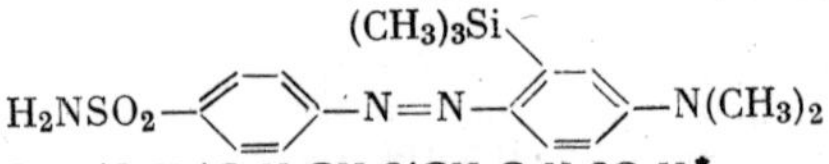

TABLE 26b (Continued)

Empirical formula	Compound	M.p., °C	Literature
	dl -acid + l -menthylamine	226—227 236—237 *	410
	dl -acid + l -menthylamine	225—226 233—235 *	410
	l -acid + d -menthylhydrindamine	175—185	408, 410, 411
	d -acid + d -menthylhydrindamine	205—207	408, 410, 411
	dl -acid + d -menthylhydrindamine	170	408, 411
	d -acid + l -menthylhydrindamine	152	408, 411
	dl -acid + dl - menthylhydrindamine	160	408
$C_{26}H_{42}O_7S_2Si_2$	$[C_2H_5(iso\text{-}C_4H_9)SiCH_2C_6H_4SO_3H]_2O$		
	d -acid + l -menthylamine . $4H_2O$	254—258	461
	l -acid + l -menthylamine . $4H_2O$	253—257	461
	dl -acid + l -menthylamine. $4H_2O$	245	460
	d -acid + d -menthylhydrindamine	207—209	461
	l -acid + l -menthylhydrindamine	207—209	461
	d -acid + cinchonidine	171—173	461
	d -acid + cinchonidine + HCl	282—284	461
	l -acid + cinchonidine	181—183	461
	l -acid + cinchonidine + HCl	229—232	461
	dl -acid + cinchonidine	177	460, 461
	dl -acid + cinchonidine + HCl	225—229	460, 461
	l -acid + d -bornylamine	210—212	461
	dl -acid + d -bornylamine	207—209	460, 461
$C_{30}H_{25}NO_2SSi$	$(C_6H_5)_3SiC_6H_4SO_2NHC_6H_5\text{-m}$	154—155	205
	$(C_6H_5)_3SiC_6H_4SO_2NHC_6H_5\text{-p}$	196—197	205

---

* The compound is dimorphous.

[205, 253, 288]:

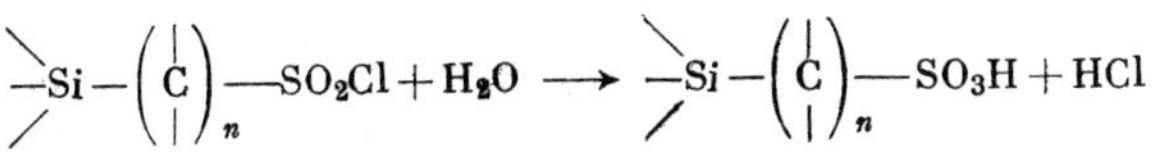

$$\text{—Si} - \left(\overset{|}{\underset{|}{C}} \right)_n \text{—SO}_2\text{Cl} + \text{H}_2\text{O} \longrightarrow \text{—Si} - \left(\overset{|}{\underset{|}{C}} \right)_n \text{—SO}_3\text{H} + \text{HCl} \qquad (2.311)$$

$$\text{—Si} - \langle\!\!\!=\!\!\!\rangle \text{—SO}_2\text{OSi—} + \text{H}_2\text{O} \longrightarrow \text{—Si} - \langle\!\!\!=\!\!\!\rangle \text{—SO}_3\text{H} + \text{HO—Si—} \qquad (2.312)$$

In the hydrolysis of $(CH_3)_3SiCH_2SO_2Cl$ there is also partial cleavage of the $Si-C$ bond and the formation of hexamethyldisiloxane [253].

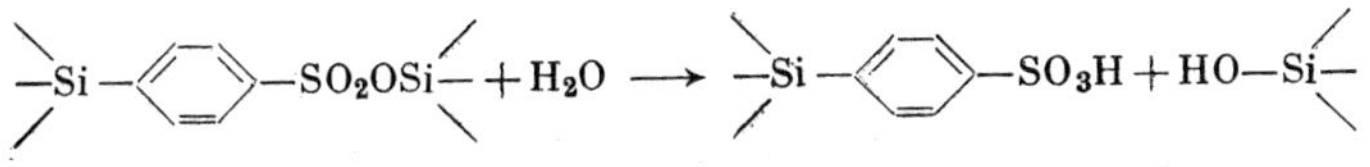

By treatment with ammonia or amines, organosilicon sulfonyl chlorides may be converted into the corresponding sulfonamides [205, 206, 233, 234, 253, 288]:

$$\diagup\!\!\!\!\diagdown\!\!-Si-\left(\begin{array}{c}|\\C\\|\end{array}\right)_n-SO_2Cl+RNH_2 \xrightarrow[-HCl]{} \diagup\!\!\!\!\diagdown\!\!-Si-\left(\begin{array}{c}|\\C\\|\end{array}\right)_n-SO_2NHR \qquad (2.313)$$

Salts with optically active bases may be used for the separation of organosilicon sulfonic acids containing an asymmetric silicon atom [243, 408, 410, 411, 413, 460, 461, 473]. Organosilico azo dyes containing sulfonic acid and sulfonamide groups are known [676]. Polysiloxanes containing a sulfonic acid group [375] have been proposed as foaming agents [544] and polycondensation catalysts [306].

8. COMPOUNDS CONTAINING THE GROUPINGS Si − (C)$_n$ − O − S AND Si − O − (C)$_n$ − S

8.1. Preparation Methods

Compounds containing the grouping Si−(C)$_n$−O−S include sulfate esters of organosilicon alcohols. They are obtained by the action of the corresponding alcohols with sulfuric and chlorosulfonic acids, toluenesulfonyl chloride, or sulfur trioxide [312, 324, 355, 386, 584, 586, 588]. In this way it is possible to obtain both full esters and acid esters of sulfuric acid. The latter may be converted into salts of amines [515] or alkali metals [683]. The reaction of trimethylsilylmethanol with concentrated sulfuric acid forms bis(trimethylsilylmethyl) sulfate in 44.4% yield [386]:

$$2(CH_3)_3SiCH_2OH + H_2SO_4 \longrightarrow [(CH_3)_3SiCH_2O]_2SO_2 + 2H_2O \qquad (2.314)$$

A side reaction in (2.314) is the liberation of methane due to the elimination of a methyl group from the starting alcohol.

Bis(trimethylsilylmethyl) sulfate is also obtained as a result of the reaction of trimethyl(iodomethyl)silane with silver sulfate [386]:

$$2(CH_3)_3SiCH_2I + Ag_2SO_4 \longrightarrow [(CH_3)_3SiCH_2O]_2SO_2 + 2AgI \qquad (2.315)$$

Compounds containing the grouping $Si-O-(C)_n-S$ include the organosilicon derivatives of oxysulfides, oxythiocyanates, thiocarboxylic acids, and O-derivatives of mercapto acids. Organosilicon derivatives of oxysulfides are obtained by using the general methods for the formation of the $Si-O$ bond by treating the appropriate alcohols with chlorosilanes [44, 138, 390, 513, 547, 816, 817]; alkoxysilanes [44]; hydrosilanes in the presence of H_2PtCl_6, Pd, and $ZnCl_2$ [809]; and hexamethyldisilazan [390]:

$$-(C)_n-S-(C)_m-OH + X-Si\equiv \rightarrow -(C)_n-S-(C)_m-O-Si\equiv \qquad (2.316)$$

$$X = Cl,\ OR,\ H,\ NHSi\equiv$$

The reaction of chlorosilanes with sodium mercaptoacetate in ethyl ether gives mercaptoacetoxysilanes in 78-83% yield [654, 655]:

$$\equiv Si-Cl + NaOOCCH_2SH \xrightarrow[-NaCl]{} \equiv SiOCOCH_2SH \qquad (2.317)$$

Mercaptoacetoxysilanes are also formed by the action of excess mercapto acids on O- and S-silylated mercapto acids [577, 933]:

$$\equiv SiSCHRCOOSi\equiv + HSCHRCOOH \rightarrow 2HSCHRCOOSi\equiv \qquad (2.318)$$

If the $Si-S$ bond in O- and S-silylated mercapto acids is cleaved by acid chlorides of carboxylic acids, then the reaction products are organosilicon esters of acylthiocarboxylic acids [578]:

$$\equiv SiSCHRCOOSi\equiv + RCOCl \xrightarrow[-\equiv SiCl]{} RCOSCHRCOOSi\equiv \qquad (2.319)$$

With cooling, ethylene oxide reacts with isothiocyanatosilanes, converting them to β-thiocyanatoethoxysilanes [564]:

$$\equiv SiNCS + CH_2-CH_2 (-O-) \rightarrow \equiv SiOCH_2CH_2SCN \qquad (2.320)$$

The reaction of triphenylsilanethiol with benzoyl chloride does not lead to the formation of S-triphenylsilylthiobenzoate. Instead the O-ester is obtained in 75% yield [282, 478]:

$$(C_6H_5)_3SiSH + C_6H_5COCl \xrightarrow[-HCl]{} (C_6H_5)_3SiOCSC_6H_5 \qquad (2.321)$$

The same compound is formed in 65% yield by the reaction of triphenylchlorosilane with thiobenzoic S-acid in the presence of triethylamine [367, 478], and in 85% yield when triphenylaminosilane is boiled with dibenzoyl disulfide in cyclohexane for 3 h [478, 479]:

$$(C_6H_5)_3SiCl + C_6H_5COSH \xrightarrow[-R_3N \cdot HCl]{R_3N} (C_6H_5)_3SiOCSC_6H_5 \qquad (2.322)$$

$$(C_6H_5)_3SiNH_2 + C_6H_5COSSCOC_6H_5 \longrightarrow (C_6H_5)_3SiOCSC_6H_5 + C_6H_5CONH_2 + S \qquad (2.323)$$

Thioacyloxyalkylsilanes are obtained from alkylchlorosilanes by scheme (2.322), and also by the reaction of hexamethyldisilazan with S-thio acids [367] and triethylsilane with dibenzoyl disulfide in the presence of zinc chloride [477]:

$$(CH_3)_3SiNHSi(CH_3)_3 + 2CH_3COSH \xrightarrow[-NH_3]{} 2(CH_3)_3SiOCSCH_3 \qquad (2.324)$$

$$2(C_2H_5)_3SiH + C_6H_5COSSCOC_6H_5 \xrightarrow{ZnCl_2} 2(C_2H_5)_3SiOCSC_6H_5 + H_2 \qquad (2.325)$$

8.2. Physical and Chemical Properties

The physical and chemical constants of compounds containing the grouping Si$-(C)_n-$O$-$S and Si$-$O$-(C)_n-$S are given in Table 27.

Thioacyloxysilanes are hydrolyzed rapidly by water even at room temperature [367]:

$$2R_3SiOCSCH_3 + H_2O \xrightarrow{ZnCl_2} R_3SiOSiR_3 + 2CH_3COSH \qquad (2.326)$$

Their ethanolysis proceeds equally readily. In contrast to the hydrolysis and alcoholysis, the reaction of thioacyloxysilanes with acetic acid proceeds at an appreciable rate only in the presence of sodium acetate [367].

TABLE 27. Compounds Containing the Grouping Si – (C)$_n$ – O – S and Si – O – (C)$_n$ – S

Empirical formula	Compound	B.p., °C (mm)	M.p., °C	n_D^{20}	d_4^{20}	Literature
$C_5H_{12}OSSi$	$(CH_3)_3SiOCOCH_2SH$	67 (10)	—	1.4478	—	577
	$(CH_3)_3SiOCSCH_3$	73–75.5 (30)	—	—	—	654, 655
		84–85	—	1.4513^{25}	0.929^{25}	367
$C_6H_{12}O_2S_2Si_2$	$(CH_3)_2Si(OCSCH_3)_2$	—	55	—	—	367
$C_6H_{12}O_4S_2$	$(CH_3)_2Si(OCOCH_2SH)_2$	110.5–112 (1.5)	—	—	—	655
$C_6H_{14}O_2SSi$	$(CH_3)_2Si\langle(OCH_2CH_2)_2\rangle S$ (cyclic)	204.5 (756)	—	1.4832	—	44
	$(CH_3)_3SiOCOCH(CH_3)SH$	61 (10)	—	1.4408	—	577
$C_7H_{14}O_3SSi$	$(CH_3)_3SiOCOCH_2SCOCH_3$	106 (10)	—	1.4589	—	578
$C_7H_{16}OSSi$	$(CH_3)_3SiOCH_2CH_2SCH{=}CH_2$	95–97 (47)	—	1.5736^{24}	0.91	390
$C_8H_{14}N_2O_2S_2Si$	$(CH_3)_2Si(OCH_2CH_2SCN)_2$	140–144 (1.5)	—	—	—	564
$C_8H_{16}O_3SSi$	$(CH_3)_3SiOCOCH(CH_3)SCOCH_3$	102 (10)	—	1.4529	—	578
$C_8H_{22}O_4SSi_2$	$[(CH_3)_3SiCH_2O]_2SO_2$	—	78.2–78.6	—	—	386
$C_9H_{14}OSSi$	$(CH_3)_3SiOCSC_6H_5$	98–99 (1)	—	1.557	1.035	478
$C_9H_{22}O_2SSi$	$(CH_3)_3Si(OCH_2CH_2)_2SC_2H_5$	81 (4)	—	1.4505	0.9380	122
$C_{10}H_{20}OSSi$	$(CH_3)_3SiOCSC_6H_5$	98–99 (1)	—	1.5570	1.0355	479
$C_{11}H_{26}O_3SSi$	$(CH_3)_3Si(OCH_2CH_2)_3SC_2H_5$	112 (3)	—	1.4525	0.9615	122
$C_{12}H_{16}O_3SSi$	$(CH_3)_3SiOCOCH_2SCOC_6H_5$	140 (0,8)	—	1.5292	—	578
$C_{13}H_{18}O_3SSi$	$(CH_3)_3SiOCOCH(CH_3)SCOC_6H_5$	137 (0,4)	—	1.5203	—	578
$C_{13}H_{30}O_4SSi$	$(CH_3)_3Si(OCH_2CH_2)_4SC_2H_5$	142 (3)	—	1.4569	0.9872	122
$C_{16}H_{26}O_3SSi$	trans-$(CH_3)_3Si$–C_6H_{10}–OSO_2–C_6H_4–CH_3	—	97–98	—	—	312
	cis-$(CH_3)_3Si$–C_6H_{10}–OSO_2–C_6H_4–CH_3	—	83–84	—	—	312
$C_{18}H_{24}O_3SSi$	p-$(CH_3)_3SiC_6H_4CH_2CH_2OSO_2C_6H_4CH_3$-p	208–212 (0.5)	—	—	—	324
$C_{24}H_{52}O_4S_4Si$	$(C_4H_9SCH_2CH_2O)_4Si$	220 (0.5)	—	—	—	547
	(iso-$C_4H_9SCH_2CH_2O)_4Si$	200 (1)	—	—	—	547
$C_{25}H_{20}OSSi$	$(C_6H_5)_3SiOCSC_6H_5$	—	96–98	—	—	367
		—	92	—	—	282, 478, 479
$C_{28}H_{28}O_3SSi$	$(C_6H_5)_3Si(CH_2)_3OSO_2C_6H_4CH_3$-p	—	103–104.5	—	—	355
$C_{28}H_{60}O_4S_4Si$	$[C_4H_9SCH_2CH(CH_3)O]_4Si$	220 (0.5)	—	—	—	547
$C_{28}H_{62}O_8SSi_2$	[(sec-$C_4H_9O)_3SiOCH_2CH_2]_2S$	194 (1)	—	1.4318	0.9636	138

9. ORGANOSILICON COMPOUNDS OF SULFUR CONTAINING INORGANIC ELEMENTS AND PHOSPHORUS

The synthesis methods and chemical properties of organosilicon compounds containing both sulfur and phosphorus atoms [7, 8, 13, 21, 37, 39, 83, 94, 181, 255, 307, 308, 384, 525, 535, 543, 741, 768, 790–792, 872, 926, 948, 953] are examined in chapter one (Sections 2 and 6).

In a series of papers there are descriptions of organosilicon compounds containing both sulfur and boron [86], sulfur and germanium [591, 592, 615], sulfur and tin [614, 638], sulfur and arsenic [469, 593], sulfur and vanadium [245], and sulfur and selenium atoms [630]. Compounds of this type have been examined in a monograph [17].

LITERATURE CITED IN CHAPTER TWO

1. Publications of Soviet Authors

1. V. Z. Alekseev, Zh. Fiz. Khim., 28:945 (1954).
2. K. A. Andrianov, Usp. Khim., 27:1257 (1958).
3. K. A. Andrianov, S. I. Dzhenchel'skaya, and Yu. K. Petrashko, Plast. Massy, No. 3, p. 20 (1960).
4. K. A. Andrianov, A. A. Zelenetskaya, N. N. Nikitina, V. I. Sidorov, and L. M. Khananashvili, Zh. Obshch. Khim., 36:1633 (1966).
5. K. A. Andrianov, B. A. Izmailov, A. M. Kononov, and G. V. Kotrelev, J. Organomet. Chem., 3:129 (1965).
6. K. A. Andrianov and A. M. Kononov, Dokl. Akad. Nauk SSSR, 156:858 (1964).
7. K. A. Andrianov and I. K. Kuznetsova, Izv. Akad. Nauk SSSR, Otdel. Khim. Nauk, 1962:456.
8. K. A. Andrianov, I. K. Kuznetsova, and I. Pakhomova, Izv. Akad. Nauk SSSR, Otdel. Khim. Nauk, 1963:500.
9. K. A. Andrianov, M. Ya. Levshchuk, S. A. Golubtsov, and T. A. Krasovskaya, Authors' cert. 127259, 1960; Byull. Izobr., No. 7, p. 17 (1960).
10. K. A. Andrianov, S. N. Leznov, and L. A. Sabun, Authors' cert. 121449, 1959; Byull. Izobr., No. 15, p. 17 (1960).
11. K. A. Andrianov, I. Khaiduk, and L. M. Khananashvili, Usp. Khim., 32:539 (1963).
12. K. A. Andrianov, I. Khaiduk, L. M. Khananashvili, and N. I. Nekhaeva, Zh. Obshch. Khim., 32:3447 (1962).
13. G. F. Bebikh, Yu. A. Pentin, and T. V. Ershova, Zh. Obshch. Khim., 33:3544 (1963).

14. Z. V. Belyakova, Yu. K. Yur'ev, and G. B. Elyakov, Proceedings of the Conference on the Chemistry and Practical Use of Organosilicon Compounds [in Russian], No. 1, TsBTI LSNKh, p. 197 (1958).
15. A. S. Berezhnoi, Silicon and Its Binary Systems, Izd. AN USSR, Kiev (1958).
16. M. A. Blokhin, A. T. Shuvaev, and V. V. Gorskii, Izv. Akad. Nauk SSSR, Ser. Fiz., 28:801 (1964).
17. S. N. Borisov, M. G. Voronkov, É. Ya. Lukevits, Organosilicon Heteropolymers and Heterocompounds, Plenum Press, New York (1969).
18. S. N. Borisov and A. V. Karlin, Izv. Akad. Nauk Latv. SSR, Ser. Khim., 1965:89.
19. S. N. Borisov, A. V. Karlin, and N. G. Sviridova, Zh. Priklad. Khim., 35:917 (1962).
20. S. N. Borisov and N. G. Sviridova, Vysokomol. Soed., 3:50 (1961).
21. A. B. Bruker, L. D. Balashova, and L. Z. Soborovskii, Zh. Obshch. Khim., 36:75 (1966).
22. A. A. Bugorkova, Effect of Unsaturated Compounds on Thiocyanation, Author's abstract of Cand. dissertation, VNII synt. i nat. dushistykh veshchestv, Moscow (1952).
23. A. A. Bugorkova, V. F. Mironov, and A. D. Petrov, Izv. Akad. Nauk SSSR, Otdel. Khim. Nauk, 1960:474.
24. A. A. Bugorkova, L. N. Petrova, and V. M. Rodionov, Zh. Obshch. Khim., 23:1808 (1953).
25. V. M. Vdovin, N. S. Nametkin, K. S. Pushchevaya, and A. V. Topchiev, Izv. Akad. Nauk SSSR, Otdel. Khim. Nauk, 1963:274.
26. V. M. Vdovin, N. S. Nametkin, E. Sh. Finkel'shtein, and V. D. Oppengeim, Izv. Akad. Nauk SSSR, Ser. Khim., 1964:458.
27. T. M. Voronkina, I. T. Strukov, and M. F. Shostakovskii, Zh. Obshch. Khim., 34:1464 (1964).
28. M. G. Voronkov, Izv. Akad. Nauk SSSR, Otdel. Khim. Nauk, 1957:517.
29. M. G. Voronkov, Authors' cert. 110968, 1957; Byull. Izobr., No. 2, p. 16 (1958).
30. M. G. Voronkov, Heterolytic Cleavage Reactions of the Siloxane Bond, Report of scientific works presented at the competition for the science degree of Doctor of Chemical Sciences, INKhS AN SSSR, Moscow (1961).
31. M. G. Voronkov and V. P. Davydov, Dokl. Akad. Nauk SSSR, 125:553 (1959).
32. M. G. Voronkov and B. N. Dolgov, Zh. Obshch. Khim., 24:1082 (1954).
33. M. G. Voronkov, B. N. Dolgov, and G. B. Karpenko, Zh. Obshch. Khim., 24:269 (1954).
34. M. G. Voronkov, V. A. Kolesova, and V. N. Zgonnik, Izv. Akad. Nauk SSSR, Otdel. Khim. Nauk, 1957:1363.
35. M. G. Voronkov, A. A. Pashchenko, E. A. Lasskaya, and K. K. Karibaev, Zh. Priklad. Khim., 39:1345 (1966).
36. N. S. Vyazankin, M. N. Bochkarev, and L. P. Sanina, Zh. Obshch. Khim., 36:1961 (1966).
37. M. L. Galashina, M. V. Sobolevskii, D. Z. Levina, and T. P. Alekseeva, Plast. Massy, No. 8, p. 16 (1964).

38. G. Herzberg, Spectra and Structure of Diatomic Molecules [Russian translation], IL (1949) [G. Herzberg, Molecular Spectra and Molecular Structure. I. Diatomic Molecules, New York (1939)].
39. V. G. Glukhovtsev, The Synthesis and Properties of Unsaturated Silanes and Disilanes, Author's abstract of Cand. dissertation, IOKh AN SSSR, Moscow (1956).
40. V. N. Gruber and L. S. Mukhina, Vysokomol. Soed., 3:84 (1961).
41. G. G. Gustavson, Zh. Russ. Khim. Obshch., 4:101 (1872).
42. G. G. Gustavson, Ber., 5:332 (1872).
43. G. G. Gustavson, Ber., 6:9 (1873).
44. V. P. Davydova, M. G. Voronkov, and B. N. Dolgov, Proceedings of the Conference on the Chemistry and Practical Use of Organosilicon Compounds, No. 1, TsBTI LSNKh, p. 204 (1958).
45. N. Ya. Derkach and N. P. Smetankina, Zh. Obshch. Khim., 34:3613 (1964).
46. B. N. Dolgov, D. N. Andreev, and V. P. Lyutyi, Dokl. Akad. Nauk SSSR, 118:501 (1958).
47. B. N. Dolgov, É. V. Kukharskaya, and D. N. Andreev, Izv. Akad. Nauk SSSR, Otdel. Khim. Nauk, 1957:968.
47a. A. N. Egorochkin, N. S. Vyazankin, G. A. Razuvaev, O. A. Kruglaya, and M. N. Bochkarev, Dokl. Akad. Nauk SSSR, 170:333 (1966).
48. D. Ya. Zhinkin, M. M. Morgunova, and K. A. Andrianov, Dokl. Akad. Nauk SSSR, 165:114 (1965).
49. D. Ya. Zhinkin, E. A. Semenova, M. V. Sobolevskii, and K. A. Andrianov, Plast. Massy, No. 11, p. 16 (1963).
50. A. V. Karlin and S. N. Borisov, Material of Conference on the Production and Use of Organosilicon Compounds, Coll. 2, MDNTP (1964), p. 40.
51. L. I. Kartasheva, N. S. Nametkin, and T. I. Chernysheva, Dokl. Akad. Nauk SSSR, 170:848 (1966).
52. É. V. Kogan, A. G. Ivanova, V. O. Reikhsfel'd, N. I. Smirnov, and V. N. Gruber, Vysokomol. Soed., 5:1183 (1963).
53. V. P. Kozyukov, V. D. Sheludyakov, and V. F. Mirnov, in: Organosilicon Compounds (Proceedings of Conference), Moscow, 1:32 (1966).
54. V. P. Kozyukov, V. D. Sheludyakov, and V. F. Mironov, in: Organosilicon Compounds (Proceedings of Conference), Moscow, 1:52 (1966).
55. N. V. Komarov and N. N. Vlasova, Izv. Akad. Nauk SSSR, Otdel. Khim. Nauk, 1963:90.
56. N. V. Komarov and N. N. Vlasova, Izv. Akad. Nauk SSSR, ser. khim., 1965:1687.
57. N. V. Komarov, N. N. Vlasova, G. I. Kagan, and G. A. Gladkova, Zh. Obshch. Khim., 35:1763 (1965).
58. N. V. Komarov, N. N. Vlasova, and Z. I. Mikhailov, Zh. Obshch. Khim., 35:1692 (1965).
59. N. V. Komarov and O. G. Yarosh, Authors' cert. 182150, 1965; Byull. Izobr., No. 11, p. 20 (1966).
60. N. V. Komarov and O. G. Yarosh, Authors' cert. 188494, 1965; Izobr., Prom. Obr., Tov. Zn., No. 22, p. 37 (1966).

61. P. A. Konstantinovich and R. I. Shupik, Zh. Obshch. Khim., 33:1251 (1963).
62. A. P. Kreshkov, V. A. Drozdov, and R. R. Tarasyants, Zav. Lab., 30:143 (1964).
63. A. P. Kreshkov, V. A. Drozdov, and R. R. Tarasyants, Authors' cert. 161948, 1964; Byull. Izobr., No. 8, p. 66 (1964).
64. A. P. Kryuchkova, Investigation of Derivatives of Organosilicon Carboxylic Acids, Author's abstract of Cand. dissertation, MINKhGP, Moscow (1966).
65. O. Kubaschewski and E. L. Evans, Metallurgical Thermochemistry, London (1951).
66. A. G. Kuznetsova and V. I. Ivanov, Plast. Massy, No. 10, p. 17 (1963).
67. N. S. Leznov, L. A. Sabun, and K. A. Andrianov, Zh. Obshch. Khim., 29:1270 (1959).
68. É. Ya. Lukevits, Izv. Akad. Nauk Latv.SSR, Ser. Khim., 1963:111.
69. É. Ya. Lukevits and M. G. Voronkov, Kh. G. S., 1965:36.
70. É. Ya. Lukevits and S. Giller, Izv. Akad. Nauk Latv.SSR, No. 4, p. 95 (1961).
71. M. Ya. Marova, M. G. Voronkov, and B. N. Dolgov, Zh. Prikl. Khim., 30:650 (1957).
72. L. A. Mai and R. P. Kalvishkis, Authors' cert. 139320, 1961; Byull. Izobr., No. 13, p. 22 (1961).
73. L. A. Mai and R. P. Kalvishkis, Izv. Akad. Nauk Latv.SSR, Ser. Khim., 1962:473.
74. L. A. Mai and R. P. Kalvishkis, Izv. Akad. Nauk Latv.SSR, Ser. Khim., 1963:240.
75. L. A. Mai, R. P. Kalvishkis, and O. Ya. Neiland, Izv. Akad. Nauk Latv.SSR, Ser. Khim., 1962:147.
76. L. A. Mai and T. Miller, Authors' cert. 162659, 1964; Byull. Izobr., No. 10, p. 59 (1964).
77. L. K. Maslyi and T. T. Razbegaeva, Zh. Obshch. Khim., 37:250 (1967).
78. V. F. Mironov, Carbofunctional Organic Compounds of Silicon and Germanium, Author's abstract of Doctor's dissertation, INKhS AN SSSR, Moscow (1961).
79. V. F. Mironov, V. P. Kozyukov, and V. D. Sheludyakov, Zh. Obshch. Khim., 36:1860 (1966).
80. V. F. Mironov and A. L. Kravchenko, Izv. Akad. Nauk SSSR, Ser. Khim., 1965:1026.
81. V. F. Mironov and N. A. Pogonkina, Izv. Akad. Nauk SSSR, Otdel. Khim. Nauk, 1956:707.
82. V. F. Mironov and N. A. Pogonkina, Izv. Akad. Nauk SSSR, Otdel. Khim. Nauk, 1957:1199.
83. V. F. Mironov and N. A. Pogonkina, Izv. Akad. Nauk SSSR, Otdel. Khim. Nauk, 1959:85.
84. V. F. Mironov and N. A. Pogonkina, Izv. Akad. Nauk SSSR, Otdel. Khim. Nauk, 1960:1998.
85. V. F. Mironov and N. A. Pogonkina, Authors' cert. 126495, 1960; Byull. Izobr., No. 5, p. 17 (1960).
86. B. N. Mikhailov and A. N. Blokhina, Zh. Obshch. Khim., 30:3615 (1960).

87. M. M. Morgunova and D. Ya. Zhinkin, Plast. Massy, No. 5, p. 16 (1965).
88. M. M. Morgunova, D. Ya. Zhinkin, K. K. Popkov, and K. A. Andrianov, Scientific Communications of the International Symposium on Organosilicon Chemistry, B/25-317, Prague (1965).
89. N. S. Nametkin, A. V. Topchiev, and L. S. Povarov, Dokl. Akad. Nauk SSSR, 117:245 (1957).
90. N. S. Nametkin, A. V. Topchiev, T. I. Chernysheva, and L. I. Kartasheva, Izv. Akad. Nauk SSSR, Otdel. Khim. Nauk, 1963:654.
91. N. S. Nametkin, T. I. Chernysheva, and L. I. Kartasheva, Dokl. Akad. Nauk SSSR, 156:608 (1964).
92. O. M. Nefedov, V. I. Shiryaev, R. A. Strazdynya, and M. N. Manakov, Authors' cert. 187796, 1965; Izobr., Prom. Obr., Tov. Zn., No. 21, p. 42 (1966).
93. A. D. Petrov, V. F. Mironov, and A. A. Bugorkova, Fette-Seife-Anstrichmittel, 62:1107 (1960).
94. A. D. Petrov, V. F. Mironov, and V. G. Glukhovtsev, Dokl. Akad. Nauk SSSR, 93:499 (1953).
95. A. D. Petrov, V. F. Mironov, and V. G. Glukhovtsev, Zh. Obshch. Khim., 27:1535 (1957).
96. A. D. Petrov, V. F. Mironov, and N. A. Pogonkina, Dokl. Akad. Nauk SSSR, 100:81 (1955).
97. A. F. Plate, N. A. Belikova, and Yu. P. Egorov, Dokl. Akad. Nauk SSSR, 102:1131 (1955).
98. A. F. Plate, N. A. Belikova, and Yu. P. Egorov, Izv. Akad. Nauk SSSR, Otdel. Khim. Nauk, 1956:1085.
99. L. S. Povarov, Synthesis and Properties of Organosilicon Compounds with Siloxane–Carbon, Silazan–Carbon, and Silthiane–Carbon Chains, Author's abstract of Cand. dissertation, IOKh AN SSSR, Moscow (1955).
100. N. A. Pogonkina, Investigation of Carbofunctional Organosilicon Compounds, Author's abstract of Cand. dissertation, IOKh AN SSSR, Moscow (1960).
101. N. A. Pogonkina and V. F. Mironov, Authors' cert. 126495, 1960; Byull. Izobr., No. 5, p. 17 (1960).
102. V. O. Reikhsfel'd and E. P. Lebedev, Zh. Obshch. Khim., 37:1412 (1967).
103. K. A. Rzhendzinskaya and I. K. Stavitskii, in: Proceedings of the Conference on the Chemistry and Practical Use of Organosilicon Compounds, No. 2, TsBTI LSNKh (1958), p. 82.
104. T. V. Samsonov, Silicides and Their Use in Technology, Izd. AN USSR, Kiev (1959).
105. V. M. Svetozarova and I. K. Stavitskii, in: Proceedings of the Conference on the Chemistry and Practical Use of Organosilicon Compounds, No. 2, TsBTI LSNKh (1958), p. 71.
106. I. K. Stavitskii, Khim. Nauka i Prom., 2:331 (1957).
107. I. K. Stavitskii, B. E. Neimark, and Z. M. Kryukovskaya, Authors' cert. 115674, 1958; Byull. Izobr., No. 10, p. 77 (1958).
108. I. K. Stavitskii, B. E. Neimark, Z. M. Kryukovskaya, V. A. Kirichenko, and V. N. Churmaeva, Proceedings of the Conference on the Chemistry

and Practical Use of Organosilicon Compounds, No. 2, TsBTI LSNKh (1958), p. 57.

109. I. K. Stavitskii, B. E. Neimark, Z. M. Kryukovskaya, V. A. Kirichenko, and V. N. Churmaeva, Proceedings of the Conference on the Chemistry and Practical Use of Organosilicon Compounds, No. 2, TsBTI LSNKh (1958), p. 203.

110. N. G. Tolstikova, Homolytic Silylation of Alkyl and Chloro Derivatives of Benzene, Naphthalene, and Thiphene, Author's abstract of Cand. dissertation, IOKh AN SSSR, Moscow (1964).

111. A. V. Topchiev, N. S. Nametkin, and L. S. Povarov, "Organosilicon compounds," Izv. Akad. Nauk SSSR, 1962:130.

112. N. S. Fedotov and V. F. Mironov, in: Organosilicon Compounds (Proceedings of conference), Vol. 1, Moscow (1966), p. 7.

113. E. Sh. Finkel'shtein, Synthesis and Some Properties of Derivatives of Silaindane and Silatetralin, Author's abstract of Cand. dissertation, INKhS AN SSSR, Moscow (1965).

114. I. Khaiduk and K. A. Andrianov, Izv. Akad. Nauk SSSR, Otdel. Khim. Nauk, 1963:1537.

115. E. A. Chernyshev, A. A. Zelenetskaya, and T. L. Krasnova, Izv. Akad. Nauk SSSR, Ser. Khim., 1966:1118.

116. E. A. Chernyshev and G. F. Pavelko, Izv. Akad. Nauk SSSR, Ser. Khim., 1966:2205.

117. E. A. Chernyshev and N. G. Tolstikova, Izv. Akad. Nauk SSSR, Otdel. Khim. Nauk, 1964:1700.

118. E. A. Chernyshev and N. G. Tolstikova, Authors' cert. 172786, 1964; Byull. Izobr., No. 14, p. 24 (1965).

119. E. A. Chernyshev, N. G. Tolstikova, A. A. Ivashenko, A. A. Zelenetskaya, and L. A. Leites, Izv. Akad. Nauk SSSR, Otdel. Khim. Nauk, 1963:660.

120. T. I. Chernysheva, N. S. Nametkin, N. A. Pritula, and L. I. Kartasheva, Plaste und Kautschuk, 10:390 (1963).

121. V. G. Shakhovskoi, Investigation of 1,3-Diyne Organosilicon Hydrocarbons, Author's abstract of Cand. dissertation, LTI im. Lensoveta, Leningrad (1966).

122. M. F. Shostakovskii, A. S. Atavin, V. M. Nikitin, B. A. Trofimov, V. V. Keiko, and V. I. Lavrov, Izv. Akad. Nauk SSSR, Ser. Khim., 1965:2049.

123. M. F. Shostakovskii, N. V. Komarov, and N. N. Vlasova, Scientific Communications of the International Symposium on Organosilicon Chemistry, A/28-21, Prague (1965).

124. M. F. Shostakovskii, N. V. Komarov, N. N. Vlasova, and G. A. Rinkus, Zh. Obshch. Khim., 36:904 (1966).

125. M. F. Shostakovskii, N. V. Komarov, and O. G. Yarosh, Izv. Akad. Nauk SSSR, Ser. Khim., 1966:101.

126. Yu. K. Yur'ev and Z. V. Belyakova, Zh. Obshch. Khim., 29:1458 (1959).

127. Yu. K. Yur'ev, Z. V. Belyakova, and V. P. Volkov, Zh. Obshch. Khim., 29:1463 (1959).

128. Yu. K. Yur'ev, Z. V. Belyakova, and V. P. Volkov, Zh. Obshch. Khim., 29:3652 (1959).

2. Publications of Foreign Authors

129. E. W. Abel, J. Chem. Soc., 1960:4406.
130. E. W. Abel, J. Chem. Soc., 1961:4933.
131. E. W. Abel and D. A. Armitage, J. Chem. Soc., Suppl. 2, 1964:5975.
132. E. W. Abel and D. A. Armitage, J. Organomet. Chem., 5:326 (1966).
132a. E. W. Abel and D. A. Armitage, "Organosulfur derivatives of silicon, germanium, tin, and lead," in: Advances in Organometallic Chemistry (ed. by F. G. A. Stone and R. West), Vol. 5, Academic Press, New York, London (1967), p. 1.
133. E. W. Abel, D. A. Armitage, and D. B. Brady, J. Organomet. Chem., 5:130 (1966).
133a. E. W. Abel, D. A. Armitage, and D. B. Brady, Trans. Far. Soc., 62:3459 (1966).
134. E. W. Abel, D. A. Armitage, and R. P. Bush, J. Chem. Soc., 1964:2455.
135. E. W. Abel, D. A. Armitage, and R. P. Bush, J. Chem. Soc., Suppl. 1, 1964:5584.
136. E. W. Abel, D. A. Armitage, and R. P. Bush, J. Chem. Soc., 1965:3045.
137. E. W. Abel, D. A. Armitage, and R. P. Bush, J. Chem. Soc., 1965:7098.
138. A. D. Abbot and R. O. Bolt (California Research Corp.), USA Patent 2776307, 1957; C. A., 51:7401 (1957).
139. A. D. Abbot, J. R. Wright, A. Goldschmidt, W. T. Stewart, and R. O. Bolt, J. Chem. Eng. Data, 6:437 (1961).
140. A. Almenningen, K. Hedberg, and R. Seip, Acta Chem. Scand., 17:2264 (1963).
141. R. Amberg, Stahl und Eisen, 29:146 (1909); C., 1909(I):798.
142. B. Anders and H. Malz, Federal Germ. Rep. Patent 1215144, 1966; C. A., 65:3908 (1966).
143. H. H. Anderson, J. Am. Chem. Soc., 67:223 (1945).
144. H. H. Anderson, J. Am. Chem. Soc., 67:2176 (1945).
145. H. H. Anderson, J. Am. Chem. Soc., 69:3049 (1947).
146. H. H. Anderson, J. Am. Chem. Soc., 70:1220 (1948).
147. H. H. Anderson, J. Am. Chem. Soc., 71:1801 (1949).
148. H. H. Anderson, J. Am. Chem. Soc., 72:193 (1950).
149. H. H. Anderson, J. Am. Chem. Soc., 72:194 (1950).
150. H. H. Anderson, J. Am. Chem. Soc., 72:196 (1950).
151. H. H. Anderson, J. Am. Chem. Soc., 72:2761 (1950).
152. H. H. Anderson, Angew. Chem., 66:714 (1950).
153. H. H. Anderson, J. Am. Chem. Soc., 73:2351 (1951).
154. H. H. Anderson, J. Am. Chem. Soc., 74:2371 (1952).
155. H. H. Anderson, J. Am. Chem. Soc., 75:1576 (1953).
155a. H. H. Anderson, J. Org. Chem., 19:1766 (1954).
156. H. H. Anderson, IUPAC Coll. Münster (1954), Silicium, Schwefel, Phosphate, Verl. Chemie GmbH, Weinheim (1955), s. 37.
157. H. H. Anderson, J. Am. Chem. Soc., 81:4785 (1959).
158. H. H. Anderson, J. Org. Chem., 26:276 (1961).
159. H. H. Anderson, Inorg. Chem., 3:910 (1964).

160. H. H. Anderson and H. J. Fischer, J. Org. Chem., 19:1296 (1954).
161. H. H. Anderson and L. R. Grebe, J. Org. Chem., 26:2006 (1961).
162. H. H. Anderson and A. Hendifar, J. Org. Chem., 26:3033 (1961).
163. H. H. Anderson and G. M. Stanislow, J. Org. Chem., 18:1716 (1953).
164. B. J. Aylett, H. J. Emeleus, and A. G. Maddock, Research, London, 6:30 S (1953).
165. H. J. Backer and W. Drenth, Rec. Trav. Chim., 70:559 (1951).
166. H. J. Backer and H. A. Klasens, Rec. Trav. Chim., 61:500 (1942).
167. H. J. Backer and F. Stienstra, Rec. Trav. Chim., 51:1197 (1932).
168. H. J. Backer and F. Stienstra, Rec. Trav. Chim., 52:912 (1933).
169. H. J. Backer and F. Stienstra, Rec. Trav. Chim., 54:38 (1935).
170. H. J. Backer and F. Stienstra, Rec. Trav. Chim., 54:607 (1935).
171. D. L. Bailey (Union Carbide Corp.), USA Patent 2955128, 1960; C. A., 55:4430 (1961).
172. D. L. Bailey (Union Carbide Corp.), USA Patent 2968643, 1961; C. A., 55:10387 (1961).
173. R. F. Barrow, Trans. Far. Soc., 36:1053 (1940).
174. R. F. Barrow, Proc. Phys. Soc., 56:204 (1944).
175. R. F. Barrow, Nature, 154:364 (1944).
176. R. F. Barrow, Proc. Phys. Soc., 58:606 (1946).
177. R. F. Barrow and W. Jevons, Nature, 141:833 (1938).
178. R. F. Barrow and W. Jevons, Proc. Roy. Soc., A169:45 (1939).
179. V. Bažant, V. Chvalovsky, and J. Rathousky, Organosilicon Compounds, Vol. 1, Prague (1965).
180. M. Becke-Goehring and G. Wunsch, Ann., 618:43 (1958).
181. M. Becke-Goehring and G. Wunsch, Chem. Ber., 93:326 (1960).
182. H. Behrens and J. Ostermeyer, Chem. Ber., 95:487 (1962).
183. R. A. Benkeser and R. B. Currie, J. Am. Chem. Soc., 70:1780 (1948).
184. R. A. Benkeser and H. Landesmann, J. Am. Chem. Soc., 71:2493 (1949).
185. R. A. Benkeser and A. Torkelson, J. Am. Chem. Soc., 76:1252 (1954).
186. P. L. D. Benneville and M. I. Hurwitz (Rohm. a. Hass Co.), USA Patent 2876209, 1959; C. A., 53:12321 (1959).
187. A. Berger and J. A. Magnuson, Anal. Chem., 36:1156 (1964).
188. E. Berger, C. R., 170:1492 (1920).
189. J. J. Berzeliuss, Pogg. Ann., 1:216 (1824).
190. J. J. Berzeliuss, Pogg. Ann., 8:411 (1826).
191. A. Besson, C. R., 113:1040 (1891); C., 1892(I):272.
192. W. Biltz, Z. Anorg. Chem., 146:289 (1925).
193. W. Biltz and F. Caspari, Z. Anorg. Chem., 71:182 (1911).
194. L. Birkofer, Belgian Patent 615401, 1962; C. A., 59:2826 (1963).
195. L. Birkofer, W. Konkol, and A. Ritter, Chem. Ber., 94:1263 (1961).
196. L. Birkofer, A. Ritter, and H. Goller, Chem. Ber., 96:3289 (1963).
197. L. Birkofer, A. Ritter, and P. Richter, Tetrahedron Letters, 5:195 (1962).
198. T. A. Bither, W. H. Knoth, R. V. Lindsey, and W. H. Sharkey, J. Am. Chem. Soc., 80:4151 (1958).
199. L. C. F. Blackman and M. J. S. Dewar, J. Chem. Soc., 1957:169.

200. F. Blazy, J. Bonastre, and G. Pfister-Guillauzo, Bull. Soc. Chim. Fr., 1966:2136.
201. M. Blix, Ber., 36:4218 (1903).
202. M. Blix and W. Wirbelauer, Ber., 36:4220 (1903).
203. B. A. Bluestein (General Electric Co.), USA Patent 2559340, 1951; C. A., 46:1581 (1952).
204. P. V. Bonsignore, C. S. Marvel, and S. Banerjee, J. Org. Chem., 25:237 (1960).
205. R. N. Bott, C. Eaborn, and T. Hashimoto, J. Organomet. Chem., 3:442 (1965).
206. R. W. Bott, C. Eaborn, and B. M. Rushton, J. Organomet. Chem., 3:448 (1965).
207. P. Bourgeois, R. Calas, and N. Duffaut, Bull. Soc. Chim. Fr., 1965:2694.
208. P. Bourgeois, R. Calas, and N. Duffaut, Bull. Soc. Chim. Fr., 1966:1171.
209. P. Bourgeois, R. Calas, N. Duffaut, and C. Hou, Bull. Soc. Chim. Fr., 1965:1255.
210. P. Bourgeois and N. Duffaut, Bull. Soc. Chim. Fr., 1965:2697.
211. G. Brauer, Handbuch der präparativen anorganischen Chemie, Bd. I, Stuttgart (1960).
212. H. Breedervald, Rec. Trav. Chim., 79:1126 (1960).
213. H. Breedervald, Rec. Trav. Chim., 81:276 (1962).
214. British Thomson-Houston Co., Ltd., British Patent 688407, 1949; C. A., 47:6976 (1953).
215. British Thomson-Houston Co., Ltd., British Patent 643941, 1950; C. A., 45:7585 (1951).
216. British Thomson-Houston Co., Ltd., British Patent 671879, 1952; C. A., 46:8146 (1952).
217. British Thomson-Houston Co., Ltd., British Patent 688408, 1953; C. A., 48:10055 (1954).
218. British Thomson-Houston Co., Ltd., British Patent 694440, 1953; C. A., 48:10775 (1954).
219. British Thomson-Houston Co., Ltd., British Patent 695461, 1953; C. A., 48:8264 (1954).
220. British Thomson-Houston Co., Ltd., British Patent 695462, 1953; C. A., 47:10894 (1953).
221. M. J. Buerger and R. D. Butler, Amer. Miner., 23:471 (1938); C., II, p. 3514 (1938).
222. G. F. Bulbenko (Thiokol Chemical Corp.), French Patent 1433372, 1966; C. A., 66:2635h (1967).
223. W. J. Burke and W. A. Hoffman (E. J. du Pont de Nemours and Co.), USA Patent 2515857, 1950; C. A., 44:8943 (1950).
224. C. A. Burkhard, J. Am. Chem. Soc., 67:2173 (1945); C. A., 40:1150 (1946).
225. C. A. Burkhard, J. Am. Chem. Soc., 72:1078 (1950).
226. C. A. Burkhard (General Electric Co.), USA Patent 2544296, 1951; C. A., 45:7786 (1951).
227. C. A. Burkhard (General Electric Co.), USA Patent 2563516, 1951; C. A., 46:1580 (1952).

228. C. A. Burkhard (General Electric Co.), USA Patent 2583322, 1952; C. A., 46:8670 (1952).
229. C. A. Burkhard (General Electric Co.), USA Patent 2604486, 1952; C. A., 47:12798 (1953).
230. C. A. Burkhard (General Electric Co.), USA Patent 2604487, 1952; C. A., 47:3334 (1953).
231. A. Buschfeld, Dissertation, Techn. Hochschule Aachen, 1962; cited from [374].
232. W. Bussem, H. Fischer, and E. Gruner, Naturwiss., 23:740 (1935); C., 1936(I):2052.
233. A. Bygdén, Diss., Uppsala, 1916; C. A., 14:1974 (1920).
234. A. Bygdén, J. Prakt. Chem., 96:88 (1917).
235. R. Calas, P. Bourgeois, and N. Duffaut, C. R., 263:243 (1966).
236. R. Calas, N. Duffaut, and P. Bourgeois, Bull. Soc. Chim. Fr., 1964:9.
237. R. Calas, N. Duffaut, B. Martel, and C. Paris, Bull. Soc. Chim. Fr., 1961:886.
238. L. Cambi, Atti. Acad. Lincei [5], 19(II):294 (1910); C., 1910(II):1863.
239. L. Cambi, Atti Acad. Linceri [5], 20(I):433 (1911); C., 1911(II):263.
240. L. Cambi and G. G. Monselise, Gazz., 66:696 (1936).
241. G. L. Carlson, Spectrochim. Acta, 18:1529 (1962).
242. F. Challenger and F. S. Kipping, J. Chem. Soc., 97:142 (1910).
243. F. Challenger and F. S. Kipping, J. Chem. Soc., 97:755 (1910).
244. G. de Chalmot, Am. Chem. J., 19:871 (1897).
245. M. M. Chamberlain, G. A. Jabs, and B. B. Wayland, J. Org. Chem., 27:3321 (1962).
246. G. Champetier, Y. Étiennes, and R. Kullmann, C. R., 234:1985 (1952).
247. A. Colson, C. R., 94:1526 (1882).
248. A. Colson, Bull. Soc. Chim. Fr., [2], 38:56 (1882).
249. G. D. Cooper, J. Am. Chem. Soc., 76:2499 (1954).
250. G. D. Cooper, J. Am. Chem. Soc., 76:2500 (1954).
251. G. D. Cooper, J. Am. Chem. Soc., 76:3713 (1954).
252. G. D. Cooper (General Electric Co.), USA Patent 2719165, 1955; C. A., 50:8709 (1956).
253. G. D. Cooper, J. Org. Chem., 21:1214 (1956).
254. G. D. Cooper (General Electric Co.), USA Patent 2789121, 1957; C. A., 51:12129 (1957).
255. G. D. Cooper (General Electric Co.), USA Patent 2811540, 1957; R. Zh. Khim., 1959:83661.
256. N. D. Costeanu, C. R., 156:1985 (1913).
257. N. D. Costeanu, Ann. Chim. Phys., [9], 2:189 (1914); C. R., 1915(II):65.
258. J. B. Culbertson, H. W. Erasmus, and R. M. Fowler (Union Carbide Carbon Corp.), USA Patent 2569455, 1951; C. A., 46:3084 (1952).
259. J. B. Culbertson, H. W. Erasmus, and R. M. Fowler (Union Carbide Carbon Corp.), USA Patent 2569746, 1951; C. A., 46:3566 (1952).
260. J. B. Culbertson, H. W. Erasmus, and R. M. Fowler (Union Carbide Carbon Corp.), USA Patent 2569747, 1951; C. A., 46:3558 (1952).
261. C. C. J. Culvenor and N. S. Ham, Chem. Comm., 1966:537.

262. C. W. N. Cumper, A. Melnikoff, and A. J. Vogel, J. Chem. Soc., 1966(A):242.

263. C. W. N. Cumper, A. Melnikoff, and A. J. Vogel, J. Chem. Soc., 1966(A):246.

264. C. W. N. Cumper, A. Melnikoff, and A. J. Vogel, J. Chem. Soc., 1966(A):323.

265. W. H. Daudt (Corning Glass Works), British Patent 585589, 1947; C. A., 45:9918 (1951).

266. W. H. Daudt (Corning Glass Works), USA Patent 2451664, 1948; C. A., 43:1803 (1949).

267. F. B. Deans and C. Eaborn, J. Chem. Soc., 1959:2299.

268. F. B. Deans and C. Eaborn, J. Chem. Soc., 1959:2303.

269. A. W. Dearing and E. E. Reid, J. Am. Chem. Soc., 50:3058 (1928).

270. Q. W. Decker, Silaorganic Polysulfides, Diss. Abstr., 18:1979 (1958).

271. Q. W. Decker and H. W. Post, J. Org. Chem., 25:249 (1960).

272. W. Dilthey, Ber., 36:3207 (1903).

273. W. Dilthey, Ann., 344:300 (1905).

274. P. Dolch, Chem. Fabrik, 8:512 (1935); C., 1936(I):2520.

275. Dow Corning Corp., Netherlands Applic. 6516388, 1966; C. A., 65:15612 (1966).

276. A. J. Downs and E. A. V. Ebsworth, J. Chem. Soc., 1960:3516.

277. N. Duffaut and P. Bourgeois, Bull. Soc. Chim. Fr., 1964:1723.

278. N. Duffaut, R. Calas, and J. Dunouges, Bull. Soc. Chim. Fr., 1960:597.

279. N. Duffaut, R. Calas, and J. Dunouges, Bull. Soc. Chim. Fr., 1961:886.

280. N. Duffaut, R. Calas, and J. Dunouges, Bull. Soc. Chim. Fr., 1963:512.

281. N. Duffaut, R. Calas, and B. Martel, Bull. Soc. Chim. Fr., 1960:597.

282. N. Duffaut, B. Martel, A. Villemiane, and R. Calas, Bull. Soc. Chim. Fr., 1962:1533.

283. Ch. H. Van Dyke and A. G. MacDiarmid, Inorg. Chem., 3:1071 (1964).

284. C. Eaborn, Nature, 165:685 (1950).

285. C. Eaborn, J. Chem. Soc., 1950:3077.

286. C. Eaborn, J. Chem. Soc., 1953:494.

287. C. Eaborn, Organosilicon Compounds, Butterworths Scientific Publications, London (1960).

288. C. Eaborn and T. Hashimoto, Chem. Ind., 1961:1081.

289. C. Eaborn and J. A. Sperry, J. Chem. Soc., 1961:4921.

290. E. A. V. Ebsworth, R. Mould, R. Taylor, G. R. Wilkinson, and L. A. Woodward, Trans. Far. Soc., 58:1069 (1962).

291. E. A. V. Ebsworth, M. Onyszchuk, and N. Sheppard, J. Chem. Soc., 1958:1453.

292. E. A. V. Ebsworth, R. Taylor, and L. A. Woodward, Trans. Far. Soc., 55:211 (1959).

293. Elektrokemisk AIS, Norwegian Patent 74969, 1949; C. A., 44:3682 (1950).

294. H. J. Emeléus, B. J. Aylett, A. G. MacDiarmid, and A. G. Maddock, IUPAC Coll. Munster, 1954, Silicium, Schwefel, Phosphate, Verl. Chemie GmbH. Wienheim (1955), p. 50.

295. H. J. Emeléus and H. G. Heal, J. Chem. Soc., 1946:1126.

296. H. J. Emeléus, A. G. MacDiarmid, and A. G. Maddock, J. Inorg. Nucl. Chem., 1:194 (1955).

297. H. J. Emeléus and M. Onyszchuk, J. Chem. Soc., 1958:604.

298. H. J. Emeléus, M. Onyszchuk, and W. Kuchen, Z. Anorg. Chem., 283:74 (1956).
299. H. J. Emeléus and L. E. Smythe, J. Chem. Soc., 1958:609.
300. Y. Étienne, C. r., 235:966 (1952).
301. Y. Étienne, Bull. Soc. Chim. Fr., 1953:791.
302. Farbenfabriken Bayer A. G., British Patent 791609, 1958; C. A., 52:19947 (1958).
303. Farbenfabriken Bayer A. G., Netherlands Applic. 6408328, 1965; C. A., 63:18153 (1965).
304. Farbenfabriken Bayer A. G., Netherlands Applic. 6408441, 1965; C. A., 63:16616 (1965).
305. Farbenfabriken Bayer A. G., Netherlands Applic. 6408329, 1965; C. A., 63:632 (1965).
306. Farbenwerke Hoechst A. G., Belgian Patent 667089, 1966; C. A., 65:7313 (1966).
307. F. Fehér and A. Blümcke, Chem. Ber., 90:1934 (1957).
308. F. Fehér and K. Lippert, Chem. Ber., 94:2437 (1961).
309. R. J. Fessenden and M. D. Coon, J. Org. Chem., 29:1607 (1964).
310. R. J. Fessenden and M. D. Coon, J. Org. Chem., 29:2499 (1964).
311. R. J. Fessenden, J. G. Larsen, M. D. Coon, and J. S. Fessenden, J. Med. Chem., 7:695 (1964).
312. R. J. Fessenden, K. Seeler, and M. Dagani, J. Org. Chem., 31:2483 (1966).
313. W. Fielding, Chem. News, 100:14 (1909); C., 1909(II):1086.
314. M. Fild, W. Sundermeyer, and O. Glemser, Chem. Ber., 97:620 (1964).
315. E. A. Flood, J. Am. Chem. Soc., 55:1735 (1933).
316. R. H. Flowers, R. J. Gillespie, and E. A. Robinson, Can. J. Chem., 41:2464 (1963).
317. G. S. Forbes and H. H. Anderson, J. Am. Chem. Soc., 67:1911 (1945).
318. G. S. Forbes and H. H. Anderson, J. Am. Chem. Soc., 69:3048 (1947).
319. W. Förster and H. Kriegsmann, Z. Anorg. Chem., 326:186 (1963).
320. W. Förster and H. Kriegsmann, Z. Anorg. Chem., 327:305 (1964).
321. J. A. Forstner and E. L. Muetterties, Inorg. Chem., 5:552 (1966).
322. E. W. Foster and P. E. Koenig, (Ethyl Corp.), USA Patent 2998407, 1956; C. A., 56:6170 (1962).
323. M. Frankel, I. Belsky, D. Gertner, and A. Zilkha, J. Chem. Soc., 1966(C):493.
324. M. Frankel, M. Broze, D. Gertner, and A. Zilkha, J. Chem. Soc., 1966(C):379.
325. M. Frankel, D. Gertner, A. Zilkha, and A. Shenhar, Israeli Patent 19902, 1966; C. A., 65:20215 (1966).
326. E. Fremy, Ann. Chim. Phys., [3], 38:312 (1853).
327. E. Fremy, C. r., 36:178 (1853).
328. C. Friedel and A. Landenburg, Bull. Soc. Chim. Fr., [2], 7:472 (1867).
329. C. Friedel and A. Ladenburg, C. r., 64:1295 (1867).
330. C. Friedel and A. Ladenburg, Ann., 145:179 (1868).
331. C. Friedel and A. Ladenburg, Ann. Chim. Phys., [4], 27:416 (1872).
332. H. Gabriel and C. Alvarez-Tostado, J. Am. Chem. Soc., 74:262 (1952).
333. J. J. Gallagher (Imperial Chemical Industries, Ltd.), British Patent 1042865, 1966; C. A., 65:20248 (1966).

334. A. Gauthier, C. r., 107:911 (1888).
335. A. Gauthier, C. r., 132:740 (1901).
336. A. Gauthier, C. r., 143:7 (1906).
337. A. Gauthier and L. Hallopeau, C. r., 108:806 (1889).
338. P. F. Gawrys and H. W. Post, J. Org. Chem., 27:634 (1962).
339. General Electric Co., British Patent 869844, 1961; C. A., 56:3515 (1962).
340. General Electric Co., British Patent 869845, 1961; C. A., 55:26497 (1961).
341. P. D. George (General Electric Co.), USA Patent 2640818, 1953; C. A., 47:9055 (1953).
342. P. D. George (General Electric Co.), USA Patent 2640833, 1953; C. A., 48:5227 (1954).
343. P. D. George (General Electric Co.), USA Patent 2645644, 1953; C. A., 48:7064 (1954).
344. P. D. George (General Electric Co.), USA Patent 2802853, 1957; C. A., 51:17992 (1957).
345. P. D. George, J. Org. Chem., 26:4235 (1961).
346. P. D. George, M. Prober, and J. R. Elliot, Chem. Rev., 56:1065 (1956).
347. M. C. L. Gerry, J. C. Thompson, and T. M. Sugden, Nature, 211:846 (1966).
348. D. Gertner and A. Schenhar, Israel J. Chem., 2:109 (1964); C. A., 61:16170 (1964).
349. R. J. Gillespie, J. Am. Chem. Soc., 82:5978 (1960).
350. H. Gilman, R. A. Benkeser, and G. E. Dunn, J. Am. Chem. Soc., 72:1689 (1950).
351. H. Gilman, L. F. Cason, and H. G. Brooks, J. Am. Chem. Soc., 75:3760 (1953).
352. H. Gilman and J. W. Diehl, J. Org. Chem., 26:2938 (1961).
353. H. Gilman, B. Hofferth, and H. W. Melvin, J. Org. Chem., 72:3045 (1950).
354. H. Gilman and G. D. Lichtenwalter, J. Org. Chem., 25:1064 (1960).
355. H. Gilman and O. L. Marrs, J. Org. Chem., 30:2096 (1965).
356. H. Gilman and F. J. Marshall, J. Am. Chem. Soc., 71:2066 (1949).
357. H. Gilman and J. F. Nobis, J. Am. Chem. Soc., 72:2629 (1950).
358. H. Gilman and M. A. Plunkett, J. Am. Chem. Soc., 71:1117 (1949).
359. H. Gilman and G. N. R. Smart, J. Org. Chem., 16:424 (1951).
360. H. Gilman and D. S. Swayampati, J. Am. Chem. Soc., 79:208 (1957).
361. H. Gilman and G. R. Wilder, J. Org. Chem., 22:523 (1957).
362. H. Gilman and D. Wittenberg, J. Am. Chem. Soc., 79:6339 (1957).
363. P. A. Di Giorgio, W. A. Strong, L. H. Sommer, and F. C. Whitmore, J. Am. Chem. Soc., 68:1380 (1946).
364. K. W. Glombitza, Kongr. Pharm. Wiss., Münster, 1963, p. 491; C. A., 62:7744 (1965).
365. K. W. Glombitza, Ann., 673:166 (1964).
366. G. R. Glowacki and H. W. Post, J. Org. Chem., 27:634 (1962).
367. G. A. Gornowicz and J. W. Ryan, J. Org. Chem., 31:3439 (1966).
368. J. Goubeau and E. Heubach, Z. Phys. Chem., 25:271 (1960).
369. J. Goubeau and W. D. Hieresemane, Z. Anorg. Chem., 290:292 (1957).

370. J. Goubeau and J. Reyhing, Z. Anorg. Chem., 294:96 (1958).
371. V. Gutmann, P. Heilmayer, and K. Utvary, Mh. Chem., 1961:942.
372. A. Haas, Dissertation, Techn. Hochsch. Aachen (1960); cited from [374].
373. A. Haas, Chem. Ber., 97:2189 (1964).
374. A. Haas, Angew. Chem., 77:1066 (1965).
375. L. A. Haluska (Dow Corning Corp.), French Patent 1413604, 1965; C. A., 64:2126 (1966).
376. E. Hayek, A. Czaloun, and B. Krismer, Monatsch., 87:741 (1956).
377. H. Hecht, Präparative Anorganische Chemie, Berlin (1951).
378. W. Hempel and V. Haasy, Z. Anorg. Chem., 23:32 (1900).
379. E. Hengge and U. Brychcy, Monatsh., 97:1309 (1966).
380. E. Hengge and K. Pretzer, Chem. Ber., 96:470 (1963).
381. J. M. Hersch (Continental Oil Co.), USA Patent 2464231, 1949; C. A., 43:8210 (1949).
382. G. W. Holbrook (Dow Corning Corp.), USA Patent 2833801 (1958); C. A., 52:16787 (1958).
383. G. W. Holbrook, A. F. Gordon, and O. R. Pierce, J. Am. Chem. Soc., 82:825 (1960).
384. O. A. Homberg and I. Hechenbleikner (Carlisle Chemical Works, Inc.), French Patent 1365375, 1964; C. A., 61:14711 (1964).
385. K. A. Hooton and A. L. Allread, Inorg. Chem., 4:671 (1965).
386. Huang Chih-tang and Wang Pao-jen, Acta Chim. Sinica, 23:291 (1957).
387. Huang Chih-tang and Wang Pao-jen, Acta Chim. Sinica, 25:330 (1959).
388. B. J. Humphrey and H. H. Wasserman (Connecticut Hard Rubber Co.), USA Patent 2673843, 1954; C. A., 48:8580 (1954).
389. D. T. Hurd, J. Am. Chem. Soc., 77:2998 (1955).
390. M. J. Hurwitz and P. L. De Benneville (Rohm a. Hass Co.), USA Patent 2865899, 1958; C. A., 53:5735 (1959).
391. J. F. Hyde (Dow Corning Corp.), USA Patent 2571039, 1951; C. A., 46:2837 (1952).
392. G. Illuminati, J. F. Nobis, and H. Gilman, J. Am. Chem. Soc., 73:5887 (1951).
393. F. M. Jaeger and D. W. Dijkstra, Z. Anorg. Chem., 143:79 (1925).
394. D. R. Jenkins, R. Kewley, and T. M. Sugden, Proc. Chem. Soc., 1960:220.
395. D. R. Jenkins, R. Kewley, and T. M. Sugden, Trans. Far. Soc., 58:1284 (1962).
396. K. A. Jensen, A. Holm, B. Kägi, and C. T. Pedersen, Acta Chem. Scand., 19:772 (1965).
397. H. W. Johnston (Union Carbide Corp.), USA Patent 3170940, 1965; C. A., 63:4332 (1965).
398. H. Jorg and I. Stetter, J. Prkt. Chem., 117:305 (1927).
399. R. M. Joyce (E. J. du Pont de Nemours & Co.), USA Patent 2485603, 1949; C. A., 44:1281 (1950).
400. M. Kanazashi, Bull. Chem. Soc. Japan, 28:44 (1955).
401. S. W. Kantor (General Electric Co.), USA Patent 2997457, 1961; C. A., 55:27939 (1961).

402. W. H. Keeber and H. W. Post, J. Org. Chem., 21:509 (1956).
403. E. Kieffer and J. Czech, Ber. Ceram. Ges., 19:54 (1938); C. 1938(I):3514.
404. K. Kimura, K. Katada, and S. H. Bauer, J. Am. Chem. Soc., 88:416 (1966).
405. F. S. Kipping, Proc. Chem. Soc., 21:65 (1905).
406. F. S. Kipping, Proc. Chem. Soc., 23:9 (1907).
407. F. S. Kipping, Proc. Chem. Soc., 23:83 (1907).
408. F. S. Kipping, J. Chem. Soc., 91:209 (1907).
409. F. S. Kipping, J. Chem. Soc., 91:717 (1907).
410. F. S. Kipping, J. Chem. Soc., 93:457 (1908).
411. F. S. Kipping, J. Chem. Soc., 95:408 (1909).
412. F. S. Kipping, J. Chem. Soc., 119:647 (1921).
413. F. S. Kipping and H. Davies, J. Chem. Soc., 95:69 (1909).
414. F. S. Kipping and G. Martin, J. Chem. Soc., 95:489 (1909).
415. H. A. Klasens, Chem. Weekbl., 38:448 (1941); C., 1942(I):984.
416. H. A. Klasens and H. J. Backer, Rec. Trav. Chim., 58:941 (1939).
417. H. A. Klasens and H. J. Backer, Rec. Trav. Chim., 61:513 (1942).
418. G. Koerner (Th. Goldschmidt A. G.), Federal Germ. Rep. Patent 1173898, 1964; C. A., 61:10707 (1964).
419. G. Koerner and G. Rossmy (Th. Goldschmidt A. G.), Federal Germ. Rep. Patent 1163818, 1964; C. A., 60:14539 (1964).
420. E. J. Kohlmeyer and H. W. Retzlaft, Z. Anorg. Chem., 261:248 (1950).
421. E. J. Kohlmeyer and X. Siebers, Germ. Patent 534984, 1931; C., 1931(II):3524.
422. K. Kojima, N. Tarumi, and S. Wakatuki, J. Chem. Soc. Japan, Pure Chem. Sect., 76:1205 (1955); C. A., 51:17369 (1957).
423. E. Krause and G. Renwanz, Ber., 62:1710 (1929).
424. H. Kriegsmann, Angew. Chem., 69:145 (1957).
425. H. Kriegsmann, Z. Elektrochem., 61:1088 (1957).
426. H. Kriegsmann, Chem. Technik, 9:508 (1957).
427. H. Kriegsmann, Z. Anorg. Chem., 294:113 (1958).
428. H. Kriegsmann, Z. Anorg. Chem., 299:138 (1959).
429. H. Kriegsmann and H. Clauss, Z. Anorg. Chem., 300:210 (1959).
430. W. Kuchen, Z. Anorg. Chem., 288:101 (1956).
431. M. Kučera, M. Lanikova, and E. Spousta, Czech Patent 110064, 1964; C. A., 61:9604 (1964).
432. M. Kučera and E. Spousta, J. Polymer Sci., A2:3443 (1964).
433. M. Kučera and E. Spousta, Makrom. Chem., 76:183 (1964).
434. M. Kumada, M. Ishikawa, S. Maeda, and K. Ikura, J. Organomet. Chem., 2:146 (1964).
435. M. Kumada, M. Ishikawa, and K. Tamao, J. Organomet. Chem., 5:226 (1966).
436. M. Kumada, K. Naka, and M. Ishikawa, J. Organomet. Chem., 2:136 (1964).
437. M. Kumada, K. Naka, and Y. Yamamoto, Bull. Chem. Soc., Japan, 37:871 (1964).
438. M. Kumada, M. Yamaguchi, Y. Yamamoto, J. Nakajima, and K. Shiina, J. Org. Chem., 21:1264 (1956).
439. Y. Kurita and M. Kondo, Bull. Chem. Soc. Japan, 27:160 (1954).

440. A. Ladenburg, Ber., 5:565 (1872).
441. A. Ladenburg, Ber., 40:2274 (1907).
442. S. H. Langer, S. Connell, and J. Wender, J. Org. Chem., 23:50 (1958).
443. E. Larsson, Trans. Chalmers Univ. Techn., Gothenburg, 79:13 (1948); C. A., 43:2929 (1949).
444. E. Larsson, Trans. Chalmers Univ. Techn., Gothenburg, 79:17 (1948); C. A., 43:2929 (1949).
445. E. Larsson, Svensk. Kem. Tidskr., 60:178 (1948).
446. E. Larsson, Trans. Chalmers Univ. Techn., Gothenburg, 115:21 (1951); C. A., 47:11124 (1953).
447. E. Larsson and C. G. Carlsson, Acta Chem. Scand., 4:45 (1950).
448. E. Larsson and R. Marin, Acta Chem. Scand., 5:964 (1951).
449. E. Larsson and R. E. I. Marin (Uddeholms Aktiebolag), Swedish Patent 138357, 1952; C. A., 48:2761 (1954).
450. A. Ledebur, Berg-hüttenmänn. Ztg., 37:321 (1878).
451. B. Lengyel, A. Prékopa, P. Révész, and F. Török, J. Phys. Chem., 280:33 (1957).
452. B. Lengyel, A. Prékopa, and F. Török, Z. Phys. Chem., 206:161 (1956).
453. G. D. Lichtenwalter, Organosilylmetallic Compounds and Derivatives, Diss. Abstr., 20:101 (1959).
454. R. V. Lindsey, J. Am. Chem. Soc., 73:371 (1951).
455. J. W. Linnett, Nature, 199:168 (1963).
456. H. R. Linton, Vibrational Spectra and Structures of Silyl Compounds, Diss. Abstr., 18:687 (1958).
457. H. R. Linton and E. R. Nixon, J. Chem. Phys., 29:921 (1958).
458. R. J. Lisanke (Union Carbide Corp.), USA Patent 3008922, 1962; C. A., 56:8744 (1962).
459. R. J. Lisanke (Union Carbide Corp.), USA Patent 3008924, 1962; C. A., 56:8745 (1962).
460. B. D. W. Luff and F. S. Kipping, J. Chem. Soc., 93:2004 (1908).
461. B. D. W. Luff and F. S. Kipping, J. Chem. Soc., 93:2090 (1908).
462. J. Lugauer, Dissertation, Techn. Hochsch. Aachen (1960), cited from [374].
463. H. Lumbroso, Bull. Soc. Chim. Fr., 1959:887.
464. A. G. MacDiarmid, J. Inorg. Nucl. Chem., 2:88 (1956).
465. A. G. MacDiarmid, J. Inorg. Nucl. Chem., 2:323 (1956).
466. A. G. MacDiarmid, Quart. Rev., 10:208 (1956).
467. A. G. MacDiarmid, J. Inorg. Nucl. Chem., 25:1534 (1963).
468. A. G. MacDiarmid and A. G. Maddock, J. Inorg. Nucl. Chem., 1:411 (1955).
469. A. G. MacDiarmid, J. J. Moscony, C. R. Russ, and T. Yoshioka, Sc. Comm. International Symposium on Organosilicon Chemistry, A/20-100, Prague (1965).
470. L. Malatesta, Gazz., 78:702 (1948).
471. L. Malatesta, Gazz., 78:753 (1948).
472. L. Malatesta, Italian Patent 436808, 1948; C. A., 43:4630 (1949).
473. H. Marsden and F. S. Kipping, J. Chem. Soc., 93:198 (1908).
474. J. Marsden and J. F. Roedel (General Electric Co.), USA Patent 2469883, 1949; C. A., 43:5624 (1949).

475. B. Martel, R. Calas, and N. Duffaut, Bull. Soc. Chim. Fr., 1964:9.
476. B. Martel, R. Calas, and N. Duffaut, Bull. Soc. Chim. Fr., 1966:1173.
477. B. Martel, R. Calas, and N. Duffaut, Bull. Soc. Chim. Fr., 1966:2134.
478. B. Martel and N. Duffaut, C. R., 263(C):74 (1966).
479. B. Martel, N. Duffaut, J. Pellaroque, and R. Calas, Bull. Soc. Chim. Fr., 1963:2007.
480. C. A. Marvel and H. N. Cripps, J. Polymer Sci., 9:53 (1952).
481. D. L. Mayfield, R. A. Flath, and L. R. Best, J. Org. Chem., 29:2444 (1964).
482. J. J. McBride, J. Org. Chem., 24:2029 (1959).
483. J. J. McBride and H. C. Beachell, J. Am. Chem. Soc., 74:5247 (1952).
484. J. A. McHard, Chem. Eng., 55:228 (1948).
485. R. N. Meals (General Electric Co.), USA Patent 3269982, 1966; C. A., 65:20337 (1966).
486. R. H. Meen and H. Gilman, J. Org. Chem., 20:73 (1955).
487. H. W. Melvin, Iowa State College J. Sci., 30:413 (1956); R. Zh. Khim., 1957:57582.
488. R. L. Merker (Dow Corning Corp.), USA Patent 2863898, 1959; C. A., 53:9060 (1959).
489. K. H. Meyer, Natural and Synthetic High Polymers, Interscience Publishers, New York (1950).
490. Z. Michalowski, Wiad. Chem., 13:543 (1957).
491. Midland Silicones Ltd., British Patent 772986, 1957; C. A., 51:9673 (1957).
492. Midland Silicones Ltd., British Patent 778272, 1957; C. A., 52:429 (1958).
493. Midland Silicones Ltd., British Patent 833142, 1960; C. A., 54:17269 (1960).
494. D. H. Miles, Low-Melting Organosilicon Monomers of High Molecular Weight, Diss. Abstr., 17:2821 (1957).
495. F. A. Miller, Pure Appl. Chem., 7:125 (1963).
496. N. E. Miller, Inorg. Chem., 4:1458 (1965).
497. A. O. Minklei, Q. W. Decker, and H. W. Post, Rec. Trav. Chim., 76:187 (1957).
498. P. Miquel, Ber., 9:852 (1876).
499. P. Miquel, Ann. Chim. Phys., [5], 11:289 (1877).
500. P. Miquel, Bull. Soc. Chim. Fr., [2], 28:103 (1877).
501. K. Moedritzer, Organomet. Chem. Rev., 1:179 (1966).
502. K. Moedritzer and J. R. Van Wazer, J. Organomet. Chem., 6:242 (1966).
503. K. Moedritzer and J. R. Van Wazer, Inorg. Nucl. Chem. Letters, 2:45 (1966).
504. K. Moedritzer and J. R. Van Wazer, J. Chem. Phys., 42:2478 (1965).
505. D. Mohler and J. E. Sellers (General Electric Co.), USA Patent 2598436, 1952; C. A., 47:3875 (1953).
506. H. Moissan, C. R., 134:1083 (1902).
507. H. Moissan and Q. Dilthey, C. R., 134:503 (1902).
508. K. R. Molt, I. Hechenbleiker, and O. A. Homberg (Carlisle Chemical Works, Inc.), USA Patent 3208966, 1965; C. A., 64:8409 (1966).
509. Monsanto Chemical Ltd., French Patent, Suppl. 81295, 1963; C. A., 60:1794 (1964).
510. G. G. Monselise, Gazz., 67:748 (1937).

511. G. G. Monselise, Atti X Congr. Int. Chim., 2:732 (1938); C., 1940(I):188.
512. E. L. Morehouse (Union Carbide Corp.), USA Patent 2938046, 1960; C. A., 55:1066 (1961).
513. R. C. Morris and J. L. Van Winkle (Shell Development Co.), USA Patent 2719860, 1955; C. A., 50:10125 (1956).
514. M. Morton and A. Deisz (US Dept. of the Army), USA Patent 2960492, 1960; C. A., 55:7884 (1961).
515. L. S. Moody (General Electric Co.), USA Patent 2567742, 1951; C. A., 47:145 (1953).
516. L. S. Moody (General Electric Co.), USA Patent 2567724, 1951; C. A., 47:7534 (1953).
517. R. Müller and C. Dathe (Inst. für Silikon und Fluorokarbon-Chemie), Belgian Patent 669338, 1965; C. A., 65:13760 (1966).
518. R. Müller and R. Koehne, Z. Anorg. Chem., 311:142 (1961).
519. J. E. Mulvaney, Addition of Mercaptans to Vinyl Silanes—Aspects of Stereospecific Polymerization, Diss. Abstr., 20:1587 (1959).
520. L. D. Nasiak and H. W. Post, J. Org. Chem., 24:492 (1959).
521. B. Neumann, Stahl und Eisen, 29:355 (1909); C., 1909(I):1615.
522. R. G. Neville and J. J. McGee, Can. J. Chem., 41:2123 (1963).
523. M. S. Newman, R. A. Craig, and A. B. Garret, J. Am. Chem. Soc., 71:869 (1949).
524. H. Niebergall, Chem. Ber., 90:1235 (1957).
525. H. Niebergall (Koppers Co., Inc.), Federal Germ. Rep. Patent 1118781, 1959; C. A., 56:11622 (1962).
526. H. Niebergall, Federal Germ. Rep. Patent 1093994, 1961; C. A., 55:20503 (1961).
527. S. Nitzsche and E. Pirson (Wacker-Chemie G. m. b. H.), USA Patent 3187033, 1965; C. A., 63:11615 (1965).
528. J. E. Noll, J. Am. Chem. Soc., 77:3149 (1955).
529. J. E. Noll (Dow Corning Corp.), USA Patent 2762826, 1956; C. A., 51:5827 (1957).
530. W. Noll; Chemie und Technology der Silicone, Verlag Chemie, Weinheim (1960).
531. D. C. Noller and H. W. Post, J. Org. Chem., 17:1393 (1952).
532. T. Nomura, M. Yokoi, and K. Yamasaki, Proc. Japan Acad., 29:342 (1953); C. A., 49:12273 (1955).
533. S. Nozakura, J. Chem. Soc. Japan, Pure Chem. Sect., 75:958 (1954); C. A., 51:14543 (1957).
534. S. Nozakura, Bull. Chem. Soc. Japan, 28:299 (1955).
535. G. Oertel, H. Holtschmidt, and H. Malz (Farbenfabriken Bayer A. G.), Federal Germ. Rep. Patent 1157226, 1964; C. A., 60:6868 (1964).
536. G. Oertel, H. Malz, and H. Holtschmidt, Chem. Ber., 97:891 (1964).
537. K. Oita and H. Gilman, J. Org. Chem., 22:336 (1957).
538. K. Olsson and C. Axen, Arkiv Kemi, 22:237 (1964).
539. B. A. Orkin (Socony Vacuum Oil Co. Inc.), USA Patent 2592175, 1952; C. A., 46:8414 (1952).

540. B. A. Orkin (Socony Vacuum Oil Co., Inc.), USA Patent 2626957, 1953; C. A., 47:6132 (1953).
541. B. A. Orkin (Socony Vacuum Oil Co., Inc.), USA Patent 2701803, 1955; C. A., 49:7236 (1955).
542. C. G. Overberger, J. E. Mulvaney, and F. M. Beringer, J. Org. Chem., 21:1311 (1956).
543. W. J. Owen and F. C. Saunders (Midland Silicones Ltd.), British Patent 1007333, 1965; C. A., 63:18154 (1965).
544. D. R. Pail (Dow Corning Corp.), French Patent 1345191, 1963; C. A., 61:5869 (1964).
545. D. J. Panckhurst, C. J. Wilkins, and P. W. Craighead, J. Chem. Soc., 1955:3395.
546. D. J. Panckhurst and C. J. Wilkins, Inorganic Syntheses, 7:28 (1963).
547. W. J. Pappel and R. S. Schiefelbein (Jefferson Chem. Co., Inc.), USA Patent 2752381, 1956; C. A., 51:2020 (1957).
548. W. I. Patnode and F. C. Schmidt, J. Am. Chem. Soc., 67:2272 (1945).
549. W. Patnode and D. F. Wilcock, J. Am. Chem. Soc., 68:358 (1946).
550. T. Perklev, Svensk Kem. Tidskr., 65:216 (1953).
551. T. Perklev, Svensk Kem. Tidskr., 65:253 (1953).
552. G. Petit and A. Seyyedi, C. R., 255:2061 (1962).
553. R. Piekoš and W. Wojnowski, Z. Anorg. Chem., 318:212 (1962).
554. I. Pierre, J. Prakt. Chem., 41:342 (1847).
555. I. Pierre, C. R., 24:814 (1847).
556. I. Pierre, Ann. Chim. Phys., [3], 24:286 (1848).
557. I. Pierre, C. R., 26:523 (1848).
558. I. Pierre, Ann., 64:259 (1848).
559. I. Pierre, J. Prakt. Chem., 46:65 (1849).
560. I. Pierre, Ann., 69:73 (1849).
561. I. Pierre and H. Kopp, C. R. Trav. Chim., 5:170 (1849).
562. E. P. Plueddemann (Dow Corning Corp.), Belgian Patent 628951, 1963; C. A., 60:16103 (1964).
563. F. H. Pollard, G. Nickless, and P. C. Uden, J. Chromatogr., 11:312 (1963).
564. J. M. Pollock (Imperial Chemical Industries, Ltd.), British Patent 923582, 1963; C. A., 60:3011 (1964).
565. B. O. Pray, L. H. Sommer, G. M. Goldberg, G. T. Kerr, P. A. Di Giorgio, and F.C. Whitmore, J. Am. Chem. Soc., 70:433 (1948).
566. C. T. Prewitt and H. S. Young, Science, 149:535 (1965); C. A., 63:9152 (1965).
567. F. P. Price, J. Am. Chem. Soc., 70:871 (1948).
568. M. Prober (General Electric Co.), USA Patent 2835690, 1958; C. A., 52:18216 (1958).
569. M. Prober (General Electric Co.), Federal Germ. Rep. Patent 1084264, 1960; C. A., 55:24567 (1961).
570. M. Prober (General Electric Co.), USA Patent 3185663, 1965; C. A., 63:8403 (1965).
571. I. G. Rankin and S. M. Revington, Proc. Chem. Soc., 24:131 (1908).

572. R. E. Reavill, J. Chem. Soc., 1964:519.
573. J. E. Reynolds, J. Chem. Soc., 89:397 (1906).
574. J. E. Reynolds, Proc. Chem. Soc., 22:17 (1906).
575. F. P. Richter and B. A. Orkin (Socony Vacuum Oil Co., Inc.), USA Patent 2590039, 1952; C. A., 46:5892 (1952).
576. R. V. Riley, J. Iron. Inst., 156:528 (1947); C. A., 41:6857 (1947).
577. M. Rimpler, Chem. Ber., 99:1523 (1966).
578. M. Rimpler, Chem. Ber., 99:1528 (1966).
579. O. Roder (Hans Heinrich Hütte G. m. b. H.), German Patent 627200, 1936; C., 1936(I):4492.
580. W. Rodziewicz, Z. Michalowski, and J. Prejzner, Roczn. Chem., 33:579 (1959).
581. W. Rodziewicz and W. Wojnowski, Roczn. Chem., 34:843 (1960).
582. R. Roquet and M. F. Ancey-Moret, Bull. Soc. Chim. Fr., 1954:1038.
583. G. Rossmy (Th. Goldschmidt A. G.), Federal Germ. Rep. Patent 1106324, 1961; C. A., 56:24721 (1962).
584. G. Rossmy (Th. Goldschmidt A. G.), Federal Germ. Rep. Patent 1157789, 1963; C. A., 60:12052 (1964).
585. G. Rossmy (Th. Goldschmidt A. G.), Federal Germ. Rep. Patent 1175643, 1964; C. A., 61:12161 (1964).
586. G. Rossmy (Th. Goldschmidt, A. G.), Federal Germ. Rep. Patent 1179937, 1964; C. A., 62:1690 (1965); C. A., 64:11249 (1966).
587. G. Rossmy and J. Wassermeyer (Th. Goldschmidt A. G.), Federal Germ. Rep. Patent 1108918, 1959; C. A., 56:1481 (1962).
588. G. Rossmy and J. Wassermeyer (Th. Goldschmidt A. G.), USA Patent 3109012, 1963; C. A., 61:7044 (1964).
589. K. Rühlmann, Chem. Ber., 94:1876 (1961).
590. K. Rühlmann and E. Ettenhuber, Chem. Ber., 98:2855 (1965).
591. I. Ruidisch and M. Schmidt, Angew. Chem., 75:1108 (1963).
592. I. Ruidisch and M. Schmidt, Chem. Ber., 96:1424 (1963).
593. C. R. Russ and A. G. MacDiarmid, Angew Chem., 78:391 (1966).
594. J. W. Ryan (Dow Corning Co.), French Patent 1345190, 1963; C. A., 60:5816 (1964).
595. P. Sabatier, C. R., 90:819 (1880).
596. P. Sabatier, Ann. Chim. Phys., [5], 22:5 (1881).
597. P. Sabatier, Bull. Soc. Chim. Fr., [2], 32:153 (1882).
598. Y. Sakata and T. Hashimoto, J. Pharm. Soc. Japan, 79:872 (1959); C. A., 54:357 (1960).
599. H. Sargent (U.S. Rubber Co.), USA Patent 3068241, 1962; C. A., 58:10238 (1963).
600. K. Sathienandan and J. L. Margrave, J. Mol. Spectr., 10:442 (1963).
601. O. J. Scherer and P. Hornig, Angew. Chem., 78:776 (1966).
602. O. Scherer and M. Schmidt, Angew. Chem., 75:139 (1963).
603. O. Scherer and M. Schmidt, Angew. Chem., 75:1115 (1963).
604. O. Scherer and M. Schmidt, Naturwiss., 50:302 (1963).
605. O. Scherer and M. Schmidt, Z. Naturf., 18b:415 (1963).

606. M. Schmeisser (Kali-Chemie A.G.), Federal Germ. Rep. Patent 1008265, 1957; C. A., 54:3886 (1960).
607. M. Schmeisser and W. Burgemeister, Angew. Chem., 69:782 (1957).
608. M. Schmeisser and A. Haas, Sc. Comm. Inter. Symposium on Organosilicon Chemistry, A/25-125, Prague (1965).
609. M. Schmeisser and H. Müller, Angew. Chem., 69:781 (1957).
610. M. Schmeisser, H. Müller, and W. Burgemeister, Angew. Chem., 69:781 (1957).
611. H. Schmidbaur, J. Am. Chem. Soc., 85:2336 (1963).
612. H. Schmidbaur, Z. Anorg. Chem., 326:272 (1964).
613. H. Schmidbaur, Chem. Ber., 98:83 (1965).
614. H. Schmidbaur, and I. Ruidisch, Inorg. Chem., 3:559 (1964).
615. H. Schmidbaur, and M. Schmidt, Chem. Ber., 94:2137 (1961).
616. H. W. Schmidt, M. Schmeisser, and M. Jenker (Karlie-Chemie A. G.), British Patent 738703, 1953; C. A., 50:16825 (1956).
617. M. Schmidt and I. Ruidisch (Wasag-Chemie A. G.), Federal Germ. Rep. Patent 1190462, 1965; C. A., 63:631 (1965).
618. M. Schmidt, I. Ruidisch, and H. Schmidbaur, Chem. Ber., 94:2451 (1961).
619. M. Schmidt and H. Schmidbaur, Angew. Chem., 70:469 (1958).
620. M. Schmidt and H. Schmidbaur, Angew. Chem., 70:470 (1958).
621. M. Schmidt and H. Schmidbaur, Angew. Chem., 70:657 (1958).
622. M. Schmidt and H. Schmidbaur, Angew. Chem., 71:384 (1959).
623. M. Schmidt and H. Schmidbaur, Chem. Ber., 93:878 (1960).
624. M. Schmidt and H. Schmidbaur, Chem. Ber., 94:2446 (1961).
625. M. Schmidt and H. Schmidbaur, Chem. Ber., 95:47 (1962).
626. M. Schmidt and H. Schumann, Z. Anorg. Chem., 325:130 (1963).
627. M. Schmidt and G. Talyky, Chem. Ber., 94:1352 (1961).
628. M. Schmidt and M. Wieber, Inorg. Chem., 1:909 (1962).
629. M. Schmidt and M. Wieber, Federal Germ. Rep. Patent 1210192, 1966; C. A., 64:17747 (1966).
630. M. Schmidt and I. Wilhelm, Chem. Ber., 97:876 (1964).
631. E. Schnell and G. Wersin, Monatsh., 92:647 (1961).
632. R. S. Schreiber (E. J. du Pont de Nemours and Co.), USA Patent 2465339, 1949; C. A., 43:6220 (1949).
633. G. E. Schroll (Ethyl Corp.), USA Patent 2914548, 1959; C. A., 54:7651 (1960).
634. W. C. Schumb and W. J. Bernard, J. Am. Chem. Soc., 77:862 (1955).
635. W. C. Schumb and W. J. Bernard, J. Am. Chem. Soc., 77:904 (1955).
636. R. Schwarz, Z. Anorg. Chem., 276:33 (1954).
637. D. W. Scott, J. Am. Chem. Soc., 68:2294 (1946).
638. D. Seyferth, J. Am. Chem. Soc., 79:5881 (1957).
639. A. G. Sharkey, R. A. Friedel, and S. H. Langer, Anal. Chem., 29:770 (1957).
640. L. M. Shorr, The Mechanism of Methyl-Silicon Cleavage of Certain Substituted Carboxylic Acids in Sulfuric Acid, Diss. Abstr., 14:766 (1954).
641. L. M. Shorr, H. Freiser, and J. L. Speier, J. Am. Chem. Soc., 77:547 (1955).
642. X. Siebers, Diss., Berlin (1930); cited from [374].

643. X. Siebers and E. I. Kohlmeyer, Arch. Erzbergbau, Erzaufber. Metallhüttenwes., 1:120 (1931); cited from [374].
644. M. S. Silverman and J. R. Soulen, Inorg. Chem., 4:129 (1965).
645. W. Simmler, Chem. Ber., 96:349 (1963).
646. W. Simmler, Angew. Chem., 75:859 (1963).
647. W. Simmler and H. Niederprüm (Farbenfabriken Bayer A. G.), Belgian Patent 634126, 1963; C. A., 61:1893 (1964).
648. W. Simmler and H. Niederprüm (Farbenfabriken Bayer A. G.), Belgian Patent 667386, 1965; C. A., 65:13759 (1966).
649. W. Simmler, H. Niederprüm, and H. Jonas (Farbenfabriken Bayer A. G.), French Patent 1378592, 1964; C. A., 63:5677 (1965).
650. W. Simmler, H. Walz, and H. Niederprüm, Chem. Ber., 96:1495 (1963).
651. B. O. H. Sjoberg and A. B. Ekstrom, Belgian Patent 628231, 1963; C. A., 61:1870 (1964).
652. S. Sliwinski, Germ. Dem. Rep. Patent 11977, 1956; C. A., 53:2681 (1959).
653. F. A. Smith (Union Carbide and Carbon. Corp.), USA Patent 2569784, 1951; C. A., 46:5073 (1952).
654. Société Monsavon-l'Oréal, French Patent 1157158, 1958; C. A., 54:18364 (1960).
655. Société Monsavon-l'Oréal, British Patent 827419, 1960; C. A., 54:19494(1960).
656. L. H. Sommer and G. R. Ansul, J. Am. Chem .Soc., 77:2482 (1955).
657. L. H. Sommer, D. L. Bailey, W. A. Strong, and F. C. Whitmore, J. Am. Chem. Soc., 68:1881 (1946).
658. L. H. Sommer, W. P. Barie, and J. R. Gould, J. Am. Chem. Soc., 75:3765 (1953).
659. L. H. Sommer and G. A. Baum, J. Am. Chem. Soc., 76:5002 (1954).
660. L. H. Sommer, W. D. English, G. R. Ansul, and D. N. Vivona, J. Am. Chem. Soc., 77:2485 (1955).
661. L. H. Sommer and F. J. Evans, J. Am. Chem. Soc., 76:1186 (1954).
662. L. H. Sommer, G. T. Kerr, and F. C. Whitmore, J. Am. Chem. Soc., 70:445 (1948).
663. L. H. Sommer, N. S. Marans, G. M. Goldberg, J. Rockett, and R. P. Pioch, J. Am. Chem. Soc., 73:882 (1951).
664. L. H. Sommer and J. McLick, J. Am. Chem. Soc., 88:5359 (1966).
665. L. H. Sommer, E. W. Pietrusza, G. T. Kerr, and F. C. Whitmore, J. Am. Chem. Soc., 68:156 (1946).
666. L. H. Sommer, E. W. Pietrusza, and F. C. Whitmore, J. Am. Chem. Soc., 68:2282 (1946).
667. L. H. Sommer, R. P. Pioch, N. S. Marans, G. M. Goldberg, J. Rockett, and J. Kerlin, J. Am. Chem. Soc., 75:2932 (1953).
668. L. H. Sommer, L. J. Tyler, and F. C. Whitmore, J. Am. Chem. Soc., 70:2872 (1948).
669. J. L. Speier (Dow Corning Corp.), USA Patent 2746956, 1957; C. A., 51:1246 (1957).
670. J. A. Sperry, U. S. Dept. Com. Office Tech. Serv., PB Rept., 145953, p. II (1959); C. A., 58:4592 (1963).

671. W. Stamm, J. Org. Chem., 28:3264 (1963).
672. A. J. Starshak, R. D. Joyner, and M. E. Kenney, Inorg. Chem., 5:330 (1966).
673. B. Sternbach and A. G. MacDiarmid, J. Inorg. Nucl. Chem., 23:225 (1961).
674. A. Stock and C. Somieski, Ber., 56:247 (1923).
675. W. Sundermeyer, Angew. Chem., 77:241 (1965).
676. S. V. Sunthankar and H. Gilman, J. Org. Chem., 15:1200 (1950).
677. S. V. Sunthankar and H. Gilman, J. Org. Chem., 18:47 (1953).
678. H. H. Szmant, O. M. Devlin, and G. A. Brost, J. Am. Chem. Soc., 73:3059 (1951).
679. Y. Takami, J. Soc. Org. Synth. Chem., Japan, 19:449 (1961).
680. M. Takeda and A. Yamada, Chem. High Polymer, 17:154 (1960).
681. T. Takiguchi and M. Abe, J. Chem. Soc. Japan, Ind. Chem. Sect., 68:679 (1965); C. A., 63:4323 (1965).
682. J. S. Thayer and D. P. Strommen, J. Organomet. Chem., 5:383 (1966).
682a. J. S. Thayer and R. West, Organometallic Pseudohalides, in: Advances in Organometallic Chemistry (ed. F. G. A. Stone and R. West), Vol. 5, Academic Press, New York, London (1967), p. 169.
683. Th. Goldschmidt A. G., British Patent 945181, 1963; C. A., 60:16069 (1964).
684. Th. Goldschmidt A. G., British Patent 1011320, 1965; C. A., 64:8237 (1966).
685. M. Thimann, Ueber die chemischen Bildungs-bedingungen des Silicium-disulfides und die Auffindung seiner Phosphoreszenzfähigkeit, Diss., Berlin (1927).
686. E. Tiede and M. Thimann, Ber., 59:1703 (1926).
687. E. Tiede and M. Thimann, Ber., 59:1706 (1926).
688. G. D. Tiers (Minnesota Mining and Manufacturing Co.), USA Patent 3141898, 1964; C. A., 61:9527 (1964).
689. G. D. Tiers and R. J. Coon, Org. Chem., 26:2095 (1961).
690. R. F. Toomey, J. Org. Chem., 17:473 (1952).
691. R. F. Toomey, Diss. Abstr., 21:1386 (1960).
692. E. E. Vago and R. F. Barrow, Proc. Phys. Soc., 58:538 (1946).
693. E. Vigouroux, C. R., 120:367 (1895).
694. W. L. Walton (General Electric Co.), USA Patent 2636895, 1953; C. A., 48:4002 (1954).
695. Wang Chi-Tao, Chou Hsui-Chung, and Hung Man-Shui, Acta Chim. Sinica, 30:91 (1964).
696. U. Wannagat, Angew. Chem., 75:173 (1963).
697. U. Wannagat and O. Brandstaetter, Monatsh., 94:1090 (1963).
698. U. Wannagat and H. Kuckertz, Angew. Chem., 74:117 (1962).
699. U. Wannagat, J. Pump, and H. Bürger, Monatsh., 94:1013 (1963).
700. L. G. L. Ward and A. G. MacDiarmid, J. Inorg. Nucl. Chem., 21:287 (1961).
701. J. R. Van Wazer, K. Moedritzer, and L. C. D. Groenneghe, J. Organomet. Chem., 5:420 (1966).
702. A. Weiss and Q. Rocktäschel, Z. Anorg. Chem., 307:1 (1960).
703. A. Weiss and A. Weiss, Z. Anorg. Chem., 276:95 (1954).
704. A. Weiss and A. Weiss, Angew. Chem., 66:714 (1954).
705. A. Weiss and A. Weiss, IUPAC Coll. Münster (1954), Silicium, Schwefel, Phosphate, Verl. Chemie GmbH, Weinheim (1955), s. 41.

706. R. West, Proc. International Symposium Mol. Struct. Spectr., Tokyo (1962), D 117/1.
707. K. Wetterlin, Acta Chem. Scand., 18:899 (1964).
708. J. White and H. Skelly, J. Iron. Inst., 156:529 (1947); C. A., 41:6857 (1947).
709. M. Wieber and M. Schmidt, Angew. Chem., 74:902 (1962).
710. M. Wieber and M. Schmidt, Angew. Chem., 75:1116 (1963).
711. M. Wieber and M. Schmidt, J. Organomet. Chem., 1:22 (1963).
712. M. Wieber and M. Schmidt, Chem. Ber., 96:1019 (1963).
713. M. Wieber and M. Schmidt, Chem. Ber., 96:1561 (1963).
714. M. Wieber and M. Schmidt, Chem. Ber., 96:2822 (1963).
715. M. Wieber and M. Schmidt, Z. Naturf., 18b:846 (1963).
716. M. Wieber and M. Schmidt, Z. Naturf., 18b:849 (1963).
717. M. Wieber and M. Schmidt, J. Organomet. Chem., 2:129 (1964).
718. C. J. Wilkins and L. E. Sutton, Trans. Far. Soc., 50:783 (1954).
719. W. Wirbelauer, Beiträge zur Chemie des Siliciums, Diss., Berlin (1904).
720. D. Wittenberg, H. A. McNinch, and H. Gilman, J. Am. Chem. Soc., 80:5418 (1958).
721. D. Wittenberg, T. C. Wu, and H. Gilman, J. Org. Chem., 23:1898 (1958).
722. W. Wojnowski, Rocz. Chem., 38:115 (1964).
723. W. Wojnowski, Rocz. Chem., 38:1263 (1964).
724. W. Wojnowski and R. Piekoš, Polish Patent 46366, 1962; R. Zh. Khim., 23H57P (1963).
725. W. Wojnowski and R. Piekoš, Z. Anorg. Chem., 314:189 (1962).
726. L. Wolinski, Univ. Microf., Pub. Nr. 2567; C. A., 46:2482 (1952).
727. L. Wolinski, H. Tieckelmann, and H. W. Post, J. Org. Chem., 16:395 (1951).
728. L. Wolinski, H. Tieckelmann, and H. W. Post, J. Org. Chem., 16:1134(1951).
729. L. Wolinski, H. Tieckelmann, and H. W. Post, J. Org. Chem., 16:1138 (1951).
730. M. Yokoi, Bull. Chem. Soc., Japan, 30:100 (1957).
731. M. Yokoi, Bull. Chem. Soc., Japan, 30:106 (1957).
732. M. Yokoi, T. Nomura, and K. Yamasaki, J. Am. Chem. Soc., 77:4484 (1955).
733. E. L. De Young and R. W. Watson (Standard Oil Co.), USA Patent 3192164, 1965; C. A., 63:8106 (1965).
734. A. Zappel (Farbenfabriken Bayer A. G.), Federal Germ. Rep. Patent 1000817, 1957; C. A., 53:13054 (1959).
735. N. Zenzen, Ark. Kemi, Mineral. Geol., 8:34 (1923); C. A., 18:922 (1924).
736. E. Zintl, W. Bräuning, H. L. Grube, W. Krings, and W. Morawietz, Z. Anorg. Chem., 245:1 (1940).
737. E. Zintl and K. Loosen, Z. Phys. Chem. (Leipzig), A174:301 (1935).
738. A. Zinzen, Z. Ver. Deut. Ing., 88:171 (1944); C. A., 39:401 (1945).

3. Supplement to Literature References for Chapter Two

739. D. N. Andreev, B. N. Dolgov, and S. V. Buts, Zh. Obshch. Khim., 32:1275 (1962).
740. D. N. Andreev and L. I. Lavrinovich, Zh. Obshch. Khim., 38:2743 (1968).

741. N. F. Baina, O. I. Kochkina, and L. Z. Soborovskii, Authors' cert. 226607, 1967; Izobr. Prom. Obr., Tov. Zn., No. 29, p. 21 (1968).
742. N. G. Bokii and Yu. T. Struchkov, Zh. Strukt. Khim., 9:722 (1968).
743. M. N. Bochkarev, Synthesis and Properties of Some Organosilicon, Germanium, and Tin Chalcogenides, Author's abstract of Cand. dissertation, Gor'kii (1968).
744. M. N. Bochkarev, L. P. Sanina, and N. S. Vyazankin, Zh. Obshch. Khim., 39:135 (1969).
745. T. D. Burnasheva, Unsaturated Organosilicon and Tin Compounds Based on Diacetylene, Author's abstract of Cand. dissertation, Leningrad (1968).
746. V. M. Vdovin, Investigation of Compounds with silicon-carbon Heterocycles, Author's abstract of Doctor's dissertation, Moscow (1968).
747. N. S. Vyazankin, M. N. Bochkarev, and L. P. Sanina, Zh. Obshch. Khim., 37:1037 (1967).
748. N. S. Vyazankin, M. N. Bochkarev, and L. P. Sanina, Zh. Obshch. Khim., 37:1545 (1967).
749. N. S. Vyazankin, M. N. Bochkarev, and L. P. Sanina, Zh. Obshch. Khim., 38:414 (1968).
750. N. S. Vyazankin, M. N. Bochkarev, L. P. Sanina, A. N. Egorochkin, and S. Ya. Khorshev, Zh. Obshch. Khim., 37:2576 (1967).
751. N. S. Vyazankin, E. N. Gladyshev, M. N. Bochkarev, O. A. Kruglaya, G. S. Kalinina, and G. A. Razuvaev, Abstracts of Reports to the Fourth Conference on the Chemistry and Use of Organosilicon Compounds, NIITÉKhIM, Moscow (1968), p. 34.
752. M. L. Galashina, G. A. Matveeva, M. V. Sobolevskii, E. A. Chernyshev, and N. G. Tolstikova, Authors' cert. 190571, 1966; Byull. Izobr., No. 2, p. 88 (1967).
753. D. Ya. Zhinkin, Preparation and Properties of Organoaminosilanes and Organosilazans, Author's abstract of Doctor's dissertation, Moscow (1968).
754. A. M. Evdokimov, E. P. Lebedev, and V. O. Reikhsfel'd, Scientific and Technological Conference of the Lensovet Leningrad Technological Institute, Section of Organic Chemistry and Technology, Brief Communications, Leningrad (1968), p. 30.
755. A. N. Egorochkin, A. I. Burov, N. S. Vyazankin, V. I. Savushkina, V. Z. Anisimova, and E. A. Chernyshev, Dokl. Akad. Nauk SSSR, 185:351 (1969).
756. A. N. Egorochkin, N. S. Vyazankin, M. N. Bochkarev, V. T. Bychkov, and A. I. Burov, Zh. Obshch. Khim., 38:396 (1968).
757. A. N. Egorochkin, A. I. Burov, V. F. Mironov, T. K. Gar, and N. S. Vyazankin, Dokl. Akad. Nauk SSSR, 180:861 (1968).
758. A. N. Egorochkin, S. Ya. Khorshev, and N. S. Vyazankin, Dokl. Akad. Nauk SSSR, 185:353 (1969).
759. A. N. Egorochkin, S. Ya. Khorshev, N. S. Vyazankin, M. N. Bochkarev, O. A. Kruglaya, and G. S. Semchikova, Zh. Obshch. Khim., 37:2308 (1967).
760. A. N. Egorochkin, Š. Ya. Khorshev, N. S. Vyazankin, E. N. Gladyshev, V. T. Bychkov, and O. A. Kruglaya, Zh. Obshch. Khim., 38:276 (1968).

761. N. V. Eliseeva, V. V. Kopylov, T. L. Krasnova, A. N. Pravednikov, and E. A. Chernyshev, Abstracts of Reports to the Fourth Conference on the Chemistry and Use of Organosilicon Compounds, NIITÉKhIM, Moscow., p. 761 (1968).
762. L. I. Kartasheva, N. S. Nametkin, and T. I. Chernysheva, Abstracts of Reports to the Fourth Conference on the Chemistry and Use of Organosilicon Compounds, NIITÉKhIM, Moscow, p. 9 (1968).
763. N. V. Komarov, Investigation of Acetylenic Organosilicon, Tin, and Lead Compounds, Author's abstract of Doctor's dissertation, Novosibirsk (1968).
764. N. V. Komarov, T. D. Burnashova, N. N. Vlasova, and Z. I. Mikhailov, Authors' cert. 191559, 1965; Izobr., Prom. Obr., Tov. Zn., No. 4, p. 34 (1967).
765. N. V. Komarov and N. N. Vlasova, Authors' cert. 178372, 1963; Byull. Izobr., No. 3, p. 23 (1966).
766. N. V. Komarov and N. N. Vlasova, Authors' cert. 202141, 1966; Izobr., Prom. Obr., Tov. Zn., No. 19, p. 33 (1967).
767. N. V. Komarov, Z. I. Mikhailov, and N. N. Vlasova, Izv. Akad. Nauk SSSR, Ser. Khim., 1968:905.
768. N. V. Komarov, V. G. Rozinov, L. P. Vakhrushev, E. F. Grechkin, and N. F. Chernov, Izv. Akad. Nauk SSSR, Ser. Khim., 1969:729.
769. N. V. Komarov, É. V. Serebrennikova, and Z. I. Mikhailov, Abstracts of Reports to the Fourth Conference on the Chemistry and Use of Organosilicon Compounds, NIITÉKhIM, Moscow, p. 44 (1968).
770. N. V. Komarov and O. G. Yarosh, Authors' cert. 188494, 1966; Izobr., Prom. Obr., Tov. Zn., No. 22, p. 37 (1966); C. A., 67:43917e (1967).
771. N. V. Komarov and O. G. Yarosh, Zh. Obshch. Khim., 38:202 (1968).
772. N. V. Komarov, O. G. Yarosh, and G. A. Kalabin, Izv. Akad. Nauk SSSR, Ser. Khim., 1967:690.
773. A. M. Kuliev, A. A. Dzhafarov, and F. N. Mamedov, Authors' cert. 196835, 1966; Izobr., Prom. Obr., Tov. Zn., No. 12, p. 35 (1967).
774. A. M. Kuliev, A. A. Dzhafarov, F. N. Mamedov, I. I. Namazov, and I. A. Aslanov, Authors' cert. 229513 (1967); Izobr., Prom. Obr., Tov. Zn., No. 33, p. 39 (1968).
775. A. M. Kuliev, G. A. Zeinalova, and N. S. Kyazimova, Authors' cert. 199884, 1966; Izobr., Prom. Obr., Tov. Zn., No. 16, p. 24 (1967).
776. A. M. Kuliev, N. S. Kyazimova, and G. A. Zeinalova, Azerb. Khim. Zh., No. 6, p. 57 (1967).
777. A. M. Kuliev, N. S. Kyazimova, and G. A. Zeinalova, Zh. Obshch. Khim., 39:557 (1969).
778. A. M. Kuliev, M. A. Salimov, N. S. Kyazimova, É. A. Agaeva, and T. Yu. Iskenderova, Dokl. Akad. Nauk Azerb. SSR, 24(7):32 (1968).
779. E. P. Lebedev and V. O. Reikhsfel'd, Scientific and Technological Conference of the Lensovet Leningrad Technological Institute, Section of Organic Chemistry and Technology, Brief Communications, Leningrad (1968), p. 30.
780. É. Lukevits, M. G. Voronkov, A. E. Pestunovich, A. A. Kimenis, S. Z. Gutberga, and Z. A. Atare, Izv. Akad. Nauk. Latv.SSR, No. 4, p. 93 (1968).

781. É. Ya. Lukevits, M. G. Voronkov, and L. M. Chudesova, Authors' cert. 196841, 1966; Izobr., Prom. Obr., Tov. Zn., No. 12, p. 36 (1967).

782. É. Lukevits, A. E. Pestunovich, and M. G. Voronkov, Kh.G.S., 1968:949.

783. É. Lukevits, A. E. Pestunovich, É. A. Reshetilova, and M. G. Voronkov, Abstracts of Reports to the Fourth Conference on the Chemistry and Use of Organosilicon Compounds, NIITÉKhIM, Moscow, p. 48 (1968).

784. L. A. Mai, Izv. Akad. Nauk Latv.SSR, Ser. Khim., 1967:506.

785. L. Mai [L. May], II Symposium International sur la Chimie des Composés Organiques du Silicium, Résumés des Communications, Bordeaux (1968), p. 129.

786. V. F. Mironov, V. P. Kozyukov, and V. D. Sheludyakov, Zh. Obshch. Khim., 37:1669 (1967).

787. V. F. Mironov, V. P. Kozyukov, and V. D. Sheludyakova, Zh. Obshch. Khim., 37:1915 (1967).

788. M. M. Morgunova, Some Reactions of Organoaminosilanes and Organosilazans with Nucleophilic and Electrophilic Reagents, Author's abstract of Cand. dissertation, Moscow (1967).

789. D. Kh. Nguen, Yu. I. Baukov, and I. F. Lutsenko, Zh. Obshch. Khim., 38:191 (1968).

790. N. F. Orlov, M. A. Belokrinitskii, and B. L. Kaufman, Authors' cert. 226604, 1967; Izobr., Prom. Obr., Tov. Zn., No. 29, p. 21 (1968).

791. N. F. Orlov, M. A. Belokrinitskii, B. L. Kaufman, and É. V. Sudakova, The Chemistry and Practical Application of Organosilicon Compounds, Leningrad (1968), p. 117.

792. N. F. Orlov, M. A. Belokrinitskii, É. V. Sudakova, and B. L. Kaufman, Zh. Obshch. Khim., 38:1656 (1968).

793. A. A. Petrov, and M. P. Forost, Zh. Organ. Khim., 2:1358 (1966).

794. A. N. Pravednikov, V. A. Shapatyi, N. V. Eliseeva, N. V. Zakatova, and V. I. Frants, Zh. Fiz. Khim., 41:118 (1967).

795. D. A. Predvoditelev and Z. A. Rogovin, Vysokomol. Soed., 9V:611 (1967).

796. V. O. Reikhsfel'd and E. P. Lebedev, Zh. Obshch. Khim., 37:1412 (1967).

797. V. O. Reikhsfel'd and E. P. Lebedev, The Chemistry and Practical Application of Organosilicon Compounds, Leningrad (1968), p. 137.

798. V. O. Reikhsfel'd and E. P. Lebdev, II Symposium International sur la Chimie des Composés Organiques du Silicium, Résumés des Communications, Suppl., Bordeaux (1968).

799. V. O. Reikhsfel'd and E. P. Lebedev, Zh. Obshch. Khim., 39:221 (1969).

800. S. I. Sadykh-Zade, I. I. Tsetlin, and A. D. Petrov, Zh. Obshch. Khim., 26:1239 (1956).

801. G. S. Smirnova, D. N. Andreev, and I. M. V'rbanova, Zh. Obshch. Khim., 37:1676 (1967).

802. B. A. Sokolov, G. M. Alekseeva, and G. V. Dmitrieva, Authors' cert., 190896, 1967; Izobr., Prom. Obr., Tov. Zn., No. 3, p. 25 (1967); R. Zh. Kh., 5H194P (1968).

803. M. D. Stadnichuk, Investigation of Unsaturated Silicohydrocarbons and Germaniohydrocarbons with Conjugated Multiple Bonds, Author's abstract of Doctor's dissertation, Leningrad (1969).

804. I. G. Sulimov, T. M. Sleta, M. D. Stadnichuk, and A. A. Petrov, Zh. Obshch. Khim., 38:202 (1968).
805. I. G. Sulimov and M. D. Stadnichuk, Zh. Obshch. Khim., 37:1906 (1967).
806. I. G. Sulimov and M. D. Stadnichuk, Zh. Obshch. Khim., 37:2329 (1967).
807. I. G. Sulimov and M. D. Stadnichuk, The Chemistry and Practical Application of Organosilicon Compounds, Leningrad (1968), p. 57.
808. D. L. Timrot, V. E. Lyusternik, and V. N. Prostov, Plast. Massy, No. 3, p. 35 (1968).
809. Yu. I. Khudobin, M. G. D'yachenko, N. P. Kharitonov, and P. A. Vasil'eva, Zh. Obshch. Khim., 38:181 (1968).
810. E. A. Chernyshev, M. E. Kurek, V. I. Savushkina, T. A. Klochkova, B. M. Tabenko, S. N. Polivanov, and V. Z. Anisimova, Abstracts of Reports to the Fourth Conference on the Chemistry and Use of Organosilicon Compounds, NIITÉKhIM, Moscow, p. 26 (1968).
811. E. A. Chernyshev, V. I. Savushkina, and V. M. Tabenko, Authors' cert. 228685, 1967; Izobr., Prom. Obr., Tov. Zn., No. 32, p. 13 (1968).
812. E. A. Chernyshev, V. I. Savushkina, N. G. Tolstikova, V. M. Tabenko, and V. Z. Anisimova, Abstracts of Reports to the Fourth Conference on the Chemistry and Use of Organosilicon Compounds, NIITÉKhIM, Moscow, p. 27 (1968).
813. B. G. Shakhovskii and A. A. Petrov, Zh. Obshch. Khim., 37:1371 (1967).
814. B. G. Shakhovskii and A. A. Petrov, The Chemistry and Practical Application of Organosilicon Compounds, Leningrad (1968), p. 134.
815. G. A. Shvekhgeimer and A. P. Kryuchkova, Zh. Obshch. Khim., 38:904 (1968).
816. M. F. Shostakovskii, A. S. Atavin, S. V. Amosova, and B. A. Trofimov, Izv. Akad. Nauk SSSR, Ser. Khim., 1968:1852.
817. M. F. Shostakovskii, A. S. Atavin, A. I. Mikhaleva, N. P. Vasil'ev, and G. A. Rinkus, Authors' cert. 225192, 1967; Izobr., Prom. Obr., Tov. Zn., No. 27, p. 18 (1968).
818. M. F. Shostakovskii, N. V. Komarov, and N. N. Vlasova, Zh. Obshch. Khim., 37:1151 (1967).
819. M. F. Shostakovskii, N. V. Komarov, N. N. Vlasova, and F. P. L'vova, Izv. Akad. Nauk SSSR, Ser. Khim., 1967:2678.
820. M. F. Shostakovskii, N. V. Komarov, N. N. Vlasova, and Z. I. Mikhailov, in: The Chemistry of Acetylene, Izd. "Nauka," Moscow (1968), p. 295.
821. M. F. Shostakovskii, N. V. Komarov, V. K. Roman, and N. I. Golovanova, Zh. Obshch. Khim., 38:2309 (1968).
822. M. F. Shostakovskii, N. V. Komarov, and O. G. Yarosh, Zh. Prikl. Khim., 38:435 (1965).
823. O. G. Yarosh, A. S. Nakhmanovich, and N. V. Komarov, Kh. G. S., 1968:642.
824. E. W. Abel and D. B. Brady, J. Organomet. Chem., 11:145 (1968).
825. E. W. Abel and O. R. Jenkins, J. Organomet. Chem., 14:285 (1968).
826. E. W. Abel and I. H. Sabberwal, J. Chem. Soc., 1968(A):1105.
827. E. W. Abel and D. J. Walker, J. Chem. Soc., 1968(A):2338.
828. E. W. Abel, D. J. Walker, and J. N. Wingfield, J. Chem. Soc., 1968(A):1814.

829. E. W. Abel, D. J. Walker, and J. N. Wingfield, J. Chem. Soc., 1968(A):2642.
830. F. P. Adams, J. B. Carmichael, and R. J. Zeman, Am. Chem. Soc., Div. Polym. Chem., Preprints, 7:960 (1966); C. A., 66:29128w (1967).
831. Astra Aktiebdag, British Patent 1073530, 1967; C. A., 68:12984 r (1968).
832. R. C. Anderson and R. A. W. Hill (Imperial Chemical Industries, Ltd.), British Patent 879439, 1959; C. A., 56:8745 (1962).
833. R. Appel, L. Siekmann, and H. O. Hoppen, Chem. Ber., 101:2861 (1968).
834. J. C. Baldwin, M. F. Lappert, J. B. Pedley, and J. A. Treverton, J. Chem. Soc., 1967(A):1980.
835. S. Barcza and C. W. Hoffman, II Symposium International sur la Chimie des Composés Organiques du Silicium, Résumés des Communications, Bordeaux (1968), p. 12.
836. J. C. J. Bart and J. J. Daly, Chem. Comm., 1968:1207.
837. L. Birkofer and K. Krebs, Tetrahedron Letters, 1968:885.
838. G. H. Birum and G. A. Richardson (Monsanto Chemical Co.), USA Patent 3113139 (1963); C. A., 60:5551 (1964).
839. L. C. F. Blackman, M. J. S. Dewar, and H. Hampson, J. Appl. Chem., 7:160 (1957).
840. F. Blazy, J. Bonastre, and G. Pfister-Guillouzo, Bull. Soc. Chim. Fr., 1968:4247.
841. P. Bourgeois and N. Duffaut, C. R., 266C:810 (1968).
842. L. A. Bromley, J. W. Porter, and S. M. Read, Am. Inst. Chem. Eng. J., 14:245 (1968); C. A., 68:98530 (1968).
843. A. G. Brook and D. G. Anderson, Can. J. Chem., 46:2115 (1968).
844. A. G. Brook and P. F. Jones, Can. J. Chem., 46:2119 (1968).
845. A. G. Brook, D. G. Anderson, J. M. Duff, P. F. Jones, and D. M. MacRae, J. Am. Chem. Soc., 90:1076 (1968).
846. A. G. Brook, J. M. Duff, P. F. Jones, and N. R. Davis, J. Am. Chem. Soc., 89:431 (1967).
847. G. F. Bulbenko (Thikol Chemical Corp.), French Patent 1439027 (1966); C. A., 66:19420 w (1967).
848. G. F. Bulbenko (Thikol Chemical Corp.), USA Patent 329473 (1967); Offic. Gaz., 834(2):683 (1967).
849. H. Büger, II Symposium International sur la Chimie des Composés Organiques du Silicium, Résumés des Communications, Bordeaux (1968), p. 28.
850. H. Bürger, Organomet. Chem. Rev., 3A:425 (1968).
851. H. Bürger, U. Goetze, and W. Sawodny, Spectrochim. Acta, 24A:2003 (1968).
852. R. Calas, P. Bourgeois, and N. Duffaut, Bull. Soc. Chim. Fr., 1967:762.
853. J. D. Citron, Stereochemistry of Asymmetric Silicon: a Study of the Silicon–Nitrogen Bond, Diss. Abstr., 28B:1851 (1967).
854. R. E. Coldwell, The Addition of Silicon Hydrides to Ferrocenylalkenes, Diss. Abstr., 29B:1300 (1968).
855. E. J. Corey, D. Seebach, and R. Freedman, J. Am. Chem. Soc., 89:434 (1967).
856. R. Corriu and J. Masse, C. R., 266C:1709 (1968).

857. R. H. Cragg, M. F. Lappert, and B. F. Tilley, J. Chem. Soc., 1967(A):947.
858. C. W. N. Cumper, A. Melnikoff, E. F. Mooney, and A. I. Vogel, J. Chem. Soc., 1966(B):874.
859. J. Diekman, J. B. Thomson, and C. Djerassi, J. Org. Chem., 32:3904 (1967).
860. B. R. Donadson and J. C. P. Schwarz, J. Chem. Soc., 1968(B):395.
861. Dow Corning Corp., Netherlands Applic., 6516641, 1966; C. A., 66:3536 v (1967).
862. Dow Corning Corp., Netherlands Applic. 6609898, 1967; C. A., 67:11578 d (1967).
863. Dow Corning Corp., British Patent 1101236, 1968; C. A., 68:60372 m (1968).
864. N. Duffaut, P. Bourgeois, and R. Calas, 3 Internationale Symposium über Metallorganische Chemie, München (1967), p. 22.
865. E. A. V. Ebsworth, F. Rockstaschel, and J. C. Thompson, J. Chem. Soc., 1967(A):362.
866. E. A. V. Ebsworth and J. C. Thompson, J. Chem. Soc., 1967(A):69.
867. F. Fehér and H. Goller, Z. Naturf., 22b:1223 (1967).
868. F. Fehér and H. Goller, Z. Naturf., 22b:1224 (1967).
869. F. Fehér and H. Goller, Z. Naturf., 22b:1225 (1967).
870. K. Friederich, A. de Montigny, and W. Simmler, Kolloid. Z. Z. Polym., 218:7 (1967); C. A., 67:91292 u (1967).
871. P. J. Gellings, Chem. Weekbl., 62:425 (1966).
872. O. Glemser and E. Niecke, Z. Naturf., 23b:743 (1968).
873. C. Glidewell and D. W. R. Rankin, J. Chem. Soc., 1969(A):753.
874. Th. Goldschmidt A. G., French Patent 1456142, 1966; C. A., 66:105452 q (1967).
875. W. G. Gondy and J. W. Keil (Dow Corning Corp.), French Patent 1659124, 1967; C. A., 67:33646 b (1967).
876. G. A. Gornowicz, J. W. Ryan, and J. L. Speier, J. Org. Chem., 33:2918 (1968).
877. B. G. Gowenlock and J. Stevenson, J. Orgnomet. Chem., 13(P):13 (1968).
878. P. A. Griffin, R. N. Haszeldine, and M. J. Newlands, II Symposium International sur la Chimie des Composés Organiques du Silicium, Résumés des Communications, Bordeaux (1968), p. 85.
879. W. C. Hammann, C. F. Hobbs, and D. J. Bauer, J. Org. Chem., 32:2841 (1967).
880. S. Herrling and H. Mückter (Chemie Grünenthal G.m.b.H.), Federal Germ. Rep. Patent 1445015, 1968; Pharm. Ind., 1969:203.
881. J. R. Horder and M. F. Lappert, J. Chem. Soc., 1968(A):1167.
882. B. K. Hunter and L. W. Reeves, Can. J. Chem., 46:1399 (1968).
883. Imperial Chemical Industries, Ltd., Netherlands Applic. 6611054, 1967; C. A., 67:44529 (1967).
884. K. Itoh, I. K. Lee, I. Matsuda, S. Sakai, and Y. Ishii, Tetrahedron Letters, 1967:2667.
885. K. Itoh, K. Matsuzaki, and Y. Ishii, J. Chem. Soc., 1968(C):2709.
886. J. F. R. Jaggard, II Symposium International sur la Chimie des Composés Organiques du Silicium, Résumés des Communications, Bordeaux (1968), p. 99.

887. W. H. Keeber, Synthesis and Reactions of Some Organosilicon Phosphonates, Diss. Abstr., 28B:1854 (1967).
888. J. F. Klebe (Comp. Fr. Thomson-Houston), French Patent 1442585, 1966; C. A., 66:85854 s (1967).
889. J. F. Klebe, J. B. Bush, and J. E. Lyons, J. Am. Chem. Soc., 86:4400 (1964).
890. G. Koerner (Th. Goldschmidt A.-G.), Federal Germ. Rep. Patent 1236521, 1967; C. A., 67:64525 r (1967).
891. G. Koerner (Th. Goldschmidt A.-G.), Federal Germ. Rep. Patent 1239046, 1967; C. A., 67:65158 k (1967).
892. G. Koerner and G. Rossmy, USA Patent 3314982, 1967; Offiz. Gaz., 837(3):938 (1967).
893. W. A. Kornicker (Monsanto Co.), USA Patent 3318932, 1967; C. A., 67:64541 t (1967).
894. G. Krüger, M. Büttner, G. Oliew, and E. Thilo, Z. Anorg. Allg. Chem., 360:70 (1968).
895. V. J. Kuck and R. W. Wright, J. Organomet. Chem., 14(P):17 (1968).
896. M. Kumada (Tokyo Shibaura Electric Co., Ltd.), Japanese Patent 6624, 1960; C. A., 55:6376 (1961).
897. M. Kumada, Organosilicon Chemistry (Special Lectures Presented at the International Symposium on Organosilicon Chemistry held in Prague, Czechoslovakia, 6-9 Sept., 1965), Butterworths, London (1966), p. 167.
898. M. Kumada and M. Ishikawa, J. Organomet. Chem., 1:153 (1963).
899. M. Kumada, K. Tamao, T. Takubo, and M. Ishikawa, J. Organomet. Chem., 9:43 (1967).
900. D. H. Lemmon, The Infrared and Raman Spectra, and Structure of $P(C\equiv CH)_3$, $As(C\equiv CH)_3$, $NC(C\equiv C)_2CN$, Cl_3SiSH, Cl_3SiCN, Diss. Abstr., 28B:632 (1967).
901. E. Louis and G. Urry, Inorg. Chem., 7:1253 (1968).
902. B. Martel and N. Duffaut, C. R., 264C:452 (1967).
903. B. Martel, N. Duffaut, and R. Calas, Bull. Soc. Chim. Fr., 1967:758.
904. B. Martel, N. Duffaut, and R. Calas, C. R., 264C:2160 (1967).
905. A. Marzocchi (Owens-Corning Fiberglass Corp.), USA Patent, 3364059, 1966; C. A., 68:50834 z (1968).
906. I. Matsuda, K. Itoh, and Y. Ishii, J. Chem. Soc., 1969(C):701.
907. R. N. Meals (General Electric Co.), USA Patent 3269982, 1966; R. Zh. Kh., 8C249 (1968).
908. M. M. Millard, K. Steele, and L. J. Pazdernic, J. Organomet. Chem., 13:P7 (1968).
909. K. Moedritzer and J. R. Van Wazer (Monsanto Co.), USA Patent 3344161, 1967; R. Zh. Kh., 6S451 P (1969).
910. K. Moedritzer and J. R. Van Wazer, Inorg. Chem., 6:93 (1967).
911. K. Moedritzer and J. R. Van Wazer, Monsanto Tech. Rev., 12:19 (1967); C. A., 67:116914 p (1967).
912. K. Moedritzer and J. R. Van Wazer, Inorg. Chim. Acta, 1:152 (1967).
913. K. Moedritzer and J. R. Van Wazer, J. Inorg. Nucl. Chem., 29:1851 (1967).
914. K. Moedritzer and J. R. Van Wazer, Inorg. Chem., 7:2105 (1968).
915. K. R. Molt, I. Hechenbleickner, and O. A. Homberg (Deutsche Advance Production G.m.b.H.), Belgian Patent 622649, 1963; C. A., 59:5325 (1963).

916. R. Müller and C. Dathe, Germ. Dem. Rep. Patent 50828, 1966; C. A., 66:95177 n (1967).
917. R. G. Neville and J. J. McGee, Inorg. Synth., 8:27 (1966).
918. H. Niederprüm and W. Simmler (Farbenfabriken Bayer A.-G.), British Patent 1072930, 1967; C. A., 67:64912 q (1967).
919. H. Niederprüm and W. Simmler (Farbenfabriken Bayer A.-G.), Federal Germ. Rep. Patent 1249530, 1968; R. Zh. Kh., 5N86 P (1969).
920. H. Niederprüm and W. Simmler, USA Patent 3370076, 1968; RAPRA Abstr., 3:622P (1969).
921. J. E. Noll, J. L. Speier, and B. F. Daubert, J. Am. Chem. Soc., 73:3867 (1951).
922. S. Nozakura and S. Konotsune, Bull. Chem. Soc., Japan, 29:322 (1956).
923. S. Nozakura and S. Konotsune, Bull. Chem. Soc., Japan, 29:326 (1956).
924. G. A. Olah, D. H. O'Brien, and C. Y. Lui, J. Am. Chem. Soc., 91:701 (1969).
925. D. J. Peterson, J. Org. Chem., 32:1717 (1967).
926. D. J. Peterson, J. Org. Chem., 33:780 (1968).
927. C. G. Pitt and M. S. Fowler, J. Am. Chem. Soc., 90:1928 (1968).
928. T. J. Pullukat, Reactive Intermediates in Organosilicon Chemistry, Diss. Abstr., 29B:924 (1968).
929. W. Rodziewicz and Z. Michalowski, Roczn. Chem., 43:31 (1969).
930. W. Rodziewicz and Z. Michalowski, Roczn. Chem., 43:465 (1969).
931. N. A. Rosenthal (Thiokol Chemical Corp.), USA Patent 3284466, 1966; C. A., 68:13015 n (1968).
932. K. Rühlmann, II Symposium International sur la Chimie des Composés Organiques du Silicium, Résumés des Communications, Bordeaux (1968), p. 159.
933. K. Rühlmann, J. Hils, and H.-J. Graubaum, J. Prakt. Chem., [4], 32:37 (1966).
934. K. Rühlmann, U. Kaufmann, and D. Knopf, J. Prakt. Chem., [4], 18:131 (1962).
935. K. Rühlmann, H. Seefluth, and H. Becker, Chem. Ber., 100:3820 (1967).
936. H. Sakurai, M. Kira, and M. Kumada, Chem. Comm., 1967:889.
937. H. Sakurai, M. Kira, and M. Kumada, Bull. Chem. Soc., Japan, 41:1494 (1968).
938. R. M. Salinger and R. West, J. Organomet. Chem., 11:631 (1968).
939. O. J. Scherer and R. Schmitt, II Symposium International sur la Chimie des Composés Organiques du Silicium, Résumés des Communications, Bordeaux (1968), p. 165.
940. O. J. Scherer and R. Schmitt, Tetrahedron Letters, 1968:6235.
941. O. J. Scherer and R. Schmitt, Chem. Ber., 101:3302 (1968).
942. O. J. Scherer and R. Schmitt, J. Organomet. Chem., 16(P):11 (1969).
943. O. J. Scherer and J. Wokulat, Z. Anorg. Allg. Chem., 357:92 (1968).
944. M. Schmeiber and H. Fruozanfar, Z. Chem., 8:254 (1968).
945. H. Schmidbaur, L. Sechser, and M. Schmidt, J. Organomet. Chem., 15:77 (1968).
946. M. Schmidt, Organosilicon Chemistry (Special Lectures presented at the International Symposium on Organosilicon Chemistry held in Prague, Czechoslovakia, 6-9 Sept., 1965), Butterworths, London (1966), p. 15.

947. M. Schmidt and H. Fischer, Z. Chem., 8:253 (1968).
948. O. Schmitz-Du Mont and W. Jansen, Angew. Chem., 80:399 (1968).
949. U. Schöllkopf and N. Rieber, Angew. Chem., 79:906 (1967).
950. G. Schott and B. Säverin, Z. Chem., 8:427 (1968).
951. G. Schott and B. Säverin, Z. Chem., 8:428 (1968).
952. W. W. Schwarz and R. D. Lowrey, J. Appl. Polym. Sc., 11:553 (1967); C. A., 67:45694 d (1967).
953. W. Siebert, W. E. Davidson, and M. C. Henry, J. Organomet. Chem., 15:69 (1968).
954. W. Simmler (Farbenfabriken Bayer A.-G.), Federal Germ. Rep. Patent 1250448 (1968).
955. W. Simmler and H. Niederprüm (Farbenfabriken Bayer A.-G.), Federal Germ. Rep. Patent 1222057, 1967; R. Zh. Kh., 6N167 P (1968).
956. W. Simmler and H. Niederprüm (Farbenfabriken Bayer A.-G.), Federal Germ. Rep. Patent 1227456, 1966; C. A., 66:38350 q (1967).
957. S. Sliwinski, Germ. Dem. Rep. Patent 56102, 1967; C. A., 67:100581 s (1967).
958. L. H. Sommer (Dow Corning Corp.), USA Patent 2589446, 1952; C. A., 47:145 (1953).
959. L. H. Sommer (Dow Corning Corp.), USA Patent 2591736, 1952; C. A., 46:6435 (1952).
960. L. H. Sommer (Dow Corning Corp.), USA Patent 2607793, 1952; C. A., 46:10680 (1952).
961. L. H. Sommer (Dow Corning Corp), USA Patent 2662910, 1953; C. A., 48:5553 (1954).
962. L. H. Sommer and J. D. Citron, J. Org. Chem., 32:2470 (1967).
963. L. H. Sommer, G. M. Goldberg, G. H. Barnes, and L. S. Stone, J. Am. Chem. Soc., 76:1609 (1954).
964. L. H. Sommer and J. McLick, J. Am. Chem. Soc., 89:5806 (1967).
965. G. C. Tesoro (J. P. Stevens and Co., Inc.), USA Patent 3278484, 1966; C. A., 66:3027 v (1967).
966. S. F. Thames and L. H. Edwards, J. Heteroc. Chem., 5:115 (1968).
967. S. F. Thames and L. H. Edwards, J. Chem. Soc., 1968(C):2339.
968. S. F. Thames and J. E. McCleskey, J. Heteroc. Chem., 4:371 (1967).
969. S. F. Thames, J. E. McClesky, and P. L. Kelly, J. Heteroc. Chem., 5:749 (1968).
970. J. S. Thayer, Inorg. Chem., 7:2599 (1968).
971. W. J. Trepka (Phillips Petroleum Co.), USA Patent 3324089, 1967; C. A., 67:33138 n (1967).
972. C. Ungurenasu, 3 International Symposium über Metallorganische Chemie, München (1967), p. 18.
973. W. Verbeek and W. Sundermeyer, Angew. Chem., 79:860 (1967).
974. N. Viswanathan and C. H. Van Dyke, J. Organomet. Chem., 11:181 (1968).
975. N. Viswanathan and C. H. Van Dyke, J. Chem. Soc., 1968(A):487.
976. R. V. Viventi (General Electric Co.), USA Patent 3346405 (1967); Offic. Gaz., 843(2):618 (1967).

977. J. T. Wang and C. H. Van Dyke, Chem. Comm., 1967:612.
978. J. T. Wang and C. H. Van Dyke, Chem. Comm., 1967:928.
979. J. T. Wang and C. H. Van Dyke, Inorg. Chem., 6:1741 (1967).
980. J. T. Wang and C. H. Van Dyke, Inorg. Chem., 7:1319 (1968).
981. U. Wannagat, K. Hensen, P. Petesch, F. Vielberg, and H. Vob, Monatsh., 99:438 (1968).
982. R. M. Washburn and K. R. Eilar (American Potash and Chemical Corp.), USA Patent 3347824, 1967; C. A., 67:117695 c (1967).
983. N. Wiberg and W. C. Joo, II Symposium International sur la Chimie des Composés Organiques du Silicium, Résumés des Communications, Bordeaux (1968), p. 196.
984. N. Wiberg, W. C. Joo, and W. Uhlenbrock, Angew. Chem., 80:661 (1968).
985. N. Wiberg, W. C. Joo, and M. Veith, Inorg. Nucl. Chem. Letters, 4:223 (1968).
986. M. Wieber and G. Schwarzmann, II Symposium International sur la Chimie des Composés Organiques du Silicium, Résumés des Communications, Bordeaux (1968), p. 197.
987. E. V. Wilkus and A. Berger (General Electric Co.), USA Patent 3326952, 1967; C. A., 67:65050 u (1967).
988. E. V. Wilkus and W. H. Rauscher, J. Org. Chem., 30:2889 (1965).

Subject Index†

†An asterisk after a page number indicates that the reference occurs in a footnote to that page.